C0-AWD-587

Biology of the Ubiquitous House Sparrow

Biology of the Ubiquitous

House Sparrow

From Genes to Populations

Ted R. Anderson

OXFORD
UNIVERSITY PRESS
2006

OXFORD
UNIVERSITY PRESS

Oxford University Press, Inc., publishes works that further
Oxford University's objective of excellence
in research, scholarship, and education.

Oxford New York
Auckland Cape Town Dar es Salaam Hong Kong Karachi
Kuala Lumpur Madrid Melbourne Mexico City Nairobi
New Delhi Shanghai Taipei Toronto

With offices in
Argentina Austria Brazil Chile Czech Republic France Greece
Guatemala Hungary Italy Japan Poland Portugal Singapore
South Korea Switzerland Thailand Turkey Ukraine Vietnam

Published by Oxford University Press, Inc.
198 Madison Avenue, New York, New York 10016

www.oup.com

Library of Congress Cataloging-in-Publication Data

Anderson, Ted R.
Biology of the ubiquitous house sparrow : from genes to populations / Ted R. Anderson.
p. cm.
Includes bibliographical references and index.
ISBN-13: 978-0-19-530411-4
ISBN 0-19-530411-X
1. English sparrow. I. Title.
QL696.P264A527 2006
598.8'87—DC22
2006016745

9 8 7 6 5 4 3 2 1

Printed in the United States of America
on acid-free paper

To my wife, Carol

PREFACE

The house sparrow has several attributes that make it an ideal subject for many types of biological inquiry. These include its accessibility as a widespread commensal of urban and agricultural communities; its ready acceptance of nest-boxes as nest sites; its status as a pest of agriculture and as a disease vector for both humans and their livestock; its highly social behavior, which results in its foraging in flocks and breeding semicolonially; its ready adaptation to laboratory conditions permitting extensive laboratory research (including captive breeding); and its lack of formal governmental protection in many places.

Denis Summers-Smith published a landmark study on the biology of the house sparrow in 1963. *The House Sparrow* is a remarkably thorough description of a species that, although it is abundant throughout much of the world, had remained relatively unknown before that date, with few ornithologists deigning to study the "common" sparrow. By summarizing the descriptive biology of the species, *The House Sparrow* stimulated many new studies that have greatly expanded understanding of the biology of the species and, indeed, of the biology of birds. The purpose of this volume is to provide a comprehensive and critical review of the results of these studies and to present an updated synthesis of current knowledge of the biology of the species.

A major thread in this work is the attempt to integrate evolutionary and mechanistic understandings of the biology of sparrows. It has long been recognized that phenotypes of organisms result from both ultimate (adaptive or evolutionary) causes and proximate (mechanistic) causes; they are the product of the interaction of the organisms' genotypes with their environments during development to shape their anatomy, physiology, and behavior. These inextricably related aspects of the biology of a species are frequently treated separately, often resulting in one-dimensional and necessarily incomplete understandings of the biology of the organism. Because the house sparrow has been so extensively studied in both laboratory and field settings, a treatment of its biology, provisional as it is, offers a unique opportunity to integrate a discussion of the adaptive and mechanistic understandings of its biology.

I have received much support, both institutional and personal, in my attempt to meet the many challenges of writing this review. I owe a particular debt of gratitude to numerous librarians who helped me track down difficult-to-obtain literature. In particular, I thank Dr. Linda Birch of the Alexander Library of the Edward Grey Institute of Field Ornithology, Oxford University; Elizabeth Vogt and Bill Haroff of Holman Library, McKendree College; and several summer librarians at the University of Michigan Biological Station (UMBS). I thank Prof. J. Pinowski for assisting in obtaining literature from former eastern bloc countries. I thank Dr. Peter Lowther, Dr. Harvey Blankespoor, Prof.Chris Perrins, and three anonymous reviewers for helpful suggestions after reading part or all of the manuscript, and I thank Mike Wilson for Russian translation. I also thank Dr. J. D. Summers-Smith for his encouragement and support. Susan Fawcett, a student in my Biology of Birds course at the UMBS, prepared the cover drawing of house sparrows on stalks of wheat.

The principal financial support came from McKendree College in the form of a sabbatical leave in 2000, and from the UMBS and its director, Dr. James Teeri, for a research fellowship during the summer of 2001. I am immensely grateful to both institutions. I also thank Prof. Perrins and Wolfson College, Oxford, for hosting me during the sabbatical. Much technical support at the UMBS was provided by the resident biologist, Bob Vande Kopple, to whom I am deeply appreciative.

Many persons at Oxford University Press were immensely important in seeing the manuscript through to publication. In particular, I thank Peter Prescott (Science Editor for Life Sciences), Kaity Cheng and Alycia Somers (Assistant Editors for Life Sciences), Rosanne Hallowell (Production Editor, Academic EDP), and Lisa Hamilton (Cover Designer, Academic EDP. I also thank the copy editor, Beverly Braunlich.

Finally, I owe an immense debt of gratitude to my wife, Carol, who not only unflaggingly supported me during the nine-year gestation period of this work, but also spent countless hours helping me proof the final version of the manuscript.

Despite all of this splendid assistance, there are no doubt errors and omissions in this work, and these remain the full responsibility of the author.

CONTENTS

Biology of the Ubiquitous House Sparrow

Chapter 1

TAXONOMY AND DISTRIBUTION

"Know anything about bird-watching, sir?"
"More than you, I shouldn't wonder."
"Beautiful little fellow, isn't he?"
"She!"
"Pardon?"
"Immature female of the species."
"*What* species?"
"*Passer domesticus*, Morse. Can't you recognize a bloody house sparrow when you see one?"

—Colin Dexter, *The Remorseful Day*, 1999

"Daddy, what is a sparrow?" asked my 6-year-old daughter one day as I was leaving for the field. Out of the mouths of babes . . . ! My answer that a sparrow is whatever a recognized avian taxonomist says it is somehow didn't satisfy either my daughter or me (she was much happier with my definition of a sparrow as a little brown bird). Implicit in my daughter's disarmingly simple question are two of the most enduring and intractable problems in systematics: (1) the definition of species, and (2) the phylogenetic relationships among higher order taxa, from the genus on up.

The father of modern taxonomy, Karl Linnaeus, introduced the Latin binomial, the first name representing the genus and the second the species. The binomial introduced the idea of relatedness among species, with the members of a genus resembling each other more closely than they resemble species of other genera. Morphological resemblance was the primary differentiating criterion distinguishing one species from another in what later came to be termed the "morphological species concept" (Mayr 1963). Although this morphological concept worked well as a first approximation, it was not long before the complexities of nature began to offer challenges to its validity. Polymorphic species, morphologically distinct geographic races (subspecies), species with markedly different developmental forms, and sibling species all posed challenges for the concept. In addition, evolutionary convergence—similarity due to adaptation to similar environmental selective pressures—particularly complicated the determination of higher-order relationships among species. For instance, when Linnaeus originally described the house sparrow in 1758, he named it *Fringilla domestica*, thus placing it with other common seed-eating birds such as the common chaffinch (*Fringilla coelebs*), to which it is not closely related. This problem of evolutionary convergence continues to plague systematists attempting to resolve the question of the phylogenetic relationship of the Old World sparrows (*Passer* spp.) to other seed-eating passerines (see later discussion).

The morphological species concept was largely replaced during the last half-century by the so-called biological species concept, first formulated by Ernst Mayr in 1940. He defined a species as "a group of populations which replace each other geographically or ecologically and of which the neighboring ones intergrade or hybridize wherever they are in contact or which are potentially capable of doing so (with one or more of the populations) in those cases where contact is prevented by geographical or ecological barriers" (Mayr 1940:256). This definition emphasizes the potential for interbreeding among members of a species, thus ensuring that a species shares a common gene pool, and it postulates that mechanisms will rapidly evolve that reduce or preclude breeding between members of two different species. Although this definition has much to recommend it theoretically, it is often difficult to apply in practice, as shall be seen later in the efforts to delineate the species *Passer domesticus*. One of the major difficulties involves how to interpret instances of hybridization between what otherwise appear to be distinct species. A second problem involves the treatment of groups that are widely separated, either geographically or temporally, and hence have no opportunity to interbreed in nature. By default, actual taxonomic decisions at the species level are still often based to a large extent on morphological similarity.

In recent years, new species definitions have been proposed, partly in response to the difficulties with the application of the biological species concept, and partly based on the development of molecular technologies that have provided new tools that permit either direct or indirect estimates of the genetic similarities between groups. Two such definitions are referred to as the "evolutionary species concept" and the "phylogenetic species concept" (see Zink and McKitrick 1995). Although each of these concepts has features that recommend them, both also have theoretical and practical problems in their application (see Bock 1992), and a discussion of the merits and difficulties associated with them is beyond the scope of this work.

Mayr presented a "descriptive" definition of the species as "a reproductive community of populations (reproductively isolated from others) that occupies a specific niche in nature" (Mayr 1982:273). This definition has both practical and conceptual difficulties, not the least of which is that it neatly and explicitly closes the implicit circularity that Cole (1960) detected in the competitive exclusion principle. I believe, however, that it actually has utility in the attempt to identify the boundaries of the species *P. domesticus*, and it is therefore the species definition that I have used in this account.

The question of the geographic distribution of the house sparrow depends on a satisfactory answer to the prior question, "What is a house sparrow?" No matter how one resolves some of the thornier and more esoteric taxonomic questions regarding the species, however, there can be little doubt that the "ubiquitous" house sparrow is the most widely distributed land bird species in the world. This is due not only to its large natural distribution in Europe, Asia, and North Africa, but also to the fact that the species has been successfully introduced to all of the continents except Antarctica, as well as to many oceanic islands (Lever 1987; Long 1981).

This chapter deals first with the taxonomic status of the house sparrow and then summarizes what is known of its current distribution.

The Genus *Passer*

The genus *Passer* has recently been treated by Summers-Smith (1988), and this account will make extensive use of his conclusions. The genus *Passer* (Old World sparrows) is a well-defined group of 20 species whose native distributions lie principally in Africa, Europe, and Asia (Table 1.1). Johnston and Klitz (1977) recognized 15 species in the genus, whereas Sibley and Monroe (1990) identified 23. The differences among the various authors depend primarily on whether morphologically and ecologically similar groups with largely disjunct distributions are assigned species status. Eleven of the species have distributions that are limited to Africa, suggesting that the genus originated on that continent. Summers-Smith (1988) identified two subgenera and three superspecies groups within the genus, and he assigned the house sparrow to a superspecies group that includes the Spanish sparrow (*P. hispaniolensis*), the desert sparrow (*P. simplex*), and the Somali sparrow (*P. castanopterus*).

"The classification of the Old World finches [= sparrows] has been a major problem for systematists from the beginnings of avian classification" (Bock 1992:66). The relationship of the Old World sparrows to the other major groups

TABLE 1.1. The Genus *Passer*

Species Name	Common Name	Range
griseus	Grey-headed sparrow	Sub-Saharan west and central Africa
swainsonii	Swainson's sparrow	East Africa
gongomensis	Parrot-billed sparrow	East Africa
suahelicus	Swahili sparrow	East-central Africa
diffusus	Southern grey-headed sparrow	Southern Africa
iagoenesis	Iago sparrow	Cape Verde Islands
motitensis	Rufous sparrow	Eastern and southern Africa, Socotra Island
melanurus	Cape sparrow	Southern Africa
domesticus	House sparrow	Europe, Asia, and North Africa
hispaniolensis	Spanish sparrow	North Africa, southern Europe, central Asia, Madeira, and Cape Verde and Canary Islands
simplex	Desert sparrow	Deserts of North Africa and central Asia
castanopterus	Somail sparrow	East Africa
moabiticus	Dead Sea sparrow	East-central Asia (Middle East)
pyrrhonotus	Sind jungle sparrow	South-central Asia
flaveolus	Pegu sparrow	Southeast Asia
rutilans	Cinnamon sparrow	Central and southeastern Asia, Japan
ammodendri	Saxaul sparrow	Central Asia
montanus	Tree sparrow	Eurasia
luteus	Golden sparrow	Arid sub-Saharan Africa
eminibey	Chestnut sparrow	East Africa

The 20 species belonging to the genus *Passer* as identified in Table 50 of Summers-Smith (1988). Scientific names, common names, and approximate natural ranges are given.

of seed-eating passerines, the fringilline finches (Old World finches), the emberizine finches (buntings and New World sparrows), and ploceine finches (weaverbirds), has been a continuing source of controversy. The dominant position throughout the last three-quarters of a century has been that they are closely related to the weaverbirds. This conclusion dated to Sushkin (1927), who based his decision on similarities between sparrows and weaverbirds in the palatal surface of the horny bill, the possession of a complete postjuvenal molt (first prebasic molt—see Chapter 5), and the construction of domed nests. Some studies comparing other characteristics among the groups have tended to support Sushkin's conclusion, but others have raised doubts about the close affinity of the sparrows and weaverbirds (see later discussion). Some workers have suggested that the Old World sparrows and their relatives (the rock sparrows, *Petronia* spp., and the snow finches, *Montifringilla* spp.) should be assigned to their own family, Passeridae (i. e., Bock and Morony 1978a; Sibley

1970). Recent results from DNA-DNA hybridization studies have helped to clarify the affinities of the group, but their taxonomic status is still not completely resolved.

Ziswiler (1965) compared the structure of the palatal surfaces of the horny sheath of the upper mandible and the seed-husking mechanisms among the various groups of seed-eating passerines. He found that there were two mechanisms of seed-husking: crushing the seed husk or slicing it. The fringilline finches slice the seed coat, whereas the emberizines and ploceines (in which he included the sparrows) crush it. Pocock (1966), however, examined the distribution of five tiny foramina around the optic foramen in the posterior surface of the orbit and concluded that *Passer* and *Petronia* were not closely related to the weaverbirds. Sibley (1970) studied the electrophoretic properties of egg-white proteins and hemoglobin and concluded that *Passer* was closer to the emberizine finches than to the Ploceidae. He also found that the egg-white proteins of *Montifringilla* were quite different from those of *Passer*, and he questioned whether these two genera were closely related. Poltz and Jacob (1974) found that the waxes produced in the uropygial gland secretions of sparrows were similar to those produced by fringilline and emberizine finches but were unlike those produced by the weaverbirds. Bock and Morony (1978a) used the possession of a preglossale (a small skeletal element in the tongue) and its associated muscle, *m. hypoglossus anterior*, to argue that the Old World sparrows, rock sparrows, and snow finches belong to a monophyletic lineage that did not arise from the weaverbirds. The preglossale is a unique skeletal element that supports the seed cup in the tongue; not only do the weaverbirds lack this element, but they also lack an anterior hypoglossal muscle. Bock and Morony argued that this muscle, which is found in fringilline and emberizine finches, is a necessary prerequisite for the evolution of the preglossale itself.

Because such comparisons of morphological traits among groups have the persistent difficulty of determining whether the traits are similar due to homology or to convergence, efforts have often been made to study complex anatomical suites of characters, such as a structure and its associated musculature (e.g., Bock and Morony 1978a), in which close similarity among the traits would be highly unlikely unless they were derived from a common ancestor. Judgments based on such traits usually remain open to question, however, and analyses of different suites of characters often lead to conflicting taxonomic conclusions. The advent of molecular systematics, particularly DNA-DNA hybridization techniques, was promoted as an objective measure of the genetic similarity between taxa that would obviate the necessity of making judgments based on the more subjective analyses of comparative morphology (Sibley and Ahlquist 1990).

DNA-DNA hybridization studies are based on the assumption that a "molecular clock" is present in the DNA molecule. The presence of this molecular clock is based on the idea that there is a constant rate of neutral base pair changes per generation within a genetic lineage. The accumulation of these changes in two genetic lineages derived from a common ancestor results in an increasing number of nonhomologous nucleotides between the DNA molecules of the two lineages. Hybridized DNA strands can be dissociated from each other by heating, and the temperature at which dissociation occurs is an inverse function of the degree of homology between the strands. DNA-DNA hybridization studies typically compare the temperatures at which 50% of the hybridized DNA is dissociated, a statistic referred to as the $T_{50}H$. The difference between the temperature at which 50% of the radioactively labeled DNA is dissociated from unlabeled DNA from the same individual and the temperature at which 50% of the same labeled DNA is dissociated from unlabeled DNA from a second individual (i.e., from another species or group) is the $\Delta T_{50}H$. The $\Delta T_{50}H$ is therefore an indirect measure of the time in generations since the two individuals shared a common ancestor. Although a number of difficulties with the underlying assumptions of the molecular clock have been identified (Houde 1987; O'Hara 1991), such as the effect of population size on the rate at which genetic drift will result in the fixation of a neutral allele, many taxonomic decisions made on the basis of DNA-DNA hybridization studies have been widely accepted (American Ornithologists' Union 1998; Cramp and Perrins 1994). Even more recently, gene sequencing technologies have enabled taxonomists to directly compare the base sequences of specific genes among species (e.g., Allende et al. 2001).

Sibley and Ahlquist (1990) placed the three genera, *Passer*, *Petronia*, and *Montifringilla*, in the subfamily Passerinae of the family Passeridae. They also included the wagtails and pipits (Motacillinae), accentors and hedge-sparrows (Prunellinae), weaverbirds (Ploceinae), and waxbills, manakins, and widowbirds (Estrildinae) in the family Passeridae. They reported that the average $\Delta T_{50}H$ for 34 comparisons between the Passerinae and other members assigned to the family Passeridae was 8.5°C, and that this constituted the oldest branching in the group. Based on the assumption that a $\Delta T_{50}H$ of 10 corresponds to a branching time of about 25 million years before present (mybp), this would suggest that the Old World sparrows (Passerinae) began their independent evolution about 21 mybp, in the early Miocene. In an earlier treatment, Sibley and Ahlquist (1985) suggested that this divergence between the Old World sparrows and the weaverbirds occurred about 30–35 mybp, a date that places the split in the mid-Oligocene. Sibley and Ahlquist (1990) reported that the genera *Passer* and *Petronia* branched at an average $\Delta T_{50}H$ of 3.4°C ($n = 2$) suggesting that their divergence occurred about 8.5 mybp (range, 7–10 mybp), near the beginning of the Pliocene. They also reported some $\Delta T_{50}H$ values within the genus *Passer*, with *griseus* branching from the lineage containing *domesticus* at a $\Delta T_{50}H$ of 1.9°C, *melanurus* at 1.7°C, *moabiticus* at 1.4°C, and *hispaniolensis* at 0.2°C. The last value, which is indicative of the close relationship between *hispaniolensis* and *domesticus* mentioned earlier, suggests that separation between the two occurred approximately 0.5 mybp, an estimate that is not far from dates indicated by the little fossil evidence known at present (see later discussion).

Although there is still controversy about the placement of the genus *Passer*, I will follow Summers-Smith (1988) in placing it, along with the rock sparrows and snow finches, in their own family, Passeridae. The relationship of the Old World sparrows to other seed-eating groups is still not clear, although the present evidence appears to favor a closer relationship to the Ploceidae than to any other group.

The Origin of *P. domesticus*

Johnston and Klitz (1977) and Summers-Smith (1988) both provided accounts of the evolution of the house sparrow, but they differed considerably in the timing of the origin of the species. Summers-Smith (1988) suggested that the house sparrow is one of several Eurasian species in the genus *Passer* that evolved during the Pleistocene, from an ancestral sparrow that colonized the eastern Mediterranean region from tropical Africa through either the Nile or the Rift Valley. He postulated that this ancestral species was one in which the male possessed a black bib, which is characteristic of all of the present-day Palearctic and Oriental species, and that this ancestral sparrow subsequently expanded its distribution both eastward and westward in the grassland belt. The repeated glacial advances and recessions during the Pleistocene resulted in periodic isolation of sparrows in grassland refugia, which in turn resulted in adaptive divergence and speciation within the group. The lineage leading to the house sparrow was one of the resulting species.

Johnston and Klitz (1977) proposed that the house sparrow, as an obligate commensal of sedentary humans, must have evolved its species status since the advent of agriculture in the Middle East, approximately 10,000 years before present (ybp). Other lines of evidence suggest, however, that the species had an independent evolutionary history that dates to long before the development of agriculture. The fossil record of the house sparrow is sparse and suggests that the species may have arisen by 400,000 ybp. The earliest fossil evidence of *P. domesticus* was found in Oumm-Qatafa Cave near Bethlehem in Palestine (Israel) (Tchernov 1962). These fossils date to approximately 0.4 mybp and were found in association with other fossils assigned by Tchernov to *P. moabiticus* (Dead Sea sparrow) and *P. predomesticus*. The latter represented a now extinct species that had close ties with both *P. domesticus* and *P. hispaniolensis* but differed from both in having a groove in the midventral surface of the premaxilla, whereas each of the others has a ridge (Tchernov 1962). Markus (1964) compared these fossils with those of other African members of the genus *Passer* and concluded that they were most similar to *P. domesticus*. The presence of fossils of *domesticus, moabiticus,* and *predomesticus* in this cave indicates that these species probably lived in association with early Paleolithic humans.

The next fossil evidence of the house sparrow was located in Kebara Cave on Mount Carmel in Israel and dates from the late Pleistocene, approximately 65,000 ybp (Tchernov 1962). This finding again places the house sparrow in close association with humans in the region where agriculture was first developed about 10,000 ybp (Johnston and Klitz 1977).

A second line of evidence suggests a more recent origin for the house sparrow. Parkin (1988) used assays of 15 polymorphic enzymes (see Chapter 2) to determine the genetic distance (D) between the Spanish sparrow (*P. hispaniolensis*) and populations of house sparrows from Norway, France, Italy, Israel, and India. Using an empirically determined estimate of the mutation rate at 14 of these isozyme loci of 1.31×10^{-7} mutations per year (based on Parkin and Cole 1984), Parkin estimated the time to divergence of the several house sparrow populations from the Spanish sparrow as being between 105,000 and 122,000 ybp (Fig. 1.1). This suggests that the house sparrow and the Spanish sparrow last shared a common ancestor near the end of the Pleistocene, but the accuracy of this estimate is based on the correctness of the estimation of the mutation rate at these isozyme loci. This estimate is based on the assumption that the house sparrow colonized the British Isles at the time of the Roman conquest (approximately 2000 ybp) (Parkin and Cole 1984), an assumption that may be unwarranted, particularly if the house sparrow spread throughout northern Europe with the horse.

More recent studies of avian speciation based on changes in mitochondrial DNA have generally suggested earlier divergences between sibling species and even within species populations (Avise and Walker 1998). Based on gene sequencing of the mitochondrial cytochrome *b* gene and its nuclear pseudogene (see later discussion), Allende et al. (2001) suggested that speciation in the genus *Passer*, including that between the house sparrow and the Spanish sparrow, occurred in the Miocene or Pliocene.

The early house sparrow was presumably closely associated with humans, particularly seminomadic, seed-gathering groups. One early glacial advance presumably resulted in disjunct ranges, with one group of house sparrows being found in the Middle East and/or North Africa, and a second group being found on the Indian subcontinent. Differentiation during this period of isolation resulted in the two major subspecific groups discussed here (the *Domesticus*

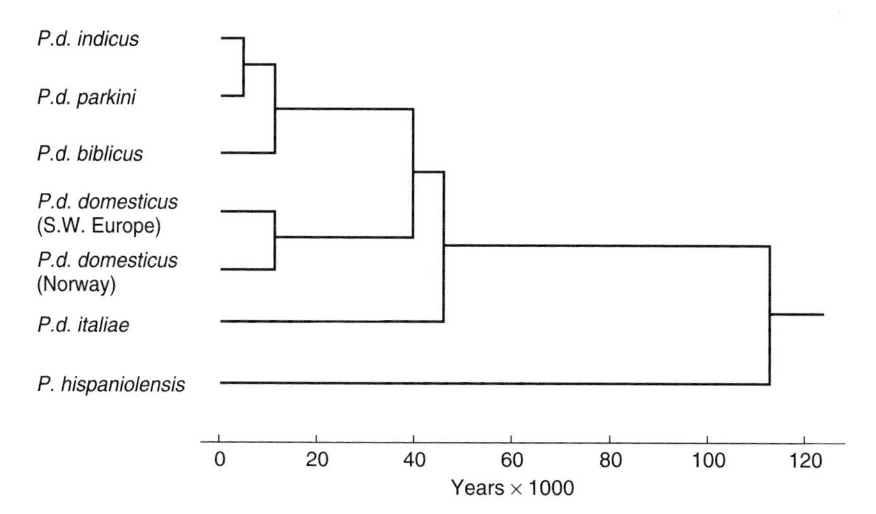

FIGURE 1.1. Dendrogram showing estimated times of divergence of several populations of the house sparrow and one population of the Spanish sparrow. Fifteen polymorphic isozyme loci (see Chapter 2) were examined in two samples each from Norway and France (*P. d. domesticus*), two samples from India (one *P. d. indicus* and one *P. d. parkini*), and one sample each from Italy (*P. d. italiae*) and Israel (*P. d. biblicus*). The one sample of the Spanish sparrow (*P. hispaniolensis*) was taken in Tunis. Note that *P. d. italiae* clusters with populations of the house sparrow rather than with the Spanish sparrow, and it has an estimated time of divergence from the latter of 113,500 ybp (Parkin 1988). Note also that *P. d. biblicus* (a member of the Palearctic group—see Fig. 1.2) clusters more closely with the two members of the Oriental group (*P. d. indicus* and *P. d. parkini*) than with the two populations of *P. d. domesticus*. From Parkin (1988: Fig. 1), courtesy of The Canadian Museum of Nature, Ottawa, Canada.

group and the *Indicus* group). If this scenario is correct, it would mean that the present-day, obligate commensalism with agricultural, sedentary humans developed independently in the two groups after the invention of agriculture in the two regions within the last 10,000 y.

Johnston and Klitz (1977) suggested that the early house sparrow was migratory, because none of the present-day members of the genus that live in temperate continental regions except the house sparrow is permanently resident. One of the major adaptations in the species as it developed its commensal relationship with agricultural humans would therefore have been the adoption of permanent residency instead of migratory behavior. Johnston and Klitz (1977) suggested that this change was facilitated by the year-round presence of stored grain in human settlements and was also advantageous because it ensured early access to nest sites. After the development of permanent residency and commensalism with early agricultural humans, sparrows presumably spread northwestward through Europe with sedentary agriculturalists who colonized the region as the climate ameliorated after the recession of the last glaciation in the region. Parkin (1988), however, suggested a second alternative, colonization of Europe from North Africa via the Iberian peninsula after the recession of the last glaciation. Sparrow bones were found at a Bronze Age site in central Sweden (50°N) that date to approximately 3000 ybp (Ericson et al. 1997). These authors concluded that sparrows probably spread into northern Europe from the Middle East along with the horse, which was used both as a pack animal and as human food at the site.

Passer domesticus

Some of the language used in the following discussion presupposes the taxonomic conclusions that I have reached with regard to the species boundaries of *P. domesticus.* Although I trust that the discussion will justify these decisions, some of them are necessarily arbitrary. I have chosen to follow Vaurie (1949) and Summers-Smith (1988), as well as others, in recognizing two major groups of house sparrows, the Palearctic (*Domesticus*) group and the Oriental (*Indicus*) group (Fig. 1.2). This means that the "Indian" sparrow and several related subspecies (see later discussion) are considered to be house sparrows. I have also chosen to consider the "Italian" sparrow as a subspecies of *P. domesticus.* This decision differs from that of Summers-Smith (1988), but it is one shared by other workers (i. e., Cramp and Perrins 1994; Sibley and Monroe 1990). As noted in Table 1.1, the Spanish sparrow (*P. hispaniolensis*) is considered to be a separate species (contra Johnston and Selander 1971).

The two major problems in attempting to delineate the species boundaries of *P. domesticus* are the "Italian" sparrow and the "Indian" sparrow. The former is the resident sparrow on the Italian peninsula; on Corsica, Sicily, Crete, Malta, and other smaller islands in the Mediterranean; and in parts of northwest Africa. However, the sparrows on the island of Sardinia, which is separated from Corsica by only 11 km, appear to be pure Spanish sparrows (Cheke 1966; Meise 1936; Steinbacher 1954). The "Italian" sparrows are permanent residents and commensals of humans, like *P. domesticus* is elsewhere, but males have distinct plumage differences from the typical house sparrow that resemble in some ways the plumage of the male Spanish sparrow (see Chapter 5). The most noticeable of these differences is that the crown is chestnut or reddish-brown rather than gray. The evolutionary origin and taxonomic relationships of the "Italian" sparrow have been controversial, and numerous hypotheses

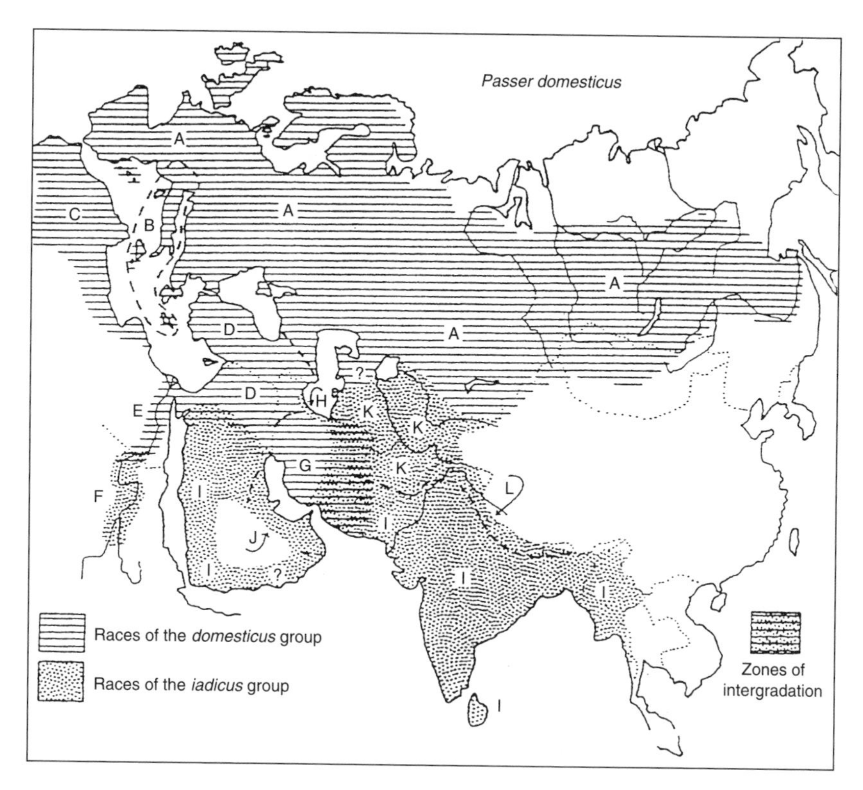

FIGURE 1.2. Natural distributions of the Palearctic and Oriental groups of the house sparrow in Eurasia (and North Africa). Subspecies in the two groups are identified by letters (see also Table 1.2): A, *domesticus*; B, *italiae*; C, *tingitanus*; D, *biblicus*; E, *niloticus*; F, *rufidorsalis*; G, *persicus*; H, *hyrcanus*; I, *indicus*; J, *hufufae*; K, *bactrianus*; L, *parkini*. From Vaurie (1956: Fig. 1), courtesy of the American Museum of Natural History.

have been proposed to account for its origin and present distribution (Baumgart 1984; Haffer 1989; Johnston 1969c; Massa 1989; Meise 1936; Lo Valvo and Lo Verde 1987; Stephan 1986; Summers-Smith 1988). Much less ink has been spilt on the question of the relationship of the "Indian" sparrow to the house sparrow, but some of the same issues are raised with respect to the "Indian" sparrow as with the "Italian" sparrow.

The "Italian" Sparrow

Meise (1936) devised a hybrid index to analyze the geographic variation in the "Italian" sparrow. He identified seven plumage characteristics that differed between the typical male house sparrow and the typical male Spanish sparrow, assigning a value of 0 for the house sparrow-like coloration and a value between 5 and 15 for the Spanish sparrow-like coloration. Specimens were scored on each trait; and a "pure" house sparrow-like male would score 0, and a "pure" Spanish sparrow-like male would score 100. Meise evaluated specimens of sparrows from throughout Italy, North Africa, and the Mediterranean islands using this index and found that there was a clinal decrease in average score from Sicily northward through the Italian peninsula. The average score for Sicilian males was 81, whereas at Rome it was 65; in Florence, 54; and in northern Italy south of the Alps, 38.5. Meise also identified a narrow hybrid zone in the Alps between "pure" house sparrows on the continental side and "Italian" sparrows on the peninsular side. He concluded that the "Italian" sparrow was the product of hybridization between *P. domesticus* and *P. hispaniolensis* and chose to consider it a well-marked subspecies of the house sparrow, *Passer domesticus italiae.* Meise also suggested that the "Italian" sparrow arose initially in North Africa, when house sparrows spreading westward from the eastern Mediterranean encountered Spanish sparrows already associated with early human settlements in North Africa. Hybridization between the two species occurred, and the hybrid sparrows subsequently colonized the Italian peninsula before the arrival of house sparrows spreading northwestward through Europe, also from the eastern Mediterranean region (see earlier discussion). The southern Alps represented a physical barrier to gene flow between typical house sparrows north of the Alps and "Italian" sparrows on the peninsula that has greatly retarded the flow of house sparrow genes southward.

Johnston (1969c) essentially repeated Meise's earlier work, collecting new specimens throughout the region and analyzing them using a hybrid index based on six of Meise's traits (all but ear coverts). His scale varied from 0 for "pure" house sparrow-like traits to 17 for "pure" Spanish sparrow-like traits. His results closely paralleled those of Meise, with Sicilian birds having an average score of 13.0; Napoli birds, 9.2; Florence birds, 7.5; and birds from five northern Italy locations, 6.8. Johnston concurred with Meise that the "Italian" sparrow was the product of hybridization between house and Spanish sparrows but suggested that it might be useful to designate this well-marked hybrid group as a separate species, *P. italiae.* Lo Valvo and Lo Verde (1987) came to a similar conclusion, designating the "Italian" sparrow an emergent interspecies.

In addition to the morphological evidence, some behavioral evidence also suggests a possible hybrid origin of the "Italian" sparrow. A recent study examined geographic variation in the structure of the advertisement vocalization of male sparrows (the "chirrup"—see Chapter 7) at 11 locations spanning the length of the Italian peninsula, including the islands of Sicily and Corsica.

Fulgione, Esposito et al. (2000) used principal components analysis to compare elements of song structure at the 11 locations with Spanish sparrow vocalizations from Sardinia and house sparrow vocalizations from France. Song structure on the Italian peninsula varied clinally, with the more southern locations showing close resemblance to Spanish sparrow song, and the most northern locations resembling house sparrow song.

Several studies have examined the hybrid zone between house sparrows and "Italian" sparrows in the mountains separating Italy from France, Switzerland, Austria, and Croatia (Johnston 1969c; Lockley 1992, 1996; Meise 1936; Niethammer 1958; Niethammer and Bauer 1960; Schifferli and Schifferli 1980; Schöll 1960; Summers-Smith 1988). These studies have tended to share a common design, that being to identify the phenotypes of male sparrows in a string of towns or villages leading up and through an alpine pass. Males with reddish-brown crowns are considered to be "Italian" sparrows; those with gray crowns, house sparrows; and those with mixed crown coloration, hybrids. The proportions of the three phenotypes are then used to determine the "hybrid zone" between house and "Italian" sparrows. Most of the studies report a hybrid zone with a width of 35–40 km, with populations on either side of this zone being either entirely house sparrows or entirely "Italian" sparrows. There is sometimes a hiatus in the distributions in the higher mountain passes, which suggests that sparrows are not able to sustain viable populations at these elevations. Although the hybrid zone appears to have remained relatively stable since it was first described by Meise (1936), some recent studies have suggested movements of either house sparrow or "Italian" sparrow phenotypes in some of the passes. Specifically, some villages at higher elevation that were previously devoid of sparrows have been colonized in recent years (Lockley 1992, 1996; Löhrl 1963; Schöll 1959). Lockley (1992) suggested that increased human development in the Alps, particularly for recreation and tourism, has contributed to the successful colonization of the higher elevations.

Stephan (1986) identified five proposals with respect to the taxonomic status of the "Italian" sparrow. In addition to those already identified—subspecies of house sparrow (Meise 1936) and its own species originating as a hybrid of house and Spanish sparrows (Johnston 1969c)—another proposal is that the Spanish and house sparrows are conspecific (Johnston and Selander 1971). A fourth alternative is that the Spanish sparrow is a subspecies of *P. italiae* (Baumgart 1984). The final alternative is that the Spanish sparrow and the "Italian" sparrow are closely related species with separate origins from a common ancestor that was not closely related to the house sparrow. Based on an analysis of feather pigmentation, Stephan actually favored the last proposal, asserting that the house sparrow was in another lineage. An additional hypothesis, proposed by J. M. Harrison (1961a), is that the house sparrow has evolved from the Spanish sparrow, and the "Italian" sparrow represents a transitional stage in that evolutionary pathway.

So, what does one do with the "Italian" sparrow? Macke (1965) established mixed pairs of German house sparrows and Sardinian Spanish sparrows in outdoor aviaries and obtained several F_1 offspring from the mixed pairs. The males were similar in plumage coloration to "Italian" sparrow males, providing some support for the hypothesis of a hybrid origin of the "Italian" sparrows. Alonso (1984a) also established mixed pairs of house and Spanish sparrows from sympatric populations of the two species in Spain and obtained F_1 hybrids from a pair consisting of a male house sparrow and a female Spanish sparrow. The one male hybrid resembled the "Italian" sparrow phenotype. Alonso's efforts to obtain an F_2 generation by pairing male and female F_1 offspring were unsuccessful. Hybridization has also been reported between house sparrows and Spanish sparrows in several areas of sympatry between the two species: North Africa and Yugoslavia (Summers-Smith and Vernon 1972), central Spain (Alonso 1985), and central Asia (Gavrilov and Stephan 1980; Stephan and Gavrilov 1980).

Parkin (1988) used data from 15 polymorphic isozyme loci to estimate genetic distances (D) and times since divergence of several populations of house sparrows and a population of the Spanish sparrow from Tunisia (see earlier discussion). Two of the populations sampled were "Italian" sparrows from northern Italy (Lake Idro and Lake Maggiore). The estimated time since divergence of the "Italian" sparrow from house sparrows of southwestern Europe was 15,300 ybp (SE = 2000), and that between the "Italian" sparrows and Tunisian Spanish sparrows was 113,300 ybp (SE = 3700). Parkin noted that the populations of "Italian" sparrows sampled were from the northern part of the peninsula, which might have influenced the estimated genetic distance due to gene flow from European house sparrows across the narrow hybrid zone. He nevertheless concluded that the "Italian" sparrow is much more closely related to *domesticus* than to *hispaniolensis*.

This conclusion was supported by recent work based on the sequencing of the mitochondrial cytochrome *b* gene and nuclear pseudogenes of the mitochondrial gene. Allende et al. (2001) isolated segments of mitochondrial and nuclear DNA from nine species belonging to the genus *Passer*, including house sparrows from Spain, Spanish sparrows from the Canary Islands, and "Italian" sparrows from Italy. After amplification of the DNA samples with the polymerase chain reaction (PCR), they sequenced both the mitochondrial cytochrome *b* gene (containing 924 base pairs) and nuclear pseudogenes of the cytochrome *b* gene that were found in five of the species. Genes from at least two individuals of each species were sequenced, and intraspecific variation was less than 0.1%. The base sequences of the mitochondrial gene and the nuclear pseudogene of the "Italian" sparrow were much more similar to those of the house sparrow (differing by 0.54% and 0.98%, respectively) than they were to those of the Spanish sparrow (2.66% and 9.46%, respectively). Dendrograms constructed on the base sequence data from both genes place the "Italian" sparrow with the house sparrow rather than with the Spanish sparrow, and the authors concluded that the "Italian" sparrow should be considered a subspecies of *P. domesticus.*

The results of the experimental hybridization studies and the evidence of hybridization in nature both support the proposition that the "Italian" sparrow is the product of hybridization between the house sparrow and the Spanish sparrow. The fact that hybridization between these two results in viable offspring raises the question of whether these two groups are really distinct at the species level. The two groups occur sympatrically over a broad band running from the Iberian peninsula in the west, through southern Europe and the Middle East, into central Asia (Meise 1937; Summers-Smith 1988). Throughout most of this region, they do not regularly interbreed. The house sparrow is a year-round resident and an obligate commensal of humans in most of the region of overlap, whereas the Spanish sparrow is migratory and less closely associated with human settlements. The Spanish sparrow usually breeds in large colonies with nests situated in trees located close to rivers or lakes, thus tending to occupy more mesic habitats than the house sparrow (Summers-Smith 1988). The clear ecological separation of these two groups, as well as the effective genetic separation found throughout most of their region of sympatry, argues in favor of their specific distinctness. Parkin's (1988) studies of the genetic distance between widely scattered house sparrow populations and the Spanish sparrow (see Fig. 1.1) and the mitochondrial DNA studies of Allende et al. (2001) also support the conclusion that the two groups are in fact specifically distinct. Despite their ability to interbreed successfully, I consider the Spanish sparrow (*P. hispaniolensis*) to be a separate species from the house sparrow.

If the house sparrow and the Spanish sparrow are different species, then one must make a decision about the taxonomic position of the "Italian" sparrow. The ecological evidence argues strongly in favor of a closer affinity to the house sparrow than to the Spanish sparrow. Like the house sparrow, the "Italian" sparrow is a permanent resident living in close association with agricultural and urban humans. The presence of a narrow and apparently stable hybrid zone between European house sparrows and peninsular "Italian" sparrows, however, suggests a degree of genetic separation between the two groups. At least one observer has suggested that, where "Italian" and house sparrows occur together in the zone of overlap, "Italian" sparrows occupy the more built-up areas, and house sparrows are more confined to the village edges (Wettstein

1959). This would suggest that there is also some ecological segregation between the two forms, with the "Italian" sparrow actually being more of a "house" sparrow than the house sparrow. Parkin's (1988) preliminary evidence of genetic distances suggests that the "Italian" sparrow is much closer to the house sparrow than to the Spanish sparrow, a position supported by the gene sequences of mitochondrial cytochrome *b* genes and cytochrome *b* pseudogenes (Allende et al. 2001). I therefore have chosen to adopt Meise's (1936) conclusion that the "Italian" sparrow is a well-marked subspecies of the house sparrow.

The "Indian" Sparrow

The taxonomic questions related to the "Indian" sparrow are somewhat different from those associated with the "Italian" sparrow. Historically, the house sparrows of India have usually been treated as a well-differentiated group within the species *P. domesticus* (Summers-Smith 1988; Vaurie 1949). There have been recent proposals, however, that these two groups are specifically distinct (Gavrilov and Korelov 1968; Yakobi 1979). The bases for this claim come from observations in central Asia east of the Caspian Sea, primarily in Kazakhstan. The nominate race of the house sparrow, *P. d. domesticus*, which was historically confined to Europe east of the Ural Mountains, colonized this region recently as it expanded its range eastward from the Urals all the way to the Pacific Ocean, following the development of cultivation in Siberia and the construction of the trans-Siberian railway (Summers-Smith 1963). This, in turn, brought *P. d. domesticus* into contact with a local subspecies belonging to the Oriental group, *P. d. bactrianus*, and proposals for species status for the "Indian" sparrow stem from observations in this region of overlap.

Gavrilov and Korelov (1968) compared the ecology and morphology of the two groups breeding in Kazakhstan. The house sparrows are permanent residents, and the "Indian" sparrows are migratory. The resident sparrows begin breeding in late March or early April and typically rear three broods. The migratory birds arrive in late April or early May, begin breeding in May, and raise only one brood. The resident sparrows are also considerably larger than the migrant sparrows (mean body mass, 28.31 g [SE = 0.28] and 24.88 g [SE = 0.15], respectively). Gavrilov and Korelov concluded that the two groups are effectively reproductively isolated from each other and hence represent two different species; they proposed the name *P. indicus* for the migratory sparrows.

Hybridization between the two groups does occur despite the temporal separation and size difference. Gavrilov (1965) described 7 apparent hybrids from a sample of 110 sparrows captured in Tajikistan. He found that the average wing length, total length, bill length, bill depth, and body mass of the 7 presumed hybrids were intermediate with values of the other 103 specimens of house and

"Indian" sparrows. He also noted that the individuals from the two groups lived side by side, often nesting in the same trees.

In the sparrows of central Asia, we therefore have a situation where two groups that have recently come into contact are partially separated both ecologically and genetically. What does this situation mean for the species status of the "Indian" sparrow? In some respects, this is a concrete example of a classic question in systematics. For a widely distributed species showing clinal variation in numerous traits over its broad geographic range, presumably as adaptations to very different local conditions (see Chapter 2), would members of the putative species from the two ends of the cline be able to interbreed and hence meet the defining criterion of the biological species concept? The traditional answer to that question is to determine whether there is a pattern of continuous breeding among variants along the clinal gradient. If this is the case, then the entire assemblage is considered to be a single species even if members from the extreme ends of the gradient would not interbreed.

In the case of the house sparrow, there are several areas of contact between subspecies of the Palearctic group and subspecies of the Oriental group. Vaurie (1956) reported that the two groups intergrade smoothly into each other in Iran, where three subspecies meet. The two groups also meet along the Egyptian-Sudanese border and interbreed, but do not intergrade smoothly (Vaurie 1956). Little information appears to be available on the situation at other sites of contact. Nevertheless, as stated earlier, I will follow the traditional view and include the "Indian" sparrow in *P. domesticus.*

Other Hybrids

Numerous instances of hybridization between the house sparrow and the tree sparrow (*P. montanus*) have been reported (Albrecht 1983; Cheke 1969; Cordero 1990a, 1990b, 1991a; Ruprecht 1967; see Cordero and Summers-Smith 1993 and Summers-Smith 1995 for recent reviews), and at least one case of hybridization with the Somali sparrow (*P. castanopterus*) has been reported (Ash and Colston 1981). The tree sparrow is a sexually monomorphic species, and it is the only species in the genus in which the female possesses a black throat patch, or badge. It is widely sympatric with the house sparrow, co-occurring throughout much of Europe, northern Asia, and parts of southeastern Asia. It is commensal with humans throughout most of its range, and in most of China and in Japan, where *P. domesticus* is absent, it is the "house" sparrow (Summers-Smith 1963, 1988). Summers-Smith (1988) suggested that plumage monomorphism may have evolved in the tree sparrow as a species-isolating mechanism in the region of sympatry in southeastern Asia and spread throughout the species subsequent to its development.

Cordero and Summers-Smith (1993) identified four hypotheses that have been advanced to explain the occurrence of hybridization between house and tree sparrows: (1) sexual imprinting (usually by cross-fostering), (2) promiscuous behavior, (3) mate restriction (too few potential mates), and (4) misidentification. The first of these hypotheses was proposed by Cheke (1969), who cross-fostered chicks of the two species by transferring eggs between nests in England. One of the cross-fostered young, a male house sparrow, paired with a female tree sparrow in the following year. Cheke suggested that either egg-dumping or nest usurpation might result in young of the two species occasionally being cross-fostered naturally. If mate selection is influenced by early experience, or imprinting, these instances of naturally occurring cross-fostering could result in hybridization. Naturally occurring cross-fostering between these two species has not been observed, however (Cordero and Summers-Smith 1993). Instead, Cordero and Summers-Smith found that in 23 (72%) of the 33 recorded instances of hybridization between the two species, one or both of the species was rare, and they concluded that the mate restriction hypothesis was the most likely explanation for hybridization between these two species.

Several reports of hybridization between house sparrows and tree sparrows in western Europe have been published since the Cordero and Summers-Smith review (Constantini 1996; Crewe 1997; Lehto 1993; Solberg and Ringsby 1996; Stepniewski 1992; Taylor 1994). Recent declines in the population sizes of the two species (see Chapter 8), but particularly in tree sparrow numbers (Marchant et al. 1990), may be resulting in an increased incidence of hybridization between the two species.

Mate restriction may account for the one observed instance of hybridization between the house sparrow and the Somali sparrow in Somalia, where the house sparrow does not normally occur (Ash and Colston 1981). The reporting authors suggested that the hybrid probably resulted from the accidental arrival of a house sparrow in Somalia, possibly by ship.

At least three studies have suggested that house × tree sparrow hybrids are fertile. Hume (1983) reported observing a male hybrid, paired with a female house sparrow, feeding nestlings in England in June 1979. Similarly, Solberg et al. (2000) reported an apparently successful mating between a male F_1 hybrid and a female house sparrow on an island in northern Norway. The male had been reared in 1995 as one of seven successful fledglings from a pair consisting of a female tree sparrow and a male house sparrow. It remained unmated in 1996, when the male:female ratio on the island was 17:8, but mated with a female house sparrow in 1997, when the sex ratio was 7:5. The pair had two clutches of five and six eggs, with the first hatching all eggs and fledging four young, and the second hatching three eggs, of which two lived to be at least 2 d old (no further observations were provided). In addition to this instance, a second male hybrid, which had been banded as an adult, was observed mated to a female house sparrow on another island in 1996; however, the nest was inaccessible, and therefore no data on its success was available. Cordero (2002) also reported a male hybrid paired with a female house sparrow at a nest containing at least two nestlings in Spain.

The fact that all of these instances involved male hybrids means that the evidence of hybrid viability is inconclusive (Cordero 2002). Extra-pair fertilizations are common in house sparrows, with 10%–20% of the offspring in several populations being sired by a male other than the pair male (see Chapter 4). It is therefore possible, although unlikely, that all of the offspring observed in these instances were the result of extra-pair fertilizations. If, on the other hand, house × tree sparrow hybrids are fertile, hybridization with the tree sparrow may provide an important source of genetic novelty for house sparrow populations (Solberg et al. 2000). Solberg et al. noted that the sparse data available to date on the viability of house × tree sparrow hybrids tends to support Haldane's rule, which states that hybrids of the homogametic sex (males in birds, see Chapter 2) survive better than hybrids of the heterogametic sex.

Subspecies of the House Sparrow

Summers-Smith (1988) recognized 11 subspecies of the house sparrow. Adding the "Italian" sparrow to this list (see earlier discussion) brings the total to 12. Table 1.2 identifies the 12 subspecies and assigns them to either the Palearctic *(Domesticus)* group or the Oriental *(Indicus)* group. Brief descriptions of the current natural ranges of each subspecies are also given in the table.

Distribution of the House Sparrow

The following description of the distribution of the house sparrow treats the natural distribution of each subspecies and also describes the introduction and present distribution in areas to which the species has been either intentionally or unintentionally introduced by humans. Much of the distributional information for the following account was obtained from Summers-Smith (1988), Cramp and Perrins (1994), Lowther and Cink (1992) and American Ornithologists' Union (1998), and the information on the history of introductions is drawn primarily from Long (1981) and Lever (1987).

P. d. domesticus

The natural distribution of the nominate race extends from the British Isles and northern Scandinavia throughout Europe (except for the Italian peninsula)

TABLE 1.2. Subspecies of the House Sparrow (*Passer domesticus*)

Subspecies	Approximate Natural Range
Palearctic (Domesticus) group	
P. d. domesticus (A)	Europe (except Italy) and northern Asia
P. d. tingitanus (C)	Northwestern Africa
P. d. italiae (B)	Italian peninsula, Mediterranean islands, and north-central Africa
P. d. biblicus (D)	Middle East eastward into Iran
P. d. niloticus (E)	Northern Egypt
P. d. persicus (G)	Central and eastern Iran
Oriental (Indicus) group	
P. d. indicus (I)	Arabian Peninsula and Indian subcontinent eastward to Burma
P. d. rufidorsalis (F)	Upper Nile Valley of Sudan
P. d. hyrcanus (H)	Northern Iran around Caspian Sea
P. d. hufufae (J)	Eastern Arabian Peninsula
P. d. bactrianus (K)	Central Asia (Kazakhstan to Afghanistan)—migratory
P. d. parkini (L)	Kashmir, Nepal

The twelve subspecies are separated into two groups, the Palearctic or *Domesticus* group and the Oriental or *Indicus* group (Summers-Smith 1988). The principal differences between the two groups, besides their distributions, are that sparrows of the Oriental group are generally smaller, have whiter cheeks, and have darker and richer chestnut coloration (Vaurie 1956). Approximate natural ranges of each subspecies are also given. The letters in parentheses after each subspecies name refers to the designations in Fig. 1.2.

and in a narrowing band across northern Asia to the Pacific coast. Small populations have also become established, possibly by natural colonization, in Iceland (Cramp and Perrins 1994). Cramp and Perrins (1994) recognized another subspecies (*P. d. balearoibericus*), with a distribution along the Mediterranean coast of Europe, which is included in the nominate race in the present treatment. Range expansion northward in Scandinavia and across Asia has occurred in historical times, primarily since the middle of the 19th century, following the expansion of agriculture in these regions and the completion of the trans-Siberian railway (Summers-Smith 1963). The northern boundary of the range in Siberia is poorly known, and it dips southward as one proceeds eastward, accounting for the narrowing distributional band on the continent. The southern border of the Asian range runs from the Caucasus in southern Russia, throughout most of Kazakhstan south to eastern Uzbekistan, across northern Mongolia (south to the Gobi Desert), and across the northeastern tip of China to the Pacific. House sparrows have also become established recently on Rishiri Island, the northernmost island in Japan (Sano 1990). This subspecies is also the source of many of the introduced populations throughout the world, including those in Australia, New Zealand, North and South America, and numerous near-shore and oceanic islands.

House sparrows were first successfully introduced into North America in 1852 or 1853, when about 100 individuals were brought from Liverpool (UK) on a steamship, appropriately named *Europa*, and released in New York City (Barrows 1889). Subsequent introductions of additional sparrows from Great Britain and Germany occurred between 1854 and 1881 at several eastern and midwestern cities, as well as on the Gulf coast of Texas. In addition, there were numerous releases of sparrows from already established populations throughout much of the eastern United States and Canada, as well as on the Pacific coast of the United States. These releases undoubtedly helped to accelerate the rate of range expansion of the established populations, and by 1915 the species had achieved a transcontinental distribution in both the United States and Canada (Robbins 1973). Today, the sparrows occur ubiquitously in urban and agricultural areas of the lower 48 states and across southern Canada from central British Columbia, southwestern McKenzie District (Northwest Territories), northwestern and central Saskatchewan, and northern Manitoba to central Ontario, southern Quebec, and Newfoundland (American Ornithologists' Union 1998). Widely scattered populations persist north of this area, primarily in cities, where they apparently survive the long, harsh winters by living indoors, such as in a railroad roundhouse or in grain elevators in Churchill, Manitoba (Weaver 1939a). Sparrows have also been recently reported at two locations in Alaska (Anchorage and Petersburg) (see Summers-Smith 1993). The highest densities of sparrows in North America, based on both Breeding Bird Survey data (Robbins 1973) and Christmas Bird Count data (Lowther and Cink 1992; Root 1988), are in areas of most intense agriculture in the central and upper midwest regions of the United States from central Nebraska and Kansas east to Ohio.

North American populations also spread southward into Mexico along both the Pacific and Gulf coasts in about 1910 (Wagner 1959). Extremely arid habitat halted the advance along the Gulf coast, but along the Pacific coast it continued southward and eventually spread across the isthmus. Range expansion through Central America continues to the present, with the species spreading rapidly through Guatemala, El Salvador, and Costa Rica and into Panama during the past 30 y (Reynolds and Stiles 1982; Thurber 1972, 1986).

Australia was apparently the next continent to which the nominate race was introduced, with three shipments of sparrows from Great Britain being released in Melbourne (Victoria) in 1863 (Lever 1987; Long 1981). Subsequent introductions of imported birds from Great Britain and western Europe occurred several times between 1864 and 1872 (Sage 1957). Sparrows were also introduced at Adelaide (South Australia) and Sydney (New South Wales) in 1863, and at Brisbane (Queensland) in 1869, although the latter proved unsuccessful. Additional introductions of birds from established Australian populations occurred at numerous locations, including Tasmania, in the years following 1863, accelerating the range expansion from the early populations. Today, sparrows occupy most of eastern and southern Australia from northern Queensland to western South Australia.

The first introduction of sparrows to South America occurred in either 1872 or 1873 with the release of 20 cages (or possibly 20 pairs) of birds in Buenos Aires, Argentina (Lever 1987; Long 1981). Additional introductions of European birds were made to Santiago, Chile, in 1904 and to Rio de Janeiro, Brazil, in 1905 (Summers-Smith 1988). Sparrows from established populations in Argentina were also subsequently released in Uruguay, Chile, and Peru. Range expansion was quite rapid in the temperate regions of South America, with sparrows spreading throughout Argentina and Chile; as far south as Ushuaia, Tierra del Fuega, the southernmost town in the world (54°50'S) (Summers-Smith 1963); and also northward into Paraguay, southern Brazil, Uruguay, and Peru. Range expansion occurred rapidly along both the Pacific and Atlantic coasts, and it slowed considerably when the species reached Amazonia (Smith 1980). Sparrows were first recorded to have crossed the equator, in Esmeraldas, Ecuador (0°55'N), in 1977 (Crespo 1977). In Brazil, the construction of roads through the rainforest apparently facilitated further range expansion, and the species has recently been extending its range both along the coast and inland along the Transamazon Highway (Smith 1973, 1980). The recent establishment of a population of sparrows in La Guaira, Venezuela, is probably the result of shipborne colonists arriving from populations in the West Indies, most likely Curacao, rather than range expansion from Brazil (Sharpe et al. 1997). The present distribution of the species in South America extends from western Colombia south through Chile along the Pacific coast, from eastern Brazil south through Argentina along the Atlantic coast, and from central Bolivia and Paraguay southward in the center of the continent.

Two subspecies of the house sparrow were introduced in South Africa, the nominate race at East London and *P. d. indicus* at Durban (see later discussion). An unknown number of birds from Great Britain were released at East London, probably sometime in the first quarter of the 20th century, but various sources give dates from 1907 to 1930 (Summers-Smith 1988). The population at East London persisted but failed to spread far from the area of introduction (see Vierke 1970), and it has apparently been swamped by the much more successful *indicus* birds (Winterbottom 1959).

The nominate race has also been introduced, either intentionally or by inadvertent shipboard transport, to numerous offshore and oceanic islands, on many of which it has become established. Documentation of some of these colonizations is scanty, and the dates given here are often only approximate. Successful introductions to islands in the Atlantic Ocean have included several in the Caribbean region: Cuba (1850, late 1890s), Bermuda (1870–1871), Jamaica (1902), and St. Thomas (1953). The house sparrow has colonized other islands in the West Indies from these original sites of introduction, including Puerto Rico (1978) and Hispanola (1978). The introduction to the Bahamas in the 1870s failed, but the species has recently colonized the islands, presumably as shipborne stowaways from Florida, USA (Lever 1987). Other successful introductions or colonizations in the Atlantic have included the Falkland Islands (1919), the Cape Verde Islands (1922–1924), and the Azores (1960). Sparrows were also introduced to Greenland (about 1880), but the population died out after a few years (Long 1981). Introductions to the islands of St. Helena (early 1900s) and South Georgia (before 1975) were also unsuccessful.

In the Pacific, the nominate race was successfully introduced into New Zealand (1859–1871) and subsequently became established on the Chatham Islands (1880 or 1910), Campbell Island (1907), New Caledonia (before 1928), Auckland (before 1935), Norfolk Island (about 1939), the Snares Islands (about 1948), and Vanuatu (before 1957). Birds from New Zealand were also successfully introduced into the Hawaiian Islands (1871), and sparrows from established South American populations were successfully introduced to Easter Island (1928) and the Juan Fernandez Islands (about 1943).

P. d. tingitanus

This subspecies is a year-round resident throughout Morocco and the northern half of Algeria, although it intergrades with *P. d. italiae* in northeastern Algeria. It also has a disjunct distribution in northeastern Libya, from Ajdabiyã to the Egyptian border (Summers-Smith 1988).

P. d. italiae

The "Italian" sparrow is a year-round resident on the Italian peninsula south of the Alps; on the islands of Corsica, Sicily, Malta, Crete, Rhodes, and Karpathos in the Mediterranean; and in northeastern Algeria, northern Tunisia, and extreme northwestern Libya (Summers-Smith 1988). Throughout most of this range, there is a fairly uniform local phenotype, intermediate between house sparrow and Spanish sparrow phenotypes, that varies clinally as described earlier on the Italian peninsula. North African populations, however, contain phenotypes ranging from the house sparrow type (*P. d. tingitanus*) to the Spanish sparrow type, constituting what Summers-Smith (1988) characterized as a "hybrid swarm." This fact further complicates the already complex issues associated with identifying the species boundaries of *P. domesticus* (discussed earlier). One alternative would be to consider all North African house sparrows as belonging to *P. d. tingitanus*, which regularly hybridizes with the Spanish sparrow in a region into which it has recently extended its range (cf. Summers-Smith and Vernon 1972). In this alternative, the subspecies designation *P. d. italiae* would be applied only to those sparrow populations that are derived from hybridization between house and Spanish sparrows sometime in the past. Neither solution is entirely satisfying.

P. d. biblicus

This subspecies, two of which were apparently "sold for a farthing" 2 millennia ago in biblical Palestine (Matthew 10:29), is resident from Turkey, Syria, Lebanon, and Israel eastward through northern Saudi Arabia and Iraq into western Iran (west of the Elburz Mountains). It merges into *P. d. persicus* in western Iran, between the southwestern Elburz Mountains and the head of the Persian Gulf, and into *P. d. indicus* in northern Saudi Arabia (Summers-Smith 1988). It is also resident on Cyprus.

P. d. niloticus

This race is resident in eastern Egypt from Alexandria southward in settled areas of the Nile valley to Wadi Haifa (Sudan), along the Suez Canal, and eastward along the Mediterranean to El 'Arish (Summers-Smith 1988).

P. d. persicus

This final subspecies of the Palearctic group is a year-round resident in eastern and southern Iran, occupying the region from north and east of the Persian Gulf and east of the Elburz Mountains eastward to the borders of Pakistan and Afghanistan (Summers-Smith 1988).

P. d. indicus

The natural range of this subspecies runs from Saudi Arabia in the west to Myanmar (Burma) in the east, occupying the Indian subcontinent south of the Himalayas as well as Sri Lanka, a region throughout which it is a year-round resident. On the Arabian Peninsula, it occupies the western and southern coastal regions, including Yemen and Oman. It also occurs from extreme southeastern Iran and southern Afghanistan south and eastward through most of Pakistan, India, Bangladesh, and Myanmar south to Rangoon (Summers-Smith 1988). This race of the house sparrow has also been successfully introduced, both intentionally and unintentionally, to numerous places.

As noted earlier, this subspecies was intentionally introduced at Durban (South Africa), with dates of introduction varying from 1890 to 1900 (Lever 1987; Long 1981). Initially this population spread slowly, but the rate of range expansion has accelerated in recent years (Vierke 1970). It now occupies most settled areas from South Africa westward and northward through much of Lesotho, Namibia, Botswana, Zimbabwe, Zambia, Mozambique, and Malawi and into the southernmost regions of the Republic of Congo and Tanzania. Birds apparently belonging to this subspecies colonized Mombasa (Kenya) about 1950 and have slowly spread inland toward Nairobi and northward along the Kenyan coast. Shipborne immigrants from southern Africa have also become established in Senegal, arriving about 1970, and have spread inland in Senegal as well as into Gambia and Mauritania (Lever 1987).

This subspecies has also become established on numerous islands, particularly in the Indian Ocean. These include Reunion Island (about 1845), Mauritius (1859–1867), the Andaman Islands (1882 and 1895), the Chiagos Archipelago (British Indian Ocean Territory) (before 1905), Zanzibar (1910), Rodrigues Island (about 1930), the Maldive Islands (1962), and the Amirante Islands (before 1969). The population that has apparently become established on Java since an introduction after 1885 (Lever 1987) is presumably derived from this subspecies. The sparrows established on the Comoros Islands (1879) apparently belong either to this subspecies or to *P. d. rufidorsalis* (Summers-Smith 1988).

P. d. rufidorsalis

The natural distribution of this subspecies is in the Upper Nile Valley of Sudan, from Wadi Haifa south to about Renk (Summers-Smith 1988), an area throughout which it is a year-round resident. This subspecies may also be the source of birds introduced into the Seychelles about 1965 (Lever 1987), which presumably came from the Comoros Islands (Gaymer et al. 1969).

P. d. hyrcanus

This subspecies is a permanent resident in north-central Iran, between the Elburz Mountains and the southern shore of the Caspian Sea (Summers-Smith 1988).

P. d. hufufae

This subspecies of the Oriental group is a permanent resident along the eastern coast of the Arabian Peninsula, from Al Hufuf (Saudi Arabia) southward through Qatar and the United Arab Emirates to the northern border of Oman (Summers-Smith 1988).

P. d. bactrianus

This migratory subspecies breeds from southern Kazakhstan (from the northern end of the Caspian Sea east to Karaganda) southward through Uzbekistan, Kyrgyzstan, Tajikistan, and Turkmenistan into central Afghanistan (Summers-Smith 1988, 1995). It spends the winter on the plains of northern Pakistan and northwestern India.

P. d. parkini

This last subspecies occupies the high Karakoram and Himalaya Mountains, typically at greater than 2000 m elevation, from Kashmir (India) eastward through Nepal (Summers-Smith 1988, 1995). It breeds at up to 4500 m but usually moves to lower elevations during the winter (altitudinal migration) (Summers-Smith 1993).

Range Expansion in the House Sparrow

Several studies have examined the rate of range expansion in sparrows subsequent to their introduction to a new continent (e.g., Barrows 1889; Vierke 1970; Wagner 1959; Wing 1943), and some studies have examined range expansion during historical times in natural populations (e.g., Lund 1956; Summers-Smith 1956). One study developed a mathematical model based on empirical observations of population growth and dispersal to estimate the rate of range expansion in invasive species and used two historical examples of range expansion in sparrows to test the predictions of the model (Van den Bosch and Metz 1992).

The rate of range expansion that occurred in North America after the introduction of *P. d. domesticus* to New York City in 1852 or 1853 was truly remarkable. Range expansion occurred most rapidly across the midsection of the country, and more slowly both northward and southward (Wing 1943). The most extreme distances indicated by the isopleths in Fig. 1B from Wing (1943) suggest that, between 1868 and 1888, sparrows had an average rate of range expansion of about 190 km/y. This range expansion was aided by repeated introductions of sparrows from the eastern United States into cities in the midwest and west (Barrows 1889), and the estimated rate is therefore undoubtedly inflated above that of unaided sparrows. Robbins (1973) used additional data and accounted for some of these introductions in his description of the rate of range expansion in North America. Based on his figures, the expansion rate between 1886 and 1910 was about 96 km/y, and the completion of transcontinental colonization by 1910 would have required an average rate of range expansion of 72 km/y for the period 1853–1910. In Mexico, Wagner (1959) reported that the rate of range expansion along the Pacific coast was initially 200 km/y, and Thurber (1986) found that the rate in Central America was 160–240 km/y. On a smaller scale, Johnston and Klitz (1977) found that the average rate of expansion to 12 towns in Kansas (USA) after the introduction of sparrows from the eastern United States at Topeka in 1874 was about 16 km/y.

High rates of range expansion have also been observed in introduced sparrows (*P. d. indicus*) in southern Africa. Assuming that the initial introduction occurred about 1900 (as described earlier), the rate of range expansion for the first 48 y was relatively low, about 7 km/y, based on Fig. 4 in Vierke (1970). This figure shows the range expansion in 4–y intervals after 1948 and indicates that the average rate of expansion (both northward and westward) varied from 30 to 101 km/y in the four periods between 1948 and 1964. Niethammer (1971), discussing Vierke's data, suggested that sparrows in southern Africa first showed a slow rate of growth, followed by an exponential increase in both range and population size.

The construction of roads through the rainforest in tropical South America has facilitated the recent range expansion of sparrows there. Smith (1973) reported that the species spread from Brasilia to Imperatriz (Brazil) in 6 y, a distance of 500 km (83 km/y); it then moved the 170 km from Salinas to Belem (Brazil) in 2 y (85 km/y) (Smith 1980). In Belem (1°30'S), the newly established population grew to about 1000 individuals in 10 y (Da Silva and Oren 1990).

The natural range expansion of *P. d. domesticus* across Siberia (Russian Republic) beginning in about 1800 was analyzed by Summers-Smith (1956). The spread of the species from Irtysh to the western edge of Lake Baikal between 1800 and 1850 suggests a rate of range expansion of about 37 km/y, and the further spread from Lake Baikal to Khabarovsk between 1850 and 1928 suggests a rate of about 30 km/y. Both of these estimates are lower than the 40 km (25 mi) per year suggested by Summers-Smith.

The mathematical model of range expansion developed by Van den Bosch and Metz (1992) is based on a modification of the diffusion equation and empirically determined estimates of dispersal distance and rate of population growth. Van den Bosch and Metz compared the predictions of rates of range expansion made by the model with the historical evidence of the rates of range expansion of sparrows in North America and Siberia. The model predicted rates of 10.2 and 23.0 km/y, respectively, for these two regions, and the authors concluded that this represented a fairly good fit with the empirically determined rates that they chose, 16.8 km/y for North America (based on Johnston and Klitz 1977) and 27.9 km/y (based on the data of Summers-Smith 1956, who actually suggested a rate of about 40 km/y [see earlier discussion]). The predicted rate of range expansion in North America appears to be almost an order of magnitude lower than the rate suggested by the bulk of the evidence. This discrepancy may be due to the assumption of a homogeneous environment made by the diffusion equation model. Because of the close association between sparrows and human settlement (see Chapter 10), most regions are clearly not homogeneous with respect to suitability for sparrows. Dispersing individuals apparently continue to move until they encounter suitable habitat before settling, so dispersal distance is affected by the distribution of suitable habitats.

Chapter 2

EVOLUTION AND GENETICS

Everything that is is adaptive.
—Richard F. Johnston, personal communication, 1967

That depends on what the meaning of the word "is" is.
—William Jefferson Clinton, Grand Jury testimony, 17 August 1998

Evolutionary biologists have been attempting to explain the adaptive bases of the morphology, physiology, behavior and ecology of organisms since the so-called Modern Synthesis of the 1930s and early 1940s. Several years ago, however, two well-known evolutionary biologists offered a strenuous critique of the tack that this research program has taken, particularly among British and American researchers, a program they referred to as the "Panglossian paradigm" (Gould and Lewontin 1979). One of their most serious criticisms concerned the approach taken by many students of evolutionary biology, that of "atomizing" an organism into "traits" and then attempting to identify the contemporary selective factors that have operated to optimize that trait. Gould and Lewontin asserted that this approach fails to give proper recognition to the organism as an organic whole and to constraints to adaptation of a particular trait based on the evolutionary history or the "architectural" structure of the whole organism. This emphasis on contemporary selection is based, according to Gould and Lewontin, on the assumption that natural selection is so powerful that it can override such constraints.

This assumption may be based in part on the paradigmatic studies that form a part of the basic instruction of all young evolutionary biologists (using paradigm in the second sense developed by Kuhn [1970]). Classic studies such as those of industrial melanism in the peppered moth (*Biston betularia*) suggest that natural selection is strong, persistent, and unidirectional, whereas, in fact, selection may more often be weak, intermittent, and multidirectional. One such paradigmatic study that has been a part of the educational background of virtually every evolutionary biologist for the last century is Bumpus's (1899) study of the selective effects of a severe winter storm on house sparrows.

Professor Bumpus and His Sparrows

The first lecture in the summer of 1896 at the Marine Biological Station at Wood's Hole, Massachusetts (USA) was presented by a professor of anatomy from Brown University, Hermon C. Bumpus, and was entitled "The Variations and Mutations of the Introduced Sparrow, *Passer domesticus*" (Bumpus 1897). In this lecture, Bumpus was probably the first to suggest that the widespread introduction of sparrows outside their natural range, and their subsequent spread from the sites of introduction, represented a huge experiment in natural selection. In this initial paper he tested the idea that rapid population growth and spread of sparrows in North America should result in a relaxation of selection on the species. He compared the size, shape, and coloration of 868 sparrow eggs from Great Britain (where most of the sparrows introduced into North America had originated) with those of 868 eggs collected in Massachusetts. Working without the benefit of modern statistical analyses, Bumpus concluded that Massachusetts eggs were shorter and more variable in size and coloration than English eggs.

In his discussion of the implications of these results, Bumpus raised the question of whether the observed changes were caused by selection (adaptive change in the hereditary structure of the house sparrow population) or were the result of phenotypic developmental responses to a changed environment (somatic change): "In brief, is the new variety merely ontogenic, or is it phylogenic?" (Bumpus 1897:13). He suggested that individuals from both populations should be reared in some neutral third locality (what we would now call a common garden experiment) to determine the answer to this question. Although much work has been done in the century since Bumpus's pioneering work, his conclusions remain surprisingly contemporary.

A severe snow, sleet, and rain storm in Providence, Rhode Island (USA), on 1 February 1898 provided Bumpus with the material for another lecture at Wood's Hole, given in the summer of 1898. This lecture, entitled "The Elimination of the Unfit as Illustrated by the Introduced Sparrow, *Passer domesticus*" (Bumpus 1899), included what was to become the most frequently reanalyzed dataset in evolutionary biology. After the storm, 136 immobilized sparrows were brought to Bumpus in his anatomy laboratory at Brown University. Seventy-two of the birds revived, and 64 died. Bumpus recorded the sex of each individual, the age for males (first-year or adult), and nine measured morphological characters of each individual. Working again without the benefit of statistical tools, Bumpus concluded that natural selection, acting through the agency of the storm, selected against females (only 21 of 49 females survived, compared with 51 of 87 males) and favored birds that were shorter and weighed less but had longer wing bones, leg bones, and sternums and greater "brain capacity" (= skull width). He also concluded that the "amplitude of variation" was less in surviving birds than in those that died (= stabilizing selection).

The published version of Bumpus's lecture (Bumpus 1899) contains his entire dataset, a fact that has permitted numerous other scientists (beginning as early as 1911) to reanalyze all or part of his data statistically. Some of the later analyses (e.g., Crespi and Bookstein 1989; Pugesek and Tomer 1996) introduced new methods of analysis that do not depend on the assumption of normality required in multiple regression analysis. Interestingly, the interpretations of those who have reanalyzed Bumpus's data vary considerably, both from Bumpus and from each other, depending in part on the portion of the dataset analyzed and the method of analysis. Table 2.1 summarizes the conclusions of the reanalyses.

When the entire nine-character dataset was used, body mass and total body length were consistently identified as the most important characters discriminating between surviving and nonsurviving sparrows. Lighter, shorter birds survived better, indicating possible directional selection for these traits (Crespi and Bookstein 1989; Harris 1911; Lande and Arnold 1983; O'Donald 1973). A principal components analysis of all nine traits showed no difference on PCI (general body size) and survivorship in either sex, but there was some evidence of stabilizing selection on both PCI and PCII (representing an inverse relationship between body size and limb length) in females (Johnston et al. 1972).

Many of the reanalyses used only a subset of the original data, however. Calhoun (1947a), for example, used only humerus length and femur length, and found no evidence of stabilizing selection in either trait, but did conclude that there was evidence for selection for longer femurs in males. Grant (1972) suggested that total body length might be subject to artifactual measurement error (but see O'Donald 1973), and Johnston et al. (1972) concluded that body mass was lower in surviving birds than in nonsurvivors because of a longer period of metabolic activity in surviving birds before measurement. Johnston et al. (1972) also ignored alar extent, because they believed that it also was subject to greater measurement error. Several studies have followed Grant and Johnston et al. in ignoring either two (body mass and body length) or all three

TABLE 2.1. Summary of Statistical Procedures and Major Conclusions of Studies Reanalyzing the Classic Data of Bumpus (1899) on Differential Survival of House Sparrows after a Severe Winter Storm in Providence, Rhode Island, on February 1, 1898

Study	Statistical Tools	Major Conclusions
Harris 1911	Mean, standard deviation, standard error	Among females, adult males, and juvenile males, shorter, lighter birds survived better; survivors in all three groups had longer humeri and femurs but showed no consistent differences in wing, sternum, or tibiotarsus length or head size; no evidence of stabilizing selection
Calhoun 1947a	χ^2 of 2×2 contingency tables	Treated only humerus and femur length; no evidence of stabilizing selection in either sex in these traits; some evidence of directional selection for longer humeri in males ($.05 < P < .1$)
Grant 1972	*t* test, F test, correlation	Tested for directional and stabilizing selection in each trait; found directional selection for smaller body mass in both adult and first-year males and for longer combined head and beak length in first-year males; found stabilizing selection for mass, humerus length, tibiotarsus length, and keel (sternum) length in females, with high correlation among the four traits
Johnston et al. 1972	Principal components analysis, discriminant function analysis	Tested nine-character set (including body mass) and six-character subset (excluding body mass, total length, and alar extent); surviving and nonsurviving males differed significantly on PCII ($P < .001$) on nine-character set (interpretation uncertain); males differed significantly on PCI ($P = .016$) on six-character set, indicating directional selection for larger size; variance of PCI scores on six-character set for females was significantly lower for survivors than nonsurvivors (stabilizing selection for size); discriminant function analysis on PCI scores (six-character set) of males and females showed no significant difference in scores of nonsurviving birds but significant difference ($P < .001$) indicating selection for sexual size dimorphism
O'Donald 1973	Discriminant function analysis, normal and quadratic models of selection	Total length and humerus length were the most important characters discriminating between survivors and nonsurvivors; variance in humerus length in females, and in both sexes combined, was significantly lower in survivors (stabilizing selection); directional selection for lower total body length in males, and in both sexes combined was found; there was no significant difference in survival by sex; approximate proportion of selective deaths for humerus length was 0.2, with an increase of fitness of about 0.1; approximate proportion of selective deaths for body length was about 0.2, with increase of fitness of about 0.07

Manly 1976	Double exponential fitness functions	Considered all variables (including sex, but not age) except body mass; significant directional selection for shorter body length and longer humerus length in males; nonsignificant selection for the same traits in females
Lande and Arnold 1983	Multivariate analysis of variance, principal components analysis, quadradic regression	Used all nine characters (including cube root of body mass) to estimate selection intensities for both directional and stabilizing selection; found no size differences in adult and first-year males; sexual size dimorphism in body mass, total length, alar length, humerus length, and sternum length; significant directional selection on body mass in both sexes (lighter birds survived better); significant directional selection for shorter body length in males; significant stabilizing selection for overall body size (PCI) in females; selection coefficients ranging from 0.27 to 0.52
Crespi and Bookstein 1989	Path analysis, factor analysis, multiple regression analysis	Analyzed nine-character set and seven-character subset (excluding body mass and total length); both multiple regression analysis and path analysis of nine-character set identified significant contributions of body mass and total length to survivorship in males and of body mass in females (concluded that these are artifacts); both multiple regression and path analysis on seven-character set identified significant effect of wing length on survivorship in males (those with shorter wings survived better)
Crespi 1990	Path analysis, factor analysis	Performed factor analysis on seven traits (excluding body mass and total length) of female sparrows and identified general body size as the first factor and residual large positive correlations between tarsus and femur length (leg size) and head length and width (head size); variance in general size factor was lower in surviving females than in nonsurvivors (stabilizing selection); selection coefficients were negative for tarsus and femur lengths; positive for head length and width, wing length, and sternum length; and negligible for humerus length
Buttemer 1992	Principal components analysis	Principal components analysis on eight metric characters identified PCI as general body size; found that nonsurviving birds had higher body mass-to-PCI score ratios and concluded that birds did not die from energy depletion but from hypothermia
Pugesek and Tomer 1996	Structural equation modeling	Used seven characters (excluding body mass and total length) to construct model using three "latent variables"—general body size, leg size, and head size; males and females did not differ in the way in which the measured variables contributed to the latent variables; larger body size contributed to survival; shorter wings contributed to survival; sex acting indirectly through larger size affected survivorship (males survived better than females)

of these traits that otherwise show the strongest selective effects. Some of these studies found evidence for stabilizing selection for body size in the female sparrows (Crespi 1990; Grant 1972; Johnston et al. 1972; Lande and Arnold 1983), and one found evidence for directional selection favoring larger males (Johnston et al. 1972). Johnston et al. also concluded that there was selection for increased sexual size dimorphism. Manly (1976), using all traits except body mass, concluded that shorter sparrows survived better. In two studies using all traits except body mass and total length, Crespi and Bookstein (1989) concluded that shorter-winged males survived better, whereas Pugesek and Tomer (1996) concluded that larger birds survived better, as did birds with shorter wings.

O'Donald (1973) used the data to estimate selection rates. He used two models to estimate the proportion of selective deaths caused by the storm and found that the proportions of selective deaths were approximately 0.2 for both the stabilizing selection on humerus length and the directional selection on total length. He also concluded that the selection event resulted in increases in mean fitness in the population of about 0.11 and 0.07 for the two characters, respectively.

Although there is much disagreement among those reanalyzing Bumpus's data about the particular targets of selection and even the direction in which selection operated, most agree that Bumpus's sparrows represent an example of a powerful selective event. No one, however, seems to have considered the fact that Bumpus's sparrows actually represent two selection events, the first being the storm itself, which presumably would have resulted in the deaths of all of the individuals collected, and the unspecified collection and laboratory conditions which resulted in the survival of some of the otherwise doomed sparrows. Despite the limitations and disagreements, Bumpus's sparrows have served for more than a century as a salient example of natural selection at work, and, through the numerous reanalyses, have generated a wealth of interesting hypotheses. However, the dataset simply has too many limitations to resolve the conflicts among the different interpretations that have emerged from these reanalyses.

Bumpus's initial insight regarding the opportunity to study natural selection in operation in the areas where sparrows have been successfully introduced has helped to stimulate many efforts to document the action of natural selection in introduced sparrow populations. Natural selection operates on phenotypic variation in a population, resulting in changes in the relative frequencies of phenotypes. To the extent that these phenotypic changes result in changes in the genetic structure of the population, natural selection acts to bring about adaptive evolutionary change. The remainder of this chapter is devoted to describing the results of studies examining geographic variation and natural selection in introduced and native populations of sparrows, as well as current knowledge about the genetics of the species and the evidence for the genetic bases of adaptive change.

Geographic Variation in Introduced House Sparrows

Most of the studies on geographic variation in introduced sparrow populations have focused on morphological characters, particularly body size and proportions, and plumage coloration. In many studies, body size has been based on external measurements such as body mass, bill length and width, wing length, tail length, and tarsus (tarsometatarsus) length. One of the major difficulties associated with these external measurements is that some are subject to seasonal, or even diurnal, change. Some also show variation due to differences in development rate. Wing and tail lengths undergo a regular pattern of seasonal change, with lengths being greatest after the prebasic (postnuptial) molt which follows the breeding season (see Chapter 5) and decreasing thereafter due to abrasion (Schifferli 1981). Bill length also shows a seasonal pattern of change in some populations, with bills being longer in the summer than during the winter (Davis 1954; Packard 1967b; Saini et al. 1989; Steinbacher 1952). This is apparently a result of different rates of abrasion of the beak due to seasonal changes in diet. Age differences in bill size may be caused by the fact that the skeletal elements of the jaw continue to grow for almost a year after birth (Ruprecht 1968). Body mass shows not only regular seasonal changes, with higher masses during the winter than during the breeding season in males (Anderson 1978; Folk and Novotny 1970), but also diurnal fluctuations. Packard (1967a) reported a 7.7% lower body mass for birds at dawn (before feeding) than for birds going to roost in the evening.

In an effort to avoid the complications resulting from seasonal and diurnal variation, many studies have used skeletal material to characterize geographic variation (e.g., Johnston and Selander 1971; McGillivray and Johnston 1987). The basic research design of these studies is one in which morphological measurements of introduced sparrows are compared with those of other introduced populations across a broad geographic and climatic spectrum, or with measurements from birds of the source populations. Morphological differences are then presumed to be the result of natural selection operating in the novel environment.

The first such study followed Bumpus's pioneering work by only 10 years. Townsend and Hardy (1909) compared wing length, tail length, bill length, and tarsus length of samples of house sparrows collected in Massachusetts between 1873 and 1886 with similar measurements from specimens collected in the same area in 1907 and from birds collected near Liverpool, UK, in 1907. No consistent differences were found among the samples. In a second early study, Phillips (1915) measured the wing, tail, and tarsus lengths of sparrows from a wide range of North American localities (including Hawaii) and concluded that the sparrows had not changed in any consistent way from English birds, although they were generally somewhat larger.

Lack (1940) compared the bill lengths, bill depths, and wing lengths of sparrows from three regions of North America (eastern, midwestern, and western) and from Hawaii with birds from England and Germany. North American birds were in general significantly larger than English birds but not German birds. The bill lengths of birds captured in southern California, Baja California, and Hawaii were greater than those of either other North American populations or the German birds, suggesting some change in these populations. Lack also found that North American populations were not more variable than those from Europe (as might be expected if relaxed selection resulted from the rapid population growth) and concluded that North American populations showed surprising stability.

In a more comprehensive study, Calhoun (1947a) measured wing length of 1877 sparrow specimens from throughout North America to examine both temporal and geographic changes since the introduction of the species. Specimens were grouped into four time periods numbered 1 through 4 (1851–1885, 1886–1907, 1908–1930, and 1931–1944, respectively) and six climate zones based on the severity and duration of the winter (with zone 1 being the most moderate and zone 6 the most extreme). The wing lengths of eastern seaboard birds collected before 1886 (all period 1, zone 4) were used as the putative ancestral birds and compared with all other groups. Only two groups showed a significant difference from the ancestral group: both males and females from period 3, zone 2 were smaller, and males from period 2, zone 5 were larger. Calhoun also attempted to determine whether there had been directional change in size over time within North American populations. Three populations had sufficient material from three or four of the time intervals for analysis. Wing length increased significantly in these three populations, with the most dramatic increase occurring between periods 1 and 2. No change occurred, however, between periods 3 and 4. Samples of 310 males and 213 females collected since 1908 east of the Rocky Mountains showed a significant increase in wing length with temperature zone (zones 3 through 6) in both sexes. Calhoun recognized the difficulty of determining whether the patterns observed were caused by natural selection or by direct environmental influences on the birds, but he concluded that the increase in wing size over time represented strong evidence that the changes were genetically based, not just phenotypic responses to cold weather.

A paper somewhat enigmatically entitled "House Sparrows: Rapid Evolution of Races in North America" (Johnston and Selander 1964) marked the beginning of a long series of studies by Richard F. Johnston, Robert K. Selander, and their students on morphological change in sparrows. The paper presented preliminary results of a study based on newly collected samples of 100–250 specimens from each of 17 localities throughout North America (from Oaxaca City, Mexico, to Edmonton, Canada), as well as from Hawaii and Bermuda. It documented pronounced, clinal differences in coloration, wing length, bill length, and body mass in North American sparrows. However, the paper did not name any new races or subspecies. The changes in coloration corresponded to Gloger's ecogeographic rule (members of a species tend to be paler in drier, hotter climates and darker in more mesic, cooler climates), and changes in size tended to conform to Bergmann's ecogeographic rule (members of a species tend to increase in size with increasing latitude) (Fig. 2.1). These changes in the North American sparrows therefore parallel patterns of clinal variation observed in numerous other species, and Johnston and Selander concluded that the changes were the result of adaptive evolutionary responses of the sparrows to North American environments that had occurred within the 111 generations since their introduction.

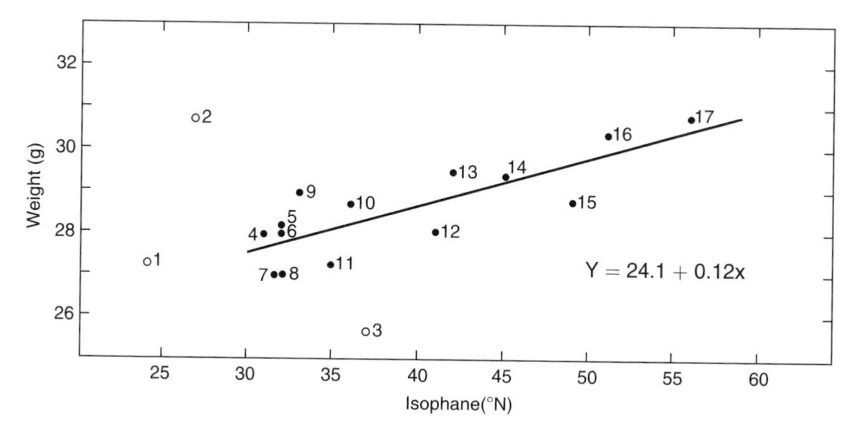

FIGURE 2.1. Body mass of adult male house sparrows plotted against isophane in North America. Isophane is a general climatic variable based on latitude, longitude, and altitude that is expressed as a latitudinal equivalent (°N). Data are means of samples from 17 North American sites from Oaxaca City, Mexico (1), to Edmonton, Canada (17). The regression equation is based on the 14 sites represented by solid dots, with 3 sites in southern Texas and Mexico (1–3, *open dots*) ignored. From Johnston and Selander (1964), with permission. Copyright © 1964 American Association for the Advancement of Science (AAAS).

The 1964 paper introduced some of the protocols that were to be used in subsequent studies in the series. These included the collection of specimens primarily in October and November, to ensure that the plumage was fresh after the prebasic molt (see Chapter 5), and to avoid the problem of seasonal changes in external morphological traits. The use of the spectrophotometer as an objective means of determining plumage coloration was also identified (see also Selander et al. 1964). Additional protocols that were generally used in the series included determination of age categories (subadult and adult) by examination of the degree of skull ossification (Selander and Johnston 1967), and the washing of specimens with detergent and white gasoline to remove dirt particles that might alter plumage coloration (Johnston 1966; Selander and Johnston 1967). Many of the studies were based on skeletal specimens rather than traditional skins, to permit the study of variation in skeletal size and proportions (e.g., Johnston and Selander 1971; Lowther 1977b). Several studies in the series examined variation in native European populations (e.g., Johnston 1969b; Murphy 1985), and at least one also examined variation in introduced South American populations (Johnston and Selander 1973). The studies also pioneered the use of multivariate statistics such as principal components analysis, discriminant function analysis, and stepwise multiple regression in analyzing morphometric data.

The use of degree of skull ossification, which refers to the development of a second layer of bone covering the cranium, to discriminate between adult and subadult sparrows requires some discussion. At least three studies have examined the process of skull ossification in sparrows. Nero (1951) found that skull ossification was usually completed between 190 and 200 days after hatch in sparrows in Wisconsin (USA). His drawings of the temporal course of skull ossification have also been used to age subadult birds (e.g., Graber and Graber 1965). J. G. Harrison (1960, 1961) described skull ossification in British sparrows, and his description closely parallels that of Nero. He did report finding first-year birds with fully ossified skulls as early as August–September, however. Niles (1973) examined the skulls of sparrows from submontane and montane populations in Colorado (USA), and found birds with incompletely ossified skulls at all seasons. The proportion of breeding adults with incompletely ossified skulls in the submontane samples (10.0%) was significantly greater than the proportion of montane breeding adults (1.3%), a fact which Niles attributed to the earlier initiation of breeding in submontane populations and the suspension of skull ossification during breeding. It is therefore apparent that the use of complete skull ossification (skull completely ossified, or SCO) to identify adults, and incompletely ossified skulls (skull not completely ossified, or SNCO) to identify subadults, may actually err in both directions, although it appears much more likely that a high proportion of early-hatched birds in many populations will have completed skull ossification by November. Conclusions based on this method of determining age in sparrows should, therefore, be treated with some caution.

The results of the numerous studies by Johnston and Selander and their colleagues have tended to converge around several broad generalizations. In North America, sparrows show an increase in body size with increasing latitude and with decreasing winter temperatures. There is also a small but consistent difference in overall body size between the sexes, with males being larger than females, and there is a tendency for this sexual size dimorphism to increase with increasing latitude or decreasing winter temperatures. The ratio of hind limb length to core body size tends to decrease with increasing latitude or decreasing winter temperatures. And finally, as noted earlier, plumage coloration tends to be paler (with greater reflectance) in xeric, hot environments, compared with mesic, cool environments. Comparisons of some of these generalizations with the patterns found in Europe have generally found correspondence on the two continents. One difference, however, is that interpopulation variation in morphological traits in North America is significantly lower than that found in Europe (Johnston 1976a; Johnston and Selander 1971).

Body Size and Latitude

The observation of increased body size with increasing latitude in North America has been reinforced by studies of both external and skeletal morphology (Johnston 1969b, 1973; Johnston and Selander 1971, 1972, 1973). Although there is a tendency for size to increase with increasing latitude in North America, the picture is actually more complicated than that. Fig. 2.2 shows the geographic distribution of wing length for North American males. The largest individuals are found in central Canada, the Rocky Mountains, and central plains of the United States, whereas birds from southwestern and south-central areas of the United States are intermediate in size, and birds from the West and Gulf coasts and Mexico are small. Johnston and Selander (1973) performed a principal components analysis on external traits of sparrows from 53 North American sites, as well as from 38 European sites and 11 South American sites. Principal component I (PCI) represented overall body size, having high positive loadings from all five of the characters measured, and it increased significantly with latitude in North American males. In Europe, however, the PCI of males decreased significantly with latitude, and in South America there was no significant relationship between size and latitude.

Similar results were obtained in studies of skeletal characteristics. Johnston and Selander (1971) performed a principal components analysis on 16 skeletal characters of sparrows collected at 36 North American sites (including Bermuda

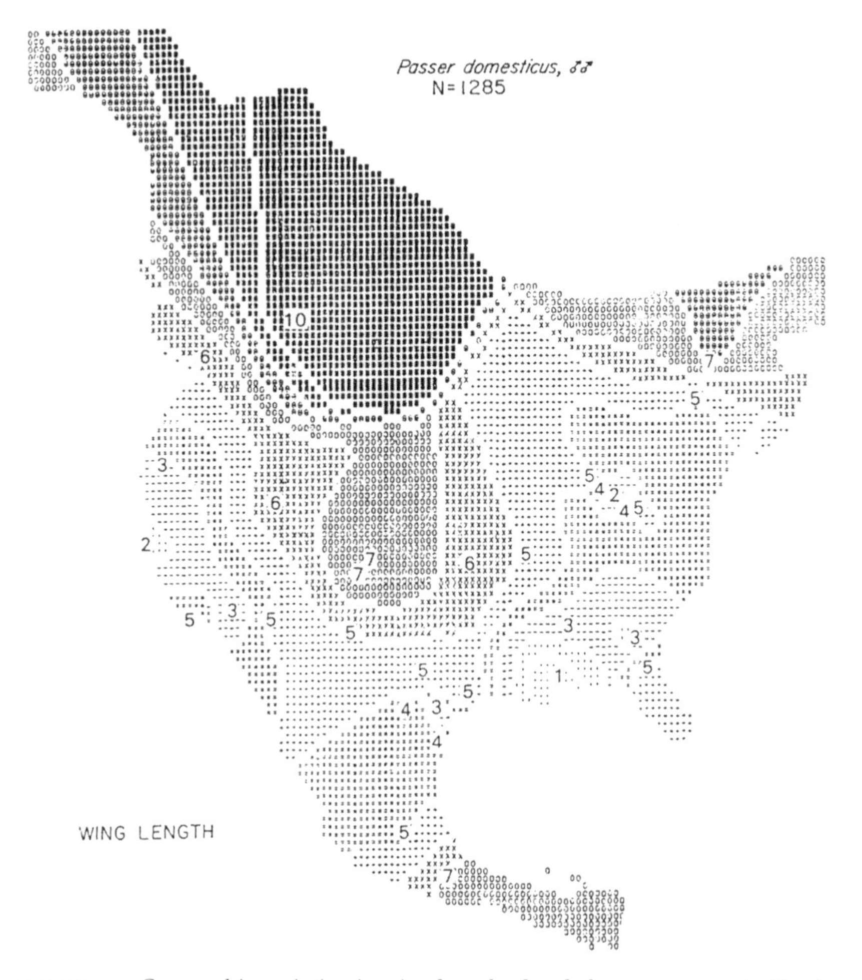

FIGURE 2.2. Geographic variation in wing length of male house sparrows in North America. A generalized contour diagram of variation in wing length of male sparrows in North America based on samples collected at 33 sites from Oaxaca City, Mexico, to Edmonton, Canada. The numbers represent equal-sized increments in flattened wing chord from 1 (75.9 mm) to 10 (79.4 mm). From Johnston and Selander (1972: Fig. 4).

and Hawaii). PCI again represented overall skeletal size and was significantly negatively correlated with mean winter temperature (Fig. 2.3), but it was not correlated with mean summer temperature (see also Johnston and Selander 1972). Johnston (1973) repeated the analysis on sparrows from an additional 10 North American sites and on birds from 30 European sites. PCI was positively correlated with latitude in the North American samples but was nega-

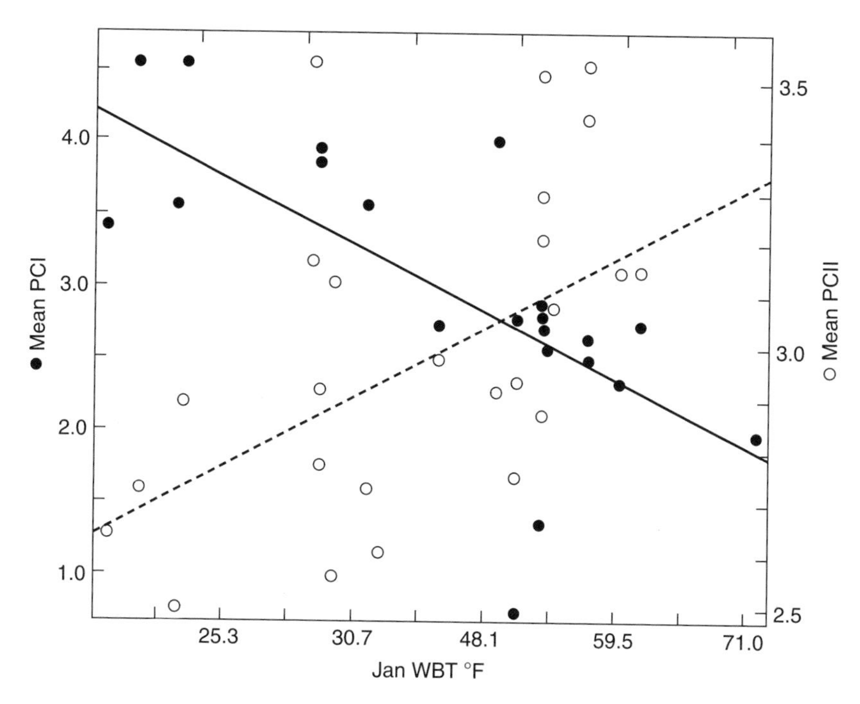

FIGURE 2.3. Relationship of overall body size or PCI (MPC1, *solid circles*) and core-to-limb ratio or PCII (MPC2, *open circles*) of North American male house sparrows with average January wet-bulb temperature (JAN WBT °F). The solid line represents the linear regression of PCI of 16 skeletal elements of male sparrows from sites in North America: $Y = 4.75\ (0.04X - P < .004)$. The dotted line is the linear regression of PCII on the same set of samples: $Y = 2.48 + 0.01X (P < .005)$. From Johnston and Selander (1971: Fig. 11), with permission of *Evolution*.

tively correlated with latitude in the European samples. Johnston (1969b) also reported a significant clinal increase in tarsus length from north to south in Europe (primarily Italy) and North Africa. In another study of geographic variation in an introduced population in New Zealand, Baker (1980) also found that overall body size (PCI on 16 skeletal characters) was negatively correlated with both latitude and isophane.

As mentioned earlier, the clinal increase in size with increasing latitude corresponds to Bergmann's ecogeographic rule, an empirical generalization that has been observed in numerous endothermic animals (Mayr 1956). The most common explanation for the rule is that it represents an adaptation to both cold and hot environments based on surface-to-volume effects. The surface of an animal is its interface with the external environment and is therefore its avenue for heat exchange with the environment. It tends to increase with the square of a linear body measurement, whereas the animal's volume or mass, and hence its potential for generating heat, tends to increase with the cube of a linear body measurement. Cold temperatures favor individuals with lower surface-to-volume ratios (larger size) because of their enhanced ability to conserve heat, and warmer climates favor individuals with higher surface-to-volume ratios (smaller size) because of their enhanced ability to dissipate excess heat.

The reversed latitudinal size clines in Europe and New Zealand are therefore puzzling. In Europe, there are at least two possible reasons for this result that are not accounted for in the original studies. The first is that many of the high-latitude samples were obtained from island or coastal sites (e.g., London and Oxford, England; Oslo, Norway; Copenhagen, Denmark), where winter temperatures are considerably ameliorated by maritime influences compared with sites at lower latitudes in central and eastern Europe. Murphy (1985) performed a stepwise regression analysis of several climate variables on PCI (body size) for samples of sparrows from 25 European localities (Fig. 2.4). Seasonality (mean July temperature − mean January temperature), which was strongly positively correlated with PCI, entered the regression first, followed by precipitation. This suggests conformity to Bergmann's ecogeographic rule in Europe despite the negative correlation between size and latitude observed in the other studies.

A second factor is the inverse relationship between size and latitude on the Italian peninsula (Johnston 1969b). This situation is potentially complicated

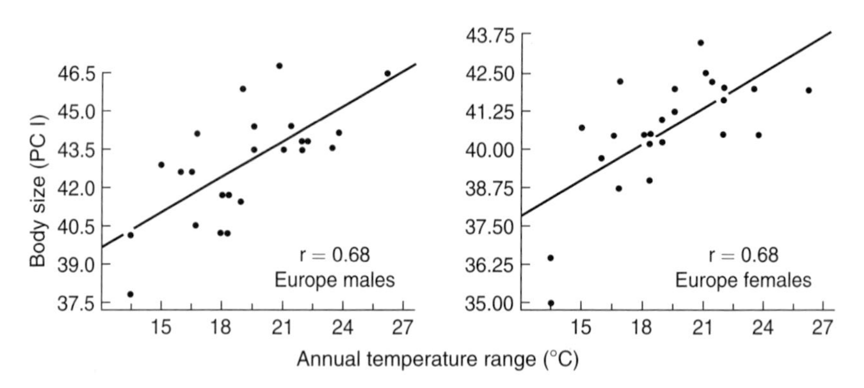

FIGURE 2.4. Body size versus seasonality in European house sparrows. Size is based on 14 skeletal measurements of specimens from 25 European localities from Italy to Sweden, and the annual temperature range is calculated as mean July temperature minus mean January temperature. From Murphy (1985: Fig. 1), with permission of *Evolution*.

by the fact that the Italian peninsula is occupied by a morphologically distinct subspecies, *Passer domesticus italiae* (see Chapter 1), which is generally considered to be the product of hybridization between the house sparrow and the Spanish sparrow. The sparrows on the peninsula show a clinal change from being more Spanish sparrow-like in the south to more house sparrow-like in the north (Johnston 1969c; Meise 1936). If the Spanish sparrows from the original source population were larger than the house sparrows with which they hybridized, then this clinal variation in size on the Italian peninsula might be a result of the degree of introgression of genes from the two species. Apparent hybrids in Spain showed a significant positive correlation between hybrid index score (0 for house sparrow to 17 for Spanish sparrow) and tarsus length ($r = 0.55$, $n = 16$, $P < .05$) (Alonso 1985). Summers-Smith and Vernon (1972) mentioned that hybrids in North Africa were noticeably smaller than coexisting Spanish sparrows. Comparative mensural data from two more recent studies of sympatric populations of the house and Spanish sparrows, one in western Spain and the other in western Algeria, were inconclusive. In Spain, the Spanish sparrow had a significantly longer tarsus and a larger beak than the house sparrow, but a significantly shorter tail (Alonso 1986). In Algeria, the Spanish sparrow had a significantly larger beak than the house sparrow, but a shorter tail (Metzmacher 1986). The two species did not differ significantly in other external traits at the two sites. The more critical question, however, concerns the size differences between the two populations that first hybridized to establish the hybrids from which the present-day populations of *P. d. italiae* are derived (see Chapter 1). These source populations are unknown, but Meise (1936) suggested that *P. d. italiae* is the result of hybridization between house sparrows spreading across North Africa from the eastern Mediterranean and Spanish sparrows from western North Africa.

Sexual Size Dimorphism

Male sparrows in North America are slightly larger than females in some characters, but not in others. Selander and Johnston (1967) found that males were significantly larger than females in body mass, tarsus length, wing length, and tail length in 24 North American samples, but that the sexes did not differ significantly in either bill length or bill width. The differences in wing and tail lengths were about 3.5%, whereas body mass was about 2.8% greater in males, and tarsus length was about 0.9% greater. In a subsequent study on skeletal samples from 33 North American localities, Johnston and Selander (1971) found that males were significantly larger than females in skull width, femur length, tibiotarsus length, and pectoral girdle size (six pectoral girdle elements). Females were significantly larger than males only in narial width, and there were no significant differences in six other elements, including tarsometatarsus (tarsus) length. Johnston and Selander (1973) used a discriminant function analysis on the external characters of sparrows from 54 North American sites and were able to correctly assign the sex of 80% of the individuals based on the analysis.

Similar sexual size dimorphism has been reported for many native sparrow populations. Significant sexual size dimorphism in body mass, wing length, and/or tail length has been reported for several large samples of sparrows obtained during programs to reduce populations in Germany, primarily by strychnine poisoning (Geiler 1959; Grimm 1954; Löhrl and Bohringer 1957; Niethammer 1953; Nordmeyer et al. 1970, 1972; Scherner 1974). In these samples, which were obtained in winter or early spring, the average body mass of males exceeded that of females by 1.0%–2.9%, wing length by 2.9%–3.7%, and tail length by 2.6%–4.2%. Johnston (1969b) also reported that wing length of male "Italian" sparrows was 3%–4% greater than that of females, but there was no sexual size dimorphism in bill length. Males were also significantly larger in sternum size, humerus length, and ulna length. Danilov et al. (1969) found no difference in tarsus or bill length between males and females in Russia but did observe significantly longer wings and tails in males. Folk and Novotny (1970) reported that males were larger than females in body mass (1.7%) and wing length (3.3%) for a large sample of sparrows collected throughout the year in the former Czechoslovakia. Cordero (1991a) reported that males were larger than females in body mass and wing length in northeastern Spain, and Solberg and Ringsby (1997) reported similar differences in body mass and wing length in northern Norway. Similar levels of sexual size dimorphism have been reported in *P. d. indicus* in India (Saini et al. 1989). Male wing length and tail length were both significantly greater than in females (4.4% and 4.8%, respectively), whereas body mass and tarsus length did not differ significantly. Flemban and Price (1997) found that males were significantly larger than females in body mass, wing length, and tail length in two populations of *P. d. indicus* in Saudi Arabia, but they found no difference in tarsus length or bill dimensions.

Sexual size dimorphism favoring males has also been reported in other introduced populations. Medeiros (1995) observed sexual size dimorphism in a study of specimens from six sites in the Azores, where sparrows were introduced from the Portuguese mainland in 1960. Wing length of males was significantly longer than that of females at all six sites, and tail length was significantly longer in males at four of the six sites (and longer, but not significantly so, at the other two sites). Of the other external traits measured (tarsus length, middle toe length, and bill length, width, and depth), only middle toe length differed significantly at two of the sites, with males having significantly longer middle toes at both sites.

The adaptive basis of this relatively small size difference between the sexes is not understood. The fact that the primary differences are in wing and tail length and in the bones of the pectoral girdle suggests that it may be related to differences in the role of flight between the sexes. Except for the migratory populations of central Asia, the species is very sedentary. Natal dispersal distances are short, and females actually tend to show greater dispersal distances than males (see Chapter 8). The fact that bill and hind limb size tend to show no differences between the sexes suggests that differences in foraging method or food type are not involved. One possibility is that social interactions are responsible for the differences. Males compete for nest sites and for access to extra-pair copulations with females (see Chapter 4). Body size or flight ability may play an adaptive role in these intrasexual contests. This conclusion does not, however, account for an apparent increase in sexual size dimorphism with latitude in North America.

Johnston and Selander (1973) used discriminant function analysis to obtain a measure of Mahalanobis distance (D^2) between the sexes and found that this measure increased significantly with latitude, particularly in midcontinental prairie regions of North America (Fig. 2.5). Additional evidence for this increase in sexual size dimorphism with increasing latitude has been provided by some of the overwinter selection studies described later (e.g., Fleischer and Johnston 1984; Johnston and Fleischer 1981).

McGillivray and Johnston (1987) attempted to determine whether there was a latitudinal or climatic gradient in sexual size dimorphism by analyzing 12 skeletal characters in approximately 3000 specimens from 56 North American sites. One of the problems with principal components analyses is that direct comparisons cannot be made between the sexes, because loadings of the various characters differ between the groups. PCI may be an indicator of overall body size in both males and females, but the individual loadings of the characters for each sex on PCI will differ from each other. In an attempt to circumvent this problem and still compare overall size differences between the sexes, McGillivray and Johnston created a size index for each specimen by summing the measurements of the 12 skeletal characters for that specimen. They also recorded three climate measures for each collecting site: mean January temperature, mean July temperature, and seasonality (July mean − January mean). Latitude was significantly correlated with each of these climate variables, but most strongly with January temperature (-0.88) and seasonality (0.71). The body size index showed significant positive correlations with seasonality in both sexes, and in both adults and subadults. However, none of

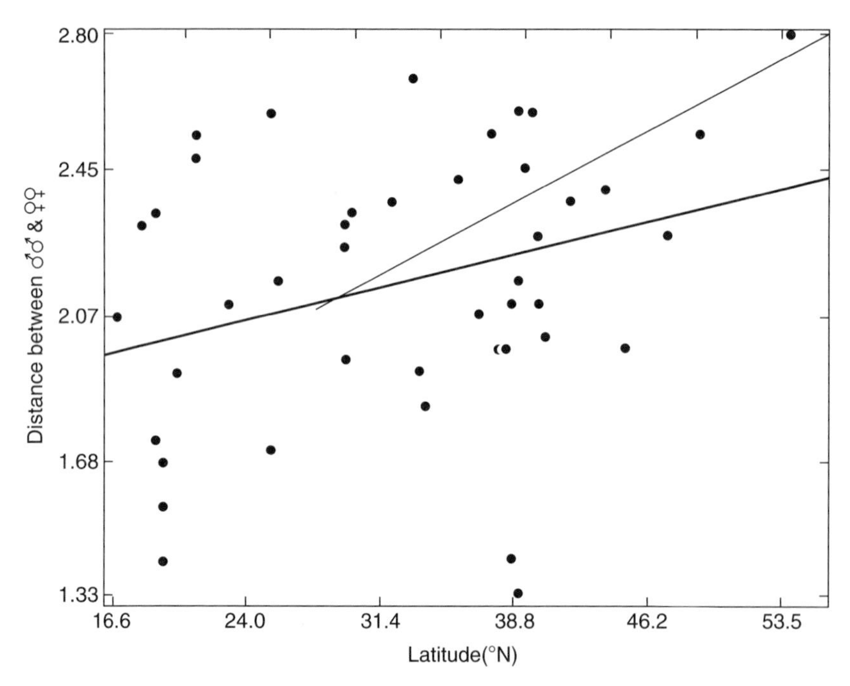

FIGURE 2.5. Sexual size dimorphism versus latitude in North American house sparrows. The distance between males and females is based on discriminant function analyses of five external traits (wing length, tail length, tarsus length, bill length, and bill depth) of 3784 sparrow specimens collected from 40 locations in North America. For the entire sample (*heavy line*) the regression equation of the measure of size dimorphism (Y) with latitude (X) is $Y = 1.79 + 0.011X$ ($P = .049$); the equation for the thin line (representing 11 midcontinental grassland sites) is $Y = 1.32 + 0.027X$ ($P = .009$). From Johnston and Selander (1973: Fig. 3), with permission of University of Chicago Press.

the differences between the regression equations for either sex or age categories were significantly different from each other, indicating no difference in sexual size dimorphism with seasonality.

The adaptive basis for increased sexual size dimorphism with increasing latitude is not clear. Fleischer and Johnston (1984) identified four hypotheses that have been suggested to explain this pattern of sexual size dimorphism in sparrows: (1) food niche splitting between males and females (the hypothesis originally put forward by Johnston and Selander 1973); (2) differential access to cavity roosts in winter (small females being able to use cavities that are unavailable to larger males); (3) differences in thermoregulatory efficiency (larger individuals may be favored in cold climates due to the surface-to-volume relationship discussed earlier, whereas small individuals may be favored during periods of food shortage by having a lower absolute food requirement; it was unclear, however, how this might result in increased sexual size dimorphism with latitude); and (4) consequences of dominance interactions (small females are obviously subdominant and save energy by fighting less, large males are obviously dominant and therefore also fight less, and intermediate-sized birds of both sexes fight more and feed less because of their uncertain social status). Fleischer and Johnston concluded that the existing evidence favored the last hypothesis (see later discussion).

Hamilton and Johnston (1978) proposed a fifth hypothesis based on an examination of skeletal samples from 14 North American sites (primarily in the central prairies) and 22 European sites. When both sexes were pooled, variance in PCI (overall size) increased significantly with latitude in both North America ($r_s = 0.807$, $P < .01$) and Europe ($r_s = 0.423$, $P < .05$). Hamilton and Johnston attributed this increase in size variation to an increase in the breadth of the foraging niche due to a reduction in the number of seed-eating competitors with increasing latitude.

Core-to-Limb Ratio

The observation of increasing core-to-limb ratio with increasing latitude or decreasing winter temperatures has been demonstrated in both North America and Europe (Johnston 1973; Johnston and Selander 1971). The second principal component (PCII) of an analysis of 16 skeletal characters of samples from 36 North American sites had significant negative loadings from core elements and positive loadings from hind limb elements (Johnston and Selander 1971). Birds with relatively large limbs tended to occur in warmer climates (PCII was significantly positively correlated with both summer and winter temperatures). Fig. 2.3, which shows the negative relationship between male body size and winter temperature, also shows the positive relationship between PCII and winter temperature. Using the same 16 skeletal elements, Johnston (1973) reported a significant decrease in relative limb size with increasing latitude in both North America and Europe. McGillivray and Johnston (1987) created an Allen score (see later discussion) as an index of the core-to-limb ratio by adding the standard scores (z-scores) of femur length, tibiotarsus length, and tarsometatarsus length for each of 3000 specimens from 56 North American sites and subtracting the standard scores for sternum length, keel length, and sternum depth. Allen scores for both sexes correlated significantly ($P < .05$) with January temperature (males, $r = -0.62$; females, $r = -0.65$) and also with seasonality.

This relationship between core-to-limb ratio and latitude or winter temperature corresponds to the tenets of Allen's ecogeographic rule (Mayr 1956). The general interpretation of Allen's rule again involves thermoregulatory considerations. Because heat is lost at a faster rate from the extremities than from the body core, individuals with a relatively larger core-to-limb ratio are able to conserve heat more efficiently in cold environments and to dissipate heat more rapidly in warm environments.

Plumage Coloration

The pattern of geographic variation in color in North American sparrows has received less attention than have the patterns of size variation. Johnston (1966) compared the spectrophotometric characteristics of dried soil samples washed from the plumage of sparrows collected along a 1600–km east-west transect in the central United States with spectrophotometric characteristics of female breast plumage from the same specimens. The hue (dominant wavelength) and chroma (saturation) of the soil samples and plumage samples were very similar at each locality, but the soil samples were lighter due to the fact that their color value (brightness) was greater. In another study involving 23 North American sites, Johnston (1972) found that the brightness of the breast of female birds showed a significant negative correlation with the average number of days of precipitation annually. He also found that mean brightness scores of the crowns of females from 17 sites on the Italian peninsula and nearby Mediterranean islands were significantly negatively correlated with both average annual precipitation and isophane, meaning that birds with darker crowns were found in cooler, moister climates.

As mentioned earlier, this pattern of color variation means that both European sparrows and their North American derivatives conform to Gloger's ecogeographic rule. The adaptive significance of Gloger's rule is poorly understood (Mayr 1963). Thermoregulatory properties again offer one possible explanation for the selective advantage of conformity to Gloger's rule. Paler coloration in more xeric, hot environments might serve to reduce heat gain from solar radiation by increasing reflectance, whereas darker coloration in cold environments might increase absorbance of solar radiation. A second possibility is that paler coloration in xeric environments and darker coloration in more mesic environments may tend to match more closely the substrates in these environments, thereby increasing crypticity and decreasing the likelihood of detection by visually hunting predators. The latter hypothesis was invoked by Johnston (1966) to explain the close color match between breast plumage of female sparrows and dried soil samples washed from their plumage.

Intrapopulation and Interpopulation Variation

The intrapopulation variation in morphological traits of North American sparrows does not differ significantly from that of European populations (Johnston 1976a; Johnston and Selander 1972; Selander and Johnston 1967). Johnston (1976a) examined variation in 14 skeletal characters of large samples collected from sites in both North America and Europe (including a subset of sites from northwestern Europe, the source area for the original North American sparrows). Intrapopulation variances in North America were similar to those in Europe, with the maximum variances being about 4% for premaxilla, dentary, keel, and sternum depth. Interpopulation variances, however, differed considerably between North America and Europe: about 10% for females in North America, compared with about 30% for the entire European sample and 27% for the northwestern Europe subset. Similar results were found in males, with North American values averaging about 14% and the entire European sample, 27%, and the subset, 21%. Johnston (1976a) concluded that the reduced interpopulation variation in North America is due to the founder effect, and that interpopulation variation in North America has been generated from the intrapopulation variation in the original founding population or populations.

One study suggested that the morphology of native sparrow populations is temporally stable. Niethammer (1969) compared the wing and tail lengths of 37 sparrows collected in 1963 and 1965 with those of 60 sparrows collected between 1815 and 1854 in the same locality in Germany and found that there were no significant differences in size in either males or females.

Other Patterns of Geographic Variation

In addition to the morphological variation documented by Johnston and Selander and their colleagues, geographic variation has also been documented in other characteristics of North American sparrows. Hudson and Kimzey (1966) determined the resting, postabsorptive metabolic rates at thermoneutrality (20°C–35°C; see Chapter 9) of captives from four sites in North America: Texas, Colorado, Illinois, and New York. The average metabolic rate of sparrows from Texas (46.3 J/g per hour) was significantly lower than that of the other three groups (55.4–58.2 J/g per hour). Hudson and Kimzey concluded that the lower metabolic rate of the Texas birds was an adaptation to warmer temperatures and reported that this lower rate could not be changed with prolonged exposure to low ambient temperatures (the length of this exposure was not specified, however). Kendeigh and Blem (1974) reported a clinal increase in existence metabolism with increasing latitude at nine localities in North America and also found that the lower limit of temperature tolerance decreased with increasing latitude. Thermal conductance, which is the inverse of insulation, was also found to vary geographically in North America, with northern populations having significantly lower average conductances than southern populations (Blem 1974). Clutch size of North American sparrows also varies geographically, in a manner that is consistent with clutch size variation in many other bird species (Anderson 1978, 1994; Murphy 1978a) (see Chapter 4).

Natural Selection in Introduced Sparrows

Collectively, the studies of Johnston, Selander, and their colleagues provide overwhelming evidence of phenotypic change in both the external and internal morphology of North American sparrows since their introduction from Europe. The patterns of variation that have developed in North America conform to those observed in numerous other species (as encapsulated in the ecogeographic rules discussed earlier). Are these patterns of variation the product of natural selection, and, if so, has this selection resulted in genetic modification in the populations?

Evidence that these morphological changes are the result of natural selection is largely indirect and relatively sparse. Two major approaches have been employed to attempt to demonstrate the action of natural selection. Both, following Bumpus's observations, assume that cold winter weather acts as a powerful selective agent on sparrow populations. The first approach is based on a comparison of the morphological traits of pre-winter samples of sparrows with those from post-winter samples (e.g., Fleischer and Johnston 1982). The second involves comparisons of the morphological characteristics of fall-captured first-year (SNCO) sparrows that have not survived a winter with those of fall-captured adult (SCO) sparrows from the same population that have survived at least one winter (e.g., Johnston 1976b). One study combined both approaches (Rising, 1972). Differences between the two groups are interpreted to be the result of overwinter selection. At least one study has attempted to compare the fitnesses of morphological variants as determined by breeding success (Murphy 1980).

Rising (1972) collected a sample of 57 sparrows (32 males) in October 1967 and a second sample of 45 sparrows (22 males) in March 1968 from the same population in Kansas (USA). He performed univariate analyses comparing skeletal traits of individuals that had not experienced winter conditions (October SNCO birds) with those that had experienced overwinter selection (October SCO birds and March birds) and found that males showed directional selection for increased premaxilla length, skull width, tibia length, tarsus length, and coracoid length. There were no significant differences in females, however. Rising concluded that the local winter conditions resulted in directional selection for larger males. A multivariate analysis of Rising's data found significant sexual dimorphism between both October and March samples, but significantly greater Mahalanobis distance (D^2) between the March birds than between the October birds, leading to the conclusion that overwinter selection had acted to increase sexual size dimorphism (Johnston and Selander 1973).

Lowther (1977b) also reanalyzed Rising's data using principal components analysis and concluded that selection intensities were greater in males than in females, and that the highest selection intensities in both sexes were on PCII (core-to-limb proportion). He concluded that overwinter selection was operating primarily on body proportions, not on body size. He also analyzed a second set of samples collected in October, November, December, and February 1962–1963 in Kansas. Selection intensities were calculated for each time interval and again were highest for PCII, but they were also significant for PCIII (bill size relative to core body size). The selection intensity for PCII was 0.1974, and for PCIII it was 0.1930. Most (68%) of the selection on core-to-limb ratio (PCII), favoring relatively shorter limbs, occurred between December and February.

Johnston and Fleischer (1981) compared the skeletal morphology of 242 sparrows collected in November 1978 with that of a second sample of 197 sparrows taken in March 1979 from the same population. Nine of 10 noncranial elements of the spring-collected females were smaller than those of fall SNCO females, with three being significantly smaller. In males, on the other hand, all 14 skeletal characters of spring birds were larger than those of fall SNCO males, with five of the differences being significant. Spring males also had a significantly larger average PCI score (overall body size) than fall SNCO males, whereas spring females had a smaller mean PCI than fall SNCO females, although the difference was not significant. Discriminant function analysis found that the D^2 between spring males and females was greater than that between SNCO males and females in the fall (Fig. 2.6). Johnston and Fleischer (1981) concluded that directional selection was taking place in both sexes. Fleischer and Johnston (1982) used the same dataset to estimate selection intensities on PCI and PCIII (core-to-limb ratio) between fall and spring. The selection intensity (0.154) for larger body size (PCI) in males was significant, whereas the selection intensity for smaller body size in females (0.056) was not. Selection favored larger core-to-limb ratios in both sexes, but not significantly so.

Fleischer and Murphy (1992) analyzed two samples of sparrows collected during February 1982, also in Kansas. The first sample was collected during a very cold spell (2–8 February, with average daytime high temperatures of

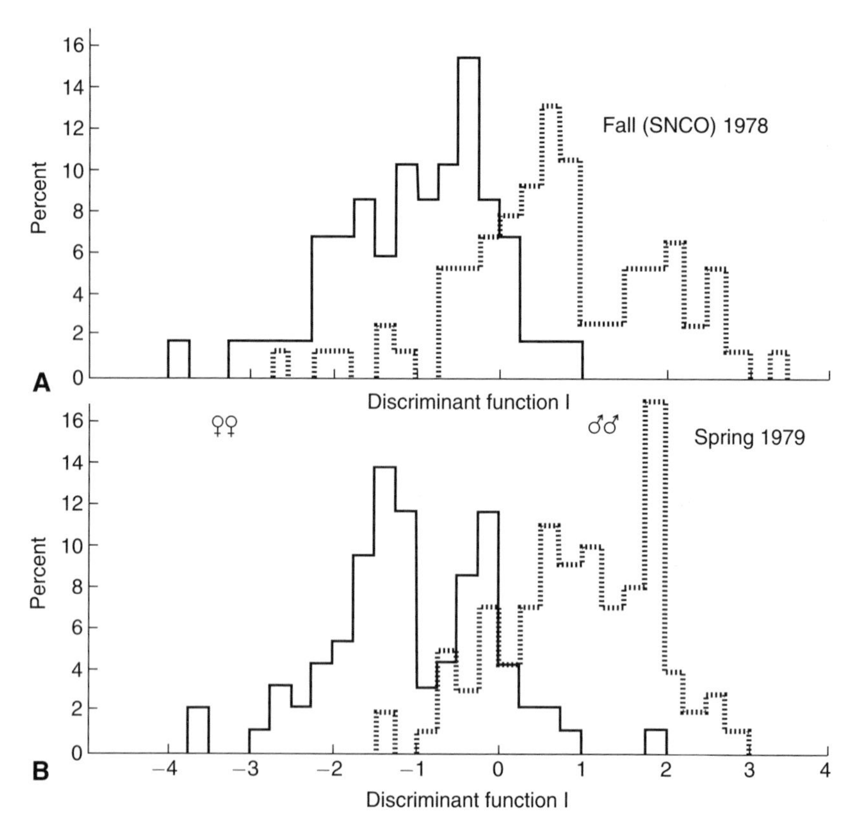

FIGURE 2.6. Change in sexual size dimorphism in house sparrows from autumn to spring in a population from Kansas (USA). Upper graph (A) is a frequency distribution on discriminant function I (overall body size) of 81 subadult male sparrows (*dotted line*) and 62 subadult female sparrows (*solid line*) collected in fall 1978. The lower graph (B) is a frequency distribution based on 116 males and 81 females collected from the same population in spring 1979. Note the greater overlap among the fall subadult birds. From Johnston and Fleischer (1981: Fig. 2), with permission of the American Ornithologists' Union.

−3.5°C and average lows of −15.2°C), and the second was collected during a subsequent warm spell (17–27 February, average highs, 13.3°C, and average lows, 0.7°C). The PCI scores (overall body size) of the latter group, based on 14 skeletal variables, were significantly greater than those of the former group in both sexes, indicating that the cold weather had selected for larger body size in both sexes.

Johnston (1976b) compared skeletal characteristics of 20 SCO and 38 SNCO males collected in October 1971 from a single population in Kansas. All skull measurements except premaxilla length were slightly larger in SCO birds, whereas all leg measurements were slightly larger in SNCO birds. A canonical correlation analysis was performed on the relationship between head variables and leg variables, and there was no significant relationship between these variables in SNCO males. There was, however, a significant relationship in SCO males, and Johnston concluded that the first-year males showed characteristics that would make them subject to selection during the winter.

Lowther (1977b) estimated selection intensities based on differences in skeletal morphology between SNCO and SCO sparrows from eight North American localities at which there were at least 20 SCO birds. He also evaluated the influence on selection intensity of climatic variables (principally temperature and precipitation). No consistent relationship between climatic variables and selection intensities on the first three principal components of sparrow morphology were detected, and Lowther concluded that local populations of sparrows had already adjusted to local climatic conditions.

Fleischer and Johnston (1984) attempted to measure overwinter selection by comparing specimen sets from the earlier studies of Johnston and Selander and their colleagues. Each set included at least 20 individuals of one sex from both fall (mid-October to mid-December) and spring (late February to May) of the same winter. Differences in both the means and variances of PCI (overall body size) and PCII (core-to-limb ratio) were examined (e.g., spring mean PCI vs. fall mean PCI). Weather variables were also analyzed for each winter, using the deviation of temperature for the period December–February from the mean for the period, and precipitation and snowcover deviations were similarly determined. For all the samples combined, PCI of males tended to increase with increased winter precipitation ($r = 0.94$, $P < .01$), whereas female size tended to decrease with lower than normal winter temperatures ($r = 0.94$, $P < .05$). The difference in variance in female size was also significantly correlated with winter temperature ($r = 0.884$, $P < .05$), suggesting that females were also under significant stabilizing selection during cold winters. Although this suggests the interesting possibility that females were under both directional and stabilizing selection, Fleischer and Johnston noted that the latter result might be partly artifactual due to the decrease in average female size. They found no relationship between any weather factor and change in either the mean or variance of PCII. They did conclude, however, that variation in winter weather conditions could influence the intensity of selection on sparrow morphology.

Murphy (1980) attempted to address the question of whether selection during the breeding season might be counteracting the apparent directional selection during winter for increased sexual size dimorphism. He captured and color-banded sparrows in Alberta (Canada) during two breeding seasons and recorded five external measurements on each bird (body mass, wing length, tarsus length, tail length, and bill length). There was no correlation between PCI (body size) and any measure of reproductive success for either males or females. There was also no tendency toward assortative mating for any size variable. Surviving and nonsurviving adults were also compared for size differences after each winter; surviving males were slightly smaller and surviving females slightly larger than nonsurvivors, although neither difference was significant.

Taken collectively, these studies provide strong circumstantial evidence that natural selection has operated on introduced populations of sparrows, and that winter weather can act as a powerful selective agent. The evidence is equivocal, however, about the actual targets of selection, and in some cases even about its direction. One study reported no evidence of directional selection for size in either male or female sparrows (Lowther 1977b), another found evidence for directional selection for larger size in males (Rising 1972), and one study detected selection for larger size in both males and females (Fleischer and Murphy 1992). Three studies (including one reanalyzing Rising's data) found evidence of directional selection for smaller size in females and larger size in males, thereby increasing the degree of size dimorphism (Fleischer and Johnston 1984; Johnston and Fleischer 1981; Johnston and Selander 1973). One study found that there was selection only for core-to-limb ratio, with winter favoring birds with relatively smaller limbs (Lowther 1977b). Although much of the variation in results among the studies was, no doubt, due to differences in the particular weather conditions in the years under study, the present data certainly leave room for a wide range of interpretations about the role of overwinter selection on sparrows. Still unanswered is the question of whether the observed changes are genetically based and therefore will result in true adaptive changes in the populations. Although there have been some attempts to link genetic change in the species to the observed phenotypic changes (see later discussion), this question remains open. The remainder of this chapter discusses the present knowledge of house sparrow genetics.

Genetics of the House Sparrow

Most of the hereditary material in sparrows is chromosomal and is transmitted by normal mendelian processes, but some is mitochondrial and is transmitted maternally in the cytoplasm of the egg cell. Most of what is known about the genetics of sparrows involves the chromosomal DNA (Parkin 1987). This section begins by discussing the karyotype of the house sparrow and then proceeds to discuss what is known about genetic variation in the species and its possible role in adaptive change. The chapter ends with a discussion of the use of DNA fingerprinting to identify individuals, as well as other new genetic technologies.

Karyotype

The karyotype of birds is particularly difficult to characterize accurately because of the numerous microchromosomes that they possess. The normal method for karyotyping is to obtain samples of tissues undergoing rapid mitotic division and stain them for the chromatin. The tissue samples are then examined for cells in metaphase, and photomicrographs are taken of these cells. A karyogram (Fig. 2.7) is prepared by cutting out the metaphase chromosomes from the photomicrograph and placing them in order based on chromosome size. The gross morphology of the chromosomes can be described based on the position of the centromere, the site of spindle attachment to metaphase chromosomes.

Four studies have suggested that the 2n chromosome number of the house sparrow is 76 (Bulatova et al. 1972; Christidis 1986; Fulgione, Aprea et al. 2000; Panov and Radjabli 1972); two studies reported that it is either 76 or 78 (Castroviejo et al. 1969; van Brink 1959), one that it is 78± (Prasad and Patnaik 1977), and one that it is 80± (Ray-Chaudhuri 1976). The number of

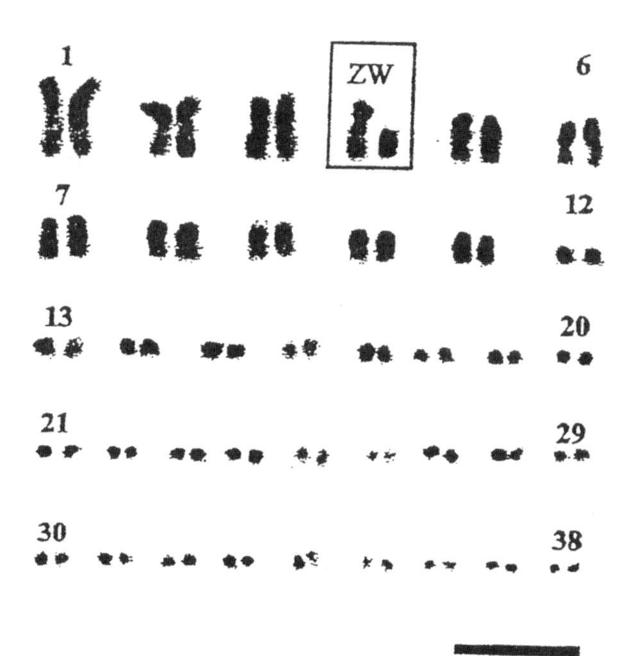

FIGURE 2.7. Karyotype of the house sparrow. Photograph of the karyotype of a female *P. d. italiae* from Italy. Chromosomes from spleen or bone marrow tissue were stained with Giemsa 24 h after inoculation with a mitogenic pokeweed solution. The bar in the lower right corner of the figure represents 10 μm. From Fulgione, Aprea et al. (2000: Fig. 1), with the permission of *Folia Zoologica*.

pairs of macrochromosomes is between 10 and 13 (Castroviejo et al. 1969; Christidis 1986; Fulgione, Aprea et al. 2000; Prasad and Patnaik 1977; Ray-Chaudhuri 1976; van Brink 1959). The discrepancies among the studies are due to the fact that the decision as to the number of macrochromosomes versus the number of microchromosomes is somewhat arbitrary. There is general agreement among the studies in the identification of the first 10 pairs of macrochromosomes. Riley (1938), one of the first to karyotype the house sparrow, identified the fourth chromosome (in order of size) as the sex chromosome, and all later studies have agreed with this identification. Fig. 2.7 shows a karyogram of a female bird from Italy (*italiae*).

Sex determination in sparrows is chromosomal, with the male being the homogametic sex (i.e., possessing two Z chromosomes), and the female being the heterogametic sex (i.e., possessing one Z chromosome and the W chromosome). The W chromosome is approximately half as long as the Z chromosome (Bulatova et al. 1972; Panov and Radjabli 1972).

Two of the studies compared the karyotype of the house sparrow with those of two congeneric species, the Spanish sparrow and the tree sparrow. The chromosome number of the Spanish sparrow matched that of the house sparrow in both studies, $2n = 76$ (Bulatova et al. 1972; Panov and Radjabli 1972), and a comparison of the morphology of the macrochromosomes suggested that they were very similar. The 2n number of the tree sparrow was reported to be 78, however, and some of the macrochromosomes differed in morphology from those of the house sparrow (Bulatova et al. 1972). Bulatova et al. (1972) also karyotyped an apparent house × tree sparrow hybrid and found a 2n number of 77. These differences reinforce the conclusions about the degree of relatedness among the members of the genus *Passer* (see Table 1.1).

In his review of avian cytogenetics, Shields (1982) concluded that the 2n chromosome number of both the house sparrow and the Spanish sparrow is 76, whereas that of the tree sparrow is 78. He also reported that the karyotype of the Spanish sparrow differed from that of the house sparrow due to fission-fusion in chromosome 7 and the W chromosome, and the tree sparrow karyotype differed due to fission-fusion in chromosomes 5 and 6 and the W chromosome.

Heritability of Morphological Traits

Intensively studied populations of sparrows on five Norwegian islands have provided the best evidence for the genetic contribution to morphological variation in the species (Jensen et al. 2003). Long-term population studies, including DNA fingerprinting of individuals using eight microsatellite loci (see later discussion), permitted the identification of both genetic parents of many individuals. This allowed Jensen et al. to construct a pedigree involving 2563 individuals, including 1228 individuals for which both parents were known, 351 for which only the mother was known, 254 for which only the father was known, and 730 that provided information on more distant relationships. Heritabilities of five morphological measurements (tarsus length, wing length, bill length, bill depth, and body mass) and of body condition index were calculated from the pedigree. Heritabilities (h^2) for both sexes combined ranged from 0.175 for body mass to 0.484 for bill length ($P < .001$ in each case).

When heritabilities were calculated separately by sex, h^2 for males ranged from 0.256 for bill depth to 0.586 for bill length; in females, h^2 ranged from 0.124 for body mass to 0.681 for tarsus length ($P < .05$ in each case) (Jensen et al. 2003). Differences in h^2 between the sexes were significant for four of the six traits, with females having significantly larger values for tarsus length, wing length, and bill depth and males having a significantly larger h^2 for body condition. Analysis of genetic covariances and correlations in each sex indicated that selection on some traits would significantly affect other traits. In some cases, these relationships indicated that the direction of change in one trait would be opposite to that in the correlated trait (e.g., wing length and both bill depth and body mass in males, bill depth and body mass in females).

Gynandromorphs

Plumage gynandromorphs have been reported in the house sparrow (Cink and Lowther 1987; Gavrilov and Stephan 1980; J. M. Harrison 1961b [specimen photographs in Figs. 5 and 6]; Lowther 1977a; McCanch 1992; Selander and Johnston 1967). The occurrence of such anomalies is apparently caused by the loss of a Z chromosome from a somatic cell line early in development. If the loss occurs at the two-cell stage of cleavage, the result could be a complete bilateral gynandromorph, because the sexual dimorphism in plumage is controlled genetically without being influenced by the sex hormones (see Chapter 5).

The results of a study on the skeletal morphology of two gynandromorphs, one collected in Iowa (USA) and the other in Kansas, suggested that sexual size dimorphism is genetically based (Lowther 1977a). Lowther measured the lengths of six bones from the right and left sides of the two specimens and compared these measurements with the average measurements of at least 20 males and 20 females collected at the same time from each location. The plumage on the left side of the Iowa bird was female, and that on the right was male; the bird had an apparent ovary on the left side but no testis on the right. The six bones on the right side of the body averaged 2.4% longer than those on the left side, which was similar to the 2.0% larger average size of males at the site. The average difference in the lengths of the six bones in five individuals of each sex (= fluctuating asymmetry) was only 0.41%, however. A discriminant function analysis successfully discriminated between males and females in most cases (except for one to three specimens, depending on the characters used) and placed the left side of the gynandromorph with the females and the right side with the males. One difficulty with these results is that the largest difference (4.6%) between the six bones was for the tarsometatarsus (tarsus), which normally does not show sexual dimorphism in sparrows (see earlier discussion). The right side of the Kansas bird, which also had female plumage on the left side, was 1.6% larger than the left side, compared with an average difference of 1.4% for males and females at the site (calculated from data in Lowther [1977a:Table 1]). Although the small sample size is obviously problematic, these results do offer some support for the genetic basis of sexual size dimorphism in the house sparrow.

Isozymes

Protein electrophoresis has been used for the last 40 years to indirectly study genetic variation in sparrow populations. Beginning with the studies of Bush (1967) and Bush and Fraser (1969), there have been numerous investigations of enzyme variation in sparrows, involving a wide range of enzymes. Some of these studies have attempted to assess the potential loss of genetic diversity in introduced populations due to the founder effect (e.g., Klitz 1972; Parkin and Cole 1985), and others have attempted to determine whether there are geographic or seasonal effects on the frequency of alleles (e.g., Bates and Zink 1992). The study of isozyme distributions and frequencies has also been used to determine the degree of inbreeding in a population (e.g., Parkin and Cole 1984) or to ascertain the relative effects of migration, selection, and genetic drift on the maintenance of observed patterns of isozyme distribution (e.g., Fleischer 1983). Other studies have used isozyme data to estimate mutation rate (e.g., Parkin and Cole 1985). Many of these studies have used population genetic models whose treatment is beyond the scope of this work, and persons interested in the details of these models should consult the original references and the sources cited therein.

I analyzed the findings of eight studies that identified the frequency of polymorphic isozyme loci in 14 native or introduced sparrow populations (see Table 2.2). I identified 64 enzyme loci, some of which may be redundant due to difficulties in determining the exact congruence of enzyme loci among studies using different tissue sources, electrophoretic techniques, and enzyme terminologies. Some of the studies sampled single populations (e.g., Medeiros 1997a), whereas others sampled multiple populations from a particular geographic region (e.g., Klitz 1972; Parkin and Cole 1984). Klitz (1972), for instance, sampled sparrows from 10 European sites from Italy to Sweden.

Eighteen of the enzyme loci appear to be monomorphic throughout the range of populations sampled. They were monomorphic at all of five or more sites at which they were examined: aconitase 2 (ACON-2), acid phosphatase (ACP), adenylate kinase (AK), esterase 2 (EST-2), guanine deaminase (GDA), gluta-mate dehydrogenase (GLUDH), glutamate oxalate deaminase B (GOT-B), α-glycerophosphate dehydrogenase (GPD), hemoglobin (three loci: Hb1–3), lactate dehydrogenase 1 and 2 (LDH-1, LDH-2), malate dehydrogenase B (MDH-B), mannose phosphate isomerase (MPI), dipeptidase 1 (PEP-D1), and superoxide dismutase 1 and 2 (SOT-1, SOT-2). An additional 24 monomorphic loci were sampled in only one study (17 loci) or two studies (7 loci), but it would be premature to suggest that they are monomorphic in the house sparrow on the basis of so few studies.

Four enzyme loci were polymorphic in all of nine or more sites at which they were characterized and are presumably polymorphic in most, if not all, sparrow populations: adenosine deaminase (ADA), esterase 1 (EST-1), isocitrate dehydrogenase A (IDH-A), and tripeptidase (PEP-T). In addition, dipeptidase 2 (PEP-D2) was polymorphic at 11 of 12 sites and showed a rare ($p < .01$) alternative allele at the remaining site. Isocitrate dehydrogenase B (IDH-B) and phosphoglucomutase 2 (PGM-2) were polymorphic in all populations except one small Australian sample based on only 16 individuals (Christidis 1987, *in litt.*), and sorbitol dehydrogenase (SDH) was polymorphic in all but another Australian sample based on 28 individuals (Manwell and Baker 1975). Transferrin (Tf) was found to be polymorphic in both studies in which it was characterized, and it was also found to be polymorphic in other populations (Fleischer et al. 1983; Johnston 1975; Wetton and Parkin 1991; Wetton et al. 1992). Both glyceraldehyde-3-phosphate dehydrogenase (GAPHD) and NADP specific dehydrogenase (NADP nDH) were polymorphic in the one population in which they were assayed. The remainder of the enzyme loci were polymorphic in some populations and monomorphic in others (numbers in parentheses following each of the enzymes are the number of populations in which the enzyme was polymorphic over the number of populations assayed for the enzyme: aconitase 1 (ACON-1), (5/8); creatine kinase (CK), (2/3); esterase 3 (EST-3), (6/7); glutamate oxalate deaminase A (GOT-A), (1/12) glucose phosphate isomerase (GPI), (2/12); glucose-6-phosphate dehydrogenase (G6PDH), (3/6); malate dehydrogenase A (MDH-A), (1/13); nucleoside phosphorylase (NP), (3/6); dipeptidase 3 (PEP-D3), (4/5); phosphoglucomutase 1 (PGM-1), (6/11); and phosphoglucomutase (PGM), (3/11).

Comparisons between native and introduced populations are difficult because different studies assayed different enzymes. Two studies made direct comparisons between native and introduced populations. Parkin and Cole (1985) compared samples from native populations in England (UK) and southwestern

Europe with samples from introduced populations in Australia and New Zealand, whereas Medeiros (1997a) compared a sample from the Portuguese mainland (native) with two samples from islands in the Azores (introduced). Parkin and Cole found that the average heterozygosity of the New Zealand populations was significantly lower than that of either of the two native populations or the Australian introduced population ($P < .05$), but that the Australia population actually had a slightly higher average heterozygosity than the two native populations. The total number of alleles over all polymorphic loci was lower in the introduced populations in both studies. For the 12 polymorphic loci in the Parkin and Cole (1985) study, there were a total of 41 alleles in England, 40 in southwestern Europe, 30 in Australia, and 29 in New Zealand. At the 14 polymorphic loci in Medeiros' (1997a) study, there were 31 alleles on the Portuguese mainland, and 28 and 24 alleles on the two islands in the Azores. Inspection of the data reveals that it was invariably very rare alleles ($p < .02$) that were lost in the introduced populations (see also Cole and Parkin 1986). This suggests that either the founder effect or genetic drift in small populations during the first years after introduction have been responsible for the loss of some genetic diversity in introduced populations.

Comparisons across the entire dataset are presented in Table 2.2 and support this conclusion. The average proportion of polymorphic loci for the seven native populations, each treated as an independent sample, is 0.379, and that for the seven introduced populations is 0.324; the mean number of alleles per polymorphic locus is 2.89 in native populations and 2.50 in introduced populations. The mean heterozygosities of the native populations also appear to generally exceed those of introduced populations (Table 2.2), but these data must be interpreted cautiously, because differences among studies are caused in part by the selection of loci (Parkin 1987).

Several studies reported that genotypic frequencies at polymorphic loci did not deviate significantly from the Hardy-Weinberg equilibrium (Bates and Zink 1992; Fleischer and Murphy 1992; Medeiros 1997a; Parkin and Cole 1984, 1985). However, in a study of the polymorphic ADA locus in 47 sparrow populations from Great Britain, Norway, southwestern Europe, Canada, Australia, and New Zealand, Cole and Parkin (1986) found that 6 of the populations deviated significantly from the Hardy-Weinberg equilibrium at $P < .05$, and an additional three deviated at $P < .01$. The latter three all showed an excess of heterozygotes, but no consistent habitat or ecological differences were apparent in the three populations.

One of the polymorphic enzyme loci is possibly located on the Z chromosome and is hence sex-linked. Baverstock et al. (1982) determined by artificial interspecific hybridization experiments that the ACON-1 locus is probably sex-linked in the guinea fowl (*Numida meleagris*). They also reported on the fre-

TABLE 2.2. Enzyme Polymorphism in Populations of the House Sparrow

	Native Populations							Introduced Populations						
	1	2	3	4	5	6	7	8	9	10	11	12	13	14
N[a]	273	57	76	988	501	354	64	32	111	16	204	39	47	28
Proportion of polymorphic loci	.278	.323	.323	.500	.393	.429	.406	.538	.393	.132	.321	.375	.312	.200
Mean number of alleles per polymorphic locus	2.6	2.5	2.3	3.6	3.6	3.3	2.3	2.0	2.6	2.8	2.9	2.2	2.0	3.0
Mean heterozygosity	—	.147	.157	.101[b]	.097	.095	.074	—	.107	—	.079	.082	.075	.027[b]

In most studies, an enzyme locus was considered monomorphic if the most common allele had a frequency of 0.99 or greater ($q > .99$). Because of differences in enzyme nomenclature in the various sources, some decisions were made about the equivalence of enzymes among the studies. Whenever possible, these were made based on information within the sources on details such as the tissue from which the enzyme was derived and whether it migrated anodally or cathodally in the electrical field. The proportion of polymorphic loci and average heterozygosities for each population are given at the bottom of the table for studies in which they were provided or for which they could be calculated or estimated based on data in the sources.

[a]Numbers in each sample varied among the enzymes assayed; the number presented is a maximum number.

[b]Calculated from data in the original reference.

Sources: 1, 10 sites from throughout Europe (Klitz 1972); 2, Nottinghamshire, UK (Cole and Parkin 1981); 3, Lincolnshire, UK (Cole and Parkin 1981); 4, 14 sites in East Midlands, UK (Parkin and Cole 1984); 5, six sites in Nottinghamshire, UK (Parkin and Cole 1985); 6, seven sites in southwestern Europe (Parkin and Cole 1985); 7, Portugal (Medeiros 1997a); 8, south Australia (Manwell and Baker 1975); 9, three sites in Australia (Parkin and Cole 1985); 10, northwestern Australia (Christidis 1987 and *in litt.*); 11, four sites in New Zealand (Parkin and Cole 1985); 12, San Miguel, Azores (Medeiros 1997a); 13, Graciosa, Azores (Medeiros 1997a); 14, Louisiana, USA (Bates and Zink 1992).

quencies of homozygotes and heterozygotes for male and female sparrows from three locations (Australia, Canada, and the United Kingdom). The pooled data showed 147 homozygotes and 72 heterozygotes among males, whereas all 203 females were homozygous at the ACON-1 locus. The authors therefore concluded that the ACON-1 gene is sex-linked in the house sparrow. However, Medeiros (1997a) reported that 2 females were heterozygous at the ACON-1 locus among 68 females from Portugal and the Azores, raising doubt about this conclusion.

Both seasonal and geographic variation in heterozygosity have been reported in sparrows. Bates and Zink (1992) compared mean heterozygosities of males and females at various seasons of the year in Louisiana (USA). Winter and spring males and winter females had significantly lower mean heterozygosities than did sparrows at other seasons ($P < .05$), but the adaptive significance of this pattern was difficult to interpret. Parkin and Cole (1984), who sampled 14 populations from the East Midlands (UK), found that mean heterozygosity decreased northward along the 100-km transect ($P < .02$). Again, the adaptive significance of this pattern was not apparent, particularly over such a short distance of relatively homogeneous habitat.

Fleischer et al. (1983) examined the relationship between morphological variation and isozyme variation in sparrows. They compared the number of heterozygous loci (0, 1, or 2+) at between two and four polymorphic enzyme loci with the variation in skeletal morphology of males and females in four populations: Alberta, Kansas (two populations), and California (USA). The average variance in PCI (overall skeletal size) was generally lower in sparrows with two or more heterozygous loci (and significantly so in two of the eight sex × site comparisons). Although none of the individual enzyme loci showed a consistent pattern with respect to degree of variation in PCI, and the study involved only a tiny minority of the loci coding for enzymes of intermediary metabolism, Fleischer et al. (1983) suggested that their data does provide support for the notion that enzyme heterozygosity may confer some developmental stability, resulting in reduced size variation in sparrows with high heterozygosity. Fleischer and Murphy (1992) found a significant positive correlation between number of heterozygous loci (0, 1, or 2+) among three polymorphic loci sampled and lipid content of the pectoral muscles of wintering sparrows in Kansas; they suggested that heterozygotes are better buffered physiologically against variation in winter weather. However, this suggestion would appear to run counter to the observed decrease in heterozygosity in more northern samples in the East Midlands (described earlier).

Isozyme data have been used to estimate both mutation rates and coefficients of inbreeding in sparrow populations. The estimates of mutation rate depend on knowledge of the history of the introduction or establishment of a population, as well as estimates of genetic distance between the source population and the introduced or derived population. Parkin and Cole (1985), for instance, estimated the mutation rate based on a comparison of isozymic frequencies in England and southwestern Europe by assuming that the arrival of the house sparrow to the British Isles coincided with the Roman invasion approximately 2 millennia ago. This yielded an estimated mutation rate of 1.76×10^{-7} mutations per locus per year. Based on better information about the time of introduction and the size of founding populations, Parkin and Cole also estimated mutation rates for Australia and New Zealand populations. These estimates were 9.26×10^{-6} and 1.03×10^{-5}, respectively, for the two introduced populations. Medeiros (1997a) estimated that the mutation rate in the introduced populations in the Azores was 5.17×10^{-5} mutations per locus per year. These estimates are similar to those for isozyme mutation rates in other species (Parkin 1987).

The coefficient of inbreeding (F_{ST}) essentially measures the probability that two alleles in the same individual are identical due to descent from one allele in a common ancestor. Several studies have used isozyme frequency data from sparrows to estimate F_{ST}. Fleischer (1983) used four polymorphic isozyme loci to compute an observed F_{ST} of 0.0076 for five subpopulations in Kansas. Demographic data from the same populations permitted him to also calculate a predicted value of F_{ST} (0.0135), which did not differ significantly from the observed value, and he concluded that migration and genetic drift were sufficient to explain the pattern of isozymic variation among these populations. There was no evidence of selection operating at this microgeographic scale. Parkin and Cole (1984) found an average F_{ST} of 0.0032 for 14 populations of sparrows in the East Midlands, and Medeiros (1997a) reported an average of 0.0022 for two populations in the Azores.

DNA Fingerprinting

The utility of DNA fingerprinting in the study of the biology of birds was first demonstrated in investigations of the house sparrow (Burke and Bruford 1987; Wetton et al. 1987). The results of these and other DNA fingerprinting studies are treated in other chapters (e.g., Chapter 4). The first studies on sparrows actually used probes developed for human DNA fingerprinting; remarkably, these were able to successfully identify individual sparrows and also to accurately assess the paternity and maternity of offspring. More recently, both minisatellite (Hanotte et al. 1992; Wetton et al. 1995) and microsatellite (Griffith, Stewart et al. 1999; Jensen et al. 2003; Neumann and Wetton 1996) probes have been developed for use in studies of the house sparrow. The four microsatellite loci (*Pdo*μ1–4) identified by Neumann and Wetton (1996) consist of restriction fragment length polymorphisms (RFLPs) of the nucleotide sequences $(G)_6(TG)_{23}$, $(T)_7(GT)_{23}$, $(TCCA)_{18}$, and $(A_nG_n)_n(GAGAGAAA)_{13}(GAAA)_{34}$, with the subscripts representing the number of repeated nucleotide sequences. The primary advantage of microsatellites over minisatellites is that they do not require high-quality DNA and can therefore be used on feathers or other sources of degraded DNA material (Neumann and Wetton 1996).

Neumann and Wetton (1996) found that the number of alleles at the four microsatellite loci varied from 7 to 37 in a sample of 40 sparrows, and they estimated the mutation rate at the most polymorphic locus to be 2.2×10^{-2} per generation. Heterozygosities at the four loci varied from 0.400 to 0.975, with the lowest value, which was unexpectedly low, apparently being a result of the presence of "null" alleles (Neumann and Wetton 1996). Griffith, Stewart et al. (1999) assayed three microsatellite loci in three sparrow populations: a native population (Nottinghamshire, UK), an introduced population (Kentucky, USA), and a recently established island population (Lundy Island, UK). Heterozygosities at the three loci varied between 0.810 and 0.923 in Nottinghamshire, between 0.792 and 0.966 in Kentucky, and between 0.864 and 0.923 on Lundy, with no significant differences among the populations. Mutation rates in the Kentucky and Lundy populations were approximately 1.8×10^{-3} and 4.9×10^{-3} per locus, respectively (calculated from data in Griffith, Stewart et al. 1999). Wetton et al. (1995) reported that heterozygosities at four minisatellite loci varied from 0.938 to 1.00 in a sample of 144 sparrows. The mutation rate at three minisatellite loci was reported to be 1.3×10^{-3} per locus (Wetton et al. 1992). Therefore, mutation rates at both minisatellite and microsatellite loci are 25–2500 times greater than those at isozyme loci in sparrows.

Nuclear Genes

The advent of the genetic technologies that were used to sequence the human genome have also permitted researchers to sequence specific genes in many other organisms, including sparrows. All or parts of at least four genes have recently been sequenced: the *pPer2* gene, the *FoxP2* gene, and the *Mhc* class I and IIB genes. A method has also been described for identifying the sex of an embryo or nestling using the sex-linked *CHDI* gene.

Part of a period gene homologous to the *Per2* gene of the Japanese quail (*Coturnix coturnix*) was described in sparrows (Brandstätter, Abraham, and Albrecht 2001). Messenger RNA (mRNA) was isolated from the brain of a sparrow sacrificed at midday, and the reverse transcriptase polymerase chain reaction (RT-PCR), using primers from the PAS region of the *Per2* gene of the Japanese quail, was used to isolate and amplify the DNA segment coding for the mRNA. The PAS region of the sparrow *Per2* gene contains 688 nucleotides and codes for a polypeptide chain in which the amino acid sequence is 96% identical to that of the corresponding gene in the Japanese quail. The amino acid sequence is also 66%, 77%, and 56% identical to the sequences in three period genes of the mouse (*Mus musculus*): *mPer1*, *mPer2*, and *mPer3*, respectively. Expression of the gene, identified by the detection of *Per2* mRNA, has been observed in sparrows in the retina of the eye, the pineal gland, and the hypothalamus (Brandstätter, Abraham, and Albrecht 2001), suggesting that it plays a role in the control of the circadian rhythm in the species (see Chapter 3).

Two amino acid substitutions in exon 7 of the *FoxP2* gene in humans have been associated with language learning (cf. Webb and Zhang 2005). Webb and Zhang (2005) sequenced exon 7 in seven bird species (including the house sparrow). Five of the species are song learning species (three passerines and two hummingbirds) and two are not song-learning species. The amino acid sequences of all seven species were identical, and were also identical to those of the American alligator (*Alligator mississippiensis*) and the mouse. The 124 nucleotides in the seven bird species were identical, except for two synonymous changes in the budgerigar (*Melopsittacus undulates*).

Two exons from genes of the major histocompatibility complex (*Mhc*) were sequenced in unrelated individual sparrows and in families (females and her nestlings) (Bonneaud et al. 2004). *Mhc* genes code for proteins that play a major role in an organism's immune response mechanism by identifying and binding to foreign antigens. The protein-binding regions (PBRs) of the two *Mhc* genes (class I and class IIB) are located on exons 3 and 2, respectively, of the genes. Twenty different sequences were identified for the PBR of exon 3 of the *Mhc* class I gene, and 13 sequences were identified for the PBR of exon 2 of the *Mhc* class IIB gene. Bonneaud et al. concluded that there were at least five gene loci for exon 3 of the class I gene, and at least three loci for exon 2 of the class IIB gene.

Sparrows, like most other bird species, possess a sex-linked gene, the chromo-helicase-DNA-binding (*CHDI*) gene (Cordero et al. 2000; Westneat et al. 2002). Two forms of the gene occur, one on the Z chromosome and one on the W chromosome. Griffiths et al. (1998) described a technique involving two primers (P2 and P8) that bind to introns associated with either the *CHDI-W* or the *CHDI-Z* gene. DNA samples from an individual can be used to sex the individual by amplifying the DNA sample with the PCR after incubation with the primers. Gel electrophoresis of the sample after the treatment results in the appearance of either one band, corresponding to the possession of homozygous *CHDI-Z* genes (male), or two bands, representing heterozygous *CHDI-W*/*CHDI-Z* genes (female).

Mitochondrial DNA

The cytochrome *b* gene of the house sparrow, located in the mitochondrial DNA, has recently been sequenced (Allende et al. 2001). A nuclear pseudogene of the cytochrome *b* gene was also identified and sequenced. The function of the pseudogene is unclear, but Allende et al. used a comparison of the sequences of cytochrome *b* genes and pseudogenes from several members of the genus *Passer* to draw inferences about the evolutionary history and relatedness within the genus (see Chapter 1).

Summary

At present, I think it can be stated with considerable assurance that any links between observed patterns of genetic variation and the patterns of morphological variation in sparrows discussed earlier are extremely tenuous. Some of the strongest evidence for a genetic basis for the patterns of phenotypic change observed in introduced sparrow populations is the evidence for a founder effect in both isozyme variation and phenotypic variation in those populations. The founder effect here is broadly interpreted to mean both the founder effect *sensu stricto* (reduction in genetic diversity among the individuals founding a new population) and the loss of genetic diversity due to random drift during the early establishment period when the population size is small. The principal evidence for a phenotypic founder effect is the reduced interpopulation variation observed in North America compared with Europe (Johnston 1976a), even though the variation in climatic and other environmental factors across the North American sites is much greater than for the European sites. Comparative studies of isozyme loci in native and introduced populations have shown a loss of rare alleles in introduced populations (Medeiros 1997a; Parkin and Cole 1985). To the extent that the isozyme loci are typical of other genetic loci in the species, this loss of genetic diversity could certainly have had an effect on phenotypic traits such as overall body size and proportions, which are undoubtedly influenced by multiple genetic loci. A second line of support for the genetic basis of size variation is the evidence for bilateral size dimorphisms in gynandromorphs that parallel the sexual size dimorphism observed in the population (cf. Lowther 1977a). The fact that Lowther's study was based on only two gynandromorphs leaves much room for doubt, however. A third line of evidence is the reversed latitudinal size cline in sparrows on the Italian peninsula (see earlier discussion).

"In brief, is the new variety merely ontogenic, or is it phylogenic?" (Bumpus 1897:13). Despite the immense amount of excellent work discussed in this chapter, I believe that the answer to that question is still not adequately resolved. Critical common garden experiments have not yet been performed except in the cases of geographic variation in clutch size (Baker 1995: see Chapter 4) and immune response (Martin et al. 2004: see Chapter 9). Interestingly, the physiological processes involved in both clutch size determination and immune response changed considerably during the 2 y in common gardens, indicating a high degree of phenotypic plasticity in birds from source populations. A second experimental protocol that also offers promise is that of reciprocal transplantation, preferably of newly laid eggs, between widely separated and phenotypically differentiated populations (see James and NeSmith 1988).

Chapter 3

ANNUAL CYCLE

> The vast majority of birds live in a nonuniform environment and . . . [their] survival hence depends on the development of an efficient timing program that permits the adjustment of physiologically important functions to favorable periods of the year.
>
> —Immelmann 1971:342

In October 1924, the Canadian physiologist William Rowan subjected dark-eyed juncos (*Junco hyemalis*) to artificially increased daylength by exposing one cage of birds held outdoors in Alberta to two 50-W light bulbs. By late December, the light-treated males had enlarged, active testes, and one female showed marked follicular development in the ovary and an enlarged oviduct; untreated birds showed complete gonadal regression. The responses of the light-treated birds occurred despite ambient temperatures as low as −45°C. Rowan concluded that "daily increases of illumination . . . are conducive to developmental changes in the sexual organs" (1925:495). Beginning shortly after these pioneering observations and continuing to the present, sparrows have served as primary research subjects in attempts to understand the control of the annual cycle in birds.

Over its enormous geographic range, the house sparrow encounters a wide variety of climatic regimes, from tropical climates with marked wet and dry seasons to arctic climates north of the arctic circle. The latter have continuous darkness during the winter, continuous daylight during the summer, and concomitant differences in ambient temperature. Although it is resident throughout most of its range, sparrows also have migratory populations (see Chapter 8). Year-round residents must be able to survive the harshest periods of the annual environmental cycle while using the more favorable periods for energetically demanding activities such as reproduction and molt; migration permits individuals to exploit favorable local conditions for breeding while avoiding harsh conditions by moving to a more equable environment.

Sparrows have three distinct annual cycles in different parts of their range: (1) the temperate resident cycle, (2) the subtropical resident cycle, and (3) the temperate migratory cycle (Fig. 3.1). Identification of the primitive annual cycle of the species depends on whether it initially arose in the subtropics and subsequently spread to temperate regions, or vice versa (see Chapter 1). The earliest fossil evidence indicates that the species may have arisen in the Middle East as a commensal of preagricultural humans (Tchernov 1962), and its close relationship with the Spanish sparrow, which remains migratory in that region, suggests the possibility that the temperate migratory cycle was the primitive annual cycle.

In temperate areas where it is resident, the breeding season lasts from early spring through middle to late summer and is followed by a complete prebasic molt in late summer and early autumn (see Chapter 5), during which the gonads are regressed and photorefractory. Throughout the rest of the year (late autumn and winter), individuals are reproductively inactive, although occasional exceptions have been noted (e.g., Bordignon 1985; Cottam 1929; Molzahn 1997; Mostini 1987; Snow 1955; Wessels 1976).

The subtropical resident cycle, which is typical of *Passer domesticus indicus* on the Indian subcontinent, differs from the temperate resident cycle in that breeding may occur throughout the year. There are, however, two distinct peaks in the breeding season, one in the spring dry season (March–May) and the other in the late monsoon season (August–October). The molt is initiated between the two breeding periods and may be interrupted at the beginning of the second breeding period (see Chapter 5).

The temperate migratory cycle, found in *P. d. bactrianus* in central Asia, involves two periods of migration: northward to the breeding grounds in March and April, and southward to the wintering grounds in October and November (Dolnik and Gavrilov 1975). The breeding season lasts from May through early July and is followed by the prebasic molt before the autumnal migration.

This chapter focuses primarily on the annual gonadal cycles (other aspects of reproduction are covered in Chapter 4) and on the mechanisms controlling the timing of these cycles, including the role of circadian time-keeping in this process. Most subdisciplines of biology possess special terminologies, but in the case of the study of annual cycles and the endogenous rhythms that control them, this terminology may be unfamiliar even to many biologists. Table 3.1 includes definitions for many terms that are used in this chapter.

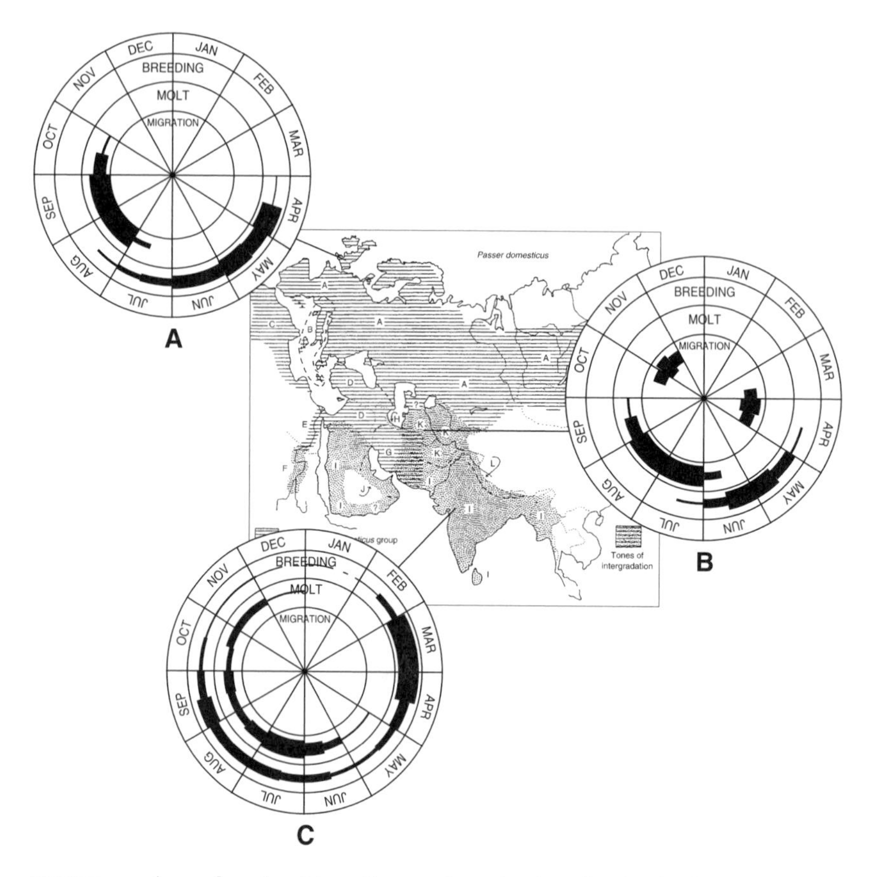

FIGURE 3.1. Annual cycle of breeding, molt and migration in the house sparrow. Pie diagrams illustrate approximate timing, duration, and intensity of the three activities in the three types of annual cycles: (A) temperate resident cycle (England, UK); (B) temperate migratory cycle (Tadjikistan); and (C) subtropical resident cycle (India). Width of black bar indicates approximate proportion of the population involved in the activity shown, in half-month increments. The breeding seasons in England and India are based on the dates of initiation of clutches, and the molt is for adult, free-living individuals only. Data for Tadjikistan are based on captive birds maintained in outdoor aviaries; the breeding season is identified based on the size of the cloacal protuberance of males, and the migratory period is based on periods of nocturnal activity. Data for preparation of the diagrams came from the following sources: (A) *P. d. domesticus* in England (Ginn and Melville 1983; Seel 1968a); (B) *P. d. bactrianus* in Tadjikistan (Dolnik and Gavrilov 1975); and (C) *P. d. indicus* in Baroda and Rajkot, India (Mathew and Naik 1986; Naik and Mistry 1980). Distributional map from Vaurie (1956: Fig. 1), courtesy of the American Museum of Natural History.

TABLE 3.1. Terms Relating to the Annual Cycle and Endogenous Rhythms in Animals

activity time (α)	Length of time of locomotor activity during the daily cycle (see Fig. 3.6)
circadian rhythm	Pattern of daily activity with a period of approximately 24 h that persists in the absence of periodic external signals
circadian time (CT)	Time calculated relative to the circadian rhythm of an individual (CT_{00} corresponds to the onset of activity)
entrainment	Synchronization of the overt rhythm of an organism with a periodic external signal (see Fig. 3.6 for example of entrainment to a photoperiodic stimulus)
free-run	Regular pattern of cyclic behavior in the absence of periodic environmental stimuli or zeitgebers
period length (T)	Time between consecutive equivalent points in an individual's cycle of activity (usually measured between consecutive onsets of periods of locomotor activity). T refers to zeitgeber period.
phase angle (ϕ)	Difference in time between the beginning of the zeitgeber stimulus and the initiation of the overt activity; usually recorded in time units with positive phase angles indicating that activity onset preceded beginning of the stimulus, and negative angles indicating that activity onset occurred after the beginning of thr stimulus (see Fig. 3.6)
phase angle shift ($\Delta\phi$)	Change in time between onsets of overt activity on consecutive days in response to manipulation in the timing of the zeitgeber (see Fig. 3.6)
photoperiod	Length of light period during a 24-h cycle; experimental photoperiods are presented as, for example, 15L:9D (240 lux) to represent 15 h of continuous light of intensity240 lux, followed by 9 h of continuous darkness (presumably 0 lux unless otherwise stated)
photorefractory	Unresponsive to stimulation by increased photoperiod
skeleton photoperiod	An experimental protocol in which two periods of light separated by darkness can be interpreted physiologically by the organism as a single period of light (e.g., a 4L:6D:2L:12D light regime could be interpreted by the organism as either a 12L:12D photoperiod or an 18L:6D photoperiod)
transients	Activity cycles observed during resetting of biological clock in response to changes in the zeitgeber (see Fig. 3.6)
transition (L/D, D/L)	Change from light to darkness (L/D = nightfall) or from darkness to light (D/L = dawn)
zeitgeber	Periodic environmental signal capable of entraining overt activity of an organism
zeitgeber time (ZT)	Time with reference to an external signal (zeitgeber) that will entrain the activity of an organism: for the typical photoperiodic signal ZT_{00} is conventionally considered to represent lights-on (the D/L transition)

The Gonadal Cycle

In temperate zone populations, individuals of both sexes go through a regular annual cycle of gonadal development and activity followed by a period of gonadal regression and quiescence. Following a detailed study of the annual cycle in New York (USA), Hegner and Wingfield (1986b) proposed that the annual cycle of the male can be divided into a short nonbreeding period (September and October), during which the birds complete the prebasic molt, and a prolonged breeding period spanning 10 or 11 months. During the nonbreeding period, the testes are fully regressed and the birds are absolutely photorefractory. The breeding period is subdivided into the preparatory phase (late October to March) and the multibrooded nesting phase (March to early September). Gonadal recrudescence occurs during the preparatory phase, which is characterized by two stages: (1) an autumnal stage lasting from October to early January, during which the gonads develop slowly, the beak darkens, and there are transitory pulses of increased production of luteinizing hormone (LH) and testosterone; and (2) a winter stage lasting from January to March, during which the testes develop rapidly (culminating in complete spermatogenesis). This results in rapid increases in the production of reproductive hormones and enlargement of the cloacal protuberance. The autumnal stage in males often coincides with increased nest site defense and sexual activity, which may function in the establishment of a pair bond or in nest site acquisition (Marshall 1952). Females show a similar pattern, with a transitory increase in LH and androgen levels during October and a slow increase in size of ovarian follicles from November through April. Threadgold (1960a) described a similar testicular cycle in a comparison of sparrows from five different temperate zone populations.

The difference in volume or mass between the inactive testes and those during full reproductive activity is approximately 100- to 200-fold (Fig. 3.2), and the left testis is usually larger and more elliptical in shape than the right (Keck 1934), although both testes are fully functional. Significant differences in size favoring the left testis have been recorded in populations from India (Rana and Idris 1989) to Denmark (Møller 1994). Table 3.2 presents the monthly minima and maxima of testes size from sparrow populations in many different areas.

Only the left ovary and oviduct develop in the female, and the difference in size between the inactive and active ovary is not as great as that of the testes, being approximately 50-fold (Witschi 1935). The oviduct also shows a seasonal cycle of regression and development; in Iowa (USA), for example, the mean mass is 3.02 mg in mid-December and 241.1 mg in late March (Keck 1934). Selander and Yang (1966) found that the mass of the oviduct varied considerably at different parts of the breeding cycle in Texas (USA), with that

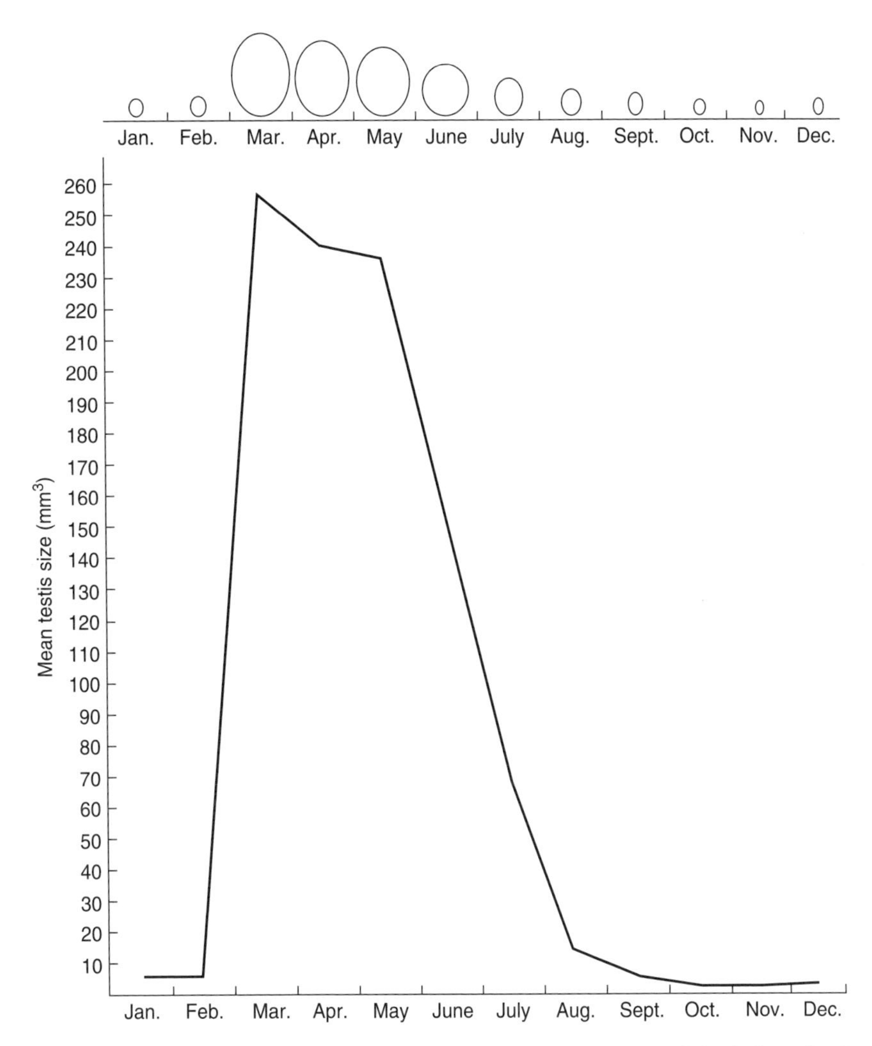

FIGURE 3.2. Graph showing monthly changes in mean volume of the left testis of the house sparrow in Iowa (USA). Volumes (V) were estimated by averaging the radii of the maximum length (R) and width (r) of the left testes of five individuals each month and using the formula $V = 4/3\pi Rr^2$. Average testis size for each month is pictured at the top of the graph (actual size). Modified from Keck (1934: Figs. 1 and 2), with permission of Journal of Experimental Zoology.

TABLE 3.2. Range of Fluctuation in Testes Mass or Volume in Various Populations of the House Sparrow

Location (latitude)	Measurement	Minimum (month)	Maximum (month)	Source
Jodhpur, India (26°N)	Left testis mass	0.225 mg[a] (Apr)	0.360 mg[a] (Sep)	1
Jaipur, India (27°N)	Combined testes mass	5.0 mg (Dec)	441.6 mg (May)	2
Baghdad, Iraq (33°N)	Left testis mass	4* mg (Dec)	312* mg (May)	3
Texas, USA (30°N)	Combined testes mass	2.58 mg (Sep)	602* mg (Apr)	4
California, USA (34°N)	Left testis volume	3.5* mm^3 (Aug)	294.4* mm^3 (Mar)	5
Oklahoma, USA (35°N)	Testis length^3	7.1* mm^3 (Oct)	1075.4* mm^3 (May)	6
New York, USA (42°N)	Left testis mass	3* mg (Oct)	500* mg (Apr–Jul)	7
Iowa, USA (42°N)	Left testis volume	2.4* mm^3 (Oct–Nov)	257.1* mm^3 (Mar)	8
Utah, USA (42°N)	Combined testes mass	2.1* mg (Dec)	582* mg (May)	9
Ontario, Canada (43°N)	Left testis volume	4* mm^3 (Sep–Dec)	359* mm^3 (Jul)	5
Minnesota, USA (45°N)	Left testis volume	1.8* mm^3 (Nov)	421* mm^3 (Apr)	10
Paris, France (49°N)	Left testis length^3	2.4* mm^3 (Jan)	1191* mm^3 (May)	11
Surrey, UK (51°N)	Combined mass	2.0* mg (Oct)	653* mg (May)	12
Cambridgeshire, UK (51°N)	Combined mass	2.4* mg (Dec)	729* mg May)	12
Gent, Belguim (51°N)	Combined mass	6.3* mg (Nov)	682* mg (Apr)	13
Leipzig, Germany (51°N)	Combined mass	2.7* mg (Jan)	414.5* mg (May)	14
Northern Ireland, UK (55°N)	Left testis volume	4.8* mm^3 (Nov)	292* mm^3 (May)	5

Several of the values in the table, indicated by an (*), represent estimates made from figures in the source. In most of the sources in which volumes were presented, they were estimated in the original source by the formula $V = 4/3\pi ab^2$ where a = ½ testis length and b = ½ testis width.

[a]Preserved in 10% formaldehyde.

Sources: 1, Rana and Idris (1989); 2, Saxena and Mathur (1976); 3, Kadhim et al. (1987); 4, Menaker (1971); 5, Threadgold (1960a); 6, Allender (1936b); 7, Hegner and Wingfield (1990); 8, Keck (1934); 9, Barfuss and Ellis (1971); 10, Kirschbaum and Ringoen (1936) (no data May–Sep); 11, Loisel (1900) (no data Jun–Nov); 12, Murton and Westwood (1974); 13, Moens and Coessens (1970) (no data Jun–Oct); 14, Etzold (1891) (no data Jun–Dec).

during prelaying averaging 514 mg; laying, 1371 mg; early incubation, 501 mg; and late incubation, 210 mg.

There is conflicting evidence concerning the gonadal cycle of subtropical populations of *P. d. indicus* in India. Clutches are initiated in every month of the year near Baroda (22°18'N), but primarily during spring (March–April) and during the late monsoon (September–October), with very few clutches being initiated in winter (November–January) (Mathew and Naik 1986; Naik and Mistry 1980). Molt normally occurs between these two periods of peak breeding (see Chapter 5). At Jodhpur (26°18'N), the highest mean monthly mass of the left testis (in September) was only 1.6 times the lowest mass (April), and active spermatogenesis was observed in all 12 months (Rana and Idris 1989). The ovary was also active throughout the year.

Studies at some Indian sites suggest, however, that the gonadal cycle is similar to that found in the temperate zone. Photorefractoriness and marked seasonal changes in the size of the ovary (Ravikumar and Tewary 1990) and complete testicular regression in males (Lal and Thapliyal 1982) have been reported at Varanasi (25°18'N). At Jaipur (26°54'), the testicular cycle is similar to that in temperate resident sparrows (Saxena and Mathur 1976). Concentrations of alkaline and acid phosphatases in the testes were 3–4 times greater during the breeding season than during the nonbreeding season at Kalyani (23°N) (Nandy and Manna 1996). At Ludhiana (30°57'N), the ovarian enzyme 3β-hydroxysteroid dehydrogenase (3β-HSD) showed peaks of activity in March–April and September, corresponding to the two peak breeding periods, whereas 3b-HSD activity is lowest during winter (November–December) (Parshad and Bhutani 1987).

Little is known about the gonadal cycles of introduced populations that have recently become established in tropical regions of Australia, southern Africa, and Central and South America. The breeding season of *P. d. indicus* in Malawi (15°44'S) is distinctly bimodal, although breeding has been observed in every month except February (Nhlane 2000). This pattern is similar to that described earlier for the source populations. A bimodal breeding season was also reported for *P. d. domesticus* in the lowlands of El Salvador (approximately 13°N) (Thurber 1986). This pattern is distinctly different from that of the temperate resident source populations from which the North American populations of *P. d. domesticus* were derived (see Chapter 1) and suggests that the mechanism controlling the annual gonadal cycle of sparrows may show considerable plasticity.

Much more work has been done on the testicular cycle than on the ovarian cycle, so the following discussion focuses primarily on the testes, with occasional comparisons to the ovarian cycle. Females do not respond as readily or as rapidly to photoperiodic manipulations as do males, and, whereas the testes become fully active, the ovary usually does not proceed to full follicular development or ovulation under experimental conditions (e.g., Bartholomew 1949; Kirschbaum et al. 1939; Riley and Witschi 1937; Tewary and Ravikumar 1989b).

Testicular Cycle

The first description of the complete testicular cycle of a periodically breeding bird species was apparently that of Etzold in 1891 on the house sparrow (Wingfield and Farner 1993). Since then, many other studies have been performed at various locations in the species' range (e.g., Allender 1936b; Davis and Davis 1954; Kadhim et al. 1987; Threadgold 1960a). Testicular activity involves two major processes: spermatogenesis, which takes place in the seminiferous tubules, and production of androgens (particularly testosterone), which occurs in the interstitial tissue between the seminiferous tubules (see also Chapter 9). Development of the interstitium generally precedes the onset of spermatogenesis in temperate zone populations (Allender 1936a).

Bartholomew (1949) developed a categorization of stages of spermatogenic activity in sparrows that has been used by a number of investigators. The system is based on histological examination of the seminiferous tubules and consists of the following six stages: I, resting spermatogonia only; II, spermatogonia dividing, but only a few spermatocytes present; III, many spermatocytes present; IV, spermatocytes with spermatids; V, spermatids with a few sperm; and VI, full spermatogenic activity with many sperm. During full spermatogenic activity, mature sperm form sperm bundles near the lumen of the seminiferous tubule (with their tails oriented toward the lumen) (Allender 1936a). Some subsequent workers (e.g., Threadgold 1960a) added another category, R, for testes that are in the process of regressing after the breeding period. Full spermatogenic activity has been observed in a testis as small as 59 mg (Kendeigh 1941).

Bill Coloration

Coloration of the bill in males is a secondary sexual characteristic of sparrows and has been used as an external indicator of testicular activity (e.g., Barfuss and Ellis 1971; Keck 1933; Pfeiffer and Kirschbaum 1943; Ringoen 1942). During the breeding season, the bills of sexually active males are a uniform deep black ("jet black" or "blue-black"), but after the breeding season, the bill becomes a pale brown or horn color, similar to that of females throughout the year. Castrated males injected with testosterone develop fully black bills within 25 d (Keck 1933), and females given twice-weekly intramuscular injections of testosterone propionate develop black bills after 2.5 wk (Pfeiffer and

Kirschbaum 1941). Fall-captured males maintained on a nonstimulatory photoperiod (8L:16D) and injected with testosterone propionate developed black bills, whereas no change occurred in controls (Haase 1975). Haase concluded that bill color responds directly to increased levels of testosterone. Lal and Thapliyal (1982) treated males that had been either castrated or thyroidectomized with one of three daily doses of testosterone (25, 50, or 100 μg), and only birds receiving the 100 μg dose showed significant darkening of the bill. Lofts et al. (1973), however, found little change in bill color in fall males injected with 0.4 mg testosterone propionate on alternate days for 4 wk, although they did observe significant darkening in males injected with both LH and follicle-stimulating hormone (FSH). They concluded that darkening of the bill is the result of synergistic effects of gonadotropins (particularly FSH) and testosterone.

Kirschbaum and Pfeiffer (1941) attempted to test whether testosterone has a direct effect on the cells responsible for pigmentation of the bill by injecting five females with yellow bills with 2 μg/d of testosterone (in alcohol). Injections were made into the skin at the angle of the bill on the right side (with the left side serving as the control). The injections resulted in consistent darkening of a narrow line on the bill near the site of injection. Kirschbaum and Pfeiffer concluded that testosterone acts directly at the local level to influence bill pigmentation. A photomicrographic study of the darkening of the bill found that the black coloration is caused by incorporation of pigmented melanocyte processes into the cornified epidermal cells of the sheath (Witschi and Woods 1936).

Of 10 male house sparrows captured in Utah (USA) in late January, 5 with yellow or gray bills had a mean testes mass of 3.01 mg (range, 1.66–4.77), whereas 5 with black bills had a mean mass of 9.30 mg (7.50–10.62), suggesting that the testes mass required for melanin deposition is between 5 and 8 mg (Fevold and Eik-Nes 1962).

Other secondary sex characteristics of the male include the cloacal protuberance, the vas deferens, and the glomus vesicles (Haase 1975; Hegner and Wingfield 1986a; Lal and Thapliyal 1982). In New York, the length of the cloacal protuberance remains low (mean, approximately 1 mm) until February, after which it enlarges rapidly to a mean length of more than 3 mm by April (Hegner and Wingfield 1986b). During the breeding season, it has a mean length of about 6 mm, but the length drops to less than 3 mm by mid-September (Hegner and Wingfield 1986a).

Control of the Testicular Cycle

Control of the testicular cycle is dependent on the interaction of internal physiological control mechanisms and external environmental cues. Wingfield and

Farner (1993) identified four categories of external agents. *Initial predictive information* includes those environmental cues that initiate gonadal development and other periodic behaviors, such as migration, that bring the organism into readiness to breed. Since Rowan's (1925) early insight, changes in daily photoperiod have been recognized as the major initial predictive information for most temperate zone breeding species, including sparrows. *Essential supplementary information* includes external cues that supplement the initial predictive information and promote completion of gonadal development and the beginning of breeding. Numerous local factors such as temperature, inclement weather, or feeding conditions can serve to fine-tune the breeding readiness of the individual to coincide with local conditions (e.g., Il'Enko 1958 for sparrows in Russia). *Synchronizing and integrating information* includes behavioral interactions between mates and with other individuals that serve to synchronize breeding readiness. Social facilitation and mating rituals are examples of this type of information. In sparrows, for instance, males housed in flocks in large cages have a significantly higher testes mass than males housed as a member of a pair in a large cage, and the latter in turn have larger testes than males from pairs held in small cages (Hegner and Wingfield 1984). Finally, *modifying information* includes events such as nest predation that interrupt the normal progression of breeding. The bulk of the work on external agents controlling the annual cycle has focused on the initial predictive information, particularly photoperiod.

Photoperiodic Control

The annual cycle of changes in photoperiod at temperate latitudes provides the primary cues in the control of the reproductive cycle of sparrows. Lengthening photoperiods after the winter solstice precede the period of rapid gonadal recrudescence in the winter stage of the preparatory phase of the cycle. Likewise, shortening days after the summer solstice precede the rapid regression of the gonads at the end of the breeding season. The resultant refractory state, during which the gonads fail to respond to normally stimulatory photoperiods, must be broken by short photoperiods.

Kirschbaum (1933) was the first to perform manipulative experiments on the effect of photoperiod on sparrows (see also Kirschbaum and Ringoen 1936). Males (both adults and juveniles) captured in Minnesota (USA) were treated with 6–7 h/d of additional light for 6 wk beginning 21 October. Treated birds showed enlarged testes and full spermatogenic activity, whereas control birds maintained on natural photoperiods at the same elevated temperatures (7°C–13°C) showed no testicular response. Juvenile males with enlarged testes also developed black bills. Conversely, treatment with only 5 h/d of light during

January–March resulted in slowed testicular development, and juveniles failed to develop black bills.

In a study that was apparently the first to demonstrate gonadal photorefractoriness in birds (see Immelmann 1971), Riley (1936b) exposed males captured in Iowa during September and November to long photoperiods (17L:7D and 14L:10D, respectively). The bills of five of the six November males gradually darkened during the 41 d on the experimental photoperiod, and the average volume of the left testis was about 100 mm^3. Five males maintained on natural photoperiods showed no change in bill coloration and had an average testis volume of 1.0 mm^3. The five adult males captured in September and exposed to 17L:7D photoperiods showed no change in bill coloration and also showed little or no increase in testis size, with a mean left testis volume of 2.2 mm^3. Three juvenile males treated in the same way, however, showed a darkening of the bill and had an average testis size of 71.5 mm^3 at the end of treatment.

In an early review on the role of light in stimulating gonadal recrudescence, Rowan (1938) concluded that longer photoperiods act indirectly to stimulate gonadal development in birds (including sparrows) by increasing the time of daily activity and feeding, one or both of which are the direct proximate causes of gonadal growth. Riley (1940) and Kendeigh (1941) tested this hypothesis by increasing the activity period of male sparrows in the dark. Both used mechanically rotating cages to enforce activity during the dark period after a natural photoperiod so that the total activity time of the treated individuals was equivalent to that of a second group maintained on increased photoperiods. Riley (1940) initially maintained two groups of birds on a 9L:15D photoperiod, followed by either gradual increases in daily photoperiod or equivalent increases in enforced activity during the dark. Males on increasing light showed a marked increase in testes size after 46 d, whereas birds with increased activity showed no testicular enlargement. Kendeigh (1941) obtained similar results, finding that males held on 15L:9D photoperiods had testes that exceeded those of control males 18- to 90-fold, whereas exercised males had testes only 60% those of controls. Females also responded to increased photoperiod, while exercised females showed no response. Kendeigh also maintained the birds exposed to 15L:9D photoperiods at a constant temperature of either 2°C or 22°C, and with or without food during the extended period of light. Neither treatment affected the development of the gonads in either sex. Both Riley and Kendeigh concluded that light alone was responsible for gonadal recrudescence in sparrows, and that periods of wakefulness, activity, or feeding and ambient temperature have no effect.

Wavelength of light has an effect on the responsiveness of the gonads to lengthened photoperiods. Ringoen (1942) tested the effects of green and red light on gonadal development in fall-captured sparrows from Massachusetts

(USA). Experimental birds were exposed to either constant red light or constant green light (24L:0D) beginning in December, and controls were maintained on natural photoperiods. Males on green light showed no testicular development after 28 d, whereas in those on red light the testes were greatly enlarged and showed complete spermatogenic activity. After 43 d the testes of males in green light were considerably enlarged compared with those of the controls and had begun to show some spermatogenic activity. In females, the ovary showed no response to continuous green light but did show some responsiveness to red light. Red light is therefore much more effective than green light in stimulating gonadal activity in sparrows.

Light intensity also affects the rate of gonadal development. In an early experiment, sparrows from Massachusetts were exposed for 25 d during the winter to 16L:8D photoperiods with light of different intensities (0.4, 7.5, 111, 564, or 2625 lux) (Bartholomew 1949). Males showed no responsiveness to the 0.4 lux light but did respond to light intensities of 7.5 lux and higher, although the response at 7.5 lux was less than that at higher intensities. The testes increased 10-fold in size at 7.5 lux, whereas the increase was 30- to 40-fold at higher intensities. Stage of spermatogenic activity was also intermediate at 7.5 lux (III–IV) and greater (V–VI) at higher light intensities. The responses of females exposed to 46 d of 16L:8D photoperiods at the various light intensities were qualitatively similar to those of males. In a second experiment, immature males held on a 16L:8D photoperiod of either 111 or 2625 lux for 25 d beginning in October showed a stronger gonadal response at the higher light intensity (Bartholomew 1949).

Length of photoperiod also has an effect on the rate of gonadal recrudescence. Bartholomew (1949) used six groups of sparrows, with five treatment groups being exposed to different photoperiods (10L:14D, 12L:12D, 14L:10D, 16L:8D, or 24L:0D) and the control group maintained on 8L:16D. Average testes mass after 18 d increased with photoperiod, with the sharpest increase occurring between 12L:12D and 14L:10D. Males on 10L:14D showed only a modest increase in testicular size, and two of the five males in this group advanced to spermatogenic stage II. Ovary size in females held for 30 d under the same conditions showed a similar pattern, except that the greatest increase occurred between 16L:8D (about 2-fold) and 24L:0D (about 4-fold). Bill color in males showed no response at the shorter photoperiods (up to 12L:12D), but at longer photoperiods bills darkened in most individuals. Two males maintained on 10L:14D for 46 d showed marked gonadal development, indicating that a 10 h photoperiod is sufficient to stimulate gonadal development in males. Because testicular development can occur at photoperiods as short as 10 h, Bartholomew concluded that in areas where the natural daylength never drops below 10 h, testicular development must begin when the photorefractory period ends. Other experimental studies have also shown a rapid growth in the testes in response to long photoperiods, whereas short photoperiods (usually 8L:16D) resulted in little or no testicular growth (Dawson 1998a; Donham et al. 1982; Farner et al. 1961, 1977).

Surprisingly, very short photoperiods can also stimulate gonadal development in sparrows. Threadgold (1958, 1960b) exposed groups of sparrows from Ontario (Canada) to either very short or short photoperiods (1L:23D, 4L:20D, or 7L:17D). The experiments were begun in mid-February with a light intensity of either 10.8 or 75 lux, and two or three males were sacrificed each month to determine the size of the testes and the stage of spermatogenesis. At 75 lux, left testis volume increased most rapidly in the 1L:23D group, attaining an average volume of 150 mm^3 after 94 d; the 4L:20D group required 125 d, and the 7L:17D group 184 d, to achieve similar growth. Spermatogenesis showed the same trend, with 1L:23D being the most stimulatory. Similar results were obtained for the number of interstitial cells in the testes. Testicular regression also tended to follow the same pattern, with the testes of 1L:23D birds regressing sooner than those of birds on the longer photoperiods. Similar results were also obtained in the two groups exposed to 10.8 lux, with the 1L:23D group showing a more rapid increase in testis size and spermatogenic activity, although the difference between 1L:23D and 4L:20D was not as pronounced as at the higher light intensity. Similar results were obtained with sparrows captured during the winter in Washington (USA) and maintained on 2L:22D, 4L:20D, or 7L:17D (Middleton 1965). Males on both shorter photoperiods had more enlarged testes and more advanced spermatogenic activity after 100–120 d than did males on 7L:17D.

Threadgold (1960b) suggested that the shape of the response curve to photoperiod in sparrows is parabolic, with the lowest responsiveness being to a 7L:17D photoperiod and greater responsiveness to both shorter and longer photoperiods. He also suggested that the mechanism controlling the testicular cycle has both stimulatory and inhibitory elements, with the former probably being associated with the anterior pituitary and the latter possibly involving a hormone produced by the pineal gland.

Skeleton photoperiods have also been found to stimulate gonadal recrudescence in sparrows, demonstrating that it is the timing of the exposure to light and not the absolute amount of light that is responsible for recrudescence. Murton, Lofts, and Orr (1970) exposed males captured in midwinter in the United Kingdom to various skeleton photoperiods (Fig. 3.3). Maximal responsiveness in terms of testes mass occurred on 6L:9.5D:1L:7.5D and 6L:11.5D:1L:5.5D; although activity records were not kept (see later discussion), birds maintained on these photoperiods presumably responded to them as they would have to 16.5L:7.5D and 18.5L:5.5D, respectively. Takahashi et al. (1978) also observed significant

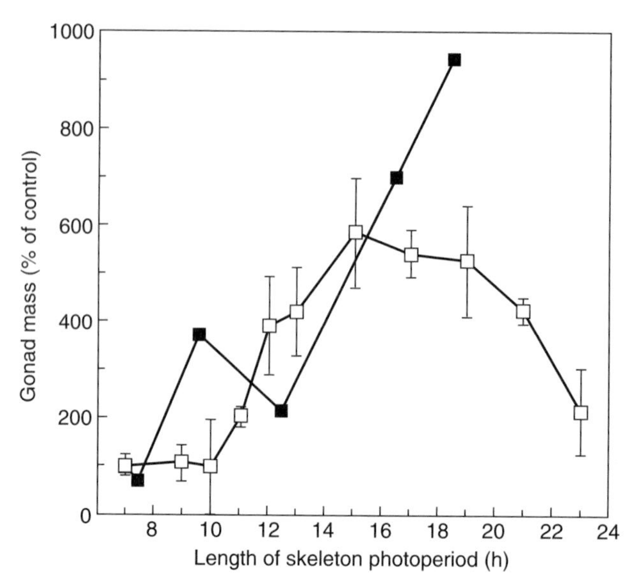

FIGURE 3.3. The effect of skeleton photoperiods on gonadal growth in house sparrows. Males captured in the United Kingdom in December and January were placed on 7L:17D photoperiod for 2 wk (Murton, Lofts, and Orr 1970). Groups were then placed on one of the following skeleton photoperiods for 19 d: 6L:0.5D:1L:16.5D, 6L:2.5D:1L:14.5D, 6L:5.5D:1L:11.5D, 6L:9.5D:1L:7.5D, or 6L:11.5D:1L:5.5D. The birds showed evidence of initiation of spermatogenesis before being transferred to 7L:17D, but control birds held for 14 d on 7L:17D showed signs of degeneration in the seminiferous tubules and had a mean testis mass of 15.4 mg. Males on the skeleton photoperiods (■) showed no increase in testes mass on 6L:0.5D:1L:16.5D but showed an increasing response with the duration of the longer skeleton photoperiods. A similar pattern was observed in the response of photosensitive females exposed to different skeleton photoperiods (□) (Ravikumar and Tewary 1991). Females were captured in February near Varanasi, India (25°18'N), and were maintained on 8L:16D for 8 wk to ensure photosensitivity. They were then exposed for 6 wk to skeleton photoperiods with 6 h of light (400 lux) followed by darkness interrupted by 1-h light pulses beginning 2, 3, 4, 5, 6, 8, 10, 12, 14, or 16 h after the transition from light to dark (L/D), with 7L:17D as control. Ovarian sizes at the beginning of the experiments were observed laparoscopically, and ovaries were fully regressed. Data from Murton, Lofts, and Orr (1970: Table 1) and Ravikumar and Tewary (1991:Table 1).

enlargement of the testes in males that were captured in December or January in Texas and exposed to 1L:11D:1L:11D or 6L:6D:6L:6D photoperiods.

Similar results were obtained in studies of females in subtropical India (see Fig. 3.4). Birds exposed to skeleton photoperiods of 12–21 h showed 4- to 6-fold increases in ovarian mass, and there was a significant increase in ovary mass at 6L:3D:1L:14D (equivalent to 10L:14D). The fact that maximum responsiveness occurred at the longer photoperiods (6L:8D:1L:9D to 6L:12D:1L:5D) is somewhat surprising, because the maximum daylength at Varanasi is only about 13.5 h.

Long photoperiods have also been shown to retard the normal onset of testicular regression during the summer. Dawson (1991) monitored the testicular cycle of a group of wild-caught males that were maintained on natural photoperiods from January to June in England (UK). On 22 June, ten birds were transferred into the laboratory and placed on 18L:6D (the natural photoperiod at the solstice); second and third groups were transferred to 18L:6D on 17 July and 31 July, respectively. Testis size was checked laparoscopically every few weeks until November. The testes of controls regressed significantly during the first 25 d after the summer solstice and reached their minimum size by late August. There was no regression in the first 25 d after the solstice in the group transferred to 18L:6D on 22 June, but regression did begin during August and continued until November. In the other two experimental groups, in which testes regression had begun before transfer, there was no enlargement after the transfer, although regression was slowed in the group transferred on 17 July.

Skeletons of long photoperiod have also been shown to retard gonadal regression during the summer. Fig. 3.4 shows the results of three studies, two in England and one in Texas, on the effect of skeleton photoperiods of various lengths on the size of the testes. In general, the findings of the three studies were similar, with retardation of testicular regression occurring with skeleton photoperiods of 12–18 h. Skeleton photoperiods of less than 12 h or more than 18 h generally resulted in decreases in testes mass, with the rate of testicular regression being inversely proportional to the effective photoperiod. Menaker (1965) also monitored the locomotor activity (see later discussion) of the males in his study. The birds exposed to 4L:2D:2L:16D through 4L:8D:2L:10D all used the beginning of the 4-h light period as subjective dawn, whereas the birds on 4L:12D:2L:6D through 4L:16D:2L:2D used the beginning of the 2-h light pulse as subjective dawn. Of the two birds on 4L:10D:2L:8D, one chose the beginning of the 4-h light pulse as subjective dawn while the other chose the 2-h light pulse.

Another experimental protocol that has been employed is the so-called T-cycle experiment, in which the zeitgeber period (T) differs from 24 h. Farner

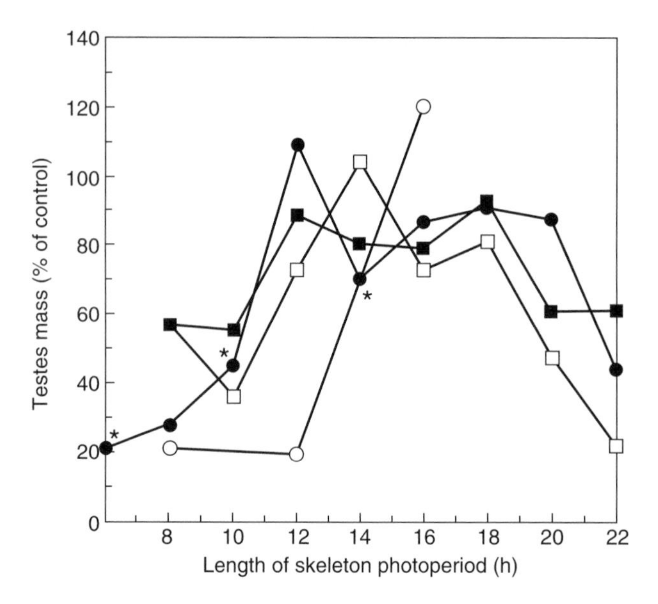

FIGURE 3.4. Retardation of testicular regression in house sparrows exposed to skeleton photoperiods during the summer. The length of the skeleton photoperiod is the time from the D/L transition at the beginning of the longer light period to the L/D transition at the end of the shorter light period (e.g., 6L:5.5D:0.5L:12D = length of 12 h). Mean testes mass of the experimental groups is plotted on the ordinate as a percentage of the mean mass of the control group. Data were obtained from Menaker (1965: Fig. 3), Murton and Westwood (1974: Fig. 2), and Murton, Lofts, and Westwood (1970). Menaker (1965) exposed two adult males captured in Texas (USA) to nine different skeleton light regimes consisting of 4 h of light followed by various intervals before a second, 2-h light pulse (e.g., 6L:18D, 4L:2D:2L:16D, 4L:4D:2L:14D) for 7 wk beginning on 13 June (●). The control group consisted of males held on 14L:10D, with the beginning of the 14-h light period coinciding with the beginning of the 4-h light period in the experimental groups. Asterisks indicate experimental groups in which only a single male survived. Murton and Westwood (1974) captured males in May in England (UK) and exposed them for either 20 (■) or 40 (□) d to one of eight skeleton photoperiods consisting of a 6-h light period followed at various intervals by a 30-min light period (e.g., 6L:1.5D:0.5L:16D, 6L:3.5D:0.5L:14D). The control group consisted of five males sacrificed at the beginning of the experimental treatment. Murton, Lofts, and Westwood (1970) captured 36 males in mid-July in England. The control group consisted of six males sacrificed in late July, when the remaining 30 males were divided into three treatment groups (○), each of which was placed on one of three skeleton photoperiods (6L:1D:1L:16D, 6L:5D:1L:12D, or 6L:9D:1L:8D) for 35 d.

et al. (1977) placed fall- or winter-captured males from Washington State on several different photocycles involving 3 h of light and varying periods of darkness (from 3L:17D to 3L:25D at 400 lux) and monitored their testicular responses by laparotomy. Birds on cycles with $T < 24$ h had testicular growth rates that did not differ from those of controls maintained on nonstimulatory photoperiods, whereas males on cycles with $T > 24$ h showed significant increases in testicular growth rate. Testicular growth rate for these birds was proportional to period length. Ravikumar et al. (1995) obtained similar results in an experiment on females from a subtropical population in India. Females exposed to 3L:23D, 3L:25D and 3L:27D (400 lux) all showed significant increases in ovarian mass compared with 8L:16D controls, whereas birds on 3L:19D or 3L:21D showed no increases. Females on 3L:17D showed a significant increase in ovarian mass, however, a result that differed from that of the Washington males.

Photoperiodic Photoreception

A truly remarkable aspect of the photoperiodic control of the gonadal cycle in sparrows is that photoperiodic light reception is extraretinal. The results of early experiments suggested that the stimulatory effects of long photoperiods could be interfered with at the level of light reception and might be extraretinal. Ringoen and Kirschbaum (1937a,b) "capped" nine males and exposed them to extended photoperiods beginning in November. The cap consisted of a double thickness of cotton stocking material that left only the bill exposed (Ringoen and Kirschbaum 1939). Six of the capped males showed no testicular response after 6 wk, and the remaining three had slightly enlarged testes with spermatocytes but no sperm. Thirteen of 15 uncapped controls had enlarged testes and active spermatogenesis. Although the cap covered the head as well as the eyes, Ringoen and Kirschbaum (1937a, 1939) concluded that retinal light reception was necessary for gonadal response to long photoperiods. Ivanova (1935) placed coverings over the eyes of males and exposed them to 17L:7D photoperiods. Birds with covered eyes showed full testicular development, including complete spermatogenesis comparable to that of controls on 17L:7D that had no eye coverings, and Ivanova concluded that photoperiodic light reception occurred through the skin of the head.

Menaker and Keatts (1968) used bilateral enucleation of male sparrows to demonstrate that an extraretinal receptor (ERR_P) is involved in photoperiodically stimulated testicular recrudescence. Birds were captured in Texas and enucleated in two surgeries, with one eye being removed in early December and the second on 1 January; birds were housed in individual cages on a 9.5L:14.5D photoperiod until the surgeries were completed. On 2 January, birds were placed on either 16L:8D or 6L:18D (500 lux) for 61 d. Controls consisted of enucleated birds maintained on 0L:24D and intact birds maintained on the two experimental photoperiods. Enucleated birds on the short photoperiod and in complete darkness failed to show testicular enlargement (mean masses of 15 and 17 mg, respectively), whereas enucleated birds on the stimulatory photoperiod had a mean testes mass of 404 mg. Control birds on the two photoperiods responded similarly to the enucleated birds on short and long photoperiods (10 and 426 mg, respectively). These results clearly demonstrate that house sparrows have an ERR_p that is linked to the system controlling the annual testicular cycle. Histological examination of the testes showed that enucleated birds on the stimulatory photoperiod had fully developed sperm (as did control birds on the long photoperiod), but that the testes of birds on nonstimulatory photoperiods typically had testes only at stage II of spermatogenesis. Menaker and Keatts concluded that the assumption that retinal light reception is necessary for photoperiodism should be suspended. Underwood and Menaker (1970) obtained similar results with blinded males exposed to a stimulatory photoperiod (16L:8D) with either bright light (460–5500 lux) or dim light (20–46 lux).

In a subsequent experiment, Menaker et al. (1970) attempted to determine whether the ERR_p is located in the brain and whether the eyes play any part in the perception of light that affects photoperiodically elicited reproductive activity. The heads of one group of males were injected subcutaneously with India ink, and a second group had the feathers of the head plucked. Based on Bartholomew's (1949) observation that a light intensity of about 7.5 lux was required to obtain a photoperiodically induced testicular response in sparrows (see earlier discussion), both groups were then exposed to 16L:8D (10 lux) beginning in early January. Birds were sacrificed on 30 January and the mass of the testes was determined. Controls sacrificed at the beginning of the experiment had a mean testes mass of 6 mg. Birds injected with India ink had a mean testes mass of 8 mg, and plucked birds had a mean of 323 mg. Tests of the effects of the two treatments on light transmission through the skin covering the braincase indicated that the India ink injection reduced light transmission to one-tenth that of untreated skin, while plucked skin permitted 100–1000 times more light to pass through. Because all of the birds were sighted, these results indicate that the eyes are not involved in photoperiodic light reception in sparrows.

To test the hypothesis that the pineal gland is the ERR_p, Menaker et al. (1970) established four treatment groups: (1) pinealectomized, plucked (PP); (2) pinealectomized, injected with India ink (PI); (3) sham-operated, plucked (SP); and (4) sham-operated, injected (SI). Birds were then exposed to 16L:8D (5–8 lux) photoperiods from 6 January to 16 February, and then sacrificed. The testes of all four treatment groups responded to the stimulatory photoperiod (mean = 281, 130, 283, and 179 for groups 1–4, respectively), and the combined mean of the plucked birds was significantly greater than that of the injected birds. Menaker et al. concluded that the pineal gland is not necessary for photoperiodically stimulated gonadal recrudescence and therefore is not the sole ERR_p.

Hormonal Control

The endogenous control of the annual gonadal cycle is mediated by hormones, particularly hormones from the hypothalamus, the pituitary gland, and the gonads. The principal hormones involved are virtually identical to the hypothalamic, pituitary, and gonadal hormones of mammals, which is attested to by the fact that the early experimental studies of sparrows described earlier used mammalian hormones in their research. The principal hormones regulating the testicular cycle are gonadotropin-releasing hormone (GnRH) and gonadotropin-inhibitory hormone (GnIH) secreted by the hypothalamus (see Bentley et al. 2003, 2004); LH, FSH, and prolactin secreted by the anterior pituitary gland; and testosterone secreted by the testes. The testes also secrete dihydrotestosterone (DHT), which is of uncertain function in the reproductive cycle. The same hypothalamic and pituitary hormones are involved in the female reproductive cycle, but the ovary secretes estradiol (E_2) as well as testosterone and DHT. See Chapter 9 for further discussion of the functions of the hormones. One effect of the gonadal steroids, particularly testosterone in males and E_2 in females, is that they exert negative feedback on the secretion of gonadotropic hormones by the pituitary (Nicholls et al. 1988).

Rapid development of the gonads in sparrows is initiated at the level of the hypothalamus (Roudneva 1970), which secretes the GnRH that stimulates the release of gonadotropins from the pituitary gland. In sparrows captured in Louisiana (USA), there were more than twice as many GnRH neurons in the preoptic area of the hypothalamus in June than in January (Meseke and Melrose 1994). Pituitary gonadotropins, in turn, stimulate the development of the testes (or the ovary in females). LH has its primary effects on the development of the testicular interstitium in males and in the secretion of androgens in both sexes, while FSH stimulates spermatogenesis in males and the maturation of ovarian follicles in females. Plasma concentration of LH increased significantly 2 d after transfer of male sparrows from 8L:16D to 16L:8D, and peaked after 7 d (Farner et al. 1977). LH levels peaked at more than 1 ng/ml, decreased to about 0.75 ng/ml after 21 d, and dropped to the levels of controls (about 0.5 ng/ml) after 45 d (Donham et al. 1982).

Photorefractoriness and gonadal regression are also presumably initiated at the level of the hypothalamus (Nicholls et al. 1988). The mechanisms responsible for turning reproduction on and off, including the interaction between the endogenous physiological components and the photoperiodic environmental cues, are still only partially understood.

Other hormones that may influence the gonadal cycle are thyroxine, produced by the thyroid glands; prolactin, produced by the anterior pituitary; and corticosterone, produced by the adrenal cortex. Thyroid activity has been found to be highest during the winter and lowest during the breeding season in sparrows in both California (USA) (Davis and Davis 1954) and Ohio (USA) (Kendeigh and Wallin 1966). Davis and Davis (1954) suggested that thyrotropin (TSH) secreted by the anterior pituitary may be antagonistic to gonadotropin secretion. Thus the demand for thyroxine, which stimulates increased metabolic activity during periods of cold stress and other periods of high metabolic demand (see Chapter 9), may indirectly act to inhibit gonadal development. Thyroxine may also play a direct role in the induction of photorefractoriness in birds (Nicholls et al. 1988). Thyroidectomy was shown to cause a rapid decrease in testis size in sparrows in India (Lal and Thapliyal 1982). In a radioimmunoassay study of plasma levels of captive sparrows in Washington, however, a significant increase in thyroxine was observed during molt, but the increase occurred after gonadal regression was complete (Smith 1982).

Prolactin plays a number of roles in avian reproduction, including the stimulation of incubation in females in species such as sparrows in which only the female incubates (see Chapter 4) and of parental care in both sexes (Wingfield and Farner 1993). In addition, prolactin may play a role in both the onset of photorefractoriness (Nicholls et al. 1988) and, in conjunction with corticosterone, the onset of reproduction (see later discussion). An intramuscular injection of 1 mg of prolactin into males captured in midsummer in England resulted in a marked decrease in size of the testes after 8 d, whereas controls showed no decrease in testes size (Lofts and Marshall 1956).

Corticosterone may have a direct role in control of the onset of reproduction in sparrows. Meier et al. (1971) captured photorefractory males in Louisiana in September and placed them in constant light (24L:0D, 300 lux) in early October. Beginning 9 d later, six groups of birds began receiving 25-µg injections of corticosterone and 25-µg injections of prolactin at intervals such that prolactin injection followed corticosterone injection by 0, 4, 8, 12, 16, or 20 h. Controls received saline injections. All birds were sacrificed in early November, and their testes were weighed. No testicular growth occurred in the groups in which prolactin injection followed corticosterone injection by 0, 12, 16, or 20 h (all testes masses about 2 mg), but significant growth did occur in the groups that received prolactin injections 4 and 8 h after corticosterone injection, with the peak growth (9 mg) in the 8 h group. Meier et al. concluded that the temporal relationship of corticosterone and prolactin is capable of stimulating testicular growth in photorefractory sparrows.

In a subsequent experiment, three groups of males were injected daily with corticosterone, followed by a prolactin injection 6 h later (Meier and Dousseau 1973). The groups were then exposed to one of the following light treatments (alternating with constant dim light, 24L:0D, 0.5 lux): (1) 6L:18D (300 lux) on alternate days with D/L 18 h after the corticosterone injection; (2) 6L:18D (300 lux) on alternate days with D/L 6 h after the corticosterone injection; or (3) constant dim light. The average testes mass of group 2 was approximately twice that of the other two groups. Meier and Dousseau concluded that plasma levels of corticosterone and prolactin must be in a certain temporal relationship for the testes of sparrows to be photoinducible. Although this is an interesting hypothesis, the possible role that temporal patterns of hormone titers may play is still largely unknown, in part because other workers have been unable to successfully replicate these experiments (Nicholls et al. 1988).

Hegner and Wingfield (1986a, 1986b, 1986c) performed the first comprehensive study of the annual cycle of reproductive hormones of a free-living bird species on a sparrow population in New York. Fig. 3.5 shows the annual cycles of several hormones and other characteristics related to the sexual cycles of both males and females. As noted previously, there are transitory increases in LH in the fall in both sexes, which apparently coincide with the breaking of photorefractoriness. In the male, this leads to a gradual increase in testicular activity, and particularly to the production of increased levels of testosterone and a concomitant darkening of the bill (Fig. 3.5a). This increase in testosterone (which is significant in December) apparently suppresses gonadotropic secretion from the pituitary until stimulatory photoperiods after the first of the year result in rapid gonadal recrudescence (cf. Nicholls et al. 1988). Significant increases in testosterone, DHT, and E_2 secretion from the ovary (all in October) apparently play a similar role in females (Fig. 3.5b). Similar patterns of seasonal change in hormonal levels and gonad size were observed in free-living populations of *P. d. italiae* in Italy (Fulgione and Milone 1998).

Testosterone can apparently play either an inhibitory or a stimulatory role in testicular development. Turek, Desjardins, and Menaker (1976) implanted testosterone-containing Silastic tubes of various lengths into male sparrows (testosterone dosage was approximately 15 μg/d for each 10 mm of tube length). Males with short tubes (<10 mm) had significantly lower testes masses than those with longer tubes, and males with tubes of 40–120 mm showed maximal testicular development. Males with 5-mm and 120-mm tubes also showed reduced levels of circulating LH, whereas controls with empty tubes showed no such reduction. In a second experiment, these authors maintained males

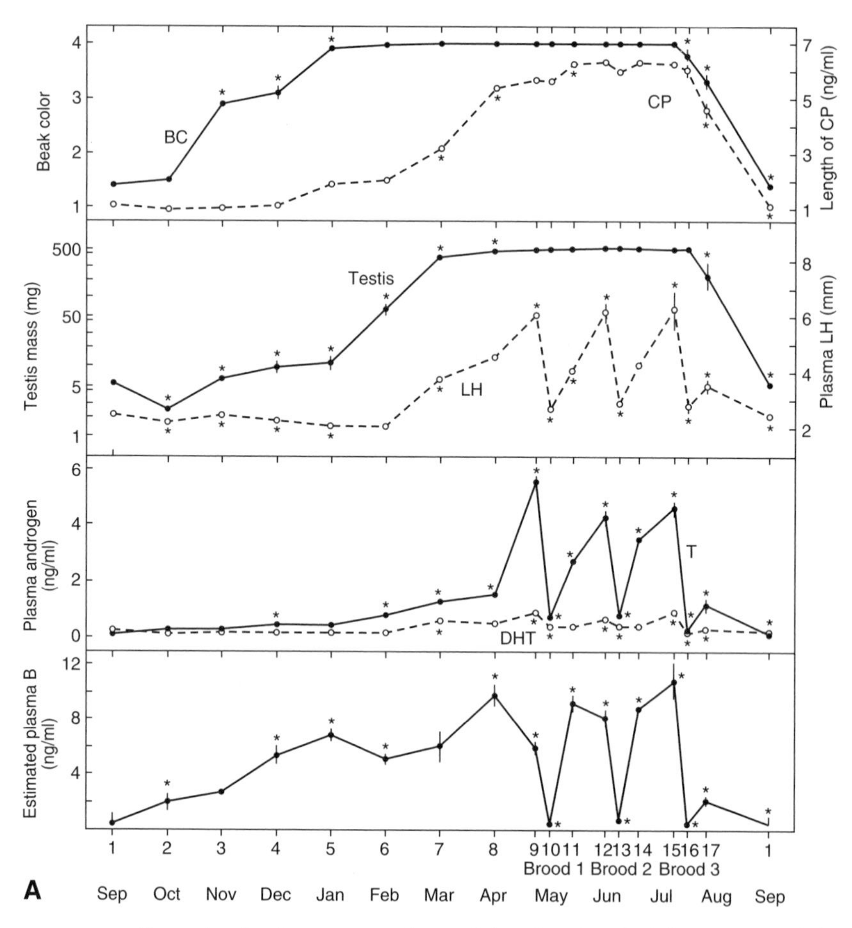

FIGURE 3.5. Reproductive hormones and gonadal cycles of male (A) and female (B) house sparrows in Dutchess Co., New York, USA (Hegner and Wingfield 1990). Birds were captured throughout the year from September 1982 to October 1984. Blood samples (300–400 μl) were taken shortly after capture and tested for testosterone (T), dihydrotestosterone (DHT), luteinizing hormone (LH), estradiol (E_2), and corticosterone (B). Corticosterone levels (in nanograms per milliliter) were positively correlated with the time (t) in minutes after capture when the blood sample was taken: B = 5.11 + 0.96t ($r = 0.55$, $P < .01$, $n = 444$). In males, the length of the cloacal protuberance (CP) was measured (to the nearest 0.5 mm) and the beak color (BC) was scored from 1 (uniform light yellow or horn color) to 4 (jet black). Laparotomies were performed to estimate left testis size and size of the largest follicle in the ovary. Adapted from Hegner and Wingfield (1990: Figs. 1–4).

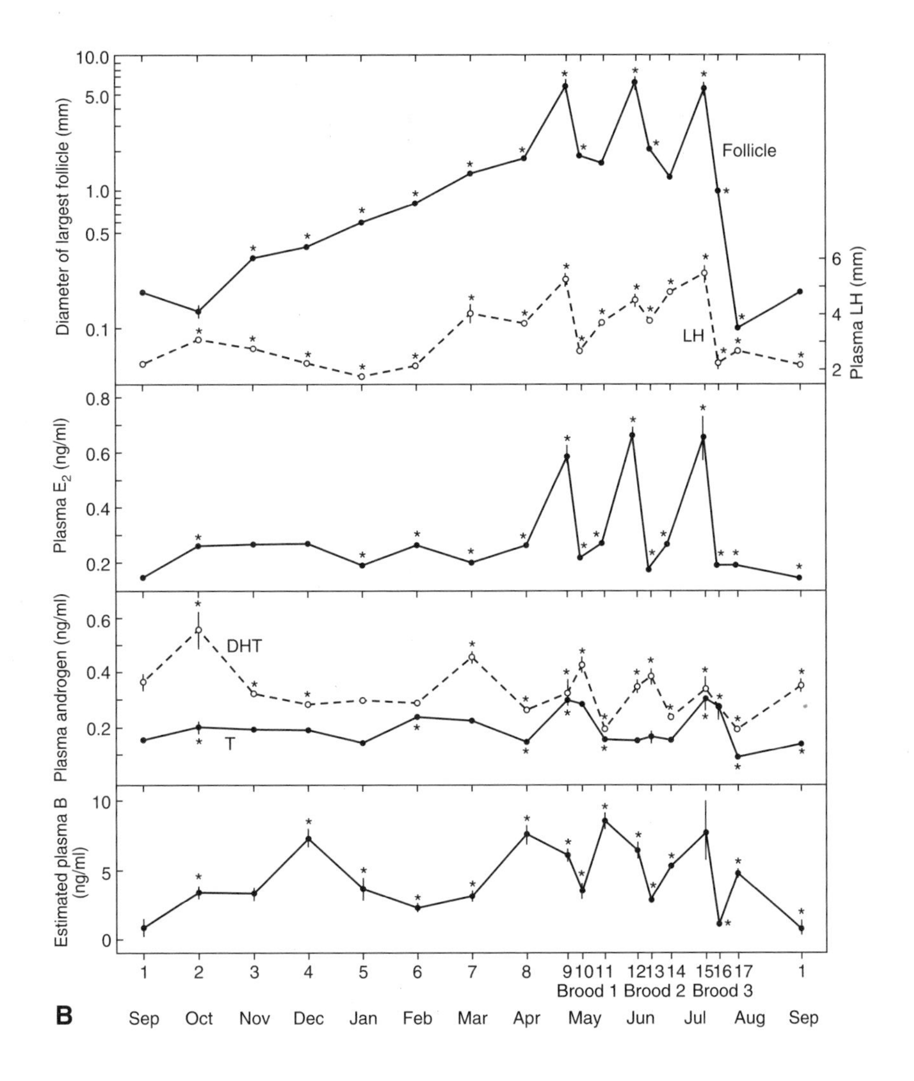

with regressed testes on a nonstimulatory photoperiod (8L:16D) and implanted 20 mm testosterone-containing Silastic tubes intraperitoneally. Controls showed no enlargement of the testes, but treated birds had fully enlarged testes after 60 d. Wolfson and Stahlecker (1950) obtained similar results with daily intramuscular injections of testosterone propionate into males being maintained on 8L:16D. Testosterone in low dosages apparently inhibits testicular development and spermatogenesis, presumably by negative feedback through the hypothalamus on pituitary gonadotropin secretion, and at high dosages it maintains testes size and stimulates spermatogenesis. These results also indicate that testosterone can stimulate testes enlargement in the absence of the normal photoperiodic cues.

Photoperiodic and Hormonal Interaction

"We remain largely ignorant of the processes, at least at the higher neural levels, by which the vertebrate reproductive system switches on and off" (Nicholls et al. 1988:133). These two events, the switching on and switching off of the reproductive system, are the pivotal events in the annual cycle of birds. The timing of molt, which is usually concurrent with the period of photorefractoriness that follows reproduction, and the timing of migration are both closely related to them. This section attempts to summarize what is known about the hormonal and photoperiodic control of these events in sparrows.

As described earlier, numerous experiments have demonstrated that long photoperiods are capable of stimulating the development of the gonads in photosensitive sparrows. Experiments with skeleton photoperiods have shown, however, that it is not the absolute amount of light but rather the timing of the light (or dark) that is effective in stimulating the response. An alternative explanation, that it is the shortness of the dark period (scotoperiod) that triggers gonadal response, is contradicted by the findings of T-cycle experiments in which birds with long scotoperiods (e.g., 3L:25D) nevertheless showed gonadal development (see earlier discussion). Two models have been proposed to explain this relationship between external cues and internal physiological mechanisms in the activation of the reproductive system, the External Coincidence Model and the Internal Coincidence Model. The former was first proposed by Bunning in 1935 and subsequently elaborated by other workers (see Bunning 1960). The Internal Coincidence Model was first proposed in 1966 by Tyscheko (cf. Meier and Russo 1985).

The External Coincidence Model postulates that the external light stimulus must occur during a photosensitive phase of some unspecified component of the internal circadian rhythm. In sparrows, this photosensitive phase usually occurs 12–16 h after dawn (D/L), which acts as an entraining zeitgeber for the internal rhythm (see later discussion). The co-occurrence of the external light stimulus with the photosensitive phase of the internal rhythm results in stimulation of the hypothalamus to secrete GnRH, thereby initiating breeding. Results of skeleton photoperiod experiments such as those described earlier represent some of the strongest evidence supporting this model. In these experiments, even a short pulse of light occurring during the photosensitive phase of the internal rhythm resulted in gonadal development, despite the fact that the birds were exposed to as little as 2 h of light during the day (Takahashi et al. 1978).

The requirement that light occur during the photosensitive period to initiate a gonadal response appears to be inconsistent, however, with the stimulation of gonadal recrudescence by very short photoperiods (described earlier). There are explanations for these results that would still be consistent with the External Coincidence Model. One possibility is that very short photoperiods do not entrain the internal circadian rhythms of the birds, which may then free-run or show only relative coordination with the periodic light stimulus. Free-running locomotor rhythms have been observed, for instance, in sparrows exposed to repeating short photocycles (see Fig. 3.8 later in this chapter). Binkley (1977) reported on the locomotor activity of sparrows maintained sequentially for 20 d each on 1L:23D, 2L:22D, 4L:20D, and so on, but the 5 d of activity presented in her study (see Fig. 3.7) do not permit one to infer whether the individual birds were free-running. If the photosensitive rhythm of sparrows does free-run on very short photoperiods, then these short periods of light would eventually coincide with the photosensitive period, resulting in photostimulation of the gonads. Some species of birds have shown gonadal development in complete darkness (cf. Meier and Russo 1985), however, and this is more difficult to rationalize with the requirement for coincidence of an external light stimulus with an internal photosensitive period in the External Coincidence Model.

A second possibility is that the birds have a circannual rhythm of gonadal activity, and that stimulation by long photoperiods acts to accelerate this endogenous rhythm to time the breeding season and other seasonal activities adaptively. Under this scenario, short photoperiods could be acting in an inhibitory way on this endogenous cycle, and very short photoperiods (e.g., 2L:22D), perhaps because they represent weaker entraining stimuli, are simply less inhibitory than longer photoperiods (e.g., 7L:17D). Full development of the testes in photosensitive sparrows stimulated with long photoperiods requires 42–45 d (Dawson 1998a, Donham et al. 1982). Testicular development in sparrows maintained on 1L:23D required 94 d to reach a volume of 150 mm^3, compared with 300 mm^3 in photostimulated males (Threadgold 1958, 1960b). On 2L:22D, the testes reached a mean mass of 23 mg after 100 d and 55 mg after 120 d (Middleton 1965). Testes masses were greater than 200 mg at 100 and 120 d with 4L:20D, but birds stimulated with a long photoperiod had a mean testes mass greater than 300 mg after only 40 d. It may therefore be inappropriate to speak of very short photoperiods as stimulatory; rather, they may be less inhibitory than short photoperiods. Although there is considerable evidence for circannual rhythms in a number of bird species (e.g., Gwinner 1986), there is as yet no direct evidence for such rhythms in sparrows (Binkley 1990).

The Internal Coincidence Model postulates that coincidence in two internal rhythms, at least one of which is photoperiodically entrained, is necessary to stimulate gonadal development. The second rhythm, which is also circadian, may have its phase relationship with the other rhythm determined by daylength (Meier and Russo 1985). If both rhythms are photoperiodically entrained, one may be entrained by dawn (D/L) and the other by sunset (L/D). The direct cause of the initiation of breeding is the coincidence in the two internal rhythms, with that coincidence in turn being a result of photoperiodic events.

The one circadian rhythm postulated by the External Coincidence Model is presumably located in the hypothalamus, as is at least one of the rhythms postulated by the Internal Coincidence Model. The finding (described earlier) that the gonads of pinealectomized sparrows respond to stimulatory photoperiods (Menaker et al. 1970; see also Yokoyama 1980) appears to eliminate the pineal gland as a possible location, despite its role in controlling other circadian rhythms in the house sparrow and its apparent sensitivity to light (see later discussion). Takahashi et al. (1978), however, did find that pinealectomized sparrows failed to show testicular responsiveness to two skeleton photoperiods (1L:11D:1L:11D and 6L:6D:6L:6D) that were stimulatory to sham-operated controls.

Photorefractoriness

The gonads of both sexes undergo rapid regression and enter a period of photorefractoriness shortly after the summer solstice. Photorefractoriness in birds is usually divided into absolute refractoriness and relative refractoriness. In the former, gonadal regression occurs during long, normally stimulatory photoperiods and results in the gonads being unresponsive to long photoperiods until they have been exposed to a period of short daylengths that breaks the refractory state. Relative refractoriness occurs in some species; the gonads remain active for an extended period after the summer solstice and then remain inducible by long photoperiods after they regress in late summer or early fall. It is unclear whether the two types of photorefractoriness represent qualitatively different phenomena or simply two extremes of a continuous spectrum (Nicholls et al. 1988). Temperate resident and temperate migratory populations of sparrows both display absolute photorefractoriness. The situation in subtropical resident populations is less clearcut, but females at both Ludhiana and Varanasi, India, show photorefractoriness under laboratory conditions (Parshad and Bhutani 1987, Tewary and Ravikumar 1989b).

The stimulus for the onset of refractoriness is apparently endogenous, although the timing and rapidity of onset may be influenced by environmental factors, particularly photoperiod (Nicholls et al. 1988). Dawson (1998a) studied the onset of refractoriness in sparrows in England. Three control groups of males were established in February, one group on 18L:6D (the photoperiod at the summer solstice at 52°N), one on 16L:8D, and one on 13L:11D. Size of the left testis was estimated by laparoscopy at the outset of the experiment and at intervals thereafter. Testis size reached its maximum after 6 wk in the 18L:6D and 16L:8D groups, whereas the 13L:11D group required 9 wk. Testis size decreased significantly after 16 wk in the 18L:6D and 16L:8D groups, but only after 26 wk in the 13L:11D group. The timing of the onset of regression was therefore inversely proportional to the photoperiod. Four experimental groups were also created from males that were initially placed on 18L:6D photoperiods for either 6 or 12 wk, after which they were placed on either 16L:8D or 13L:11D photoperiods for the duration of the experiment. Both of the two groups transferred after 6 wk (when testis size was at its peak) showed significant regression after 15 wk, and the rate of regression was similar to that in the control group maintained on the 18L:6D photoperiod. In the groups transferred after 12 wk, the decrease in daylength resulted in an immediate increase in the rate of regression with the 13L:11D group showing significant regression after 14 wk and being fully regressed after 16 wk. Similar results were obtained in the group transferred to 16L:8D, but the rate of regression was not as rapid as in the 13L:11D group. Dawson concluded that at some time during the first 6 wk of exposure to 18L:6D the timing of gonadal regression was determined to begin after 9 wk, and that subsequent transfer to a shorter photoperiod only served to accelerate the rate of regression. The fact that gonadal regression begins while the birds are being maintained on a constant photoperiod suggests that there are endogenous components in the mechanism responsible for the termination of breeding and the onset of refractoriness, but the facts that photoperiod affects the timing of that regression and that a reduction in photoperiod accelerates the rate of regression indicate that there are exogenous elements as well.

Hahn and Ball (1995) obtained similar results with 16 males captured in early June in Maryland (USA). One group of eight was maintained throughout the experiment on a 16L:8D photoperiod, and a second group of eight was maintained on 13L:11D until 2 August, when four birds were transferred to 8L:16D while the other four remained on 13L:11D. All birds were sacrificed at the end of August for immunocytoassay of GnRH in the hypothalamus. The testes of half of the 16L:8D males had regressed at the end of the experiment, and half remained active. One of four 13L:11D males had regressed testes, but all four of the males exposed to 13L:11D followed by 8L:16D were regressed or regressing. GnRH was found primarily in the preoptic and septal hypothalamic areas. Unregressed birds showed much more GnRH in both cell bodies and fibers, and the 8L:16D birds showed much less GnRH in the cell bodies, although some remained in fibers. Males on 16L:8D that had spontaneously regressed showed very low GnRH in either cell bodies or fibers. The number of cells decreased in the regressed birds on both the 16L:8D and 13L:11D followed by 8L:16D treatments.

These results support the proposition that photorefractoriness in sparrows occurs at the level of the hypothalamus and that it involves the destruction and/or inactivation of GnRH-secreting cells. The mechanism by which photoperiodic cues are transduced to affect the photorefractory or photosensitive state of the hypothalamus is still not understood. There is little doubt, however, that it involves the circadian time-keeping system.

The Ovarian Cycle

The ovarian cycle of free-living sparrows is similar to the testes cycle except that the ovary typically does not develop as rapidly in the spring (Davis and Davis 1954). The ovary also does not respond as fully as the testes to experimental photoperiodic stimulation (Bartholomew 1949, Tewary and Ravikumar 1989b), which is apparently indicative of the need for additional cues for full gonadal response in females. These cues, which presumably fall into the categories of essential supplementary information or synchronizing and integrating information (cf. Wingfield and Farner 1993), presumably include such things as temperature, male courtship behavior, and nest site (Il'Enko 1958). Once it is fully stimulated, the ovary requires 18 d to go from an inactive to an active state (Davis and Davis 1954).

Circadian Rhythmicity

The house sparrow has been one of the principal organisms used in the study of circadian rhythms, a fact attested to by Sue Binkley's book on endogenous rhythms entitled *The Clockwork Sparrow* (Binkley 1990). Circadian rhythmicity is one of the truly remarkable phenomena of nature, and it has been observed in virtually all organisms, from the simplest to the most complex, in which it has been investigated. Although some scientists still hold the agnostic view that observed rhythmicities are the result of an organism's response to as yet unidentified external environmental cues, the preponderance of evidence favors the view held by most investigators that the rhythms observed in organisms held under constant conditions are endogenous in origin (Binkley 1990). In 1935, the German botanist Erwin Bunning proposed that the internal time-keeping mechanism underlying overt circadian rhythms enable an organism to respond appropriately to the seasonal pattern of change in photoperiod to regulate its annual cycle. An understanding of the mechanism underlying this circadian pacemaker is therefore necessary if control of the annual cycle of sparrows is to be fully described.

Sparrows have been shown to have endogenous rhythms of approximately 24-h periodicity in locomotor activity (Eskin 1971), resting metabolic rate (Hudson and Kimzey 1966), body temperature (Binkley et al. 1971; Hudson and Kimzey 1966), uric acid excretion (Menaker et al. 1978), and feeding (Chabot and Menaker 1987). Although these rhythms normally become entrained to the daily cycle of light and darkness, each has been shown to continue to cycle with a period of approximately 24 h when the birds are experimentally deprived of the normal environmental cues. These rhythms are the overt manifestations of an internal time-keeping mechanism often referred to as a biological clock. Binkley (1990), however, called it a "biological watch" because, unlike a clock, it is carried around with the organism. Overt rhythms such as the locomotor rhythm are sometimes said to be "slave processes" because their periodicity is determined by the biological clock mechanism without affecting the periodicity of the mechanism itself.

The most commonly monitored circadian rhythm is that of locomotor activity, which can be continuously recorded electronically by placing microswitches on perches in the cages in which individual sparrows are kept. Printouts of the data, which are usually recorded on an event recorder or computer, are traditionally cut into 24-h segments and mounted chronologically from top to bottom to permit visualization of the activity pattern. An example of the resultant actogram is presented in Fig. 3.6, which illustrates 35 d of locomotor activity of a sparrow maintained for the first 18 d on a 12L:12D photoperiod with lights on at 0600, after which the 12L:12D photoperiod was abruptly shifted 6 h to lights on at 1200, an event referred to as a phase shift in the zeitgeber. Heigl and Gwinner (1995) suggested that the feeding rhythm is more reliable than the locomotor rhythm for determining period and activity length in sparrows but found that the two rhythms tend to coincide. Chabot and Menaker (1992) and Hau and Gwinner (1997) also concluded that the two rhythms are tightly coupled in sparrows.

Laboratory Studies of Activity of Intact Sparrows

The locomotor activity patterns of intact sparrows have been studied in the laboratory using a number of experimental manipulations. These include changes in photoperiod, changes in light intensity, skeleton photoperiods, light cycles with a period (T) differing from 24 h, periodic acoustic stimulation, periodic food availability, and constant darkness. This section summarizes some

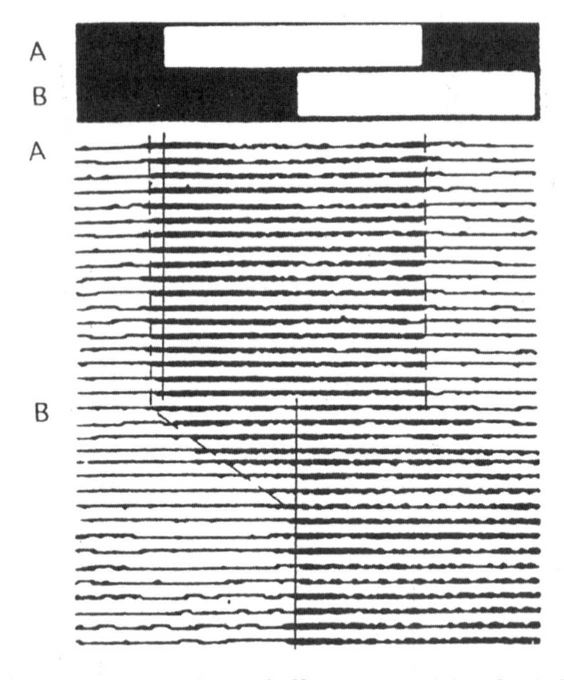

FIGURE 3.6. Actogram representing 35 d of locomotor activity of an individual house sparrow maintained in the laboratory under constant conditions (25°C and 90 dB white noise) and a 12L:12D (240 lux:0 lux) photoperiod. For the first 18 d, the lights were turned on from 0600 to 1800, after which the lights were on from 1200 to 2400 (indicated by bars A and B, respectively, at the top of the graph). Each line of the figure represents the activity record for a 24-h period, and the record is chronological from top to bottom. The two solid vertical lines represent the times of lights-on (D/L), and the vertical dashed lines at the beginning of the record indicate the average time of activity onset (*left*) and cessation (*right*). The elapsed time between these two dashed lines is the average activity time (α) of approximately 12.8 h, and the difference between the dashed and vertical line at the upper left of the record is the time by which activity onset anticipates D/L, or the phase angle (ϕ), which in this case is about 0.8 h ($\phi = +0.8$ h). The dashed diagonal line in the lower left part of the figure represents the approximate onset of activity during the period of transition to the 6-h phase delay, and the time difference between the intercepts of the vertical dashed line and the diagonal line on the day following the phase shift is the phase angle shift ($\Delta\phi \approx -0.85$ h) caused by the phase shift in the photoperiod. Note the apparent bimodality in the activity of this sparrow during the light period, with higher rates of locomotor activity for about 6 h after activity onset and for about 2 h before lights-out (L/D), a pattern that is typical in many sparrows on longer photoperiods ($\geq$12 h) (Binkley and Mosher 1985a). Modified from Binkley (1977: Fig. 7), with permission of the University of Chicago Press ©1977.

of the principal findings from these studies. These findings represent the phenomenology of the circadian locomotor rhythm and provide important insights into the mechanism of the biological clock.

Sparrows readily entrain their activity rhythms to coincide with the period of light in light/dark cycles of 24–h duration (cf. Fig. 3.6). Binkley (1977) placed sparrows sequentially on 12 different photoperiods (1L:23D, 2L:22D, 4L:20D, 6L:18D, 8L:16D, 10L:14D, 12L:12D, 14L:10D, 16L:8D, 18L:6D, 20L:4D, and 22L:2D). Fig. 3.7 shows the record of one bird on some of these photoperiods. Although birds showed a strong concentration of locomotor activity during the light periods, birds on photoperiods of 8 h or shorter showed both anticipatory activity before lights-on (D/L) and trailing activity after lights-off (L/D). Birds on photoperiods longer than 8 h showed a strong coincidence of the end of activity with the L/D transition but continued to anticipate lights-on until about 14L:10D (Binkley 1977). Fig. 3.8a shows the activity record of a sparrow maintained on a series of shorter and shorter photoperiods, followed by a period of complete darkness. Locomotor activity was again confined primarily to the light periods, including those of short duration. In complete darkness, locomotor activity free-ran with a period longer than 24 h. Fig. 3.8b shows the activity of an exceptional sparrow on a similar sequence of shorter and shorter photoperiods; although this bird was also active primarily during the light periods, it free-ran during the shorter photoperiods.

Eskin (1971) showed that sparrows entrain to light/dark cycles with T different from 24 h. When T was 15.8 h (6L:9.8D at 200–300 lux), only 25% of birds entrained their activity to the light cycle, whereas 80% entrained to 17.8–h cycles (6L:11.8D). Using 75% entrainment as the criterion, Eskin concluded that the lower limit for entrainment was between 15.8 and 17.8 h, and he also determined that the upper limit was between 28.0 and 28.7 h.

Sparrows entrain to light/dark cycles with light of very low intensity and also to cycles of light of different intensities (Fig. 3.9). McMillan, Keatts, and Menaker (1975) placed 33 wild-caught sparrows of both sexes on either a 12L:12D or an 8L:16D photoperiod with pale green light of 0.03 lux. Twenty-five birds entrained their locomotor activity to the dim light, and the other eight free-ran. Menaker (1971) reported that all sighted sparrows entrain to light periods with a light intensity of 0.1 lux (approximately equivalent to full moonlight). Hau and Gwinner (1992) found that sparrows on 12L:12D (100 lux:0.3 lux) entrained to the period of higher light intensity, and Hau and Gwinner (1994) observed the same on 12L:12D (13 lux:0.3 lux).

Numerous studies have shown that sparrows free-run in constant darkness (see Fig. 3.8a) (e.g., Binkley 1974a; Binkley et al. 1972; Eskin 1971; Gaston and Menaker 1968). A long-term study of activity rhythms of sparrows in 0L:24D showed that initially about half the birds had free-running periods of less than

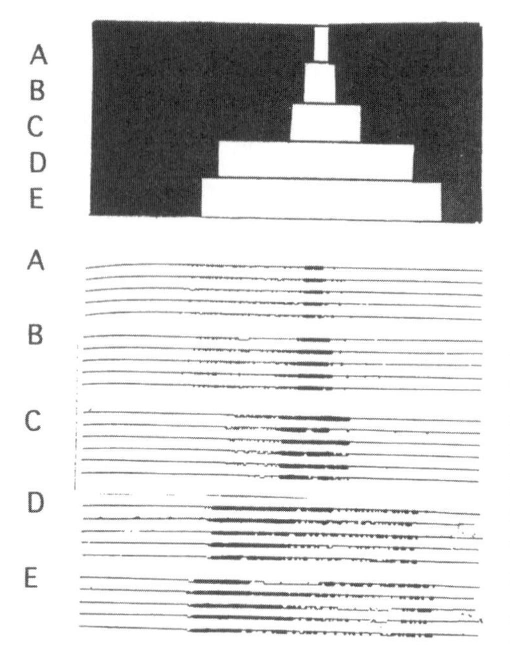

FIGURE 3.7. Actogram of locomotor activity of a house sparrow maintained sequentially for 20 d each on photoperiods of 1L:23D, 2L:22D, 4L:20D, 6L:18D, 8L:16D, 10L:14D, 12L:12D, 14L:10D, 16L:8D, 18L:6D, 20L:4D, and 22L:2D. Only the record of the last 5 d on photoperiods of 1L:23D, 2L:22D, 4L:20D, 12L:12D and 14L:10D is reproduced here. Note that, although activity was most intense during the light, activity preceded lights-on (D/L) and followed lights-off (L/D) during the shorter photoperiods. Adapted from Binkley (1977: Fig. 4), with permission of the University of Chicago Press ©1977.

24 h, and about half had periods greater than 24 h (mean period on day 10 of constant darkness = 24.08 h) (Eskin 1971). Later, however, changes in the free-running period resulted in about 95% of the birds having periods longer than 24 h (the maximum period during the first 120 d of 0L:24D had a mean of 24.87 h) (Eskin 1971). Fifteen individuals were run a second time in complete darkness after entrainment to a photoperiodic cycle, and the correlation for the maximum period values was very high ($r = 0.90$), with the mean difference between the two values being 0.02 h (mean absolute difference, 0.26 h). Eskin concluded that the peak free-running period is characteristic of an individual. Other researchers have also found that the average free-running period of sparrows held in constant darkness is longer than 24 h (Binkley and Mosher 1987b; Cassone and Menaker 1985; McMillan, Elliott and Menaker 1975b).

Sparrows also have free-running rhythms in constant dim light, but they become arrhythmic in constant bright light. Different studies report quite different results regarding the threshold light intensity between the free-running rhythm and arrhythmia. Binkley (1977) reported that at 870 lux (24L:0D) 16 of 20 sparrows were arrhythmic, while at 240 lux only 2 of 17 were arrhythmic. Binkley and Mosher (1985a) also reported that sparrows showed circadian rhythms on 24L:0D (240 lux). Several studies have found that sparrows

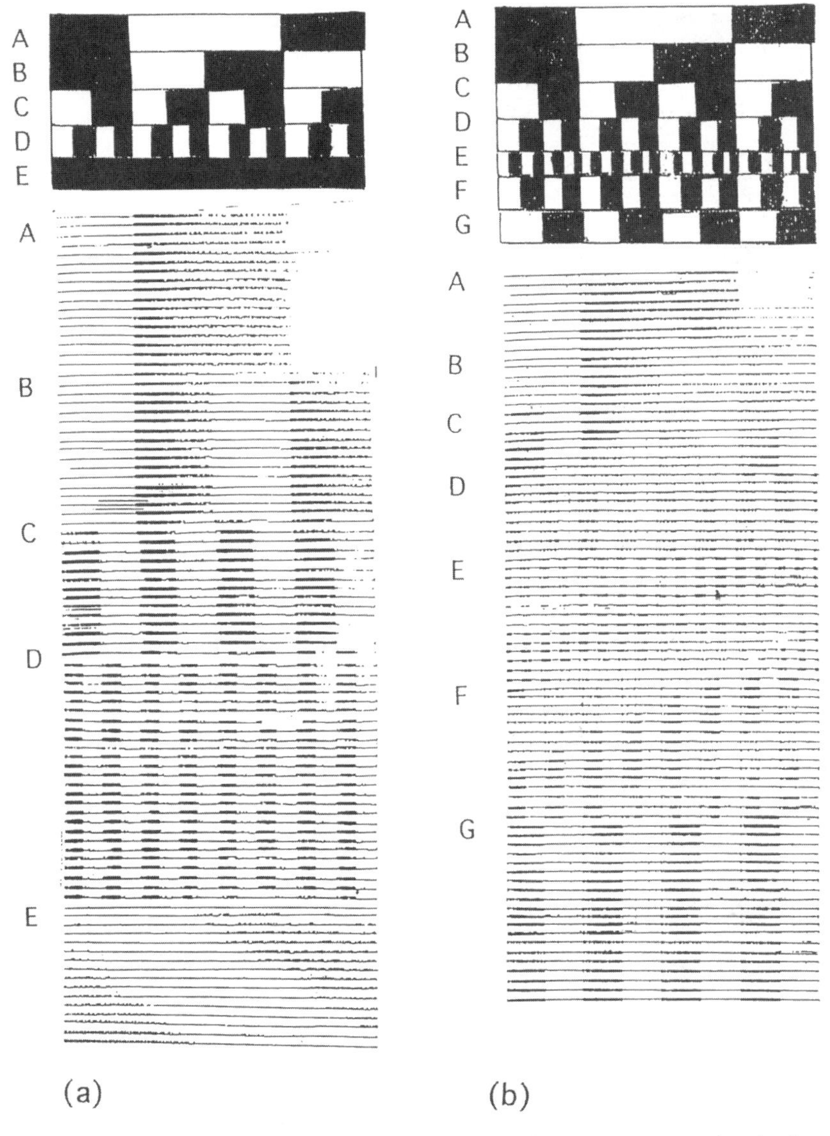

FIGURE 3.8. Actograms of locomotor activity of two house sparrows maintained on a series of different photoperiods (870 lux) followed in one case (a) by a period of complete darkness (0L:24D). Note that activity in (a) is restricted to the light periods, and that the bird free-runs in complete darkness. In (b), however, although activity is also restricted to the light period in 12L:12D, the bird tended to be active in only one of the two daily 6-h light periods and began to free-run during the shorter photoperiods. Adapted from Binkley and Mosher (1985a: Figs. 5 and 6), with permission from Elsevier.

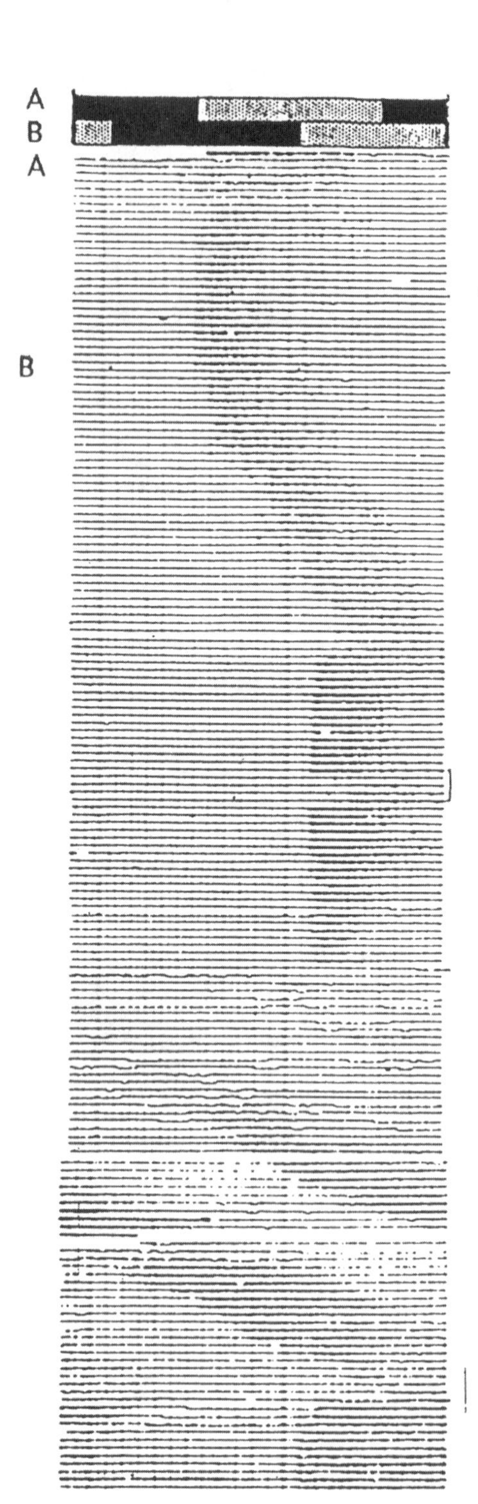

FIGURE 3.9. Actogram of 173 d of locomotor activity of a house sparrow maintained on a 12L:12D photoperiod with dim green light (≈0.03 lux at perch height). On day 20 (a) a carbon-black suspension was injected under the skin of the skull, but the bird remained entrained to the period of dim light. On day 30 (B) the photoperiod was phase-delayed by 6 h, after which the bird showed 36 d of transients before entraining to the new light period. The lights were out from days 82 to 86 (b). On day 108 (c), the bird was blinded by bilateral enucleation, after which it free-ran with $\tau > 24$ h. The subcutaneous carbon-black deposits were removed on day 162 (d), after which the bird re-entrained its activity to the light period. Adapted from McMillan, Keatts, and Menaker (1975: Fig. 3), with the kind permission of Springer Science and Business Media.

free-ran in constant light of 0.1–1.0 lux (Hau and Gwinner 1992; McMillan, Elliott, and Menaker 1975a; Menaker and Eskin 1966). McMillan, Elliott, and Menaker (1975a) reported, however, that 13 of 15 birds became arrhythmic on constant light of 9–12 lux, and the two remaining birds were rhythmic to a light intensity of 250 lux. Different authors have suggested various threshold light intensities for the onset of arrhythmia in constant light, McMillan, Elliott, and Menaker (1975a) concluding that it is normally between 1 and 10 lux, Menaker (1971) that it is between 50 and 100 lux, and Binkley et al. (1972) that it is between 100 and 500 lux.

The differences reported in the various studies are difficult to reconcile, but they may be due in part to differences in light source or in method of measuring light intensity, as well as to the obvious differences among individuals in their responses to constant light. Some studies recorded light intensity at the cage top (e.g., Binkley 1977, Binkley and Mosher 1985a), whereas some measured light intensity at perch height (McMillan, Elliott, and Menaker 1975a). Sparrows used in the former studies were captured in Pennsylvania (USA), whereas those in the latter were from Texas. This would seem to eliminate one possible adaptive explanation for the observed differences, that birds from higher latitudes (which receive lower solar light intensities than those at lower latitudes) would show greater sensitivity to light of low intensity.

Sparrows free-running under constant conditions (complete darkness or constant low-intensity light) obey Aschoff's rule, an empirical generalization relating period length and activity length to light intensity. Aschoff's rule states that (1) the free-running period (τ) is inversely proportional to light intensity, and (2) activity length (α) is directly proportional to light intensity. McMillan, Elliott, and Menaker (1975b) reported that sparrows in 0L:24D (0 lux) free-ran with $\tau > 24$ h, whereas birds on 24L:0D (0.1 lux) free-ran with τ of approximately 24 h, and birds on 24L:0D (1.0 lux) free-ran with $\tau < 24$ h. Chabot and Menaker (1992) reported a significant difference between sparrows free-running in 0L:24D (mean $\tau = 24.33$) and those free-running in constant light of low intensity (mean $\tau = 23.39$).

In the absence of photoperiodic cues, sparrows entrain to other periodic environmental stimuli. Menaker and Eskin (1966) demonstrated that sound can entrain locomotor activity. They exposed 10 sparrows that were free-running in constant pale green light to 4.5 h (during each 24-h period) of tape-recorded sparrow vocalizations (from dawn or dusk roosts—see Chapter 7). Three of the birds showed activity patterns that satisfied rigorous criteria for entrainment to the taped vocalizations (i.e., they assumed the same frequency as the signal and showed phase shifting of the activity pattern during entrainment). Two other birds met some of the entrainment criteria, three birds showed period and phase control by the sound stimulus but then spontaneously returned to circadian periodicity in the presence of the signal, and two birds were unaffected by the sound stimulus. Eskin (1971) reported that periodic white noise was as effective as the taped vocalizations in entraining sparrows. Similar results were obtained by Reebs (1989), using both taped playback of sparrow vocalizations and mechanically produced sound in 0L:24D. Although his sample size was very small, there was also the suggestion that males might entrain more readily to vocalizations than females. Reebs also examined the phase response curve of males to pulses of the mechanically produced sound and found that the greatest response was in the late subjective night.

There is also evidence for social synchronization in the locomotor rhythms of sparrows. Binkley and Mosher (1988) tested birds singly and in pairs on both 24L:0D (dim light) and 0L:24D. Under constant dim light, 50% of the pairs showed either bimodal rhythms or occasional "breakaway" rhythms (dual rhythms), whereas 21% showed only one rhythm and 29% were arrhythmic. In 0L:24D periods, 60% of the pairs had only a single activity period, while the remaining 40% were bimodal or showed occasional breakaways. These results demonstrated both social synchronization, in some cases, and that paired individuals may continue to show their own activity rhythms.

Sparrows in constant darkness have also shown entrainment of their locomotor rhythms to 24-h cycles of food availability (Hau and Gwinner 1992). Birds held on a restricted food regime with $T = 23.5$ h (11.75 food:11.75 no food), however, free-ran in constant darkness and showed only relative coordination with the periodic food stimulus (Hau and Gwinner 1992). Relative coordination occurred when the onset of a free-running rhythm is affected by a periodic stimulus only if the activity onset was close to the time of onset of the periodic stimulus. Birds did entrain, however, to a restricted feeding cycle with $T = 25$ h (12.5 food:12.5 no food). After the 25-h zeitgeber, the birds were returned to *ad libitum* food conditions in constant darkness, and they free-ran with a significantly longer τ than they had when on the 23.5-h restricted feeding cycle (25.01 vs. 24.11, $P < .05$). Hau and Gwinner concluded that food availability acts as a zeitgeber for the circadian rhythms of locomotor activity and feeding. This conclusion was further reinforced by subsequent studies that included varying the duration of food availability to see whether its strength as a zeitgeber was dependent on the degree of food restriction (Hau and Gwinner 1996, 1997). Eskin (1971) also showed that daily temperature cycles of large amplitude (6°C:38°C) entrain locomotor activity of sparrows.

Most studies on the effects of photoperiod on locomotor rhythms of sparrows have attempted to eliminate the potentially confounding effects of other known zeitgebers by using a fairly standard protocol. Birds are housed in individual cages that are kept in light-tight boxes or rooms. Constant temperature, usually between 20°C and 25°C, is maintained, and birds are acoustically isolated in soundproof cages or by constant white noise of about 90 dB. Food and water are available *ad libitum* and are replenished at irregular, long intervals to avoid the possibility of introducing a periodic stimulus. The studies described in the following paragraphs have used this protocol unless otherwise noted.

Binkley and Mosher (1985a) used a number of artificial light-dark cycles to study the direct effects of light and its indirect effects (acting on the bird's circadian system) on locomotor activity patterns of sparrows. In one experiment, birds were given 1L:11D:1L:11D (1169 lux) photoperiods after pretreatment with 12L:12D, with the first 1-h light pulse coming 2, 5, 8, 12, 18, or 24 h after the last L/D transition (circadian time CT14, CT17, CT20, and so on). All birds entrained to this skeleton photoperiod as if it were a 13L:11D photoperiod and achieved entrainment by phase-shifting up to 6 h to either of the first two pulses of light as dawn, whichever required the least phase shifting. Binkley and Mosher concluded that this demonstrated that the choice of dawn was dependent on the bird's circadian clock.

In one particularly novel experiment, Binkley et al. (1983) devised an *ad libitum* light experiment in which they placed 23 sparrows in individual activity cages with an additional perch that controlled the lights, so that birds could turn lights on or off by landing on this perch. Seven of the 23 birds used the switch to maintain an approximately circadian rhythm, with τ 0.9 h shorter than when they were free-running in 0L:24D. Five of the birds used the switch rarely and left themselves in either 0L:24D or 24L:0D for long periods, sometimes several days, four birds free-ran in 0L:24D, and five birds never used the switch. The "selected" photoperiod by those birds that used the switch in a circadian manner ranged from 8.2 to 10.0 h, averaging 9.2 h.

In another novel experiment, Binkley and Mosher (1992) examined the effect of natural photoperiod on the locomotor activity of sparrows maintained in the laboratory in Pennsylvania. Group activity onsets usually preceded sunrise, and cessation of activity preceded sunset (Fig. 3.10A). From January to June (when the photoperiod was increasing), onset of group activity preceded sunrise by a longer period (mean = 0.318 h) than when the photoperiod was decreasing (July–December, mean = 0.082 h; $P = .046$). Group activity onset was actually after sunrise for the months of July to September. Cessation of group activity preceded sunset by a longer period when days were shortening (mean = 0.317 h) than when days were lengthening (mean = 0.155 h), although the difference was not significant ($P = .125$). The same trends were observed when individual onsets and cessations of activity were compared, and both differences were highly significant ($P < .001$). The period of activity (α) was significantly greater than the photoperiod when the photoperiod was increasing, and shorter than the photoperiod when it was decreasing. Individual birds tended to be either early risers/late roosters or late risers/early roosters. No

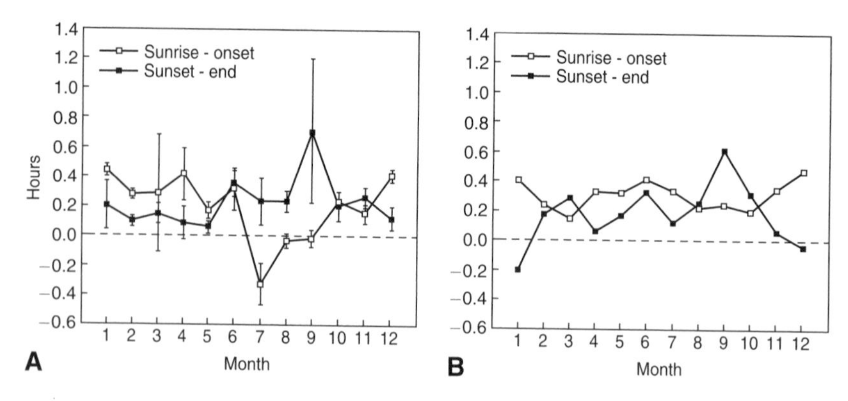

FIGURE 3.10. Two graphs showing the difference between the onset of activity and sunrise ("sunrise – onset") and the cessation of activity and sunset ("sunset – end") in the house sparrow. (A) Mean onsets and cessations of locomotor activity of 10 sparrows maintained in the laboratory for 1 y under natural lighting conditions (eastern exposure) and at temperatures between 25°C and 30°C in Pennsylvania (USA) (Binkley and Mosher 1992). Birds were held in individual cages in which their activity was automatically monitored. No white noise was used. To determine the effect of photoperiod on locomotor activity, the data from the 20th day of each month was used (1 = January). The birds displayed sharp onsets and cessations of activity, similar to those of birds on artificial photoperiods (which have no twilight). From Binkley and Mosher (1992: Fig. 5), with the permission of Taylor & Francis, Ltd. (B) Onsets and cessations of activity relative to sunrise and sunset of sparrows at a communal roost near Slupsk, Poland (Gorska 1990). Onset of activity was noted at the departure of the first individual or group from the roost, and cessation was the arrival of the last individual or group at the roost. Observations were made from December 1974 to December 1977. Adapted from Gorska (1990: Fig. 2).

effects of lunar cycles or cloudiness on sparrow activity were observed. Binkley and Mosher concluded that the results were consistent with a one-oscillator model of circadian locomotor rhythm in sparrows (see later discussion).

Field Studies of Activity

Although most studies of activity patterns have been performed in the laboratory, there have also been some studies of free-living sparrows. Gorska (1990, 1991) monitored the times of onset and end of daily activity by observing birds for 3 y at a communal roost in Poland. Sparrows began their activity earlier relative to sunrise during midwinter (November–February) and during the breeding season (April–July) and tended to cease activities earlier before sunset in the autumn (September–October) (Fig. 3.10B). Sparrows began their daily activity at lower light intensities than those at which they ended activity except during early winter (Gorska 1990). There was a highly significant correlation between the duration of daily activity and daylength ($r = 0.919$, $P < .001$) (Gorska 1991).

Considering the major differences in the way in which the two datasets represented in Fig. 3.10 were collected, there is a surprising amount of congruence in the results. In most months, activity onset preceded dawn and cessation of activity preceded sunset. Onset of activity tended to precede sunrise most during the winter months of December and January and during the breeding season months, April–June. Cessation of activity preceded sunset most in October in both studies (>0.6 h in both). There are two main differences between the two studies. Onset of activity followed sunrise in the late summer/early autumn months (July–September) in Pennsylvania, whereas it continued to precede sunrise in Poland; on the other hand, cessation of activity followed sunset in the winter (December–January) in Poland, whereas it preceded sunset in Pennsylvania. The latter difference can be accounted for by the facts that (1) free-living birds in Poland were experiencing the cold ambient winter temperatures, whereas the Pennsylvania birds were being maintained at a fairly constant temperature (25°C–30°C) and that (2) the winter photoperiod in Poland (54.5°N) is much shorter than in Pennsylvania (40°N).

Haftorn (1994) recorded the activity periods of sparrows during the winter night north of the Arctic Circle. He observed birds at two feeding stations in Pasvik, Norway, in early December, when ambient temperature fluctuated between −1°C and −18°C, and also measured light intensities. The mean arrival time of sparrows was 0754, and the mean departure time was 1247. The average light intensity at which sparrows arrived at the stations (4.5 lux) was significantly lower than the average light intensity at departure (8.9 lux) ($P < .01$).

Rana (1989a) also reported on the departure and arrival times of sparrows at a roost in India. Sparrows departed from the roost singly 10–15 min after sunrise and arrived back at the roost in flocks of 40–50 individuals about 30 min before sunset. The results of all of these field studies suggest that free-living sparrows initiate daily activity at lower light intensities than those at which they cease activity.

Phase Response Curve

The phase response curve plots the responsiveness of an organism to a pulse of either light or darkness at various times in its circadian rhythm. Responsiveness is normally determined by the phase angle shift ($\Delta\phi$) observed in response to a light or dark pulse. The curve provides an indication of the sensitivity of the organism to a zeitgeber at different times in its circadian rhythm. Binkley (1990) described seven different experimental protocols that can be used to generate a phase response curve.

Fig. 3.11 shows two phase response curves for sparrows. In one, large phase advances were observed with light pulses beginning in the late subjective night (CT16–CT23), whereas large phase delays occurred when the light pulse began during late subjective day or early subjective night (CT06–CT14) (Eskin 1971).

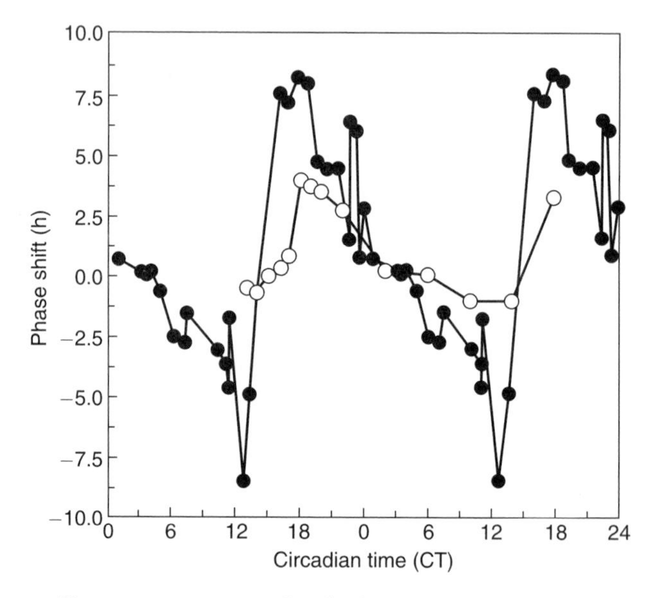

FIGURE 3.11. Phase response curves for the house sparrow redrawn from the data of Eskin (1971: Fig. 12) and Klein et al. (1985: Fig. 2). Solid circles (●) represent the individual phase shifts (Δϕ) in hours of 26 sparrows that had been maintained in constant darkness (0L:24D) for 5–7 mo before being given a single 6-h pulse of light (Eskin 1971). Responses are plotted against the individual's circadian time (CT) at the beginning of the light pulse (CT00 = activity onset of a free-running individual). Data have been duplicated on the time axis to better illustrate the shape of the curve and for comparison with the data of Klein et al. Open circles (○) represent the mean phase shifts of 4–15 sparrows exposed to a single 4-h light pulse (800 lux) at various intervals during the first 36 h of complete darkness after at least 2 wk of entrainment to a 12L:12D photoperiod. For comparative purposes, time of onset of the 4-h light pulse is plotted relative to the "anticipated" D/L in the 12L:12D entraining photoperiod (CT00 = "anticipated" lights-on). The fact that 20 control sparrows maintained in constant darkness after the entrainment had an average phase shift of +0.1 h (SE = 0.1) (Klein et al. 1985) is ignored in the construction of the phase response curve.

Small phase advances occurred in the early subjective day (~CT00), but in general the birds showed little responsiveness to light pulses occurring between CT01–CT05. In the other study, the largest phase advances again occurred during the late subjective night (CT18–CT00), and short phase delays occurred during the early subjective night (CT11–CT13) (Klein et al. 1985). In general, the shapes of the two curves are similar, but the amplitudes are quite different, with that following 5–7 mo in constant darkness having a much larger amplitude than that following entrainment to 12L:12D (see later discussion). Binkley and Mosher (1987a) replicated the latter experiment with similar results (see Fig. 3.13).

Klein et al. (1985) also developed a phase response curve for sparrows given a single 4-h dark pulse (0 lux) during the first 36 h after transfer from 12L:12D to 24L:0D with low intensity (40 lux). Sparrows were insensitive to the dark pulse if it occurred during the middle of the subjective day or during the middle of the subjective night; advances were most evident for pulses that occurred at the beginning of the subjective night, and small delays happened when the pulse occurred near the end of the subjective night. The phase response curve for the single dark pulse resembled a "mirror image" of the curve for 4-h light pulses.

Binkley and Mosher (1986) tested the effects of two different photoperiodic pretreatments (8L:16D and 16L:8D) on the phase response curves of sparrows exposed to 4-h light pulses at various intervals after transfer to 0L:24D. Sparrows were maintained on the pretreatment photoperiod for at least 2 wk before initiation of the treatment. After lights-out at the normal time, birds were exposed to a single 4-h pulse of light (800 lux) beginning at one of 12 2-h intervals during the next 24 h, and then were kept at least 1 wk in 0L:24D. Phase shifts occurred in both groups after the light pulses. Birds held on 8L:16D showed phase delays to light pulses beginning 2, 4, 6, and 22+ h after L/D, whereas birds on 16L:8D showed no such delays. Peak advances occurred at 10 h after L/D in the 8L:16D birds and at 6 h after L/D in the 16L:8D birds. In both groups this coincided, however, with the same circadian time (about CT21) (Fig. 3.12). Binkley and Mosher concluded that prior photoperiod affects the timing of the phase response curve, as well as its amplitude.

In a subsequent experiment, Binkley and Mosher (1987a) used two light pulses (doublets) in an attempt to determine whether there is instantaneous resetting of the circadian clock. Birds were maintained on 12L:12D (800 lux) for at least 2 wk and then transferred to 0L:24D. Experimental treatments consisted of exposure to either one 4-h light pulse or two 4-h light pulses (doublets) (Fig. 3.13). The first doublet protocol involved a phase-advancing initial light pulse (beginning 8 h after L/D) and resulted in a phase response curve that closely resembles the single-pulse curve (see Fig. 3.11) if a phase advance

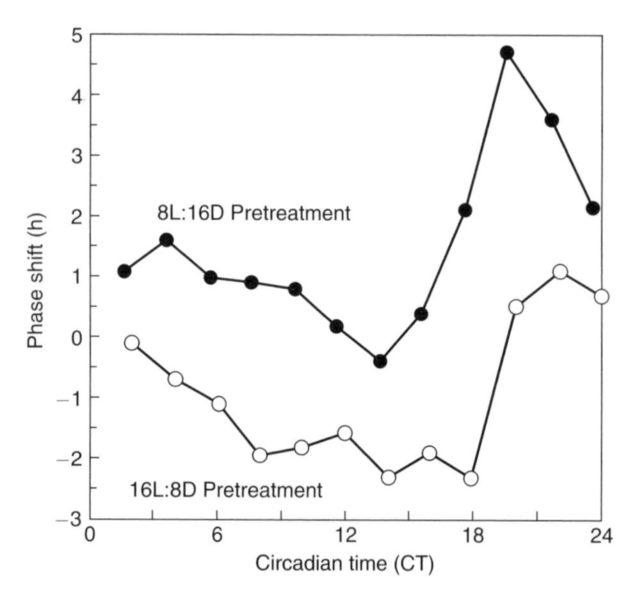

figure 3.12. Phase response curves of the locomotor activity of house sparrows maintained on two different pretreatment photoperiods before exposure to a single 4-h light pulse (800 lux) during the first 28 h of constant darkness (0L:24D) after the pretreatment. Birds were maintained on one of two pretreatments, either 16L:8D or 8L:16D, for at least 2 wk before the experimental treatment. Open circles (6) represent mean values of birds on the 16L:8D pretreatment, and closed circles (1) are means for birds on the 8L:16D pretreatment. Time on the abscissa represents the midpoint of the 4-h light pulse, and the two curves have been aligned by the bird's circadian time (CT). Phase shifts (Df) of the experimental birds were measured from electronically monitored perch-hopping records of at least 1 wk in 0L:24D and were corrected for the phase shift of control birds held in 0L:24D with no light pulse (16L:8D and 8L:16D controls initiated activity approximately 2.3 and 1.7 h after normal D/L, respectively). Phase shifts were determined for 193 trials, with 7% of records being discarded when they could not be interpreted. During the pretreatment, 8L:16D sparrows anticipated D/L by an average of 1.6 h, whereas 16L:8D sparrows began activity at D/L. Activity times (a) of birds maintained on the two treatments were also different during the first 18 d of 0L:24D, with 8L:16D birds initially having a mean α of 9.9 h that lengthened for the next 3 d after application of the light pulse. For 16L:8D birds, mean α was 13.7 h, but it shortened for the next 8 d. A difference of about 1 h still persisted after 18 d. Mean period length (t) also differed in the two treatments: 24.1 h for 8L:16D and 23.7 h for 16L:8D ($P <$ .05). Activity onset after transfer to 0L:24D also differed in the two groups: for 8L:16D, the mean onset was 14.3 h (SE = 0.2) after L/D, and for 16L:8D, it was 10.3 h (SE = 0.7) after L/D. From Binkley and Mosher (1986: Fig. 4), with permission from the American Physiological Society.

of +2.7 h is considered. In the second doublet experiment (with the initial light pulse in the phase-delaying period, beginning in the early subjective night 2 h after L/D), the phase response curve is shifted to the right compared with the single-pulse curve, and its amplitude is somewhat damped. It resembles a delayed single-pulse curve delayed −0.8 h. Binkley and Mosher concluded that the phase response to the second light pulse could be explained by assuming that the internal clock is reset by the initial light pulse before the second pulse begins.

The phase response curve of sparrows helps to explain several aspects of their circadian mechanism. It has a larger advancing portion than delaying portion (see Fig. 3.11), accounting for the advancing effect of constant light in the absence of a zeitgeber, thus shortening τ in constant dim light (Aschoff's rule). It also explains the response of sparrows to skeleton photoperiods. In the experiment described earlier in which Binkley and Mosher (1985a) presented sparrows with 1L:11D:1L:11D skeleton photoperiods, the choice of "dawn" corresponded to the predicted direction of phase-shifting from the phase response curve.

The amplitude of the phase response curve can also explain the range of periods to which sparrows entrain. Sparrows entrain to photocycles as short as 17.8 h and as long as 28.7 h (Eskin 1971). These periods correspond closely to the maximum phase angle advance (mean $\Delta\phi = +6.3$ h) and phase angle delay (mean $\Delta\phi = -4.4$ h) calculated from Eskin's (1971) phase response curve (see Fig. 3.11).

Phase response curves have also been used to explain entrainment to photoperiod (Binkley 1990). Pittendrigh (1960 cited in Binkley 1990) proposed that entrainment to a light and dark cycle represents an equilibrium between the τ of the individual in 0L:24D and the advancing or delaying effect of light. However, Binkley (1990) suggested that both light and darkness are involved in photoperiodic entrainment. This is based on the fact that sparrows have phase response curves to both light pulses and dark pulses (see earlier discussion). Binkley (1990) referred to this as the "theory of entrainment by photoperiod and scotoperiod." According to this theory, increases in photoperiod in the north temperate spring induce increases in activity period (α) of the sparrows by acting both at dawn and sunset to lengthen α, as described by the phase response curve to light. During the north temperate autumn, however, later duration of darkness in the morning and earlier onset of darkness in the evening act to shorten α, as the phase response curve to dark pulses predicts.

Locomotor Activity of Blinded Sparrows

Menaker (1968a) pioneered the study of the effects of blinding (bilateral enucleation) on the locomotor rhythm of sparrows. He removed one eyeball at a

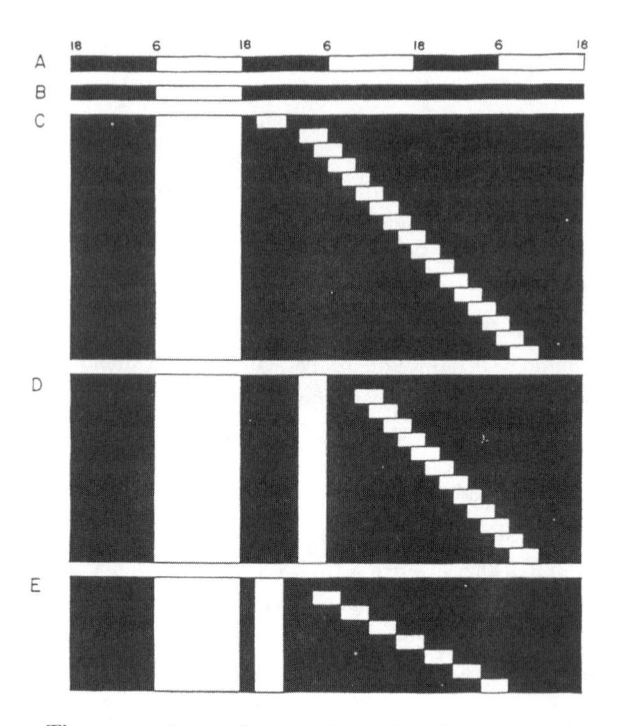

FIGURE 3.13. Three experimental protocols used by Binkley and Mosher (1987a) to determine whether the internal clock of the house sparrow is reset immediately. Sparrows (n = 186, 46% male) captured in Pennsylvania (USA) were maintained in individual cages under constant conditions (19°C–25°C, 90–100 dB white noise), and locomotor activity was electronically monitored. After at least 2 wk in 12L:12D (800 lux) (A), controls were transferred to 0L:24D (B) and experimental individuals were given one of three different light treatments during the first 42 h of darkness: protocol C, one 4-h pulse of light beginning either 2 h after L/D or at 2-h intervals beginning 8 h after L/D; protocol D, two 4-h pulses (doublets) with the first beginning 8 h after L/D and the second beginning at 2-h intervals (e.g., 16, 18, 20 h) after L/D; or protocol E, doublets with the first 4-h pulse beginning 2 h after L/D and the second beginning at 4-h intervals (e.g., 10, 14, 18 h) after L/D. Phase shift was determined based on activity onset in the first 5 d after the treatment and consisted of the horizontal displacement (in hours) of activity onset on the first day after treatment from the time of activity onset during pretreatment. Birds transferred from 12L:12D to 0L:24D showed very little phase shift (mean = +0.1 h). Protocol C replicated an experiment performed by Klein et al. (1985), and the phase response curve obtained from the results closely paralleled their phase response curve (see Fig. 3.11), with a maximum phase advance of +3.8 h resulting from a light pulse in the late subjective night (beginning 30 h after L/D) and a maximum delay of −1.7 h after a light pulse in the early subjective night (beginning 26 h after L/D). The results of protocols D and E are discussed in the text. From Binkley and Mosher (1987a: Fig. 1), with the permission of Sage Publications.

time in two successive surgeries (separated by 3–5 d). Of 53 blinded sparrows tested with 12L:12D (500 lux) photoperiods, all entrained their locomotor activity to the light period. Twenty-two blinded birds were also exposed to 12L:12D (0.1 lux, green light) photoperiods, and 12 entrained their activity rhythms to the dim light period. The remaining 10 birds free-ran as if in constant darkness. Menaker performed several experiments to eliminate the possibility that other periodic cues associated with the experimental setup, such as temperature, vibration, or noise, could be entraining the activity of the blinded sparrows, and all proved negative. Blinded sparrows treated for external parasites also entrained to light, indicating that any activity pattern of these parasites did not serve as a cue. Menaker concluded that sparrows possess an extraretinal light receptor that is coupled to the circadian mechanism controlling its locomotor rhythm.

Other experiments showed that the extraretinal light reception associated with the locomotor rhythm occurs in the brain (Menaker 1968b). Six enucleated sparrows maintained in the laboratory in individual cages free-ran on 12L:12D (0.02 lux, pale green light). They continued to free-run after a patch of feathers was plucked from the back but entrained to the light period when a comparable-sized patch of feathers was plucked from the head. Subsequent subcutaneous injection of India ink beneath the skin of the skull resulted in loss of entrainment and a return to free-running activity. Scraping of the India ink from the skull resulted in resumption of entrainment. Menaker concluded that the extraretinal receptor was located in the brain and not in the skin.

This conclusion was corroborated in experiments using blinding, carbon black injections and feather plucking to determine the relative contributions of the eyes and the extraretinal light receptors to entrainment of the circadian locomotor rhythm (McMillan, Keatts, and Menaker 1975; McMillan, Underwood et al. 1975). In the former study, as mentioned earlier, 25 of 33 sighted sparrows placed on either 12L:12D or 8L:16D (0.03 lux, pale green light) entrained to the dim photoperiod, and eight free-ran. After the head feathers of the free-running birds were plucked, they also entrained their activity to the dim light period. On the other hand, when the heads of 23 entrained birds were injected subcutaneously with a carbon black suspension, 10 of the birds began to free-run. The 13 birds that remained entrained after carbon black treatment resynchronized their locomotor activity to a 6-h phase delay. Six of these birds were then blinded by bilateral enucleation, after which they free-ran. The carbon black was then removed, and they again entrained to the dim photoperiod (Fig. 3.10 illustrates the results from one of these six birds). McMillan, Keatts, and Menaker concluded that both the eyes and extraretinal light receptors are involved in entrainment of the locomotor rhythm, and that the effects of the two photoreceptive organs are additive. McMillan, Elliott, and Menaker (1975a) concluded that plucking of the head feathers increased effective light intensity 10-100 fold, whereas injection of carbon black decreased effective light intensity 10-100 fold.

McMillan, Elliott, and Menaker (1975b) used combinations of bilateral enucleation, plucking of head feathers, subcutaneous injection of the head region with carbon black plus hooding of the head with collodion saturated with Sudan black, and exposure to different light intensities to test whether both retinal and extraretinal light receptors conform to Aschoff's rule. They estimated that the combination of injection and hooding reduced effective light intensity 100–1000 fold. Blinded birds on 24L:0D (1.0 lux) had longer free-running periods than when they were sighted. Blinded sparrows free-running in constant dim light (1 or 10 lux) had longer periods after being injected with carbon black and hooded. Blinded birds whose head feathers were plucked had shorter free-running periods in constant light of 1.0, 10, or 2000 lux, as did blinded, carbon black-injected, and hooded birds after the carbon black and hoods were removed (1.0 and 10 lux). Sighted birds whose heads were plucked had shorter periods and sighted birds injected with carbon black and hooded had longer periods in constant light (0.1 lux). McMillan, Elliott, and Menaker concluded that both the retinas and the extraretinal photoreceptors obey Aschoff's rule.

Activity of Pinealectomized Sparrows

The first major breakthrough in the search for an internal pacemaker in sparrows occurred when Gaston and Menaker (1968) surgically removed the pineal glands of individuals that had been maintained in 0L:24D. Before the surgery, the sparrows showed a circadian rhythmicity in locomotor activity. The locomotor activity of pinealectomized sparrows kept in 0L:24D became arrhythmic 0–9 d after the surgery, whereas sham-operated birds retained their circadian pattern of locomotor activity (Fig. 3.14). Pinealectomized birds transferred to 8L:16D entrained their activity rhythms to the light period and actually showed a positive phase angle of about 1 h (initiating activity about 1 h before D/L). Histological examination of the fixed brains of pinealectomized sparrows after the experiments showed that 4 of 16 birds had signs of residual pineal tissue, and 2 of these birds had shown some continued rhythmicity in 0L:24D.

Numerous subsequent studies have confirmed that pinealectomized sparrows are arrhythmic in constant darkness or dim light (e.g., Binkley et al. 1972, 1973; Brandstätter et al. 2000; Gwinner 1989; Lu and Cassone 1993a, 1993b;

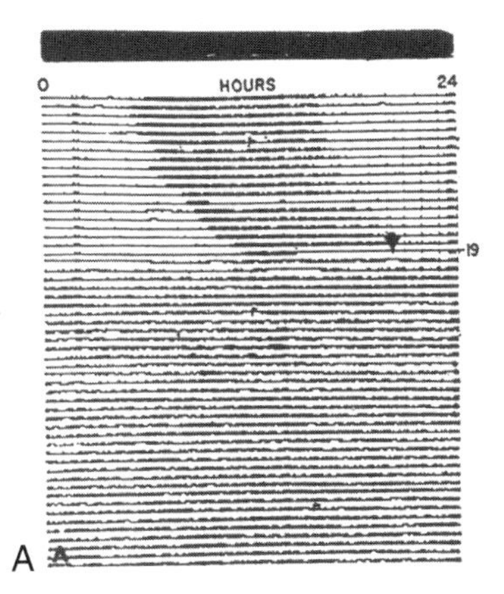

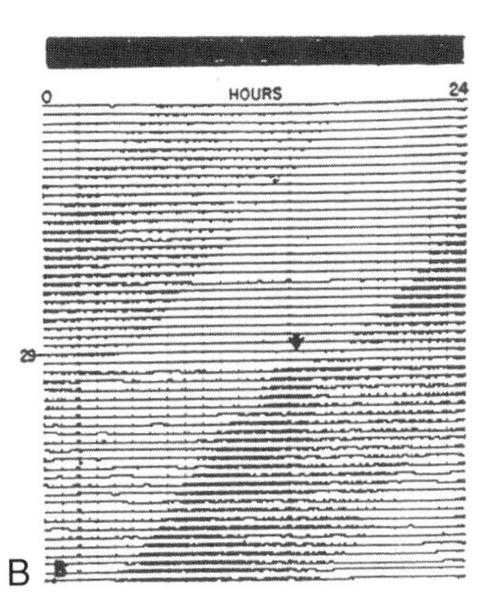

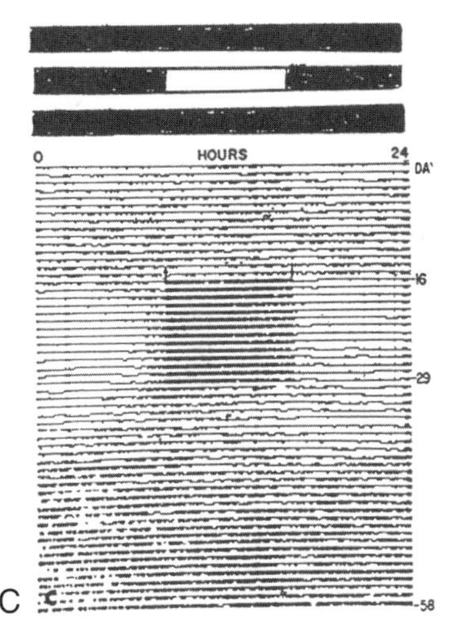

FIGURE 3.14. Effect of pinealectomy on free-running locomotor rhythm of the house sparrow. (A) Actogram of sparrow being maintained in 0L:24D. On day 19 (*arrowhead*), the bird was surgically pinealectomized, after which it became arrhythmic. (B) Actogram of a second bird free-running in 0L:24D that was sham-operated on day 29 (*arrow*). Note that the free-running rhythm continued with a small phase advance and approximately the same period after the sham operation. (C) Actogram of a pinealectomized sparrow that was arrhythmic for 15 d while being held on 0L:24D, became rhythmic when exposed to 8L:16D from days 16 to 29, and then returned to arrhythmia when returned to 0L:24D on day 30. From Gaston and Menaker (1968), reprinted with permission. Copyright © 1968 American Association for the Advancement of Science (AAAS).

Takahashi and Menaker 1982a). The feeding rhythm of pinealectomized sparrows is also abolished (Chabot and Menaker 1992; Gwinner 1989; Heigl and Gwinner 1999). Gwinner (1989) reported, however, that not all pinealectomized sparrows became arrhythmic in 0L:24D, and some birds that were rhythmic in constant darkness might become arrhythmic in dim light (24L:0D, 1 lux). He also reported that the free-running τ of pinealectomized sparrows that retained rhythmicity in 0L:24D was always shorter than their free-running τ before pinealectomy. Binkley (1974a) and Heigl and Gwinner (1995) also reported cases of pinealectomized sparrows that continued to show rhythmic activity in either 0L:24D or constant dim light (24L:0D, 0.2 lux). None of these studies reported that the brains of their sparrows had been examined subsequent to the experimental treatment to ensure that the pinealectomy had been complete. In another study, however, Heigl and Gwinner (1999) reported rhythmic locomotor or feeding activity in some pinealectomized sparrows under constant conditions (24L:0D, 0.3 lux and *ad libitum* food) and also found no evidence of residual pineal tissue in brains after the experiments.

Pinealectomized sparrows entrain their activity rhythms to very short photoperiods. Laitman and Turek (1979) placed pinealectomized or sham-operated birds on 3L:21D for 43 d and then transferred them for 26 d to 1L:23D, 5L:19D, or 7L:17D. Pinealectomy resulted in both earlier onset (compared with D/L) and later cessation of daily activity, by about 100 min in each case, comparing birds on 3L:21D before and after surgery. The birds also showed significantly increased variance in the times of onset and cessation of activity compared with controls. Cessation time was significantly lengthened for pinealectomized birds on 1L:23D, and variances of onset and cessation for pinealectomized birds were both significantly greater on 1L:23D. There were no significant differences on the other two photoperiods, however.

Binkley et al. (1971) showed that pinealectomy also abolished the free-running rhythm of body temperature of sparrows held in 0L:24D. Body temperature was monitored telemetrically (with telemeters inserted into the abdominal cavity). Sham-operated birds continued to show a circadian rhythm in body temperature in 0L:24D. Pinealectomized sparrows transferred from 0L:24D to 12L:12D entrained to the external light stimulus in both locomotor activity and body temperature, but the body temperature response required 3 or 4 d, whereas the change in locomotor activity occurred on the first day (suggesting that control of these two circadian rhythms may be independent and not simply the result of an effect of locomotor activity on body temperature) (Binkley et al. 1971).

Zimmerman and Menaker (1975) attempted to determine whether the pineal gland of sparrows sends neural signals, sends hormonal signals, or requires neural input for the control of its circadian rhythmicity. In one experiment, the pineal stalk (which contains the only known neural output from the pineal) was surgically deflected in 12 sparrows that were free-running on 0L:24D. The sparrows showed no interruption of circadian rhythmicity. In a second experiment, sparrows were given either single or multiple doses of 6-hydroxydopamine (6-OHDA), which is believed to abolish sympathetic nervous stimulation, the only known innervation of the sparrow pineal (see Ueck 1970). The 6-OHDA-treated birds continued their free-running circadian locomotor rhythms in 0L:24D. Both pineal stalk deflection and 6-OHDA treatment were performed on another 12 birds, 11 of which were unaffected. The other bird was arrhythmic for 11 d but then reestablished a circadian pattern.

Zimmerman and Menaker (1975) also transplanted pineal glands, including the stalks, into the anterior chambers of eyes of previously pinealectomized sparrows. Before receipt of the transplanted pineal, the birds were arrhythmic in 0L:24D. Circadian rhythms were reestablished in about 30% (12) of the recipients. Surgical removal of the eye containing the implanted pineal resulted in resumption of arrhythmia in 0L:24D. These results were more convincing than those of Gaston (1971), who placed pineal implants into the pineal regions of 30 previously pinealectomized, arrhythmic sparrows. Only two of those birds developed activity rhythms in 0L:24D.

In a subsequent study, Zimmerman and Menaker (1979) transplanted pineal glands from 42 sparrows that had been maintained on one of two 12L:12D photoperiods (one with D/L at 0900 and the other with D/L at 2300) into anterior chambers of the eyes of pinealectomized birds that were showing arrhythmic locomotor activity in 0L:24D. Eighteen birds established unambiguous activity rhythms within 3 d after the surgery (which took place during the overlap in the light phases of the two donor groups, 0900–1100). These birds all showed a phase of activity onset that corresponded to the entrainment schedule of the donor bird, and there was no overlap between the two groups of recipient birds. Recipient birds showing rhythmic activity were then given a single pulse of 6 h of light, and their phase shifts in response to this stimulus were compared with those of normal birds given such a light pulse while free-running in 0L:24D (see Fig. 3.8). The phase shifts of the recipient birds were very similar to those of intact birds exposed to the same treatment. Zimmerman and Menaker concluded that the sparrow pineal is an autonomous circadian oscillator.

The Hypothalamus

The suprachiasmatic nuclei (SCN) of the hypothalamus are the primary structures responsible for the control of circadian rhythmicity in mammals (Branstätter and Abraham 2003). Several nuclei in the suprachiasmatic region of the hypothalamus, including the SCN, have also been implicated in the maintenance of circadian rhythms under constant conditions in sparrows (see Fig. 9.5 in Chapter 9). Other hypothalamic nuclei that have been identified as possible components of the circadian control system include the lateral hypothalamic nuclei (LHN) and the preoptic nuclei (PON). Lack of consistency in the terminology used to describe these nuclei has contributed to some confusion about the roles of the various nuclei in the circadian control system of the species (Abraham et al. 2002; Brandstätter and Abraham 2003).

The SCN are paired nuclei located adjacent to the ventral wall of the third ventricle immediately above the rostral end of the optic chiasma (Hartwig 1974; Takahashi and Menaker 1982b). The diameter of these nuclei increases longitudinally from the rostral end to its midpoint, and then decreases again caudally (Brandstätter and Abraham 2003).

The LHN are also possible sites of circadian rhythmicity in sparrows. Cassone and Moore (1987), using immunohistochemistry techniques, found that these nuclei are quite small (75–100 μm mediolaterally, 100 μm dorsoventrally, and 700 μm rostrocaudally). Rostrally they are located lateral to the PON, and caudally they are lateral to the dorsal supraoptic decussation and medial to the ventral lateral geniculate nucleus (see Fig. 9.5).

Takahashi and Menaker (1982b) performed a series of experiments on the role of the SCN in circadian locomotor activity in sparrows. Brain lesions were produced electrically in the experimental birds by placing them in a stereotaxic instrument and inserting coated insect-pin electrodes through a hole bored in the skull into predetermined positions approximating the positions of the SCN. A current of 25–35 mA was then applied for 30–60+ s. Two control groups were used, one in which no operation was performed and one in which the birds were sham-operated (i.e., holes were drilled in the skull and electrodes inserted into the brain, but no current was applied). Three experimental groups were used, one in which birds were maintained on 12L:12D before surgery and for 2–8 wk after surgery before being transferred to 0L:24D for up to 3 mo, one with birds on 6L:18D before and after surgery before transfer to 0L:24D, and one with birds in 0L:24D both before and after surgery. After the experimental birds were sacrificed, their brains were removed and examined histologically to determine the extent of damage to the SCN. Four birds in which SCN damage was extensive (90+%) had no circadian locomotor rhythm in 0L:242D. These birds did entrain normally to 12L:12D, except that they showed no phase lead. Fourteen of 15 birds with medium SCN lesions (25%–90% destruction) continued to show locomotor rhythms in 0L:24D, and 13 birds with small lesions (<25%) and 7 with no SCN damage showed continued rhythmicity. All control birds also showed normal circadian rhythms in 0L:24D. Experimental birds showed reduced activity levels compared with controls, and there was a negative correlation between the extent of damage to the SCN and activity ($r = -0.45$, $P < .001$).

The SCN show circadian rhythmicity in the expression of the *pPer2* gene, a sparrow gene that is homologous to the PAS domain of the *qPer2* gene of the Japanese quail (see Chapter 2). Sparrows captured in Germany were maintained in the laboratory on 12L:12D (85:0.03 lux), constant temperature, and *ad libitum* food and water for at least 2 wk before the expression of the *Per2* gene was measured at different times of the day (Abraham et al. 2002). The birds were divided into two groups, one of which was maintained on 12L:12D until sampled, and the other transferred to 0L:24D (0.03 lux) before sampling. Locomotor and feeding activities of the birds were monitored electronically, and birds on 12L:12D entrained their activities to the light period, whereas birds transferred to 0L:24D began to free-run. The temporal patterns of *pPer2* gene expression varied in three regions of the SCN (Fig. 3.15). Elevated expression occurred in the rostral SCN of the 12L:12D birds at ZT24 (i.e., just before D/L), with peak expression in that region occurring at ZT06; ZT12 and ZT18 birds had background levels of expression. The medial SCN of the 12L:12D birds also had peak expression at ZT06, and slightly elevated expression at ZT12, but background levels at ZT18 and ZT24. There was little temporal variation in expression in the caudal SCN, with levels near background throughout the daily cycle. Patterns of *pPer2* gene expression in the three regions of the SCN of 0L:24D birds were similar to those of 12L:12D birds, except that the amplitudes were smaller (see Fig. 3.15). The LHN of both groups of birds showed patterns of gene expression similar to those of the medial SCN. The persistence of rhythmic *pPer2* gene expression in the SCN and LHN through three cycles in 0L:24D suggests that these hypothalamic regions may be involved in the control of circadian rhythmicity in sparrows. The fact that the difference in amplitude of gene expression in the rostral SCN between ZT24 and CT24 in the two groups was less than the differences observed in the medial SCN, rostral SCN, and LHN between ZT06 and CT06, suggests that the rhythm of gene expression in the rostral SCN may be relatively independent of the pineal gland (Abraham et al. 2003). The rostral SCN may therefore act as a second circadian oscillator in sparrows (see later discussion).

Additional evidence of circadian rhythmicity in the LHN of sparrows is provided by studies of its metabolic activity. Cassone (1988) used the uptake of radioactive glucose ([^{14}C]2-deoxyglucose, or 2-DG) to study the metabolic activity of the sparrow LHN. Twelve sparrows captured in New York were maintained in the laboratory on 12L:12D (150 lux) for 6–8 wk before the experiment, at which time the birds were transferred to 24L:0D (dim red light). Four birds each were injected with 2-DG at 12-h intervals (noon, midnight, and noon the next day (i.e., at mid-subjective day = CT06), after which they

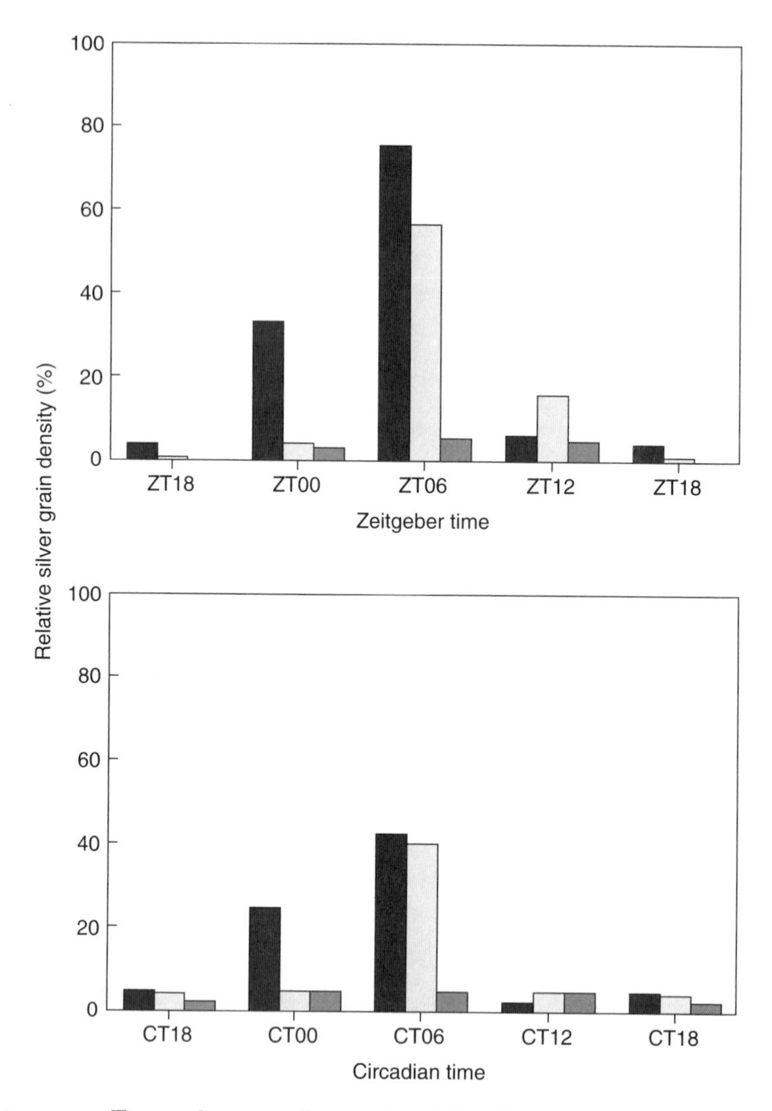

FIGURE 3.15. Temporal pattern of expression of the *pPer2* gene in rostral (*solid bar*), medial (*light gray*), and caudal (*dark gray*) parts of the suprachiasmatic nuclei (SCN) of the house sparrow. One group of sparrows was maintained on 12L:12D before the assay, and a second group was transferred to 0L:24D 3 d before assay (see text for further details). Birds in the 12L:12D group were sacrificed at ZT06, ZT12, ZT18, or ZT24 (ZT00 = D/L), and 0L:24D birds were sacrificed at CT06, CT12, CT18, and CT24 (CT00) during the third circadian cycle following transfer to 0L:24D. After sacrifice, the brains were removed and sectioned coronally, and mounted sections were incubated with labeled RNA probes. Expression of the *pPer2* gene was detected and quantified by visualizing silver grains marking labeled sites using dark-field microscopy. Relative silver-grain densities in the three regions of the SCN of the 12L:12D birds are depicted in the upper histogram, and those of 0L:24D birds are depicted in the lower histogram (values for ZT18 and CT18 are double-plotted for better visualization of the patterns of expression). From data in Abraham et al. (2002: Fig. 2).

were permitted to remain active for 1 h before being sacrificed. Brains were then removed and sectioned coronally at 20 μm, sections were exposed to x-ray film for 12 d, the resulting autoradiographs were digitized, and the label was interpreted as nanocuries per gram of tissue (nCi/g). Uptake of 2-DG by the LHN was high during the day (noon first day vs. midnight, $P < .01$) and also during midsubjective day (noon second day vs. midnight, $P < .05$). Cassone suggested that it would be very interesting to determine whether the rhythmic changes in 2-DG uptake by the LHN persist in pinealectomized sparrows.

Interaction Between the Pineal Gland and the Hypothalamus

It appears that either pinealectomy or destruction of the hypothalamic SCN abolishes free-running circadian rhythms in sparrows. What are the roles of these two structures in controlling circadian rhythmicity, and how do they interact?

The pineal gland of sparrows produces the hormone melatonin (*N*-acetyl-methoxytryptamine) using the following biosynthetic pathway: (1) tryptophan is hydroxylated by tryptophan hydroxylase to form 5-hydroxytrytophan; (2) 5-hydroxytrytophan is then decarboxylated by aromatic amino acid decarboxylase to form 5-hydroxytryptamine (5HT or pineal serotonin); (3) 5HT is converted at night to *N*-acetyl serotonin (NAS) by serotonin-*N*-acetyltransferase (NAT); and (4) NAS is 5-methylated to melatonin by hydroxyindole-*O*-methyltransferase (HMOT) (Cassone 1990). Levels of plasma melatonin are rhythmic, with high circulating levels of the hormone being found at night and low levels during the day (Fig. 3.16). Pinealectomy abolishes this rhythm, with levels remaining near the detection level (approximately 50 pg/ml) throughout the day (Janik et al. 1992). Janik et al. concluded that the pineal is the only source of circulating melatonin in sparrows, although the retinas of the eyes have been found to produce melatonin rhythmically in some other bird species (Gwinner and Brandstätter 2001).

Two studies have reported that pineal gland tissue maintains rhythmic melatonin production *in vitro*. Takahashi (1981, cited in Menaker 1982) found that there are 24-h rhythms of melatonin release during the dark period of a 12L:12D photoperiod in cultured sparrow pineal gland tissue. The rhythmic output continued at reduced amplitude for at least 2 d after the transfer of the culture to 0L:24D (Fig. 3.17). Because cultured pineal tissue receives no neural or hormonal inputs, these results indicate both that pineal cells are photosensitive and that light acts directly to inhibit melatonin synthesis and release. Brandstätter et al. (2000) obtained similar results but also found that the period and amplitude of melatonin release differed in cultures of pineal gland

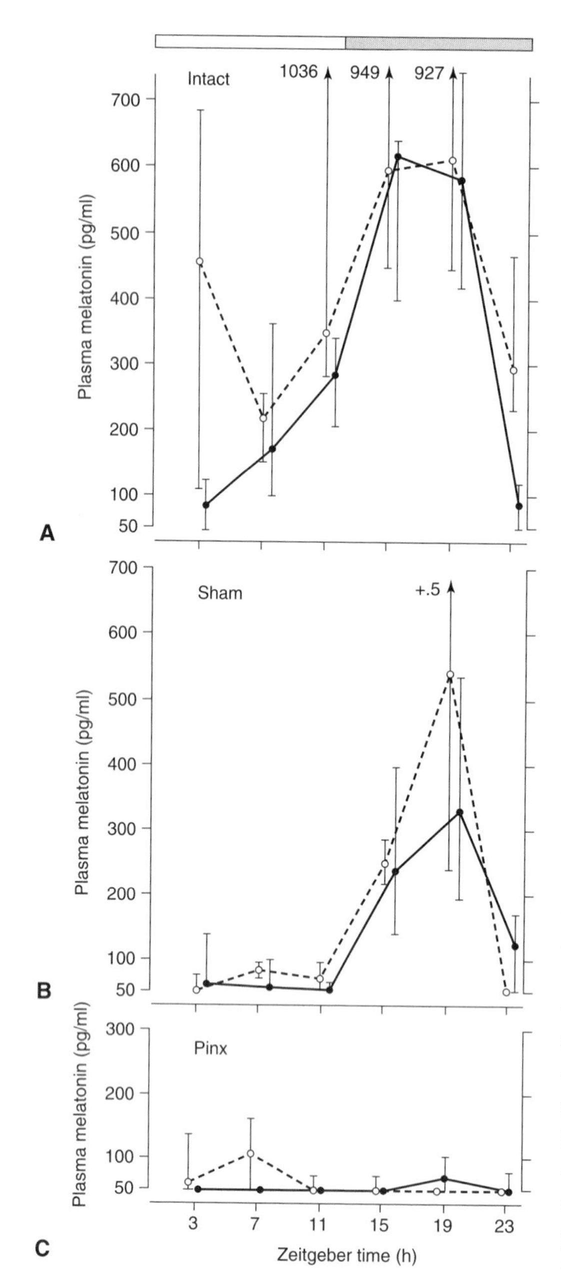

FIGURE 3.16. Daily rhythm of plasma concentration of melatonin in house sparrows obtained by radioimmunoassay (Janik et al. 1992): (A) intact sparrow, (B) sparrow with sham pinealectomy, and (C) pinealectomized sparrow. Values are medians with error bars representing quartiles, and dashed lines represent birds maintained on photoperiods of 12L:12D (400 lux); solid lines repre-sent birds maintained on 0L:24D. Data were based on 33 sparrows captured in Germany and maintained in outdoor aviaries until mid-October, when they were transferred to cages (two or three birds in an 80 × 50 × 45 cm cage) in lightproof boxes in the laboratory. Blood samples (200 µl) were taken from the wing veins of birds at weekly intervals and at various times after the onset of light (ZT03, ZT07, ZT11, ZT15, ZT19 and ZT23). Birds were also sampled on the same schedule on the second day of 0L:24D. After sampling of intact sparrows, birds were pinealectomized (*Pinx*, $n = 21$) or sham-operated ($n = 12$) and permitted to recover for 2.5 wk. Birds were then sampled again at weekly in-tervals, three times each for birds held in 12L:12D and 0L:24D. From Janik et al. (1992: Fig. 1), with the permission of Sage Publications.

cells from sparrows held on 16L:8D (short night) versus 8L:16 D (long night) before tissue explantation. The period of melatonin release was significantly longer in 0L:24D culture from the long-night birds for two circadian cycles, and peak melatonin production was significantly greater in the short-night birds for the first cycle. These results indicate that the pineal has a "memory" for the preceding photoperiod. Another study reporting similar results for "house sparrow" pineal gland cell culture in Japan (Murakami et al. 1994) presumably refers to work done with tree sparrows, because the house sparrow does not occur regularly in Japan (see Chapter 1).

Brandstätter, Kumar et al. (2001) observed similar patterns of change in the amplitude and period of circulating melatonin levels in intact sparrows maintained on natural photoperiods in Germany (Fig. 3.18). Cell cultures of pineal gland tissue from 12 of the sparrows maintained on the natural photoperiods had patterns of duration and amplitude of melatonin production in 0L:24D similar to those found by Brandstätter et al. (2000).

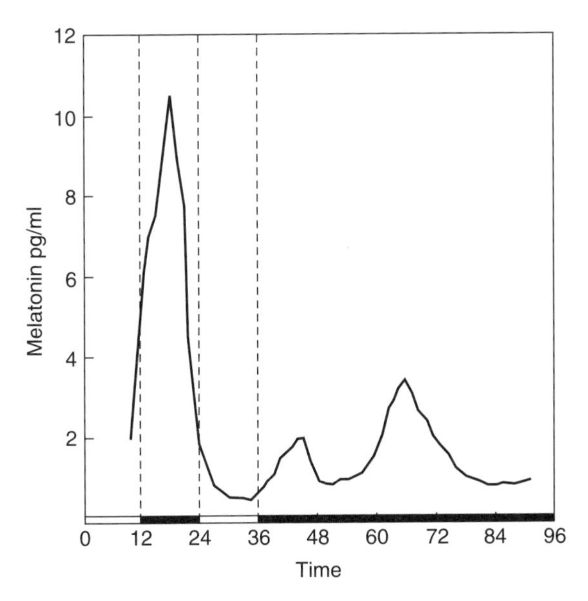

FIGURE 3.17. Melatonin production by house sparrow pineal tissue *in vitro*. Light regimes in which the tissue cultures were maintained are indicated by the bars at the bottom of the graph. Melatonin was assayed by radioimmunoassay, and the assay included some cross-reactants that are also of pineal origin (Menaker 1982). From Menaker (1982: Fig. 2), with the kind permission of Springer Science and Business Media.

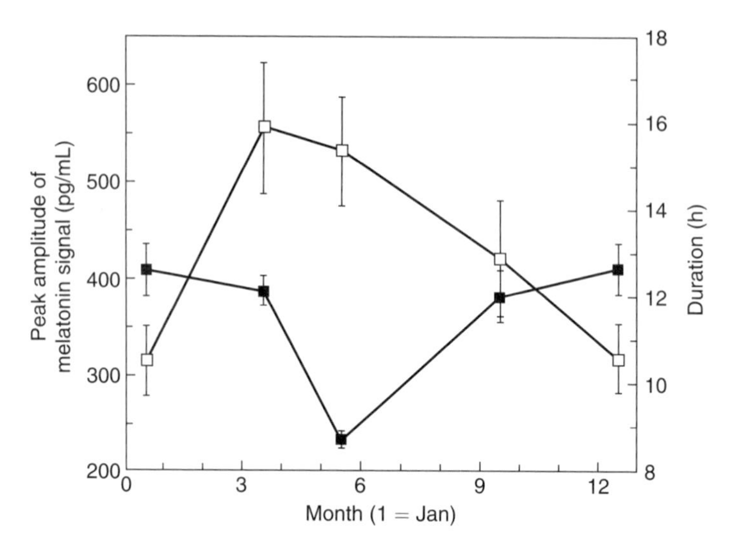

FIGURE 3.18. Annual variation in duration (■) of elevated plasma melatonin and peak amplitude (□) in male house sparrows maintained under natural photoperiods in Germany (plotted values are means ± SE). Blood samples were collected during four periods of the year (Dec 20–Jan 4, Mar 21–Apr 5, May 28–Jun 12, and Sep 24–Oct 9), and plasma concentrations of melatonin were measured by radioimmunoassay. From data in Brandstätter, Kumar et al. (2001: Fig. 2).

The nature of pineal gland photosensitivity is not known. Ralph and Dawson (1968) were unable to detect any electrical response to changes in illumination in isolated sparrow pineal glands. The damping of the peak in melatonin production after transfer to 0L:24D (see Fig. 3.17) indicates that a population of damped oscillators may gradually lose synchrony when deprived of a periodic entraining stimulus.

Binkley (1976) performed assays for the activities of the enzymes NAT and HMOT in pineal glands removed from sparrows being maintained on 12L:12D and sacrificed either 2 h before L/D or 4 h after L/D. NAT activity was higher in the dark than in the light (≈3.6 and ≈0.1 nM/h per pineal, respectively); HMOT activity did not differ between the two groups (756 and 810 pM/h per pineal, respectively). The rhythmic production of melatonin by the pineal may therefore be controlled by rhythmic changes in the availability of one of the enzymes (NAT) in its biosynthetic pathway. In the chick (*Gallus gallus*), NAT synthesis is initiated during the early subjective night and can be interrupted by light during this period (Binkley 1983).

Together with the effects of pinealectomy described earlier, these results suggest that the pineal may be a primary pacemaker in the circadian system of sparrows, and that it may exercise its control over overt circadian activities via its rhythmic production of melatonin. This hypothesis has stimulated considerable research on the effects of melatonin on circadian periodicity in sparrows, as well as its effects on other possible components of the circadian control system.

In an early experiment, Binkley (1974b) injected melatonin into sparrows during the light period of a 12L:12D photoperiod. Cloacal temperatures were monitored for the experimental group and two control groups (saline injection and ethanol injection). Body temperatures in the experimental group dropped an average of 4.7°C within 30 min, whereas they remained unchanged in the saline group and dropped by only 1.4°C in the ethanol group ($P < .02$). Birds injected with melatonin also appeared to be asleep. (A recently reported experiment on the "house sparrow" with similar protocols and results performed in Japan [Murakami et al. 2001] apparently refers to the tree sparrow [see previous discussion].)

In another early experiment, Turek, McMillan, and Menaker (1976) implanted Silastic tubules containing melatonin into sparrows that were free-running in 0L:24D. Controls implanted with empty Silastic tubes were unaffected by their presence, whereas 26 of 30 melatonin-treated birds showed one of two responses to the implant: the period of activity was shortened in 10 birds, and arrhythmia occurred in the other 16. Binkley and Mosher (1985b) obtained similar results with both intraperitoneal implants of melatonin-containing Silastic tubes and oral administration of melatonin (in the drinking water). Some sparrows in both groups became arrhythmic in 0L:24D, while others continued to free-run or were intermittently arrhythmic.

Hendel and Turek (1978) implanted Silastic tubes containing crystalline melatonin into sparrows that were being maintained in 3L:21D. Before implantation, the birds had begun activity 1–2 h before D/L (positive phase angle) and continued for 5–6 h after L/D. Total locomotor activity was reduced in all of the birds with melatonin implants ($P < .001$), whereas activity was not reduced in control birds receiving an empty capsule. The periods of intense activity immediately before D/L and after L/D were abolished In about 75% of the melatonin-treated birds. Hau and Gwinner (1995) obtained similar results in sparrows with melatonin implants being maintained on 12L:12D (4 lux:1 lux). Locomotor activity was greatly reduced and the positive phase angle was abolished in most sparrows with melatonin implants.

These results strongly suggest a role for melatonin in the control of the circadian mechanism governing locomotor and body temperature rhythms in sparrows. This conclusion is further supported by the results of two experiments performed on pinealectomized sparrows by Heigl and Gwinner (1994). In both experiments, sparrows were maintained under constant dim light (0.3 lux) and were arrhythmic after pinealectomy. In one experiment, 10 birds were given water containing 200 μg/ml of melatonin for 8-h periods daily for 3 wk, after which the birds served as their own controls by being administered tap water for 8-h periods. The sparrows tended to become inactive during the period of melatonin administration, with activity beginning to decline 30–120 min before melatonin administration began. Plasma melatonin levels of the experimental birds were higher during the treatment period (mean = 26.2 μg/ml) than during the times when tap water was being given (mean = 4.8 μg/ml); both levels were much higher than those of intact birds but had similar amplitude. The shift back to arrhythmia in the birds after return to tap water required up to 7 d. In the second experiment, 10 sparrows were given melatonin in the drinking water for 7-h periods (0800–1500) for 3 wk, after which the administration of melatonin was phase-delayed by 7 h (to 1500–2200) for another 3 wk. In this experiment, the decrease in activity gradually shifted over a period of several days in response to the phase shift of melatonin administration. The results prompted Heigl and Gwinner to conclude that melatonin acts on another oscillator, possibly the SCN.

In another experiment, Heigl and Gwinner (1995) administered melatonin in the drinking water to pinealectomized sparrows with T ranging from 21 to 27 h and monitored feeding and locomotor activity of the birds. Melatonin-water was available to the birds for 8-h intervals, and birds were considered to be synchronized with the zeitgeber if τ differed from the T by 0.3 h or less. All but one of the pinealectomized sparrows were arrhythmic in constant dim light (24L:0D, 0.2 lux), but when melatonin administration began with T = 24 h they became rhythmic, with inactivity occurring during the time of melatonin administration. A decreasing number of birds remained synchronized when T increased to 25, 26, and 27 h. All birds resynchronized their feeding rhythms when they were returned to T = 24 h, but decreasing numbers of birds then remained synchronized when T was decreased to 23, 22, and 21 h. Birds that were not synchronized on short or long T were either arrhythmic or free-running. Heigl and Gwinner concluded that these results, along with earlier findings, show that a plasma melatonin rhythm is a prerequisite for sparrows to maintain an activity rhythm in constant dim light.

Radioactive isotopes of melatonin (2[^{125}I]iodomelatonin = IMEL) and 2-DG have been used to study the physiology of the interaction between the pineal gland and specific hypothalamic nuclei, including the SCN and LHN. IMEL can be used to identify sites of melatonin binding, and 2-DG uptake can be employed to identify temporal patterns and sites of metabolic activity.

In one experiment, the brains of 11 female sparrows that had been maintained on 12L:12D (D/L at 0600) were removed and quick-frozen at 1200 (Cassone and Brooks 1991). Brains were then sectioned at 20-μm intervals, and adjacent sections were mounted and incubated either with IMEL (50 or 500 pM) or with IMEL (50 or 500 pM) plus 1 μM unlabeled melatonin (controls). Quantitative analysis of IMEL binding was performed spectrophotometrically, and nonspecific binding was determined using the IMEL plus melatonin controls. IMEL binding occurred at numerous locations related to visual processing at several levels in the brain (see also Cassone et al. 1992, 1995). In the hypothalamus, these included the LHN, which were referred to by Cassone et al. as the visual suprachiasmatic nuclei (vSCN) (cf. Brandstätter and Abraham 2003), but not to areas in the SCN (referred to as the periventricular preoptic nuclei, or PPN). IMEL also did not bind to the hypothalamo-hypophyseal structures related to regulation of reproduction, suggesting that the role of melatonin, and therefore of the pineal, in the control of reproduction is at most an indirect one. Cassone et al. did not state at what time of year females were collected or describe their reproductive state, which might affect their responsiveness to melatonin.

Cassone and Brooks (1991) also examined 2-DG uptake in the brains of female sparrows after melatonin injection. Sparrows maintained on 12L:12D (D/L at 0600) were divided into two groups, and at 1600 one group was injected intramuscularly with melatonin , while the control group was injected with saline. All birds were then injected with about 5 μCi of 2-DG and maintained for 45 min before being sacrificed. Brains were sectioned and exposed to x-ray film to determine rates of 2-DG uptake. Melatonin inhibited uptake of 2-DG in most of the same structures that showed significant IMEL binding, indicating that the sites of cerebral binding of IMEL correspond to the sites of the physiological effects of melatonin. Because of the number of cerebral sites affected, Cassone and Brooks suggested that the visual world of sparrows is modulated by the circadian secretion of melatonin by the pineal gland.

Lu and Cassone (1993b) performed two experiments to further elucidate the interaction between the pineal and the LHN (referred to as the vSCN). In the first experiment, 30 pinealectomized and 30 sham-operated sparrows were placed in 0L:24D after 1 wk on 12L:12D. The sham-operated birds free-ran in complete darkness, whereas the pinealectomized birds were arrhythmic. On the first, third, and tenth days in 0L:24D, five birds in each group were injected with 2-DG at CT06 or CT18. After 60 min, birds were sacrificed and their brains were removed, frozen, and sectioned at 20 μm. One set of sections was assayed for 2-DG by autoradiography, and two sets of sections were incubated either in IMEL or in IMEL plus melatonin and then assayed autoradiographically. In the sham-operated birds, 2-DG uptake by the LHN was significantly higher during the subjective day (CT06) than during the night on all three days. In the pinealectomized birds, rhythmic uptake of 2-DG was evident on days 1 and 3, but not on day 10. IMEL binding in sham-operated birds was also rhythmic and higher at CT06 than at CT18, indicating higher levels of endogenous melatonin during the subjective night. IMEL binding was elevated 2–4 times in pinealectomized birds (above the level in sham-operated birds) at both times on days 1 and 3 and was higher during the subjective day. Neither elevation of IMEL binding nor the rhythmic pattern of binding occurred on day 10. Rhythmic binding of IMEL was also observed in numerous other visually oriented structures of the brain, prompting Lu and Cassone to suggest that rhythmic melatonin production by the pineal may affect visual integration (as well as LHN activity) in sparrows. These results indicate that rhythmic metabolic activity continues in the LHN of pinealectomized sparrows for at least 3 d after termination of an entraining photoperiodic stimulus. Rhythmic binding of melatonin also persisted in pinealectomized sparrows for at least 3 d in complete darkness, which suggests the presence of another circadian oscillator whose rhythm slowly damps out in constant darkness in pinealectomized sparrows.

Lu and Cassone (1993b) performed a second experiment in which birds were maintained on 12L:12D and four pinealectomized and four sham-operated birds were sacrificed at ZT00 (D/L), ZT04, ZT08, ZT12 (L/D), ZT16, and ZT20. Binding of IMEL at six different concentrations (10, 50, 75, 100, 200, and 500 pM) was assayed as described earlier. IMEL binding was rhythmic in both sham-operated and pinealectomized birds, with the peak of binding in the late day (ZT08) and binding was 2–4 times higher in pinealectomized birds than in sham-operated birds.

In yet another experiment, Lu and Cassone (1993a) administered melatonin in the drinking water to pinealectomized and sham-operated sparrows and examined the uptake of 2-DG and binding of IMEL by the LHN and other brain structures associated with vision. Before the experiment, birds were treated for both ectoparasites and endoparasites to preclude possible circadian cues from these organisms. After pinealectomy or sham operation, birds were placed on 12L:12D for 10 d before being transferred to constant dim light (24L:0D, 0.1 lux). Pinealectomized birds entrained their activity rhythms to the 12L:12D photoperiod but became arrhythmic in constant dim light. Birds were then placed for 14 d on a schedule of either 12 h of melatonin in the drinking water or 12 h of ethanol in the drinking water, after which they were injected with 2-DG at either 6 or 18 h after the beginning of the melatonin or ethanol treatment. Birds were then sacrificed, and their brains were removed and sectioned. In series of three adjacent sections, one section was exposed to Kodak film for autoradiography of 2-DG uptake, and the remaining two were incubated for 1 h either with IMEL or with IMEL plus melatonin. Pinealectomized birds held on the melatonin regime had rhythmic locomotor activities, with inactivity occurring during the time there was melatonin in the drinking water. Ethanol controls were arrhythmic. Uptake of 2-DG was significantly lower 6 h after the beginning of melatonin administration than 6 h after the return to pure water in several brain structures, including the LHN, but there were no differences in 2-DG uptake between the two times in the ethanol controls. Significant differences in IMEL binding also occurred in birds on the melatonin regime, with higher levels of IMEL binding in the LHN and several visual system areas 6 h after the return to pure water. Lu and Cassone concluded that the entraining effect of melatonin in pinealectomized sparrows probably results from entrainment of residual circadian structures rather than direct suppression of activity by melatonin, and that the binding of melatonin to the LHN and to visual centers in the brain acts to downregulate the receptors.

Circadian System of the House Sparrow

The circadian system of sparrows comprises internal self-sustaining oscillators that interact with external zeitgebers to regulate daily rhythms of both behavioral and physiological attributes in the species. The oscillators are actually composed of populations of oscillators, probably individual cells. Light acting primarily at L/D and D/L transitions serves as the principal zeitgeber, although the system can use a number of secondary cues (e.g., food availability, auditory cues, temperature). The major components of the internal time-keeping system—the pineal gland, the SCN of the hypothalamus, and the retinas of the eyes—are all of diencephalic origin (Menaker 1982). Although the physiology of this biological clock is still not fully understood, it is possible to begin to suggest some of the workings of its major components.

Takahashi and Menaker (1982b) suggested three hypotheses to explain the relationship between the SCN and the pineal gland in the control of circadian activity in sparrows. In the first, the SCN could serve as components in the output pathway located between a circadian pacemaker (presumably the pineal gland) and the locomotor activity. This hypothesis is an example of a one-oscillator model (cf. Binkley 1990) and suggests that rhythmic production of melatonin by the pineal would persist in birds with SCN lesions. In the second hypothesis, also a one-oscillator model, the SCN play a permissive role and sustain pineal rhythmicity, which suggests that pineal rhythmicity would be abolished in SCN-lesioned birds. In the third hypothesis, the SCN and the pineal interact to function as a complex pacemaker (a two-oscillator model) (cf. Binkley 1990). Under this hypothesis, pinealectomized birds would be arrhythmic due to the uncoupling of numerous damped oscillators located in the SCN. At present, the data necessary to evaluate the merits of these three hypotheses are not available, although most workers seem to favor the two-oscillator model.

Two specific hypotheses have been advanced to describe the interaction of the pineal gland and the SCN in the control of circadian activity in birds. Cassone and Menaker (1984) proposed the Neuroendocrine Loop Model (Fig. 3.19), which involves the pineal gland, the SCN, the retinas of the eyes, extraretinal photoreceptors, and the superior cervical ganglia (SCG) of the sympathetic nervous system. Light (acting directly) and sympathetic nervous stimulation (indirectly activated by light) inhibit melatonin secretion by multiple oscillators in the pineal gland. Light acting directly, or indirectly via the retinohypothalamic tract, stimulates serotonin production by oscillators in the medial region of the SCN, which results in increased metabolic and electrical activity in these nuclei. This activity then stimulates overt activity rhythms (e.g., locomotor activity, body temperature, feeding), presumably via the autonomic nervous system. SCN activity can be inhibited by melatonin produced by the pineal, which passes through the blood vascular system under dark conditions, but SCN activity is apparently not inhibited by melatonin produced by oscillators in the eye.

Gwinner (1989) developed the Internal Resonance Model to explain the patterns observed in several species of birds as quantitative rather than qualitative differences. This model includes the pineal and a second extrapineal

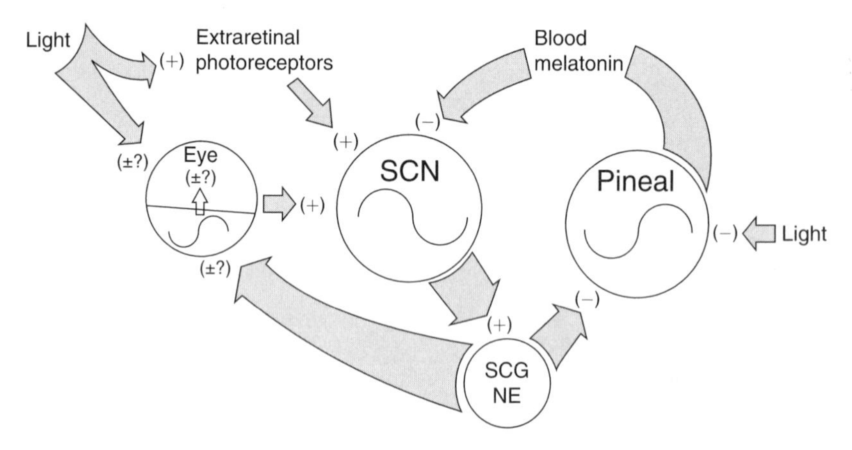

FIGURE 3.19. The Neuroendocrine Loop Model of the circadian control of locomotor rhythm in the house sparrow, proposed by Cassone and Menaker (1984). Arrows indicate either neural or hormonal pathways between organs, and (−) and (+) signs indicate inhibition and stimulation, respectively, of the target organs. ~ indicates a population of damped oscillators. Note that environmental light can act on the rhythm via both retinal and extraretinal routes. SCG NE represents norepinephrine release by the sympathetic nervous system via the superior cervical ganglia (SCG). SCN, suprachiasmatic nuclei. From Cassone and Menaker (1984: Fig. 8), with permission of the Journal of Experimental Zoology.

pacemaker system (EPPS)—probably the SCN, but possibly including other components—which both receive input from the eyes (and, in some cases, from an extraretinal light receptor). The pineal gland and the EPPS are pacemaker systems whose rhythms are mutually coupled and reinforced or amplified by each other through resonance. The EPPS has a shorter natural period, τ, than the pineal (which accounts for the universal shortening of τ in pinealectomized birds that retain rhythmicity, as discussed earlier). The EPPS in turn acts on two weak suboscillators associated with feeding and locomotor rhythms (with the feeding suboscillator more tightly linked to the EPPS than the locomotor suboscillator). Abolishment of one of the pacemaker systems may or may not result in locomotor or feeding arrhythmia in constant conditions (e.g., 0L:24D), depending on the extent to which maintenance of synchrony in the population of oscillators that make up the remaining pacemaker is dependent on the reinforcing signal from the other pacemaker. These differences in degree of coupling and resonance account for the different responses of individuals within a species and for the differences in response among bird species. One problem with the model, which might be only semantic except that it is reflected in Gwinner's depiction of the model, is that resonance implies that oscillators cycle together. In fact, as described earlier, the two principal oscillators in the circadian mechanism controlling the locomotor rhythm in sparrows appear to be active at different times, with *pPer2* gene expression in the SCN (and LHN) occurring during the daytime and the pineal gland secreting melatonin at night.

Abraham et al. (2000) attempted to test predictions of the Internal Resonance Model. Silastic tubes containing melatonin were implanted into sparrows that were then entrained to periodic food availability as a zeitgeber in constant dim light (24L:0D, 0.3 lux). Feeding activities of the sparrows were monitored with infrared sensors at the entrance to the food tray. Phase shifts in the zeitgeber (8-h phase advances and 8-h phase delays) were then performed, and birds with the melatonin implants entrained to the new zeitgeber significantly more rapidly than controls (i.e., birds with empty Silastic tubes). Abraham et al. attributed this difference to the disruption of the normal melatonin rhythm in sparrows with intact pineal glands, with the consequence that the second oscillator could re-entrain more rapidly to phase shifts in the restricted-food zeitgeber. These results, which imply the existence of a second independent oscillator, do not, however, distinguish between the Neuroendocrine Loop Model and the Internal Resonance Model.

An example of a one-oscillator model for the control of circadian rhythmicity is the mathematical model proposed by Wever (1966). The essence of a one-oscillator model is that the circadian pacemaker is located in a single structure, which then drives numerous other circadian activities in the organism. These activities are based on populations of damped oscillators whose synchrony derives either from entrainment by a rhythmic external zeitgeber or from the self-sustained rhythmicity of the internal pacemaker. Wever (1966) represented the activity of the internal pacemaker as a sinusoidal wave in which the mean value of the wave can change with respect to some fixed threshold, resulting in changes in the level of activity. Although the pineal gland is the most likely candidate for a single oscillator in the house sparrow, the fact that some pinealectomized sparrows are apparently able to retain rhythmicity in the absence of external zeitgebers, and the fact that some other bird species, such as the European starling (*Sturnus vulgaris*) and the Japanese quail, do not become arrhythmic when pinealectomized, argue in favor of a more complex avian chronometer. In a phylogenetic analysis of IMEL binding patterns in the brains of several bird species (including the house sparrow, European starling, and Japanese quail), Cassone et al. (1995) found that the four passerine species clustered together but three of the passerines (including sparrows) clustered separately from the starling within the passerine clade. This finding, as well as the differences among the other species included in the analysis, suggests that there is considerable diversity among birds in the nature of the circadian chronometer, a diversity that is probably better explained by variation in a two-oscillator system.

Two-oscillator models are also better able to account for changes in the relative timing of two parameters in response to changes in photoperiod (Binkley 1990). The one-oscillator model is, however, more useful in attempting to explain phenomena such as Aschoff's rule (the increase in τ with increases in light intensity) and changes in the duration of oscillations and the shape of the phase response curve in response to changes in photoperiod (Binkley 1990).

Much progress has been made in understanding the physiology of the circadian system that is involved in controlling daily and seasonal activities of sparrows, but there is still much that is not understood. In particular, the location of the extraretinal light receptors that mediate photoperiodic control of the annual cycle has not yet been identified, and the physiological relationships between the circadian time-keeping mechanism and the stimulation of gonadal recrudescence, regression, and photorefractoriness are also unknown. The "memory" in the pineal gland for recently experienced photoperiods (Brandstätter et al. 2000) and the seasonal variation in amplitude and length of melatonin production by the pineal (see Fig. 3.18) provide some clues as to how the system may detect changes in photoperiod and transduce that information into appropriate physiological responses (Brandstätter 2003). It is also unclear, however, how the mechanism regulating the annual cycle of temperate populations of the house sparrow also regulates the quite different annual cycle of subtropical resident populations (see Fig. 3.1).

Chapter 4

BREEDING BIOLOGY AND REPRODUCTIVE STRATEGY

Be fruitful and multiply, and fill the earth.

—Genesis 1:28

Two species that have been remarkably successful at fulfilling this command are *Homo sapiens* and the house sparrow, and for the past half-century or so the former has been attempting to discover the secrets employed by the latter in accomplishing this feat. Those attempting to unravel this mystery are known as evolutionary ecologists, and the secret they are attempting to uncover is called the reproductive strategy of a species. The reproductive strategy of a species is the integrated suite of life history characteristics that have evolved to maximize the individual's reproductive output. Characteristics frequently considered to be part of this strategy in birds include age at first reproduction, mating system, nest location and dispersion, timing of the breeding season, clutch size, juvenile development type and growth rate, egg and juvenile survivorship, number of broods per season, and reproductive lifespan. This chapter deals with these elements of the reproductive strategy of sparrows and with related components of their reproductive biology.

Age at First Reproduction

Sparrows are reproductively mature in the first breeding season following their year of birth, and most individuals either breed or attempt to breed during this first breeding season. First-year males have smaller bibs ("badges of status") than older males do (see Chapter 5), and they may have difficulty obtaining good nest sites in good colonies. This means that they may have greater difficulty attracting a mate and therefore may not breed in their first year. First-year females may also have difficulty finding and successfully mating with a male holding a good nest site in a good colony, and they often initiate breeding later than older females or become unmated floaters in the population (see Anderson 1990). In one 3-y study in England (UK), first-year females ($n = 19$) initiated their first clutches an average of 7.6 d later than 2-y-old or older females ($n = 35$) ($P < .01$) (Dawson 1972b).

Davis (1953) reported on the precocial sexual development in juvenile males in California (USA). Three males from approximately 100 juveniles trapped during June and July had enlarged testes that proved to be actively engaged in spermatogenesis when examined histologically. Davis concluded that this precocial sexual development was the result of photostimulation of the pituitary gland (see Chapter 3) in young birds that were hatched early in the breeding season and hence fledged when days were still lengthening for several weeks. Because the birds had not undergone the first prebasic molt and were still in juvenal plumage, it is unlikely that they were actually breeding. Ely and Bowman (1969) also reported on two juvenile males with enlarged testes collected in July in Kansas (USA). It is possible that individuals could breed in their natal year, and such breeding by young of the year should be looked for in subtropical populations that have a second, late-season breeding period, such as in India (Naik and Mistry 1980) or El Salvador (Thurber 1986). Thurber (1986), for instance, found females with incompletely ossified skulls and in juvenal plumage in breeding condition (one with three enlarged ovarian follicles) in May in El Salvador. These birds were presumably less than 6 mo old, apparently the product of late-season clutches in the previous year. Birds of the year may also be involved in the small number of cases of fall or winter breeding at temperate latitudes (see Chapter 3).

Mating System

Pair Bond

Sparrows are socially monogamous, meaning that each breeding attempt normally involves a mated pair and the pair bond persists throughout the attempt. Pair bonds often remain intact throughout a single breeding season, and sometimes between years (Summers-Smith 1958). Long-term persistence of the pair bond is probably due primarily to the tendency for individuals to utilize the same nest site repeatedly (nest site fidelity), rather than continuous maintenance of the pair bond itself. Individuals that change nest sites either within or between breeding seasons often change mates as well.

Formation of the pair bond is initiated by the male bird, who selects a potential nesting site, often during the autumnal period of sexual activity (see Chapter 3) or in late winter, and advertises vigorously with repeated vocalizations and displays (Summers-Smith 1958). Unmated males increase the intensity of chirping if a female appears, and they also display by dropping and shivering their wings. Females usually select a nest site based on their response to vocalizing and displaying males, and both birds may roost in the nest site once it has been selected. When one member of a pair disappears, the other bird tends to remain at the nest site until a replacement mate is acquired. Widowed females sometimes vocalize, although not as intensely as unmated males (Summers-Smith 1958).

Polygyny occurs frequently in sparrows (Cordero et al. 1999a; Griffith, Owens, and Burke 1999b; Kohler 1930; Matuhin 1994; North 1980; Pearse 1940; Schwagmeyer et al. 2002; Summers-Smith 1958; Veiga 1990b, 1990c;). Veiga (1990c) defined polygyny as a male mated to two or more females whose nesting cycles overlap (from mating to fledging of young). In some cases, two females lay their eggs in the same nest and both participate in the incubation and care of the young (Kohler 1930; Pearse 1940). However, in most cases, polygynously mated females use different nest sites—often, but not always, close to each other (Veiga 1990b). Veiga (1992c) studied a color-banded population in central Spain in 1986–1989 and found that 9.6% of males mated polygynously during the 4 y of the study, with the annual rate varying from 7.5% to 15.8%. One male mated with three females.

Polygynously mated males participate in the breeding activities at the nests of both females, but they tend to favor one nest, the primary nest, over the other. North (1980) reported that a polygynously mated male participated in the feeding of nestlings at both nests, but he made almost three times as many trips to the primary nest as to the secondary nest (7.2 vs. 2.5 trips/h). Veiga (1990c) reported that feeding rates by polygynous males at their primary nests were similar to those of monogamous males, but at secondary nests they were significantly lower ($P < .001$). In 11 of 12 cases, the primary nest was that of the female that laid first. For the years 1985–1987, polygynous males averaged 3.94 breeding attempts/y, whereas monogamous males averaged only 1.99 attempts/y ($P < .001$). Polygynous males also produced more fledglings per year (9.5) than did monogynous males (5.12), ($P < .001$) (Veiga 1990c). Griffith, Owens, and Burke (1999b) also found that most of the more productive males on Lundy Island (UK) were polygynously mated.

Why are sparrows primarily monogamous despite the obvious reproductive advantage for polygynously mated males? Veiga (1992c) attempted to answer this question by testing two hypotheses that had been advanced regarding the choice between monogamy and polygyny. The Polygyny Threshold Hypothesis is based on the assumption that territories or other resources held by males vary in quality and suggests that females should choose to mate polygynously if the resources still available in a mated male's control exceed those held by an available unmated male. The Female-Female Aggression Hypothesis proposes that, because polygyny would result in reduced reproductive success for a monogamously mated female, polygyny is prevented by aggression between females. Although the two hypotheses are not mutually exclusive, Veiga (1992c) tested a set of predictions derived from them by studying both naturally occurring and experimentally induced polygyny in a population breeding in nest-boxes in central Spain. Between late March and early May in the years 1987–1989, he removed 5, 10, and 15 mated males from the colony, usually shortly before initiation of egg-laying. A neighboring male took possession of 16 of the 30 nest-boxes from which males had been removed and, in 11 cases, mated polygynously with the female. Breeding success of polygynously mated females (including both naturally occurring and experimentally induced polygyny) was lower than that of monogamous females, primarily because of a significantly lower mean clutch size (Fig. 4.1). Before the breeding season, Veiga also added nest-boxes in close proximity to previously occupied nest-boxes (entrances 15 cm apart). The proportion of males that controlled more than one nest-box increased significantly with increased nest-box density, but the proportion of males mating polygynously did not. He also placed a female in a cage on the nest-box or on a nearby nest-box of 25 resident females, and monitored aggressive behavior of each female during the prelaying, incubation, and nestling periods. Female aggression toward the caged female decreased significantly between prelaying and incubation but was higher during incubation if the female had been previously exposed to the experimental "intruder." Veiga concluded that monogamy was maintained as the predominant mating system in the population primarily by female-female aggression, even though the polygyny threshold was apparently often met.

Copulation

Beven (1947) reported that copulation is usually initiated by the male, who approaches the female by hopping rapidly along a branch or building roofline. The male uses a high-pitched, continuous vocalization ("tee-tee-tee . . .") that, according to Beven, is not used at other times (this call apparently belongs to the "quee" family of calls—see Chapter 7). If the female is receptive, she adopts a solicitation posture by assuming a crouching position with tail slightly depressed and wings quivering. The male mounts with his wings fluttering, con-

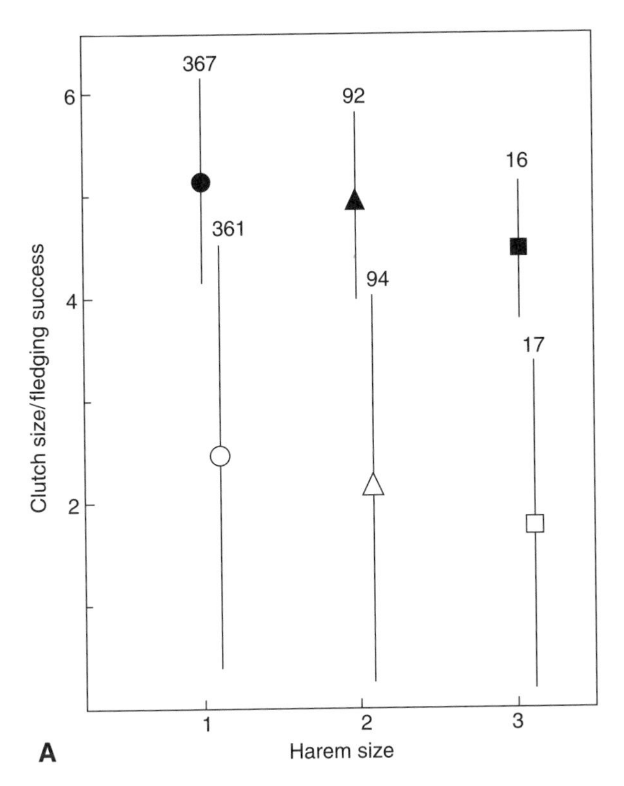

FIGURE 4.1. Mating status and breeding success of female house sparrows. Data are based on 478 breeding attempts in 1987–1989 at Collado Villalba, Spain. Mean clutch sizes and mean number of young fledging are shown for monogamously mated (●, ○), bigamously mated (▲, △), and trigamously mated (■, □) females, with vertical lines being standard deviations and numbers being sample sizes. Differences in clutch size and fledging success were both significant ($F_{2,472} = 17.35$, $P < .001$, and $F_{2,469} = 6.41$, $P < .01$, respectively). From Veiga (1992a: Fig. 1a), reprinted with permission from Elsevier.

tinuing to call. Mounting may be rapidly repeated up to 10 or 11 times with short dismounts, a bout of copulation. However, Summers-Smith (1955) stated that successful copulation is usually initiated by the female, who solicits copulation by adopting the posture described and turning her tail toward the male. The female also uses the "tee-tee-tee . . ." vocalization, perhaps as a solicitation signal (Summers-Smith 1963). If the female is unreceptive, she may fly away when approached by a displaying male or behave aggressively toward him (see Cooper 1998).

Although copulation may occur as long as 3 mo before egg-laying, it is most frequent in the 7–10 d before initiation of a clutch and during egg-laying

(Summers-Smith 1963, Wetton and Parkin 1991). However, Berger (1957) reported frequent copulation more than 1 mo before the first egg was laid. Because copulation begins before the fertile period that occurs just before and during egg-laying, it apparently plays a role in pair formation and maintenance of the pair bond, as well as insemination (Wetton and Parkin 1991). North (1980) reported that the copulation rate during the egg-laying period was about 2–4 bouts/h, and Birkhead et al. (1994) reported that bouts of copulation have an average of 4.11 copulations per bout, whereas Nyland et al. (2003) reported an average of 3.96. Vaclav and Hoi (2002b) found that copulation frequency decreased with brood number during the prolonged breeding season in Austria. In that population, the copulation rate was not affected by supplemental feeding (2.64 copulation bouts/h with supplemental food, vs. 2.00 copulation bouts/h with no extra food; $P = .71$) (Vaclav et al. 2003).

There is a significant directional bias in copulation position, with males mounting from the left side approximately 75% of the time (Nyland et al. 2003). This bias may represent an adaptation that is based on more frequent entry of sperm into the oviduct, the entrance of which is located on the left side of the proctodeum (see Chapter 9).

Sperm production in sparrows occurs only at night, and the sperm are stored in the seminal glomera for use during the daytime (see Chapter 9). Birkhead et al. (1994) estimated that the average rate of sperm production in males during egg-laying is about 31×10^6 sperm/d. They concluded that male sparrows use most of their sperm production each day during the fertile period.

Mixed Mating Strategies

Trivers (1972) suggested that selection might be different for the two members of a monogamous pair, and that this might lead to the evolution of mixed mating strategies in both sexes. Recent work using molecular techniques has confirmed that socially monogamous sparrows often participate in extra-pair copulations that frequently result in successful fertilization. These findings have helped to provide explanations for two sparrow behaviors that have long been recognized but for which there were inadequate explanations. The first of these, the communal sexual display or "sparrow party," is a very prominent feature of sparrow behavior (Simmons 1952, Summers-Smith 1954b). The second, pecking of the female's cloaca by males (Conder 1947; Cooke 1947; Hardy 1932; Simms 1948), is often observed in association with these communal displays.

The communal display always involves one female and two or more males, with as many as 11 participating in a display (Summers-Smith 1954b). The males chase the female in frantic, headlong flights, often to a dense bush, and then display vigorously to her, sometimes pecking at her cloaca. Cloacal pecking often involves several males, including the female's mate. Summers-Smith also reported that the mate of the female was always present and participated in the display. During the display, the male throws his head back, elevates and fans the tail, lowers the wings, fluffs out the feathers of the breast, and hops stiffly around the female (see Fig. 5.1 in Chapter 5). Males also chirp repeatedly. Females often react aggressively toward the males. Attempts at copulation are sometimes associated with the displays, but these were not observed to be successful. Summers-Smith recorded the timing of the displays by half-months and found that peaks occurred in late March/April and in late May, near the initiation times of first and second clutches in the population. In New Zealand, the frequency of the displays was highest in September, with slightly fewer being recorded in August and October (corresponding to the beginning of the breeding season in the southern hemisphere) (Secker 1975). Summers-Smith (1954b) proposed that the function of the displays was to bring the female into readiness to mate, thereby synchronizing her sexual cycle with that of the male.

Møller (1987a) studied communal displays in a small, marked colony of sparrows in Denmark. He found that the displays involved an average of 3.1 males and lasted an average of 35.9 s. The rate of communal displays was low during the prefertile period (generally less than 0.1 display/h) but increased dramatically during the fertile period (to about 0.5 display/h). Møller suggested that the timing of the displays and the participation of only one female in a display argued against the synchronization hypothesis and proposed instead that the displays represent attempts by males to obtain extra-pair copulations. In this context, cloacal pecking by males can be interpreted as an attempt to induce the female to expel semen that she had obtained in earlier copulations. In a follow-up experimental study on the same population, Møller (1990) reported that males initiated communal displays with females that were released after being implanted with Silastic tubes containing estradiol (to induce rapid follicular development) but failed to initiate displays with control females (empty Silastic tubes). This result suggests that males can identify fertile females and that they direct communal displays preferentially toward them.

Although extra-pair copulations had been reported in the house sparrow (e.g., Burrage 1964; Summers-Smith 1958), confirmation that such copulations resulted in successful fertilizations was first obtained using DNA fingerprinting on pairs and their putative offspring (Burke and Bruford 1987; Wetton et al. 1987). These initial fingerprinting studies used minisatellite probes from humans (see Chapter 2), and both identified offspring that were unrelated to the male, indicating that the female had participated in a successful extra-pair copulation. In an intensive 5-y study of a breeding population of sparrows in

England, Wetton and Parkin (1991) obtained DNA fingerprints of 536 nestlings and their putative parents. All of the nestlings were related to the female of the pair, indicating that there were no cases of egg-dumping or intraspecific brood parasitism. However, 73 nestlings (13.6%) were not genetically related to the male. The percentage varied annually from 10.0%–19.0% (not significant) but did vary inversely with brood size ($r = -0.29$, $P < .001$). The percentage of extra-pair offspring was also related to hatching success. Of 310 nests in which all eggs hatched, 30 (9.7%) contained extra-pair offspring, whereas 21 (19.6%) of 107 nests in which one or more eggs failed to hatch contained extra-pair young ($P < .01$).

In a subsequent study on the same population, Wetton et al. (1995) used four minisatellite probes to obtain DNA fingerprints of 144 males in order to assign paternity of offspring. A total of 47 extra-pair offspring were identified, 3 of which did not have sufficient material to test for male parentage. Matches were obtained for 37 of the 44 remaining extra-pair fledglings. In seven of nine cases in which there were two extra-pair offspring in the same nest, both were sired by the same male. In four cases involving three extra-pair offspring, one male sired all three in two cases, and in the other two cases the young were sired by two different males. Successful extra-pair copulations tended to occur during the fertile period of the siring male's own breeding cycle (prelaying and laying periods). Older males were more likely to sire extra-pair offspring than were first-year males. In a population with approximately equal numbers of first-year and older males, first-year males sired only 4 of 33 extra-pair young ($P < .0001$). Older males achieved 20% of their productivity through extra-pair fertilizations, whereas first-year males achieved only 2.5% in that manner. Older males produced on average 6.00 fledglings/y, which was a 36% higher productivity rate than that of yearlings, who produced 4.41 fledglings/y ($P < .01$). Wetton et al. suggested that younger males may be less successful at extra-pair fertilizations either because they lack time to pursue extra-pair copulations (inexperience) or because females prefer older males. Females may choose older males if there is no other reliable cue to male quality. The nestboxes at the study site were located in two areas, and the cuckolding male was, with only one exception, one of the males in the area occupied by the female.

Other studies using either minisatellite or microsatellite probes (see Chapter 2) have found similar rates of extra-pair fertilization in house sparrows. Cordero et al. (1999a) reported that 10.1% of 109 young were fertilized by a male other than the pair male in eastern Spain, whereas a study in central Spain found that 7.0% of 171 young were the result of extra-pair fertilizations (Veiga and Boto 2000). In the latter study, 1 of the 171 young was also apparently the result of intraspecific brood parasitism (see later discussion) but was apparently fathered by the pair male. In two North American studies, extra-pair fertilization rates were reported to be 10.3% in Kentucky (Griffith, Stewart et al. 1999) and about 16% in Oklahoma (Hankinson 1999). In Austria, Vaclav and Hoi (2002b) found that the percentage of extra-pair fertilizations did not differ significantly among broods (first broods, 6.7%; second broods, 28.8%; third broods, 24%) ($P = .15$). One interesting exception to the findings of extra-pair fertilization rates of 7% or higher was in the small population on Lundy Island, where the rate was only 1.3% (4 of 304 offspring) (Griffith, Stewart et al. 1999). A small sample from England also had a low rate of extra-pair fertilization (3.8%, 2 of 53 young) (Cordero et al. 1999a). The percentage of broods containing one or more extra-pair offspring varied from 9.3% in central Spain (Veiga and Boto 2000) to 28.6% in eastern Kentucky (Griffith, Stewart et al. 1999), except on Lundy Island, where only 3.6% of the 112 broods contained extra-pair offspring (Griffith, Stewart et al. 1999).

As in the case of polygyny, the advantage to the male of participating in extra-pair copulations is apparent. Males that successfully sire extra-pair offspring produce more offspring during the breeding season than do males with no extra-pair offspring, and they do so with no costs of nest defense or food provisioning. But what is the selective advantage for females, particularly if there is any chance that the pair male may reduce his effort in supporting the breeding attempt if he has reduced assurance of paternity? Two hypotheses that have been proposed to explain the selective advantage of extra-pair mating for females are the Good Genes Hypothesis (Lifjeld 1994; Trivers 1972) and the Fertility Assurance Hypothesis (Wetton and Parkin 1991). The Good Genes Hypothesis proposes that females mated to low-quality males could increase the fitness of their offspring by mating with high-quality males, those possessing "good genes." The Fertility Assurance Hypothesis, on the other hand, proposes that females enter into extra-pair matings to provide insurance against the potential infertility or low fertility of their mates. This hypothesis argues that, because of the high energetic and nutritional costs associated with the production of a clutch of eggs (see later discussion), selection would favor females that adopt mating strategies that minimize the risk of infertility. Distinguishing between the relative merits of these two hypotheses in explaining the mating strategy of female sparrows is difficult, in part because the two hypotheses are not mutually exclusive.

Cordero et al. (1999b) examined patterns of both infertility and extra-pair paternity in sparrow clutches in eastern Spain. Eggs that showed no signs of development when inspected by trans-illumination ("candling") were classified as infertile if subsequent inspection showed no signs of embryonic development. Eggs laid early in the clutch (i.e., those laid before the median) had a higher rate of infertility (12.4%) than did late-laid eggs (1.4%) ($P < .01$). Early viable eggs had a higher rate of extra-pair fertilization (19%) than did late eggs

(2.0%) (P = .01). Taken together, these results indicate that females may participate in extra-pair matings to reduce the number of infertile eggs. One difficulty with this study, however, is that examination of the perivitelline membrane of apparently infertile eggs from a population in central Spain showed that a majority of such eggs had definite signs of fertilization (Birkhead et al. 1995). Only 6 of 22 apparently infertile eggs actually showed no signs of penetration of the perivitelline membrane by sperm and were therefore truly infertile. Eggs laid early in the sequence may be subject to early embryonic mortality due to chilling (see Veiga 1992a) but nevertheless may be indistinguishable from infertile eggs on superficial inspection. The low rate of apparent infertility and extra-pair paternity in late-laid eggs observed by Cordero et al. (1999b) suggests that females do not enter into extra-pair matings to guard against infertility due to sperm depletion in the pair male late in the laying sequence.

Tests of the Good Genes Hypothesis have tended to focus on the badge size of males involved in extra-pair matings, with the assumption being that large badges are a signal of male quality (but see Chapter 5). Cordero et al. (1999a) used minisatellite DNA fingerprinting to determine extra-pair paternity in sparrow populations in both England and eastern Spain and also measured the badge size of males in the populations. Badge area of cuckolded and uncuckolded males did not differ significantly at either location (P = .26 and .61), although males with larger badges tended to be cuckolded more frequently than small-badged males in eastern Spain. Cordero et al. concluded, however, that, because badge size was positively correlated with age, females could obtain enhanced fitness for their offspring by choosing larger-badged (i.e., older) males. This conclusion is similar to that of Wetton et al. (1995), based on their study in England described earlier. In central Spain, the incidence of extra-pair fertilization in a clutch was not related to the number of broods reared by the female, the maintenance of the pair bond throughout the breeding season, or the age of the putative parents (Veiga and Boto 2000). It was also not related to male badge size.

A possible cost to both sexes associated with participating in extra-pair copulations is the risk of infection by a sexually transmitted disease. Stewart and Rambo (2000) attempted to determine whether microbes occupying the cloaca could be transmitted by copulatory behavior, by using cloacal swabs to sample the cloacal orifices and cloacae of both members of eight pairs in Kentucky. Samples were incubated on eight different media to permit specific identification of microbes, which included anaerobic bacteria, *Staphylococcus* spp., gram-negative enteric bacteria, fungi, *Salmonella* sp., *Lactobacillus* sp., and *Yersinia* sp. The average correlation coefficient of numbers of microbes per individual between members of a pair, 0.63, was significantly greater than zero ($P < .01$). Fungi and anaerobic bacteria occurred in all 16 individuals, and the incidence of the other microbes ranged from 31% to 88%. Samples collected from the testes of five males and the ovaries of five females that were not mated were also incubated, and the testes of two males and the ovaries of three females proved to be positive for microbial infection. Thirty-eight eggs that failed to hatch and appeared to be infertile were also sampled, and seven were contaminated (four with *Salmonella* sp., two with *Yersinia* sp., and one with *Salmonella* sp. and fungi). Stewart and Rambo concluded that there is evidence for the transfer of microbial infections between members of a pair during copulation, suggesting that there is a potential cost of extra-pair copulations.

Mate-guarding by the male is one strategy for minimizing the risk of being cuckolded. Hegner and Wingfield (1986a) found that mate-guarding by males in New York (USA) peaked during the prebreeding and egg-laying periods but was low during the incubation and nestling periods. Hankinson (1999) studied mate-guarding in a population in Oklahoma in which the rate of extra-pair paternity was about 16%. She defined mate-guarding operationally as the presence of the male within 10 m of the female and then recorded time spent mate-guarding during a 1–h period that included the time of egg-laying and during the hour beginning 1 h after the end of the earlier period. The male spent significantly more time mate-guarding during the egg-laying period than during the period beginning 1 h later ($P = .002$), but there was no significant difference in mate-guarding between two comparable periods during incubation. Hankinson concluded that males mate-guarded during a "fertilization window" that occurs at the time of egg-laying, before the presence of a developing egg in the oviduct, which would effectively block subsequent fertilization.

Mate-guarding is apparently successful in reducing the incidence of extra-pair fertilization. Vaclav et al. (2003) provided supplemental food to sparrow pairs in Austria and found that pairs with additional food spent significantly more time together at the nest than did control pairs without additional food. The extra-pair fertilization rate in food-supplemented pairs (0.08 chicks per brood) was significantly lower than that of control pairs (0.33 chicks per brood) ($P = .032$).

Intraspecific Brood Parasitism

Although there is some evidence of intraspecific brood parasitism in sparrows, several studies using DNA fingerprinting techniques have indicated that it occurs only rarely in the species. Using protein electrophoresis on variable albumen proteins in sparrow eggs, Manwell and Baker (1975) found two eggs among 32 clutches in Australia that differed from other eggs in the clutch and concluded that these eggs had been laid by a female other than the pair female. Similarly, Kendra et al. (1988) observed five cases of a nonmatching albumen protein (transferrin) in 41 clutches in Delaware (USA). Among several recent DNA fingerprinting studies discussed earlier, however, only two found instances of offspring that failed to match the pair female (Vaclav et al. 2003; Veiga and Boto 2000). The results of an experimental study of egg recognition and egg rejection in a captive breeding population suggest that female sparrows may recognize and attempt to eject foreign eggs (Moreno-Rueda and Soler 2001).

Egg-dumping by female sparrows may also lead occasionally to interspecific brood parasitism. A cliff swallow (*Petrochelodon albifrons*) nest in New York that contained one sparrow and three swallow eggs resulted in the successful fledging of the sparrow (Stoner 1939).

Division of Labor

There is much overlap in the contributions of males and females to breeding, but there are also some major differences. Both participate in nest building, in defense of the nest and its contents, and in feeding the nestlings. The male selects a potential nest site and advertises both his willingness to defend the site against other males and his availability for mating by displaying and chirping vigorously at the site. These nest site acquisition and advertisement behaviors may begin during the autumnal period of sexual activity (see Chapter 3), and they often continue during fair weather throughout the winter. The male often builds a rudimentary nest at the site and roosts at the site during the winter. When the male is paired, both members of the pair may roost at the site (Summers-Smith 1958). Both members of the pair participate in the building of the functional nest, usually shortly before egg-laying begins, although the female does the majority of the actual nest construction. In one study in Austria, however, the male brought nesting material to the nest an average of twice as often as the female during the 5 d preceding the onset of egg-laying ($P = .003$), although variation among males was great (range = 0%–100%) (Hoi et al. 2003). Both sexes continue to bring lining material such as feathers and grass throughout the egg-laying and incubation periods (North 1980).

Incubation is performed only by the female, because only she develops a brood patch (see Chapter 5). The female also broods the young during the first week of the nestling period. The male usually covers the eggs or small young when the female is absent from the nest, and this may perform the dual functions of retarding heat loss and protecting the eggs or young from predation or from destruction by other sparrows prospecting for nest sites. The male also spends about 62% of his perching time at the nest entrance during incubation

(Johnston 1965), presumably to defend against intrusion by other sparrows prospecting for nest sites and against potential nest predators. The male generally contributes about half of the feedings to the nestlings during the first 10 d of the nestling period, but feeding frequency drops significantly during the final third of the nestling period (Hegner and Wingfield 1986a; North 1980; Seel 1969). Nest sanitation is performed by both sexes, with both parents often eating the fecal sacs during the first days of the nestling period (Seel 1960, 1966). Assimilation efficiency of food during the first 2 d of nestling life in Poland was only about 60%, whereas by day 5 it had increased to 90% (Myrcha et al. 1972). The activities of both intestinal and pancreatic enzymes increased significantly between hatch day and day 3 or day 6 in nestling sparrows in Wisconsin (USA) (Caviedes-Vidal and Karasov 2001). This means that fecal sacs during the first few days of nestling life have food value that can be utilized by the adults. Females have been observed stimulating defecation by grasping the enlarged cloacal regions of nestlings in their beaks (Seel 1966).

Courtship feeding by the male occurs infrequently and probably represents an inconsequential contribution by the male to the breeding attempt. Summers-Smith (1955) reported that he had occasionally observed courtship feeding at the time of copulation and during the incubation and early nestling periods.

Recent theory has suggested that a female mated to a high-quality male should allocate more reproductive effort to her reproductive attempt than when she is mated to a low-quality male; this is termed the Differential Allocation Hypothesis (see Mazuc, Chastel, and Sorci 2003). Three studies have found little evidence to support this hypothesis in sparrows. In the first, conducted in Oklahoma, nestling feeding rates of females were not affected by experimental handicapping of their mates (by attaching lead weights at the base of their tail feathers) (Schwagmeyer et al. 2002). In fact, females showed a high repeatability of provisioning rate (0.375) regardless of the status of their mate (handicapped or control), even if they changed mates after the initial experiment (Schwagmeyer and Mock 2003). In the second study, an explicit test of the Differential Allocation Hypothesis conducted in France, no differences in female reproductive effort (represented by clutch size, yolk size, yolk testosterone, brooding time, and nestling provisioning rate) were found based on either badge size (see Chapter 5) or testosterone level of the pair male (Mazuc, Chastel, and Sorci 2003). The third study, in Austria, examined the contributions of males to successive stages of the breeding cycle and found that early contributions of males (nest building, nest guarding) were not related to their contributions later in the breeding cycle (brooding, provisioning) (Hoi et al. 2003).

Cooperative Breeding

One study, conducted in Mississippi (USA), reported a high incidence of helpers-at-the-nest during the nestling period (Sappington 1977b). Sappington reported that 63.4% of 254 nests under observation had one or more non-pair helpers. The helpers provided 12.4% of the total feeds at these nests, which accounted for most of the difference in average provisioning rate between nests with helpers (254.1 feeds/d) and those without (206.6 feeds/d). Although the presence of non-pair birds at a nest during the nestling period has been observed in at least one other study (Berger 1957), and a female sparrow was even observed feeding nestlings of the western kingbird (*Tyrannus verticalis*) in Kansas (Gress 1985), no other study has reported helping behavior as being common in the species. Further documentation of the behavior should be sought. It is possible that it is common in populations with large numbers of floating individuals capable of breeding (see Anderson 1990) and that it serves as a strategy for obtaining a mate and nest site, with the helpers replacing the pair female after the fledging of the brood.

Timing of Breeding

Timing of reproduction in the house sparrow varies geographically, with two major patterns being evident, one for birds breeding at temperate latitudes and the other for birds breeding in subtropical and tropical areas. Resident birds at temperate latitudes have a single breeding season that extends from early or middle spring through late summer at most localities. Migratory populations breeding at temperate latitudes also have a single breeding season that lasts from middle or late spring to midsummer (e.g., Stephan 1982). In some tropical or subtropical areas, clutch initiation has been reported in every month of the year (Naik and Mistry 1980, Nhlane 2000). These populations also show marked seasonality, however, with peaks of clutch initiation during two periods, one preceding and one after the summer rainy season (Naik and Mistry 1980; Nhlane 2000; Rana and Idris 1989). A similar pattern was observed in the lowlands of El Salvador (Thurber 1986), a particularly intriguing finding because the Central American sparrows are members of the nominate subspecies, *Passer domesticus domesticus*, and are descendants of birds introduced into North America from Europe (see Chapter 1). Chapter 3 deals at length with the proximate mechanisms controlling the timing of reproduction; this section focuses on patterns of variation among sparrow populations.

The beginning of the breeding season is negatively correlated with latitude in resident sparrows breeding at temperate latitudes, at least for populations occupying regions with continental climates. Dyer et al. (1977) examined this relationship based on 30 populations, including one subtropical population from India and one temperate-zone, maritime population from England, and found a highly significant relationship between the time of initiation of the first clutch in a population (T) and latitude (L): $T = 2.48 + 1.94L$ ($r = 0.882$). Anderson (1994) found a similar relationship for initiation of both first and second clutches in continental North America. The two studies illustrate one difficulty associated with identifying the beginning of the breeding season. The first study marked it by the initiation of egg-laying in the first clutch in a population, whereas the second used measures of central tendency (mean, median, or mode) of initiation of all first and second clutches in each population. Although the first method is subject to potential error due to outliers, Dyer et al. (1977) reported that the correlation between date of initiation of first clutch and average date of initiation of first clutches for 14 y in one population in Poland was highly significant ($P < .01$). The regression coefficients from both studies indicate that the breeding season of sparrows is retarded by about 2 d for each degree of latitude. Interestingly, this retardation rate is only half that observed in plants and insects (Hopkins 1938), which suggests that one advantage of endothermy is some degree of freedom from the constraints imposed on ectothermic organisms by ambient temperature. An alternative explanation is that the difference is due to the mitigating effects of the house sparrow's close commensal association with humans (see Chapter 10).

Il'Enko (1958) studied the timing of gonadal development in both male and female sparrows during two springs in Moscow (Russia). The timing of testes enlargement showed a close relationship to both ambient temperature and amount of sunshine, with each acting independently. Ovarian development was retarded compared to testicular development and was more closely associated with ambient temperature. During a short cold spell in April, developing follicles were resorbed in some females. Seel (1968a) examined the effects of ambient temperature, rainfall, and hours of sunlight on the timing of first clutches during 3 y in England and concluded that only temperature showed a possible relationship with the initiation of breeding. He found a significant relationship between ambient temperature and initiation of early clutches in all 3 y, with the temperatures of 4–6 consecutive days centered 4 d before clutch initiation having the strongest effect. The mean ambient daily temperature at the start of the breeding season, which he arbitrarily defined as the date on which the third clutch in the population was initiated, was 10°C. Similar values have been found in other sparrow populations. Ion and Ion (1978) reported that the mean temperature for the 2 wk preceding the initiation of breeding in Romania was 12°C, and Pinowska (1979) found that the average temperatures in the week preceding initiation of breeding in 2 y in Poland were 7.0°C and 8.5°C. Seel (1968a) noted that sparrows in

England initiated breeding at temperatures in April that did not result in initiation in March and suggested that there is a declining temperature threshold for the initiation of breeding. Similar results were obtained in a 14-y study in Poland. In that study, the date of initiation of first clutches was positively correlated with photoperiod for clutches initiated after weeks in which the average temperature was lower than 8°C, but it was not correlated with photoperiod if average temperatures were 8°C or higher (Pinowska and Pinowski 1977).

Temperature may also play a role in determining the initiation of breeding in subtropical populations. Naik and Mistry (1980) reported that an increase in daytime temperatures to about 24°C preceded the initiation of breeding in India. In 1972 the breeding season began late, however, and the average temperature for the 5 d preceding the initiation of breeding in late February was 20°C, which again suggests that there is a declining temperature threshold for the initiation of breeding.

In a study in Illinois (USA), Will (1969) found that sparrows with nest sites in a heated building initiated egg-laying earlier than did birds nesting in nearby unheated buildings. The mean initiation date in the heated building was 28 March in 1967 and 5 April in 1968, whereas in unheated buildings the mean dates were 6 April and 22 April, respectively. Because females often roost at their nest sites, the energy conserved by roosting in the heated building presumably enabled females to develop the energetic reserves required for initiating a clutch earlier than females nesting in unheated buildings.

Transplant experiments have also demonstrated the role of proximate cues in addition to photoperiod in determining the timing of breeding in sparrows. Krogstad et al. (1996) transplanted sparrows from three mainland populations in Norway (one coastal and two inland populations) to a nearby offshore island. The timing of both first and second clutches of the transplanted sparrows from the inland populations was significantly retarded compared with that in the source populations ($P < .001$ and $P < .05$), whereas the timing for transplanted coastal birds did not differ significantly from that of the source population. Krogstad et al. conclude that transplanted females were able to adjust their timing of breeding to local conditions.

Nest Sites and Nests

Nest Sites

The house sparrow is primarily a secondary cavity nester, nesting most commonly in holes or crevices in buildings or other human structures, holes in trees, and nest-boxes. It also often builds a domed nest in the branches of trees, frequently conifers, or in thick ivy covering the exterior wall of a building. It also usurps the nests of other species, and it sometimes builds nests in the lower part of the nests of birds of prey or other large birds. Occasionally it excavates its own nest cavity in rotted wood (e.g., Philipson 1938) or in an earthen bank (i. e., Penhallurick 1993; Pitman 1961). The latter site is common in Turkmenistan, where both sexes participate in excavating holes in loess banks (Ivanitzky 1996). The house sparrow also occasionally nests in holes in rocky cliffs (Schmidt 1966). Some truly unusual nest sites have also been observed. One pair built a nest in the leg of a pair of pants hanging in a shed in India (Sharma 1995), and in Kansas several sparrows built nests in the constantly moving heads of oil pumps in an oil field (Tatschl 1968).

Sparrows tend to nest in loose colonies, virtually always in close association with humans, with sites usually located within 400 m of the nearest human dwelling (Łącki 1962; Lumsden 1989). In one suburban area of about 5 km^2 in England, there were about 25 such breeding colonies, with nests located on only about 25% of the houses (Summers-Smith 1954a). The colonies had central nest sites, which were generally all located within 10 m of each other, and peripheral sites located within about 25 m of the colony center. Central sites were preferred, and Summers-Smith (1958) noted that, although peripheral sites were virtually always available in the colonies, males that chose those sites often remained unmated.

Because the preferred nest sites are usually located in crevices on buildings or other human structures, the specific nest site preferences at a particular location are dependent on the types of structures available. Kulczycki and Mazur-Gierasinska (1968) studied the location of 271 nests in three habitats (rural, suburban, and urban) in southern Poland and identified five major types of sites. Over all habitats, the most common sites were on rafters in buildings (28.8%); in vines on the exterior walls of buildings (26.5%); and in holes or crevices of buildings, trees, rocks, or nest-boxes (22.5%). Nest height varied from 1.5 to 13 m, with 74.9% falling in the 3–7 m range. Indykiewicz (1991) studied 1232 nests from the same three habitat types, also in Poland, and described 19 different types of sites. In urban environments, the three characteristic nest sites, in order of preference, were (1) on clips holding gutter spouts to houses, (2) in ventilation holes under flat roofs of houses, and (3) inside street lights. In the suburban environment, the characteristic sites were (1) in recesses between the wall and roof of houses and (2) in hollows in closed wooden cornices; in the rural environment, they were (1) on the clips holding gutters to houses and (2) in recesses between the wall and roof of houses. Nest height varied from 2.05 to 32.2 m, with an overall average of 5.7 m and the highest percentage (44.4%) being in the range of 3.0–4.9 m. Nests were placed higher in the urban habitat than in either the suburban or the rural habitat. Among about 1000 nest sites in rural and suburban habitats in eastern Spain, almost 50% of the sites were located in cavities under roof tiles, about 20% in cavities in trees, and about 15% in cavities in walls (Cordero and Rodriguez-Teijeiro 1990).

Indykiewicz (1990, 1991) examined the directional orientation of nest sites in Poland and concluded that sparrows prefer eastward-facing, spacious niches in wood, 3–5 m above the ground. McGillivray (1981) found that the directional orientation of open tree nests in Alberta (Canada) depended on the season, with early nests tending to be oriented southward and late nests northward. McGillivray attributed this tendency to the potential deleterious effects of cold northerly winds early in the breeding season, whereas, later in the season, cool northerly breezes could actually help to prevent nest overheating. In eastern Spain, Cordero and Rodriguez-Teijeiro (1988) reported that the directional orientation of open tree nests was random.

One particularly interesting site used by sparrows is the interior of occupied nest platforms of birds of prey or other large birds (Table 4.1). The frequency and wide geographical range over which this behavior has been observed imply that selection of such sites has adaptive significance. Several authors have suggested that use of these sites confers an advantage by reducing the risks of nest predation by reptiles, birds, or mammals, which would be driven from the nest area by the larger birds (Ewins et al. 1994; McGillivray 1978; Petretti 1991). Petretti (1991) suggested two additional hypotheses to explain nesting by sparrows in nests of birds of prey: (1) they offer appropriate nest site structure, and (2) they provide an opportunity for foraging by sparrows (on food scraps and detritivorous insects). Petretti studied the distribution of sparrow nests among seven black kite (*Milvus migrans*) nests in central Italy. Four pairs of kites nested each year at the site, and over the 3 y of the study there were 12 active nests and 7 inactive nests. Sparrows nested only in active nests, except for one nest located in an inactive nest that was used as a feeding platform by the kites ($P = .002$). Sparrows were observed feeding on the nest platforms. Although these results effectively negate the nest site structure hypothesis, they do not permit discrimination between the other two hypotheses.

Sparrows also frequently usurp the nests of other species and construct their nests either in or on the usurped nests. It is not always clear whether the sparrows took over the nest while it was being constructed or used by the other species or after the other species had completed their nesting cycle. The instances listed here are in addition to the instances of competition for nest cavities with other secondary cavity nesters, which are discussed in Chapter 8. Some species for which nest usurpation has been recorded are rufous hornero (*Furnarius rufus*) in Argentina (Burger 1976; Fraga 1980); cliff swallow in North America (Buss 1942); tree martin (*Hylochelidon nigricans*), fairy martin (*Hylochelidon ariel*), and pardalote (*Pardalotus* sp.) in Australia (Favaloro 1942); barn swallow

TABLE 4.1. Species of Birds of Prey and Other Large Birds in Whose Nests House Sparrows Have Been Reported to Nest

Subspecies	Host Species	Location	Source
P. d. domesticus	White stork, *Ciconia ciconia*	Poland	1
	Osprey, *Pandion haliaetus*	USA, Germany	2,3
	Mississippi kite, *Ictinia mississippiensis*	USA	4
	Common buzzard, *Buteo buteo*	Germany	3
	Red-tailed hawk, *Buteo jamaicensis*	USA	5
	Swainson's hawk, *Buteo swainsoni*	USA, Canada	4,6
	Sea eagle, *Haliaeetas* sp.	Germany	3
	Rook, *Corvus frugilegus*	UK	7
P. d. italiae	Back kite, *Milvus migrans*	Italy	8
P. d. indicus	Common pariah kite, *Milvus migrans govada*	India	9

Sources: 1, Indykiewicz (1991); 2, Ewins et al. (1994); 3, Nicolai (1971); 4, Parker (1982); 5, Wilson and Grigsby (1979); 6, McGillivray (1978); 7, Witherby et al. (1943); 8, Petretti (1991); 9, Waghray and Taher (1993).

(*Hirundo rusticus*) in North America (Werler and Franks 1975) and Europe (Nankinov 1984); house martin (*Delichon urbica*) in Europe (Balat 1973; Indykiewicz 1991; Kulczycki and Mazur-Gierasinska 1968; Nankinov 1984; Witherby et al. 1943) and Asia (Sahin 1996); American robin (*Turdus migratorius*) in North America (Werler and Franks 1975); fieldfare (*Turdus pilaris*) and chaffinch (*Fringilla coelebs*) in Poland (Kulczycki and Mazur-Gierasinska 1968); and song thrush (*Turdus philomelos*) and greenfinch (*Carduelis chloris*) in Great Britain (Summers-Smith 1963).

Cink (1976) tested for the effect of early experience on nest site selection in two colonies of sparrows breeding in Kansas in 1973. He transferred 150 nestlings (usually <1 wk old) between nests of three different types (nest-boxes, crevices, and open branches of trees), using 150 nestlings that were not transferred as controls. The nest-site choices of 83 of the birds (37 experimentals and 46 controls) were observed in 1974, and there was no significant tendency for birds to choose their natal nest-site type. Males showed no apparent preference for any of the three types, but females appeared to show a preference for cavity nest sites (nest-boxes and crevices) rather than open sites ($P = .054$). Cink concluded that early experience does not affect nest-site choice in sparrows, but that females show an innate preference for cavity sites.

McGillivray (1981) compared breeding success in open nests located in blue spruce (*Picea pungens*) trees with success in nest-box nests in Alberta. Clutch size, egg mass, and fledging mass did not differ between the two types of sites, but number of breeding attempts and total number of eggs laid per nest site were both greater in nest-boxes than in the open nests. The differences were due primarily to the fact that first clutches in nest-box nests were initiated earlier than those in open nests, with the first clutches in open nests actually coinciding with the initiation of second clutches in the nest-boxes. McGillivray (1983) also reported that the interbrood interval (the period between fledging of the last chick in a brood and laying of the first egg in the next brood) was significantly longer in open nests than in nest-boxes and attributed the difference to greater energetic costs (presumably lower insulative quality) associated with open nests.

Territorial defense is usually confined to the immediate vicinity of the nest site. Summers-Smith (1958) reported that each sex defends the nest site against conspecifics of the same sex, and that both defend it against other species that are potential competitors for the nest site, such as tits (*Parus* spp.) and the spotted flycatcher (*Musicapa striata*) (see also Chapter 8). Males in Illinois and Wisconsin defended an area around the nest site that ranged from about 0.5 m (18 in) to 6 m (20 ft), but the size of the defended area could vary with the position of the female (Owen 1957). Nest sites were an average of 3.6 m apart in a dense breeding colony in the Czech Republic (Novotny 1970). Females living in dense colonies in New York, in which the mean distance between nest sites was about 5 m, deposited significantly more testosterone in the yolks of their eggs than did females living in a more dispersed colony (mean distance between nest sites, about 15 m) (Schwabel 1997). Schwabel attributed this difference to differences in the intensity of territorial defense by females in dense and dispersed colonies, with consequent effects on the females' levels of circulating testosterone. In France, plasma concentration of testosterone in females was positively correlated with density as determined by the number of occupied neighboring nest-boxes (Mazuc, Bonneaud et al. 2003). The latter study found, however, that the residual of the yolk testosterone level was negatively correlated with female plasma testosterone.

In some situations, nests are tightly clumped with entrances only a few centimeters apart, and sometimes compound structures actually contain two or more nests (Jackson and Schardien Jackson 1985; McGillivray 1980a; Menon and Pilo 1983). McGillivray (1980b) examined the effects of nest clumping on the reproductive success of sparrows nesting in rows of blue spruce trees in Alberta. As many as 14 nests were located in a single tree, and in one case 10 nests were located within a spherical volume of 1 m radius. In many cases, the nests shared walls or were coalesced into compound structures that did not show individual boundaries. The mean nearest-neighbor distance in one row of spruce was 0.66 m, with an average of 3.43 nests within 1 m; in the other row, the mean nearest-neighbor distance was 1.52 m, with 0.76 nests within 1 m (both differences were significant, $P < .01$). The number of clutches per season was significantly greater in dispersed nests than in clumped nests, and the number of hatched young per season was also significantly greater. In general, there tended to be greater seasonal success at dispersed nests, but clumped nests produced a larger number of fledglings per attempt than dispersed nests did. Parental efficiency increased with increased nest clumping and decreased with number of attempts per season. Nesting synchrony also tended to increase with increased nest clumping. McGillivray suggested that nest clumping may be more prevalent in areas with cold winters, because compound nests may have greater insulative qualities.

Sparrows show considerable nest site fidelity, as there is a tendency for individuals to remain at the same nest site throughout a breeding season and to return to the same site in successive years (Summers-Smith 1963). Dawson (1972b) reported that, at 51 nests in England at which the females were caught on successive broods, the same female was present in 72% of the instances. In the same population, he also reported that 76% of the males were the same at 25 nests at which he captured males on successive broods. Although the presence of different birds in successive broods at the same nest site could be due to the replacement of individuals that have terminated breeding or died, there is also evidence that individuals do change sites between successive broods. Anderson (1998) found that, among 50 females that initiated late-season broods after successful midseason broods in Michigan (USA), 30% moved to a different nest site. The distance between successive sites varied from 2 to 89 m, but most birds moved to a site less than 10 m from their previous site, and the median distance moved was 5.5 m. In Russia, Il'Yenko (1965) found that individuals from pairs that separated, usually after a nest failure, moved greater distances to renest than did pairs that remained intact. Some of these movements were between breeding colonies. Sappington (1977) reported that nest site fidelity was greater in males than in females in Mississippi, with 86.0% of males retaining a nest site throughout the breeding season, compared to 45.0% of females.

Nest Structure

The nest located in the open branches of trees is thought to be the primitive nest type in the house sparrow (Heij 1986; Kulczycki and Mazur-Gierasinska 1968). The typical tree nest is a more-or-less spherical structure composed primarily of dry grass, straw, and some twigs. The entrance into the nestcup is usually located on the side, but infrequently it is near the top or even below the main body of the nest (McGillivray 1981). The nestcup is lined with feathers, hair, and/or other soft material, either natural or synthetic (Heij 1986; Kulczycki and Mazur-Gierasinska 1968), which contribute to the insulative quality of the nest. Nests located in crevices or cavities resemble tree nests in general structure but often have a reduced roof, and in some cases no roof.

Nest structure appears to be dictated by the size and shape of the cavity (Heij 1986).

O'Connor (1975d) demonstrated the insulative effect of the nest in the laboratory. An 8-d-old nestling expended 36% less energy when held in the nest than when held at the same temperature (20°C) without the nest. Ivanov (1987) reported that sparrows in Bulgaria regularly removed insulation from the nest as nestlings grew, presumably because nestlings required less insulation with increasing age.

Nests sometimes contain fresh green leaves or sprigs of plants producing potent defensive secondary compounds. In Slovakia, sparrows brought green sprigs, primarily *Artemisia absynthum*, to a nest almost daily during the early nestling period (Turcek 1972). Seven nests in Tennessee (USA) contained sprigs of wild carrot (*Daucus carrota*) (Pitts 1979). In India, green leaves of the margosa tree (*Azadirachta indica*) were found in several nests (Sengupta 1981). Both sexes brought margosa leaves to the nest, and when the margosa leaves were removed twice daily from two nests, egg-laying was delayed in both nests. Sengupta reported that Indians have long used margosa leaves as an insect repellant in their clothes and suggested that the adaptive significance of their use in sparrow nests was to discourage insects and parasites from occupying the nest. Sengupta and Shrilata (1997) reported that all of 13 nests examined in Calcutta, India, in September and October 1994 were lined with green leaves from the Krishnachura tree (*Caesalpinia pulcherrima*), and that four of the nests also contained margosa leaves. There was a local outbreak of malarial fever during the period, and the leaves of the Krishnachura tree contain quinone, which can act as an antidote to malaria. Pitts (1979) suggested that the wild carrot leaves might have a chemical effect similar to anting (see Chapter 5).

Such defenses may be effective in reducing the negative effects of the many mites and ticks that infest sparrow nests. Bhattacharyya (1995), for instance, identified 70 species of mites and ticks belonging to 40 different families in sparrow nests in West Bengal (India) (see also Chapter 8). Six species of predaceous pseudoscorpions were also found in sparrow nests in West Bengal (Bhattacharyya 1990), and five species of beetles occurred commonly in nests in Slovakia, including two carnivorous and three detritivorous species (Sustek and Kristofik 2003). This suggests the presence of a well-developed food chain in sparrow nests.

Nest dimensions vary considerably, particularly in cavity nest sites, where the shape and size of the nest depend on the dimensions of the cavity. Kulczycki and Mazur-Gierasinska (1968) reported that about 25 nests from Poland had an average outer diameter of 21.3 cm and an average height of 21.9 cm; nestcup diameter averaged 8.9 cm, and nestcup depth, 6.4 cm. Indykiewicz (1991) measured 518 nests, also in Poland, and reported that the mean length was 25.1 cm; mean width, 13.3 cm; mean height, 14.1 cm; and mean depth of the nestcup, 4.7 cm. Heij (1986) weighed 145 nests in the Netherlands from urban and suburban habitats. Nest mass varied from 20 to 500 g, but the average masses in the two habitats were very similar (urban, 115 g; suburban, 118 g). The nests from both habitats combined were composed of 77.7% vegetable material (primarily straw), 12.8% animal material (mainly feathers), and 9.5% artificial material (including string, paper, wool, and sundry other substances). The materials most frequently used in Poland were hay (in 95% of nests) and straw (in 89.4%), and feathers were found in the nestcups of 96.7% of the 96 nests examined (Kulczycki and Mazur-Gierasinska 1968).

Sparrows frequently engage in both intraspecific and interspecific kleptoparasitism while gathering the prodigious quantities of material needed for their bulky nests (Jackson and Schardien Jackson 1985; McGillivray 1980a; Suffern 1951). Sparrows breeding in close proximity to each other regularly stole nest material from nests in nearby trees, but not from other nests in the same tree (McGillivray 1980a). Another form of kleptoparasitism of nest material involves the plucking of feathers from living birds to use as lining material. In New Zealand, a male was observed to repeatedly pluck feathers from above the tail of a Barbary dove (*Streptopelia risoria*) (Sitdolphi 1974a), and several sparrows (both male and female, but mostly female) were observed removing contour feathers at a rate of 6–7 feathers/h from the rump region of an incubating rock dove (*Columba livia*) while standing on the back of the bird (Bell 1994). Sparrows have also been observed chasing flying rock doves and turtle doves (*Streptopelia decaocto*) and attempting to pluck feathers from the breast region (Bell 1994; Heij 1986).

Roost Nests

Sparrows often roost at their nest sites throughout the winter, and in some cases they build nests specifically as roost nests in sites that are too small for nesting (Mayes 1927, Summers-Smith 1963). Janssen (1983) described two such nests built by females in September in Minnesota (USA) that were used for roosting until they were blown down by a windstorm in November.

Egg-Laying and Clutch Size

Clutch size is one of the fundamental elements of the reproductive strategy of a species, because it represents the number of offspring that an individual (or pair) endeavors to produce in a reproductive attempt. Since Lack's (1947) pioneering paper on the significance of clutch size in birds, it has been more-or-less agreed that individuals should attempt to maximize the number of offspring that they can produce in a single attempt, unless there are decreases in future reproductive potential due to the effort associated with raising the maximum number of young possible ("costs of reproduction") (Williams 1966). For altricial species such as the house sparrow, Lack (1947) suggested that the primary factor limiting the number of young that could be successfully reared was the amount of food that the parents could provide to the rapidly developing chicks. Clutch size in the house sparrow varies with several factors, however, and reconciling these variations with their adaptive value relative to the food supply available for provisioning the young has not usually been possible. The most common clutch sizes are from three to six eggs in most populations, but rarely clutches contain only a single egg (Naik and Mistry 1972; Will 1969) or as many as nine (Gil-Delgado et al. 1979). Relatively little is known about the proximate mechanisms involved in determining clutch size in a particular set of circumstances, and therefore one can only guess at the potential effects of physiological constraints on clutch size.

Egg-Laying

Sparrows typically lay their eggs shortly after sunrise, but determining the precise timing of egg-laying is difficult. Seel (1968b) used the midpoint between the time a female entered the nest and the time she left after laying an egg to estimate the time of egg-laying and found that the time varied from 35 to 97 min after sunrise (mean = 55.9 min). Anderson (1997) used thermocouples to monitor nestcup temperatures during egg-laying in six nests in Michigan and estimated the timing of egg-laying using a spike in nestcup temperature shortly after sunrise. Egg-laying occurred about 45 min after sunrise in the six nests (mean = 44.9 min).

Schifferli (1979) proposed that eggs are laid at dawn because shell development takes place during the night to avoid damage to the developing shell that could be caused by normal daytime activities (see later discussion). Laying of the egg shortly after dawn allows the female to have the entire day to forage without risk of damage to a developing shell.

Seasonal Variation in Clutch Size

Clutch size varies seasonally in a predictable fashion in most temperate-latitude populations where it has been studied (Anderson 1978; Brichetti 1992; Novotny 1970; Seel 1968b; Vaclav and Hoi 2002b; Veiga 1993b; Will 1969). Fig. 4.2 illustrates the change in mean clutch size with season in Michigan. There is an increase in clutch size to a midseason maximum, followed by a

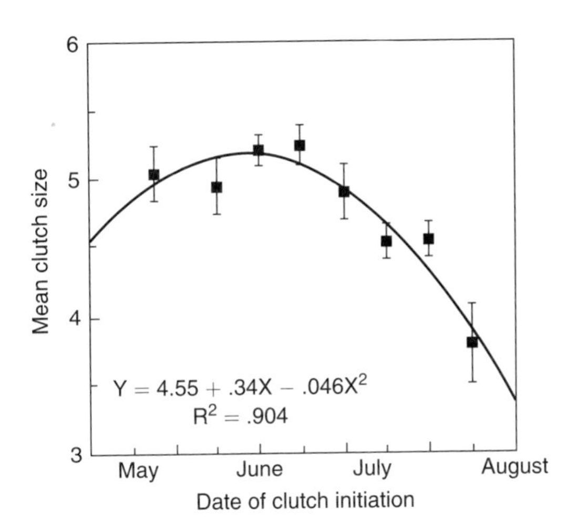

FIGURE 4.2. Seasonal change in clutch size of the house sparrow. Data are represented as means (±SE) for 10-d intervals, with interval 1 = 11–20 May, and are based on 340 clutches from 1986–1991 in Michigan (USA). From Anderson (1994: Fig. 2), reproduced with permission of *Wilson Bulletin*.

decline that is particularly pronounced for late-season clutches. Seel (1968b) noted that this seasonal pattern of change in clutch size corresponds to the seasonal change in daylength in England, with the peak in clutch size preceding the peak in daylength by about 3 wk. The timing of the largest number of chicks requiring provisioning by the parents would therefore coincide with the longest days of the year, which suggests that the seasonal pattern of change in clutch size may be related to the length of time available for the adults to search for food (see Lack 1947). One exception to the general pattern of seasonal clutch size variation was reported on an archipelago off the coast of northern Norway, where the second clutch was significantly smaller than the first and third clutches (Ringsby et al. 2002).

The late-season decline in clutch size observed in most populations is due to individuals laying fewer eggs than in their earlier clutches. In Michigan, the average clutch size of 47 marked females that produced late-season clutches after successful fledging of a midseason brood was 0.79 egg lower than that of their midseason clutch ($P < .001$) (Anderson 1998). First-year females lay smaller clutches than older females (see later discussion), and they also begin laying later than older females do (Dawson 1972b; Seel 1968b). Dawson (1972b) reported that first-year females in England initiated their first clutches of the season an average of 7.6 d later than 2-y-old or older females did ($P < .01$). If the late-season clutches are laid by first-year females, then the decline in clutch size might be due to age rather than season. In the study in Michigan, however, 2-y-old and older females were more likely to produce a late-season clutch than younger birds were (Anderson 1998).

Clutch size in tropical and subtropical resident populations of sparrows also varies seasonally, with peaks in clutch size coinciding with both of the main breeding periods, before and at the end of the rainy season (Naik and Mistry 1972; Rana and Idris 1989). Fig. 4.3 shows the pattern of monthly change in mean clutch size in two subtropical populations in India. This pattern is obviously not related to daylength, because the longest days of the year occur during the rainy season, when most birds have interrupted breeding, and the largest clutches, at least at Baroda, are initiated in the period from mid-August to mid-October, when daylength is shortening.

Geographic Variation

The nearly pan-global distribution of sparrows, and the many studies that have been conducted on their reproductive biology mean that there are probably more data on geographic variation in the reproductive parameters of the house sparrow than of any other avian species. Table 4.2 contains clutch size data from many of these studies, some of which are comprehensive, involving data from hundreds of clutches collected throughout several breeding seasons; others report data from only a small number of clutches, often collected during only part of an extended breeding season. The latter data may not accurately represent the average clutch size in the area, because of the vagaries of small

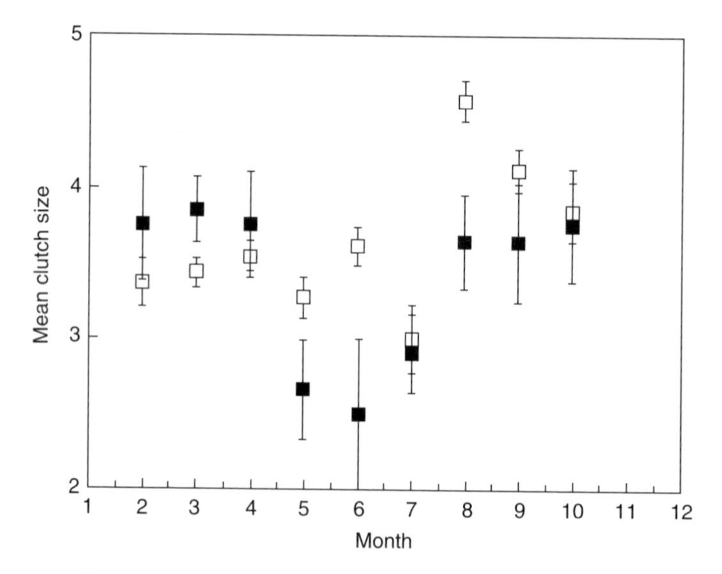

FIGURE 4.3. Seasonal change in clutch size of the house sparrow *(P. d. indicus)* in two populations in India. Values plotted are monthly means (± SE) for 1360 clutches from Baroda (□) (22°18'N) for 1969 (Naik and Mistry 1972) and 94 clutches from Jodhpur (■) (26°18'N) in 1985 (Rana and Idris 1989).

TABLE 4.2. Clutch Size of the House Sparrow

Location (latitude)	Years	N	Clutch Size: Range	Mode	Mean (SD)	Source
Natural Range						
P. d. domesticus						
Spain (39°41'N)	2	328	2–9	5	4.98 (1.17*)	1
Spain (39°41'N)	1	108	n.a.	n.a.	4.92* (1.08*)	2
Spain (40°38'N)	2	73	n.a.	n.a.	4.89 (1.06)	3
Bulgaria (42°41'N)	2	115	n.a.	5	4.94* (0.74*)	4
Bulgaria (42°50'N)	4	352	n.a.	5	4.76* (0.84*)	4
Kazakhstan (43°16'N)	n.a.	23	3–6	5	4.61* (0.89*)	5
France (46°09'N)	1	30	n.a.	n.a.	4.80* (n.a.)	6
Romania (46°05'N)	2	29*	n.a.	n.a.	4.66* (n.a.)	7
Romania (47°10'N)	2	75*	n.a.	n.a.	4.37* (n.a.)	7
Romania (47°30'N)	2	27*	n.a.	n.a.	4.70* (n.a.)	7
Austria (48°12'N)	1	~82	n.a.	n.a	4.6 (n.a.)	8
Czech Republic (49°19'N)	3	221	2–6	5	4.57 (0.92)	9
Czech Republic (50°13'N)	3	439	2–7	5	4.33 (0.745)	10
Poland (49°18'N)	1	9	n.a.	n.a.	4.90 (1.5)	11
Poland (49°28'N)	2	14	n.a.	n.a.	4.50* (1.69*)	11
Poland (49°28'N)	2	19	2–6	5	4.37* (1.16*)	12
Poland (50°04'N)	2	45	2–6	5	4.60* (0.78*)	11
Poland (50°04'N)	2	60	2–6	4	4.42* (0.72*)	12
Poland (52°15'N)	5	222	n.a.	n.a.	4.51* (1.3)	13
Poland (52°20'N)	14	1168	n.a.	n.a.	4.68* (1.50*)	14
Poland (52°21'N)	2	71	2–7	5	4.56* (1.04*)	11
Poland (52°25'N)	1	119	n.a.	n.a.	4.33* (n.a.)	15
Poland (52°25'N)	2	286	n.a.	n.a.	4.55* (1.0)	16
Poland (54°28'N)	2	31	2–6	5	4.39* (0.83*)	12
Poland (54°37'N)	1	137	n.a.	n.a.	4.47* (n.a.)	15
Poland (54°37'N)	2	111	n.a.	n.a.	4.53* (1.1)	16
UK (51°45'N)	3	831	2–7	4	3.98 (0.83)	17
UK (51°45'N)	3	630	n.a.	n.a.	4.19 (0.76*)	18
UK (52°57'N)	2	n.a.	n.a.	n.a.	4.05* (n.a.)	19
UK (53°22'N)	5	14	n.a.	n.a.	3.79* (n.a.)	20
Germany (52°31'N)	1	83	4–7	5	4.90* (0.74*)	21
Germany (53°50'N)	1	93	4–7	5–6	5.32* (0.71*)	21
Germany (53°50'N)	5	152	n.a.	n.a.	4.59* (n.a.)	22
Denmark (57°12'N)	7	61	n.a.	n.a.	4.80* (n.a.)	23
Norway (62°30'N)	2	42	n.a.	n.a.	5.14* (n.a.)	24
Norway (63°47'N)	2	44	n.a.	n.a.	5.05* (n.a.)	24
Finland (61°27'N)	10	190	3–8	5	5.32* (0.86*)	25
Finland (65°01'N)	7	44	2–7	6	5.43 (1.06)	26
P. d. italiae						
Italy (~42°10'N)	2	44*	3–7	5	5.20* (n.a.)	27
Italy (45°20'N)	1	230	2–8	5	5.27* (1.12*)	28

(continued)

TABLE 4.2. *(Continued)*

Location (latitude)	Years	*N*	Clutch Size			Source
			Range	Mode	Mean (SD)	
P. d. biblicus						
Israel (32°21'N)	2	39	3–7	5	5.05 (0.94)	29
Iraq (33°18'N)	2	127	2–7	5	4.79 (0.96*)	30
Turkey (38°24'N)	4	10	3–8	5	5.10* (1.37*)	31
Turkey (39°56'N)	10	75	2–7	5	4.68*(0.92*)	32
Turkey (39°56'N)	1	11	3–6	5	4.55* (1.13*)	33
P. d. indicus						
Yemen (≈14°30'N)	n.a.	8	n.a.	n.a.	3.25 (n.a.)	34
India (17°19'N)	1	107	2–5	3	3.79* (0.87*)	35
India (22°18'N)	4	1508	1–7	4	3.62 (n.a.)	36
India (22°18'N)						
Rural	1	82	n.a.	n.a.	3.72 (0.75)	37
Urban	1	49	n.a.	n.a.	3.78 (0.87)	37
India (26°18'N)	1	94	2–6	3	3.60* (1.19*)	38
Pakistan (31°30'N)	2	n.a.	3–5	n.a.	4.3 (n.a.)	39
P. d. bactrianus						
Kazakhstan (43°16'N)	n.a.	53	5–8	6	6.53* (0.89*)	5
Kazakhstan (~44°N)	8	114	4–8	6	6.16 (1.07)	40
Introduced Range						
P. d. domesticus						
New Zealand (43°31'S)	n.a.	109	2–5	4	3.90 (0.61)	41
Mississippi (USA) (33°28'N)	3	229	2–5	4	3.93* (n.a.)	42
Texas (USA) (34°10'N)						
Rural	1	160	1–6	n.a.	4.32* (n.a.)	43
Urban	1	97	1–6	n.a.	4.09* (n.a.)	43
Oklahoma (USA) (36°07'N)	2	80	2–7	5	4.58* (0.96*)	44
Tennessee (USA) (36°21'N)						
Rural	2	39	3–6	5	4.87* (0.89*)	45
Suburban	2	33	2–7	5	4. 42* (1.03*)	45
Kentucky (USA) (38°00'N	5	724	1–8	5	4.70* (n.a.)	46
Illinois (USA) (38°05'N)	3	337	1–6	5	4.46 (0.91*)	47
Missouri (38°55'N)	5	620	2–8	5	4.64 (0.87)	48
Kansas (USA) (39°02'N)	4	1423	1–8	5	5.14 (0.93)	49
Illinois (USA) (41°35'N)	8	220	1–6	5	4.61 (0.76)	50
New York (USA) (42°27'N)	1	38	3–6	5	4.74* (0.55)	51
Wisconsin (USA) (43°00'N)	2	103	1–6	5	4.81* (0.84*)	52
Michigan (USA) (45°32'N)	6	340	2–8	5	4.96 (1.11)	53
Washington (USA) (46°44'N)	2	41	n.a.	n.a.	4.76* (n.a.)	54
Alberta (Canada) (51°05'N)	2	734	n.a.	n.a.	5.03* (0.82*)	55
Alberta (Canada) (51°05'N)	2	1174	n.a.	n.a.	4.88* (n.a.)	56

(continued)

TABLE 4.2. *(Continued)*

Location (latitude)	Years	*N*	Clutch Size Range	Mode	Mean (SD)	Source
P. d. indicus						
Malawi (15°44'S)	2	154	2–7	4	3.93* (0.92*)	57
South Africa (29°07'S)	1	114*	n.a.	n.a.	3.88* (n.a.)	58

This list is a compilation of data from single studies on clutch size in the house sparrow, with regional data on clutch size (i. e., state- or country-wide) being ignored. In some instances, indicated by an asterisk (*), the data were calculated by the author from information in the source. Data that were unavailable in the source are indicated by the abbreviation n.a. (not available).

Sources: 1, Gil-Delgado et al. (1979); 2, Escobar and Gil-Delgado (1984); 3, Veiga (1990a); 4, Ivanov (1987); 5, Gavrilov and Korelov (1968); 6, Chastel et al. (2003); 7, Ion and Ion (1978); 8, Vaclav and Hoi (2002b); 9, Balat (1974); 10, Novotny (1970); 11, Mackowicz et al. (1970); 12, Pinowski and Wieloch (1972); 13, Luniak et al. (1992); 14, Pinowska and Pinowski (1977); 15, Wieloch and Fryska (1975); 16, Wieloch and Strawinski (1976); 17, Seel (1968b); 18, Dawson (1972b); 19, Wetton et al. (1995); 20, Craggs (1967); 21, Encke (1965b); 22, Wendtland *in litt.*; 23, Møller (1991); 24, Krogstad et al. (1996); 25, Rassi *in litt.*; 26, Alatalo (1975); 27, Sorace (1993); 28, Brichetti et al. (1993); 29, Singer and Yom-Tov (1988); 30, Al-Dabbagh and Jiad (1988); 31, Siki (1992); 32, Kiziroglu et al. (1987); 33, Erdogan and Kiziroglu (1995); 34, Al-Safadi and Kasparek (1995); 35, Kumudanathan et al. (1983); 36, Naik and Mistry (1972), Naik (1974); 37, Mathew and Naik (1998); 38, Rana and Idris (1991b); 39, Mirza (1972); 40, Gavrilov et al. (1995); 41, Dawson (1964); 42, Sappington (1975); 43, Mitchell et al. (1973); 44, North (1968); 45, Pitts (1979); 46, Westneat et al. (2002); 47, Will (1973); 48, Anderson (1978); 49, Lowther (1983); 50, Lowther (1996); 51, Weaver (1943); 52, North (1972); 53, Anderson (1994); 54, Strasser and Schwabel (2004); 55, Murphy (1978a); 56, McGillivray (1983); 57, Nhlane (2000); 58, Earle (1988).

sample size or the biases resulting from sampling from only a portion of the breeding season, and/or the seasonal change in clutch size observed in most populations. The discussion here includes analyses of the entire dataset but also includes analyses of just the larger datasets ($n \geq 50$), to ensure that conclusions are not unduly influenced by the small, incomplete samples.

The largest average clutch size was the 6.53 reported for the migratory subspecies *bactrianus* in Kazakhstan, which exceeded the next largest average clutch size (5.43, in northern Finland) by more than one egg (see Table 4.2). *Bactrianus* females typically lay only one clutch per year (Gavrilov and Korelov 1968; Gavrilov et al. 1995), unlike females of permanent resident subspecies, which are typically multibrooded. This suggests that there may be markedly different reproductive strategies for permanent resident sparrows and migratory sparrows, involving a tradeoff between clutch size and number of broods attempted. In Kazakhstan, for instance, females of the sympatric resident subspecies *domesticus* are typically double- or triple-brooded and have a mean clutch size of 4.61 (Gavrilov and Korelov 1968). The clutch size data from *bactrianus* are ignored in the analyses that follow.

Clutch size tends to increase with latitude poleward in sparrows (Anderson 1978, 1994; Murphy 1978a), as it does in many other species (Klomp 1970; Lack 1947; Murphy and Haukoija 1986). Cursory inspection of Table 4.2 tends to support this generalization, and the correlation of mean clutch size with latitude is significantly positive for the sample as a whole ($r = 0.500$, $P < .001$), as well as for the subsample of studies based on 50 or more clutches ($r = 0.582$, $P < .001$). The proportion of variation in mean clutch size explained by latitude ($R^2 = 0.339$ for the latter correlation) is small, however, and other factors are clearly also contributing to this variation.

Some of the lowest mean clutch sizes are found in populations on high-latitude islands. For instance, several studies in England (Dawson 1972b; Seel 1968b; Summers-Smith 1963; Wetton et al. 1995) and one in New Zeland (Dawson 1964) reported average clutch sizes of approximately 4. Likewise, the average clutch size for an island off the coast of central Norway (5.14) and that of near-shore populations in the same region (5.05) were considerably lower than the clutch size of two populations at similar latitudes in Finland (5.32 and 5.43, respectively) (see Table 4.2). This suggests that the ameliorating effect of maritime climates might have some effect on clutch size.

Fig. 4.4 shows the relationship between mean clutch size and seasonality (represented by the absolute value of the difference between mean high July temperature and mean high January temperature at or near the study site). The correlation between mean clutch size and seasonality is highly significant ($r = 0.617$, $P < .001$), and that for the subsample of larger studies is even higher ($r = 0.702$, $P < .001$). The latter relationship explains approximately half of the variation in average clutch size in the subsample ($R^2 = 0.493$). Slightly higher correlations were found with mean January (or July for southern hemisphere sites) temperatures ($r = -0.634$ and -0.711, respectively).

A stepwise multiple regression was performed with mean clutch size as the dependent variable and latitude, temperature seasonality (ΔT), mean January (or July) high temperature (T_h), annual precipitation, and precipitation seasonality (represented by the coefficient of variation of mean monthly precipitation) as the independent variables. Only T_h entered the model significantly ($P < .05$) for the complete sample, but for the subsample of 47 studies, both T_h ($P = .026$) and ΔT ($P = .046$) entered the model, and the proportion of variation explained ($R^2 = 0.548$) was somewhat higher than that for either of the variables alone. The multiple regression equation is Y (clutch size) $= 4.17 - 0.020T_h + 0.021\Delta T$. This suggests that both winter temperatures and the difference between winter and summer temperatures influences clutch size in sparrows (see later discussion).

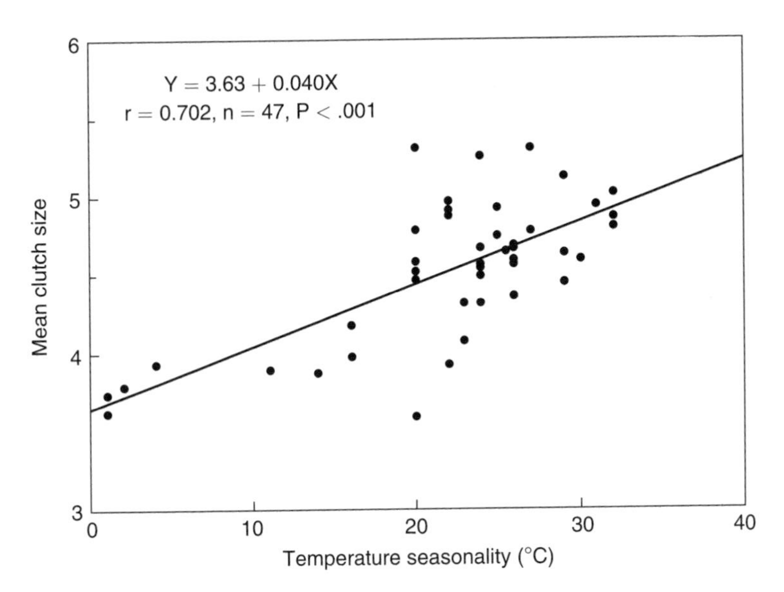

FIGURE 4.4. The relationship between mean clutch size and seasonality (absolute value of difference between mean high temperatures for January and July). Temperature data were obtained from Pearse and Smith (1990) and Universal Reference Book (2001). In three instances for which weather data were not available for sites close to the populations studied, temperature data were estimated by interpolation from data at two sites in two directions from the study sites.

Habitat Variation

Several studies have examined the effect of habitat on clutch size in sparrows. At least three studies suggested that clutch size is lower in urban populations than in rural populations in north temperate areas. Encke (1965b) reported that the average clutch size in Berlin (Germany) was 4.90, but in a nearby village it was 5.32 (see Table 4.2). In the United States, Mitchell et al. (1973) reported that the average clutch size in an urban population in Texas was 4.09, and in a nearby rural population it was 4.32. Pitts (1979) found that the average clutch size in a suburban population in Tennessee was 4.42, and in a nearby rural population it was 4.87. In South Africa, however, Earle (1988) found that the average clutch size was significantly lower in a rural population (3.1) than in suburban (4.2) and urban (3.7) populations. Mathew and Naik (1998) found no significant difference in mean clutch size between rural (3.72) and urban (3.78) populations in India. Some apparent habitat differences may be related to differences in breeding density (see later discussion).

Sorace (1993) studied clutch size in four woodland habitats located adjacent to agricultural land near Milan (Italy), including pine woods, beech woods, and two sites with Mediterranean scrub. Average clutch size in the pine woods (4.9) was lower than that in either beech woods (5.4) or Mediterranean scrub (5.1 and 5.3), but the sample sizes were too small to demonstrate statistical significance.

Density

Data on breeding density are frequently not reported, and studies that have reported breeding density often have not used comparable methods for measuring density, making comparisons among studies impossible. There does seem to be some indication that clutch size decreases with increasing density, but this is based primarily on anecdotal observations. Average clutch sizes at two farms in Bulgaria were 4.76 and 4.94, and breeding density at the first site (4200–4900 pairs/ha) was about three times greater than at the second site (approximately 1400 pairs/h1) (Ivanov 1987). Novotny (1970) presented figures showing the breeding sites of the sparrows that he studied in the Czech Republic, and these suggest that breeding density was high and increased during the 3 y of the study. This might account in part for the unusually low average clutch size for this population, 4.33 (a deviation of 0.32 egg from the 4.65 clutch size predicted for the site by the multiple regression equation described earlier) and for the very low average clutch size in the third year of the study, 4.23. The effect of density may, however, be difficult to separate from that of habitat. High-quality habitats may have superior resources for reproduction and may therefore have both higher breeding densities and larger clutch sizes than low-quality habitats (see discussion of the Ideal Free Distribution in Chapter 6).

Female Age

First-year females tend to have smaller clutches than older birds do, at least in their first breeding attempt. Dawson (1972b) reported that first-year females laid clutches that averaged 0.15 egg smaller than those of older females in England, but the difference was not significant. This fact may account for the frequent observation that newly established nest-boxes often have lower clutch sizes during the first year of a multiyear study, because the nest-site fidelity of older birds means that a high proportion of females occupying new sites are first-year birds (North 1969, 1972). It might also account for the unusual report of a modal clutch size of 4 in Ontario (Canada) by Krementz and Ankney (1986): not only was the nest-box colony newly established, but the repeated removal of females for carcass analysis meant that replacement females were probably also first-year birds.

Heritability of Clutch Size

I found no published accounts of the heritability of clutch size in sparrows, but my recent work on a population in Michigan (Anderson 1994, 1998) provided sufficient material to calculate estimates of heritability based on mother-daughter comparisons, as well as repeatability estimates. For the latter, which represent estimates of the upper limit of the heritability of a trait, I used the method of Lessells and Boag (1987) to calculate repeatability of clutch size in 42 females for which data were available on two or more midseason clutches (see Anderson 1998) and in 28 females for which data were available on two or more late-season clutches. Separating midseason and late-season clutches presumably reduced some of the phenotypic variation caused by the seasonal decline in clutch size described earlier. The two estimates, 0.295 and 0.298, respectively, suggest a low heritability of clutch size in the house sparrow and are within the range of repeatabilities observed in other passerines (0.19–0.51) (Boag and van Noordwijk 1987). The two estimates of repeatability are not independent, however, because 23 individuals were represented in both estimates.

Heritability, estimated as twice the regression coefficient of average daughter clutch size on average maternal clutch size, was calculated for 26 mother-daughter pairs. The estimate of heritability, 0.200, is lower than the two repeatability estimates, and it is considerably lower than two estimates of heritability of clutch size in the great tit (*Parus major*) based on mother-daughter data, 0.37 and 0.48 (Boag and van Noordwijk 1987).

The results of a common garden experiment also suggest that there is considerable phenotypic plasticity in clutch size in sparrows. Baker (1995) captured 51 sparrows during the summer in Costa Rica and transported them to New York, where he maintained them in outdoor aviaries along with a similar-sized group of sparrows captured locally. Initiation of egg-laying began at the same time in both groups of sparrows in two successive breeding seasons, and in the first year the average clutch size of the Costa Rican birds (3.50) was significantly lower than that of New York birds (4.38) ($P > .001$). During the second breeding season, however, the average clutch size of Costa Rican birds (4.62) did not differ significantly from that of New York birds (4.87) ($P > .15$). Although there has been no intensive study of the clutch size of house sparrows in Costa Rica, preliminary work suggests that the normal clutch size in Central America is 2 or 3 eggs (Fleischer 1982; Thurber 1986). The average clutch size around Ithaca, New York, however, was determined by Weaver (1943) to be 4.74. Based on the results of this common garden experiment, it is clear that

Costa Rican sparrows adjusted their clutch size within 2 y to correspond to that of sparrows living in the area, and Baker (1995) concluded that environmental factors play a major role in determining clutch size in the house sparrow.

Determinism

Birds have often been categorized as either determinate or indeterminate layers based on whether the addition or removal of eggs from a nest during egg-laying has an effect on the number of eggs laid by the female (Cole 1917; see Haywood 1993 and Kennedy 1991 for reviews). The usual protocols for studying determinism in birds have been either to add eggs to a nest after the laying of the first egg, or to remove the second and subsequent eggs from a nest as they are laid (always leaving one egg in the nest). The results of several such experiments in the house sparrow have yielded equivocal results, with some authors concluding that sparrows are indeterminate (Kendra et al. 1988; Witschi 1935), and others that they are determinate (Anderson 1989; Brackbill 1960; Schifferli 1976).

In his review of determinism in birds, Haywood (1993) concluded that the house sparrow is an indeterminate layer, suggesting that if each egg were removed as it was laid (leaving the nest empty), the female would continue to lay indefinitely (S. Haywood, personal communication). An experiment in Michigan confirmed this prediction, with one female laying 18 eggs in 19 d, including 11 on successive days (Anderson 1995). Analysis of egg components (yolk, albumen, and shell) in the six nests in which indeterminate laying occurred revealed that the components of the supernumerary eggs tended to weigh more than those of the first six eggs.

Significance of Clutch Size

More has probably been written about the significance of clutch size than about any other single topic in avian biology, but the present discussion must of necessity be very circumscribed. As mentioned earlier, Lack's (1947) pioneering paper on the significance of clutch size in birds suggested that, for most altricial species, clutch size should correspond to the amount of food available for provisioning the young. This hypothesis predicts that the most common clutch size in a population should be the most productive. Some studies on sparrows have found that the modal clutch size tends to be the most productive (e.g., Ivanov 1987; Seel 1970, Will 1969). Murphy (1978b), however, found that clutches of 6 and 7 eggs consistently produced more fledglings than those with the modal clutch size of 5 in both Kansas and Alberta. Brichetti et al. (1993) also found that clutches of 6 and 7 eggs produced more fledglings than did those with the modal clutch size of 5 in northern Italy.

Manipulation of brood size has been one of the experimental methods of testing Lack's hypothesis. Brood manipulation alters brood size by the addition or subtraction of nestlings to or from a nest, usually shortly after hatching. Lack's hypothesis predicts that the average number of young fledging from nests with the normal brood size (corresponding to the most common local clutch size) will be higher than that of either reduced or enlarged broods.

The results of brood manipulation studies in sparrows have been equivocal, as have similar studies in many other altricial species (Klomp 1970; Murphy and Haukioja 1986). Schifferli (1978b) added one nestling to each of 14 nests at hatching in England. Enlarged broods had lower survival than control broods, and the surviving chicks weighed less at 10 d than did nestlings from unmanipulated broods. On the other hand, two studies in the United States found that sparrows can rear enlarged broods successfully. Hegner and Wingfield (1987a) manipulated 17 second broods in New York by the removal ($n = 9$) or addition ($n = 8$) of two young when the young were 4–6 d old. Enlarged broods fledged significantly more young (5.7) than reduced (2.2) or control (3.5) broods ($P < .05$). Anderson (1998) manipulated 41 midseason broods in Michigan. Hatchlings were added either on hatching day or on the next day to establish 30 enlarged broods (each containing one to three more young than the clutch size in the nest). Twenty-seven unmanipulated broods and 11 reduced broods served as controls. Two of the 30 enlarged broods and 5 of the 38 controls failed to produce fledglings, and the percentage of young fledging from successful enlarged broods (89.5%) did not differ from that of control broods (90.0%). Both growth rate and asymptotic weight of nestlings from the enlarged broods were significantly lower than those of control broods, however ($P < .01$ in each case) (see later discussion). The results of these latter two studies suggest that sparrows are able to successfully rear more young than the number of eggs they normally lay, at least in North America.

The provision of supplemental food to breeding birds is a second method of testing Lack's hypothesis (Crossner 1977). Predictions based on Lack's hypothesis depend on the extent to which clutch size in a population is genetically determined. If there is a large genetic component, supplemental food might have little, if any, effect on clutch size. If, on the other hand, females can respond facultatively to increased food availability, clutch size should increase with the provision of supplemental food. The low heritability of clutch size in sparrows (discussed earlier) suggests that the latter may be the case.

One study has examined the effects of supplemental food on sparrow mating behavior (see earlier discussion), but I know of no study that has examined its effect on clutch size in the species. The reproductive responses of sparrows to a local emergence of periodical cicadas (*Magicicada* spp.), which represents a short, natural pulse of superabundant food during the breeding season, were examined in Missouri (Anderson 1977). There was not a significant increase in clutch size during the emergence, although there was a tendency for the interval between successive clutches to decrease. This result is consistent with those of numerous food supplementation studies on several other passerines, which typically found that supplemental food results in earlier onset of breeding but has little effect on clutch size (Crossner 1977; Desrochers 1992).

The numerous brood manipulation studies that have demonstrated that many altricial species, including the house sparrow, can successfully rear more young than the number of eggs in the normal clutch have led to the construction of various alternatives to or modifications of Lack's hypothesis (e.g., Cody 1966; Mountford 1968). One modification that has proved to be particularly influential is the "cost of reproduction" hypothesis (Williams 1966). Williams suggested that each reproductive attempt imposes certain costs, particularly on the female, and that these costs might have a negative effect on the future reproductive potential ("residual reproductive value") of the individual. If the costs in terms of future reproductive potential outweigh the benefits of rearing additional young in the current reproductive attempt, trying to rear additional young (i.e., increasing clutch size) would be selected against. This "cost of reproduction" hypothesis therefore results in a modification of Lack's hypothesis to state that the clutch size should be selected to maximize the total reproductive potential of the individual.

A few studies have attempted to identify or quantify such costs of reproduction in sparrows. In Alberta, McGillivray (1983) found that the number of young fledged in second broods was negatively correlated with the number of young fledged during the first brood at that nest site ($r = -0.85$, $P < .05$) and that the number fledged in third broods was negatively correlated with the total number of young fledged in first and second broods ($r = -0.81$, $.05 < P < .10$). Interbrood interval was also positively correlated with number fledged in the earlier brood. One difficulty with this study was that most individuals were not marked, which meant that the same individuals may not have been involved in consecutive breeding attempts at a site (see earlier discussion).

The two brood manipulation studies in the United States also attempted to identify the costs of rearing enlarged broods using marked birds. In New York, most of the pairs initiated third broods, and the average clutch size of females with reduced second broods (5.4) was significantly greater than that of control females (4.3) or enlarged-brood females (3.6) (Hegner and Wingfield 1987a). In Michigan, however, the rearing of enlarged midseason broods significantly reduced the probability of initiation of late-season broods ($P < .025$), but late-season clutch sizes did not differ between enlarged-brood females (4.83) and controls (4.92) (Anderson 1998). In addition, females rearing enlarged broods were just as likely to survive to the next breeding season as control females.

In summary, the evidence concerning the ultimate causation of clutch size in the house sparrow remains equivocal. In some cases, the average clutch size in an area appears to correspond to the maximum number of young that the parents can provide for, but in other areas, birds appear to be able to rear more young than the average clutch size. There do appear to be costs associated with increased reproductive effort, but the identity and magnitude of these costs in terms of future reproductive potential have not been clearly established or quantified.

Mechanism of Clutch Size Determination

Baker (1938) identified the distinction between ultimate and proximate causation in the explanation of biological phenomena. The former refers to the selective factors that shape a biological trait, whereas the latter refers to the mechanisms, both intrinsic and extrinsic, that contribute to the actual expression of the trait. As noted earlier, clutch size in the house sparrow varies with female age, habitat, season, density, geography, and possibly other factors. The low heritability of clutch size suggests that individual females show considerable phenotypic plasticity in clutch size, with the possibility that such plasticity may enable them to optimize each reproductive attempt in an environment with predictable variation in food resources (cf. Perrins and Moss 1975). Such a scenario would require that information from both the external and internal environments be transduced into the laying of an appropriate-sized clutch. How would such a mechanism operate?

Drent and Daan (1980) proposed that there are two possible ways in which environmental information could be utilized by birds to adjust their laying date and clutch size to optimize reproduction. Using economic terms, they described these as the "income" and the "capital" models. Both models are based on the amount of energy acquired by the female in the period leading up to egg-laying. The "income" model suggests that it is the rate of energy acquisition that determines laying date and clutch size, and the "capital" model suggests that it is the amount of energy reserves stored by the female that is determinative.

Evidence that such a mechanism operates in the determination of clutch size in the house sparrow is equivocal. In Poland, Pinowska (1979) found that the fat content in females was highest on the day before the first ovulation, and that clutch size (Y) was positively correlated with fat content (X) in grams on that day: $Y = 2.1967 + 1.207X$ ($r = 0.6683$, $P < .01$). No such relationship was found in similar studies conducted in England (Schifferli 1976) and Ontario (Krementz and Ankney 1988), however, although both studies did find that fat reserves of females increased rapidly just before the time of rapid yolk deposition in the first developing follicles of a clutch. Both studies also found that there was a significant decrease in fat reserves during the period of egg formation and laying, with Schifferli (1976) estimating that females expend 75% of their fat reserves during this time.

The fact that the house sparrow is an indeterminate layer (discussed earlier) means, however, that clutch size is not constrained by the fat reserves, or other reserves, accumulated by the female before the egg-laying period. Indeterminate laying in sparrows occurs when there is a lack of tactile stimulation from eggs in the nest (see Anderson 1995), suggesting that such stimulation also plays a role in the proximate determination of clutch size. Anderson proposed the following mechanism for clutch size determination: (1) female condition at some point before the initiation of egg-laying determines both a prospective clutch size (within some genetically determined range) and an accompanying schedule for the onset of incubation; (2) tactile feedback from the presence of eggs in the nest causes increased prolactin secretion (see Chapter 9); and (3) rising levels of circulating prolactin suppress continued follicular development, halting egg production at the predetermined number. Either the rate of accumulation of fat reserves ("income") or the amount of fat reserves ("capital") may be the relevant indicator of female condition, although accumulation of protein reserves may also be involved (Jones 1990; Krementz and Ankney 1988). No direct tests of the model have been performed, although the requisite presence of short periods of incubation early in the laying cycle has been observed (Anderson 1997).

The relative size of the liver may also play a role in clutch size determination. The liver is the site of synthesis of yolk lipids and egg proteins. In Russia, both the absolute and the relative size of the liver in males are greater during the winter and decrease during the breeding season (Danilov et al. 1969) (see Chapter 9). The size of the liver in females is similar to that in males during the winter but does not decrease during the breeding season. Increased liver size during the winter is undoubtedly related to the increased metabolic demands of thermoregulation (see Chapter 9), and the amount of increase is presumably negatively correlated with winter temperatures. The fact that female liver size does not decrease during the subsequent breeding season may help to explain the relationship between mean clutch size and both winter temperature and seasonality described earlier (see Fig. 4.4).

Egg Size and Composition

Egg Size

Egg size is usually measured in the field as fresh egg mass or as the linear measurements, length and maximum breadth, and egg volume is occasionally measured. Egg mass decreases at a rate of approximately 1.1%–1.3% per day during incubation (Dawson 1964; Weaver 1943), which means that egg masses obtained during incubation are not comparable to fresh egg mass unless they are corrected for length of incubation. Egg size varies geographically, primarily in parallel with female body size. There is also considerable variation in egg size within a population, with egg size tending to be characteristic of a particular female and showing some tendency to vary with female size within a population as well as between populations. Egg size also tends to vary with laying order within a clutch. The adaptive significance of egg size differences is poorly understood, although hatchling size does have a significant relationship with egg size.

Fresh egg mass is closely related to the linear dimensions of the egg. The linear regression equation describing this relationship for 1038 eggs that were weighed and measured on the day laid during a 5-y period in Michigan is $Y = 0.167 + 0.51LB^2$, where Y is the fresh egg mass in grams, L is egg length in centimeters, and B is maximum egg breadth in centimeters (Anderson, unpublished data). This regression explains 93.3% of the variation in fresh egg mass, and the regression coefficient of 0.51 is highly significant ($P < .001$). It is also similar to the value obtained for 217 fresh eggs measured in New Zealand, 0.543, especially considering that the latter equation was forced through the origin (Dawson 1964), and it is identical to the coefficient obtained by Hoyt (1979) in a multispecies comparison of the relationship between LB^2 and egg volume in birds. Paganelli et al. (1974) reported that the average surface area of sparrow eggs with an average mass of 2.62 g was 9.18 cm^2.

Some studies have found that egg size within a population does not vary seasonally (Anderson 1998; Sorace 1992; Veiga 1990a), although Novotny (1970) reported that egg mass was significantly higher in first clutches than in subsequent clutches in the Czech Republic. Egg size does not differ with clutch size (Dawson 1972b; Lowther 1990; Marcos and Monros 1994; Sorace 1992; Veiga 1990a). Marcos and Monros (1994) also reported that there was no annual difference in egg size within a population in Spain, and Earle (1988) found no difference in egg size among three habitats (rural, suburban, and urban) in South Africa.

Egg size does vary among individual females in a population, however. In 49 female sparrows in Michigan for which data were available on two or more clutches, more than 80% of the variation in average fresh egg mass of the eggs in a clutch was explained by differences between females ($F_{48,89} = 7.533$, $P < .001$) (Anderson, unpublished data). Repeatability of average fresh egg mass among the 49 females (calculated according to Lessells and Boag 1987) was 0.707, which is comparable to egg size repeatabilities observed in other passerines (0.59–0.82, Boag and van Noordwijk 1987; Hendricks 1991). These results indicate the possibility of a high heritability of egg size in the house sparrow. In the same population, 25 females banded as nestlings were captured as adults on their own nests, permitting a mother-daughter comparison of average egg size. The average fresh egg mass of the daughters was positively correlated with that of their mothers, and the correlation approached significance ($r = 0.373$, $.05 < P < .10$). The regression relating average fresh egg mass of mothers (X) with that of daughters (Y) is $Y = 1.33 + 0.49X$. Heritability (h^2) can be estimated as twice the regression coefficient of the offspring trait on that of a single parent, assuming that both parents contribute equally to the genetic basis of the trait, although estimates based on maternal-offspring regressions may be too high due to maternal effects (Falconer 1981).

Egg size shows a weak, but significant, relationship to female size. The average fresh egg mass of 290 clutches in Michigan was positively correlated with female tarsus length ($r = 0.183$, $P < .005$) (Anderson, unpublished data). A stepwise multiple regression of egg size on female mass, female tarsus length, clutch size, and date of clutch initiation resulted in identification of tarsus length only as having a significant effect on egg size (multiple $r = 0.200$, $P < .005$).

Egg size tends to increase with position in the laying sequence, at least in larger clutches (Marcos and Monros 1994; Murphy 1978a). Fig. 4.5 shows intraclutch egg size variation in Michigan. Egg size tended to increase with laying sequence, at least through the penultimate egg, in 5- and 6-egg clutches, but in 4-egg clutches egg size peaked in the second egg and decreased in the third and fourth eggs. Similar results were obtained in a population in Kansas (Lowther 1990).

A recent study has suggested that egg size may vary with the sex of the embryo. Cordero et al. (2000) determined the sex of embryos from 34 sparrow clutches in eastern Spain by isolating DNA from the embryos and testing for the presence of the sex-linked *CHDI* gene (see Chapter 2). They also measured egg length and breadth and calculated egg volume for each egg. The sex ratio of the eggs did not differ significantly from 50:50 (binomial test, $P = 1.0$), and there was no difference in sex ratio of eggs based on position in the laying sequence. Cordero et al. concluded that male eggs are 1.3% larger than female eggs, but their use of ANOVA and MANOVA tests, which require independence of the data, appears to be inappropriate, because the eggs within a clutch are literally "nested." Results of paired *t* tests, which are appropriate, do, however, show

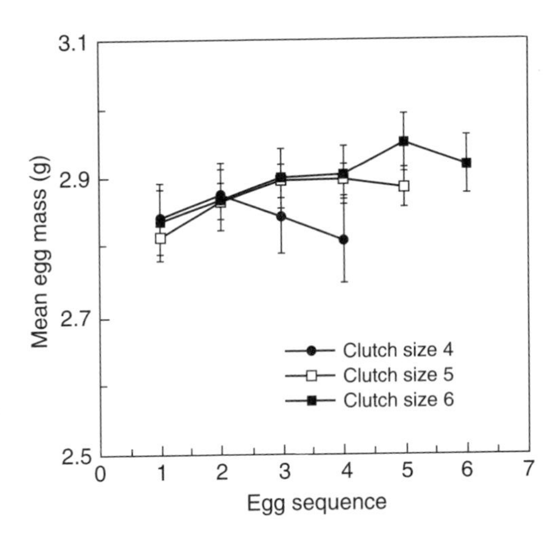

FIGURE 4.5. Change in egg size with position in the laying sequence in the house sparrow. Data are means (± SE) of fresh egg mass in Michigan (USA). Data are from 1986–1991 and are for 4-egg (n = 31), 5-egg (n = 85) and 6-egg (n = 36) clutches in which each egg was weighed on the day laid (Anderson, unpublished data).

that relative egg length is significantly greater in male eggs ($t_{30} = 2.58$, $P = .015$) and that relative egg volume approaches significance ($t_{30} = 1.98$, $P = .06$).

The adaptive significance of egg size variation in sparrows is difficult to identify. Egg size may affect hatching probability. In England, Dawson (1972b) found that infertile eggs and eggs in which the embryo died early were smaller than eggs that hatched. However, two studies have found that egg size has no effect on nestling survivorship or fledgling size, although there does appear to be a positive relationship between egg size and hatchling weight (Anderson, unpublished data; Veiga 1990a). In one population in Spain, for instance, body mass of newly hatched chicks was strongly correlated with egg mass ($R^2 = 0.841$, $P < .001$) (Veiga 1990a). In England, O'Connor (1975b) found that nestling mass remained significantly positively correlated with hatchling mass for the first 7 d of nestling life but was not related to hatchling mass thereafter.

Development of the Egg

The period from the beginning of rapid yolk deposition until the egg is laid requires approximately 4 d in sparrows (Fig. 4.6). Egg development begins with the deposition of yolk and the concomitant enlargement of the associated follicle in the ovary. Yolk deposition and follicular enlargement continue until the egg is ovulated approximately 24 h before it is laid (Schifferli 1980a). Rapid yolk deposition requires at least 3 d (Krementz and Ankney 1986; Schifferli 1980a), although Pinowska (1979) observed small increments in average follicular diameter beginning 6 d before ovulation in Poland. Albumen is added as the egg passes down the oviduct, and the shell and shell membranes are

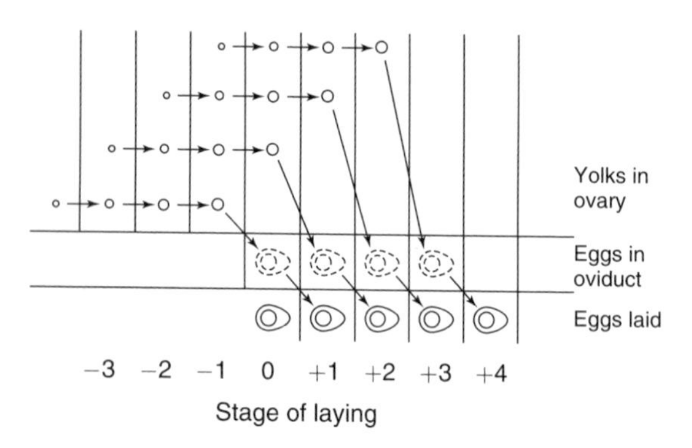

FIGURE 4.6. Egg development in the house sparrow. The figure depicts the temporal pattern of development of a 4-egg clutch of the house sparrow in England (UK). Days in the laying cycle are represented, with 0 representing the day of ovulation of the first ovum from the ovary. Follicular enlargement of the first egg begins 3 d before its ovulation (−3), and enlargement of subsequent follicles begins on successive days thereafter. From Schifferli (1980a: Fig. 1), with permission of Deutsche Ornithologen-Gesellschaft.

produced in the shell gland during the night before the egg is laid. Schifferli (1979) removed eggs from the oviducts of 50 laying female sparrows collected from their nests during the evening and night in England. The shells (including the shell membranes) were dried to constant mass at 85°C, and the equation relating dry mass of the shell (Y) to hours before sunrise (X) was $Y = -0.011X + .185$ ($r = -0.82$, $P < .001$). Development of the eggshell membranes begins about 10–11 h before sunrise, and the shell itself develops over a 9-h period after dark, completing development about 2 h before sunrise.

Egg Composition

The avian egg can be divided into three parts: yolk (the true egg), albumen, and shell. Ar and Yom-Tov (1978) reported that the yolk comprised 25% (0.7 g), the albumen 68% (1.9 g), and the shell, 7% (0.2 g) of the fresh mass of nine house sparrow eggs. Almost 80% of the fresh mass of the egg is water, however. In England, Schifferli (1976) found that 78.9% of the egg was water, whereas the dry yolk was 8.0%, the dry albumen 6.9%, and the dry shell 6.2% of total egg mass. Comparable values from Michigan were 77.8%, 8.1%, 7.9%, and 6.2%, respectively (Anderson 1995). Studies on Irish and Dutch sparrows both found that the shell made up 6.5% of fresh egg mass (Foster 1917; Hellebrakers 1950). Schifferli (1976) also reported that the lean dry mass of the yolk was 3.3% of fresh egg mass, indicating that lipids make up 59% of the dry mass of the yolk. The remainder presumably consists primarily of proteins, which also make up virtually all of the dry mass of albumen (see Krementz and Ankney 1986). Pinowska and Krasnicki (1985a) found that the relative amounts of nitrogen (N) and phosphorus (P) in sparrow eggs in Poland were 5.9 and 5.95 mg/g, respectively. Dawson (1964) reported that the specific gravity of New Zealand sparrow eggs was 1.07 g/ml, and Paganelli et al. (1974) found a value of 1.036 g/ml in New York.

The shell is comprised primarily of calcium carbonate, with small amounts of other minerals. In Poland, the egg, including shell, contained 106.1 mg calcium (Ca) per gram of dry mass, 1.02 mg/g of magnesium (Mg), and 1.65 mg/g of copper (Cu) (Pinowska and Krasnicki 1985a). Ca and Mg are found primarily in the shell, whereas Cu is located primarily in the shell membranes. Dry shell mass does not vary with position in the laying sequence (Krementz and Ankney 1995). Ar et al. (1979) tested the strength of shells of sparrow eggs from Israel using a vice-like instrument that applied pressure on the two ends of the egg until it cracked. They found that the average yield point (= strength) was 252 g. Both Ar et al. (1979) and DeGraaf and Maier (2001) reported that the average shell thickness of sparrow eggs (from Israel and Massachusetts, respectively) was 102 μm.

Egg Coloration

There is considerable variation in egg coloration in sparrows. The ground coloration varies from pure white to pale green, blue, or pink, and eggs are liberally spotted with small dark-brown or gray spots or streaks, most liberally at the blunt end of the egg (Novotny 1970). Dawson (1964) reported that egg coloration is similar among eggs laid by the same female. The last-laid egg is usually different from other eggs in the clutch, however; the spotting is larger and more diffuse, giving the last egg a lighter appearance than the other eggs in the clutch (Dawson 1964; Kendra et al. 1988; Lowther 1988). Lowther (1988) reported on known last-laid eggs from a 3-y study in Kansas and found that the frequency of such pale last eggs tended to increase with clutch size: in 4-egg clutches, they occurred in 45 (88%) of 51 cases; in 5-egg clutches, 164 (94%) of 175 cases; in 6-egg clutches, 96 (96%) of 100 cases; and in 7-egg clutches, 15 (100%) of 15 cases. Lowther also reported that there were two pale eggs in about 10% of the clutches in his study, and that in most cases the second pale egg was the penultimate egg in the laying sequence. He suggested two possible explanations for the changed spotting pattern: (1) depletion of pigment in the pigment gland at the end of egg-laying and (2) physiological changes in the female at the termination of egg-laying that affect pigment gland function.

Energetics of Egg Production

Krementz and Ankney (1986) attempted to determine the total and daily energetic costs of egg production of a 4-egg clutch in Ontario. They collected female sparrows during three breeding seasons and removed both ovaries and oviducts. Day in the prelaying or laying period of each female was determined by examination of developing and/or postovulatory follicles in the ovary, as well as the presence of eggs in the oviduct (see also Pinowska 1979; Schifferli 1976). The dry masses of developing follicles at various stages of development were obtained, as were dry masses of the oviduct throughout the prelaying and laying periods. The energy budget for egg formation of a 4-egg clutch was constructed based on the assumptions that the dry yolk mass was composed of 59% fat (see earlier discussion), that the albumen was 100% protein, that the conversion efficiencies of fat and protein were 77% for exogenous sources and 100% for endogenous sources, that 15% of the protein came from endogenous sources, and that either 0% or 70% of the fat came from endogenous sources. In the production of a 4-egg clutch, the maximum energy demand occurred on the day the first egg was ovulated, with the estimated energy requirements for egg production varying from 16.5 to 17.6 kJ/d. They estimated that this amounted to 44%–47% of the standard metabolic rate (SMR) (see Chapter 9) of the average female. The estimate of the total energetic cost of laying a 4-egg clutch was 65.8–70.6 kJ, with protein demands accounting for 58.8%–63.1% of the total demand and fat demands accounting for 36.9%–41.1%. Krementz and Ankney concluded that the energy demands of egg production in sparrows are not excessive and that clutch size in the species is not constrained by the acquisition of energy for egg production (see earlier discussion).

Incubation and Hatching

Incubation is the supplying of heat to eggs to initiate and sustain embryonic development, and it is achieved by application of the naked, highly vascularized brood patch of the female to the eggs (see Chapter 5). Although male sparrows often spend time on the nest during the incubation period (North 1980; Summers-Smith 1963), the fact that males do not develop a brood patch suggests that they are not truly incubating the eggs, but instead are merely retarding heat loss during the absence of the female and protecting the eggs from predators or from sparrows prospecting for nest sties. The role of males in incubation is not inconsequential, however. Schifferli (1978b) found that clutches incubated solely by the female (after experimental removal of the male) had a lower hatching success than controls. In an intensive study of time-budgets of breeding sparrows in Wisconsin, North (1980) found that females typically spend about half of the daylight hours incubating, with the average incubation bout lasting 9.2 min, and incubate continuously through the night. The average incubation temperature of sparrows, measured in Ohio (USA), was 34.2°C (Huggins 1941).

Continuous incubation often begins before the completion of the clutch (Dawson 1964; Seel 1968b; Weaver 1943), although there are short periods of incubation beginning on the day the first egg is laid in most clutches (Anderson 1997). Seel (1968b) reported that continuous incubation began after clutch completion in 2- and 3-egg clutches in England, but that it usually began after the laying of the penultimate egg in larger clutches.

Incubation Period

Incubation period is usually operationally defined in field studies as the period from the laying of the last egg to the hatching of that egg (Lowther 1979c; Seel 1968b). Numerous studies spanning a broad geographic range have found that the average incubation period of sparrows falls between 11.4 and 12.2 d: 11.45 d in England (Seel 1968b); 11.5 d in Malawi (Nhlane 2000); 11.55 d in Michigan (Anderson 1994); 11.7 d in Wisconsin (North 1972); 11.76 d in Kansas (Lowther 1979c); 11.8 d in Tennessee (Pitts 1979); 11.9 d in Italy (Brichetti et al. 1993); 11.9 d in Illinois (Lowther 1996); and 12.2 d in Mississippi (Sappington 1977). Mitchell and Hayes (1973) reported an average incubation period of 11.3 d for sparrows breeding in captivity in Texas (USA). The average incubation period for eggs incubated at 37°C in the laboratory was found to be 11.75 d (Weatherbee and Weatherbee 1961).

Different definitions of incubation period have also been used, with some recording the interval from the laying of the last egg to hatching of the first young (Anderson 1978; Erdoan and Kizirolu 1995; Murphy 1978a), and others recording the interval from the laying of the first egg until hatching of the first young (Singer and Yom-Tov 1988) or the period from the laying of the penultimate egg to the hatching of the first egg (Lowther 1983).

Some studies have reported average incubation periods outside of the range described. These include periods of 12.5 d in India (Rana and Idris 1991b); 12.98 d in the Czech Republic (Novotny 1970); 13–14 d in Pakistan (Mirza 1972); 13.7 d in India (Sengupta 1981); and 13.84 d in India (Kumudanathan et al. 1983). Possible explanations for extended incubation periods include social interference in dense breeding colonies (e.g., the Czech Republic colony) and human interference resulting in frequent interruption of incubation. Sappington (1977) reported one unusually long incubation period (17 d), in a nest located in a barn occupied by a barn owl *(Tyto alba)*, which apparently resulted from long interruptions in incubation due to the presence of the owl. Another possibility is that there is a latitudinal effect on incubation period, with longer periods in subtropical areas such as India and Pakistan. Tropical populations have lower clutch sizes than temperate-zone sparrows do (see earlier discussion), and, if this is caused by lower food resources for provisioning of the chicks, it may also mean that females must spend more time foraging during incubation, resulting in retarded embryonic development. No relationship was found between incubation period and latitude in continental North America, however (Anderson 1994).

Lowther (1979c) reported that incubation period decreased with season for 5-egg clutches in Kansas. He suggested that this might be due to higher ambient temperatures in middle and late summer, with resultant passive incubation during the female's absences from the nest.

Hatching Asynchrony

House sparrow eggs hatch in the order in which they were laid (Veiga 1990a) and usually asynchronously (Anderson 1994; Nhlane 2000; North 1980; Seal 1968b; Weaver 1943). In New York, Weaver (1943) found that the first three eggs hatched at about the same time, but the remaining eggs hatched in the next 24–48 h (mean = 32.0 h). Veiga (1990a) reported an overall average hatching interval of 21.3 h in Spain, with the interval increasing from 10.7 h in 3-egg clutches to 25.5 h in 7-egg clutches. Seel (1968b) also found that hatching interval increased with clutch size in England, and Anderson (1994) reported that hatching interval increased with number hatching in Michigan. Veiga (1992a) also found that hatching interval changed seasonally in Spain, with the interval being shorter in early-season clutches than in late-season clutches. As noted for the seasonal decrease in incubation period discussed earlier, higher ambient temperatures in midsummer may result in passive incubation of early-laid eggs and greater hatching asynchrony.

Both adaptive and nonadaptive hypotheses have been suggested to explain hatching asynchrony in birds. Probably the most widely accepted adaptive hypothesis, first proposed by Lack and Lack (1951), is the Brood Reduction Hypothesis, which suggests that the primary adaptive function of asynchronous hatching is to facilitate brood reduction when unpredictably variable feeding conditions for the young are poor. Good feeding conditions for the young at hatching can result in adequate provisioning of all of the young, but poor feeding conditions at hatching will result in distribution of the food primarily among the larger, first-hatched chicks and the consequent starvation of late-hatched young, with minimal energy being wasted on young doomed to die anyway. Other adaptive hypotheses include the Egg Viability Hypothesis and the "Chick-heater" Hypothesis. The former (Arnold et al. 1987) suggests that the early onset of incubation in larger clutches is due to the reduced viability of first-laid eggs if the onset of incubation is deferred until the clutch is complete. The "Chick-heater" Hypothesis (Veiga 1993b) suggests that the presence of an additional chick is a low-cost means of increasing the thermal efficiency of the entire brood, permitting the female to spend less time brooding during the chick's ectothermic period. The female can therefore spend more time foraging, increasing the rate of development of the young, and thereby increasing nestling survivorship, particularly in populations exposed to high rates of nest predation.

Veiga (1992a) used two experimental protocols to test the Egg Viability Hypothesis during three breeding seasons in Spain. Eggs were removed from nest-box nests on the day they were laid and replaced with dummy eggs or with sparrow eggs from other nests. In one protocol, the eggs were placed in nearby nest-boxes that sparrows could not enter; in the second, eggs were refrigerated at 3°C–5°C. The eggs were replaced into the nests after the clutch was complete (ensuring the simultaneous initiation of incubation of all eggs in the clutch), and the hatching success of experimental and control nests (in which parents were permitted to begin incubation at the normal time) were compared. The Egg Viability Hypothesis predicts that the hatching success of first-laid eggs in the experimental nests will be lower than that of first-laid eggs in control nests. Hatching success of eggs held in adjacent nest-boxes did decrease with a delay in onset of incubation, compared with control eggs. Eggs for which incubation was withheld for 4 or more days had a hatching success of about 60%, compared with almost 80% for the same eggs in the laying sequence in control nests; in contrast, hatching success of last-laid eggs was comparable between experimental and control nests. Eggs that were chilled had uniformly lower hatching success.

Nonadaptive hypotheses for hatching asynchrony include suggestions such as that described earlier for the proximate mechanism for the determination of clutch size. If this mechanism depends on increasing prolactin levels that are triggered by tactile stimulation from the brood patch, then asynchronous hatching may be simply a nonadaptive consequence of the periods of incubation producing these tactile stimuli.

Hatching Success

Hatching success is usually recorded as the percentage of eggs laid that hatch, or sometimes as the percentage of incubated eggs that hatch, ignoring eggs that are either abandoned or lost before the onset of incubation. Usually, only a small number of eggs fit into the latter category, so reports of hatching success can be usefully compared across studies despite this difference in method. An additional difficulty that probably has a greater impact on comparative analysis of different studies is the effect of the frequency of nest checks on recorded hatching success. Different studies have nest visitation protocols that vary from daily (e.g., Anderson 1994) to weekly (e.g., Møller 1991), with most occurring at 3- or 4-d intervals. When hatching occurs between nest checks, eggs present at the first check sometimes cannot be fully accounted for by the number of nestlings present at the second check. It is possible that eggs hatched and the hatchlings died and were removed from the nest by the parents during the intercheck interval. At least two studies (Lowther 1983; Murphy 1978a) have attempted to account for this by estimating both minimum hatching success (counting only the number of nestlings known to have been alive) and maximum hatching success (assuming that all eggs disappearing between nest checks represented nestlings that died and were removed before the check).

The major causes of egg failure include infertility; embryonic death, including death due to chilling or to microbial infection (Kozlowski et al. 1991; Pinowski et al. 1988, 1994); inclement weather (particularly high wind and heavy rain); nest predation; and destruction of eggs by sparrows seeking nest sites (Veiga 1990b). Desertion by the parents also results in nest failure, and desertion rates are high if females are captured at the nest by an observer during egg-laying or incubation (Anderson, personal observation; North 1969; Pinowski et al. 1972). Ivanov (1987) also recorded an instance of desertion in Bulgaria due to takeover of the site by *Bombus armeniacus*. Table 4.3 lists species that have been identified as nest predators of sparrows, with most taking either eggs or nestlings (and occasionally adults, see Table 8.2 in Chapter 8).

A study in Poland reported that infection rates in eggs that failed to hatch (in nests where at least one egg hatched) were significantly higher than in eggs from abandoned nests (Pinowski, Mazurkiewicz et al. 1995). Eggs infected with *Escherichia coli* were particularly vulnerable, and the authors suggested that about 25% of egg mortality may be due to infection.

At least two studies, one in Alberta (McGillivray 1981) and one in Spain (Escobar and Gil-Delgado 1984), found that hatching success is higher in nest-box nests than in open nests. This is presumably a result of the greater vulnerability of the latter sites to predation and/or weather (see also Ivanov 1987). In Austria, Vaclav and Hoi (2002b) found that hatching success was higher in colonies in which nesting attempts were synchronized than in colonies with low synchronization, and they attributed this improvement to a reduction in interference and nest destruction by sparrows seeking nest sites in the more synchronized colonies.

Table 4.4 summarizes data on hatching success from numerous studies. Recorded hatching success rates varied from a low of 43.7% (Turew, Poland)

TABLE 4.3. Nest Predators of the House Sparrow

Predator	Location	Source
Reptiles		
Ladder snake (*Elaphe scalaris*)	Spain	1
Black rat snake (*Elaphe obsoleta*)	USA	2,3,4,5
Montpellier snake (*Malpolon monspessulanus*)	Spain	1,6
Wall lizard (=wall gecko?) (*Hemidactylus prashadi*)	Pakistan	7
Yellow-bellied house gecko (*Hemidactylus flaviviridis*)	India	8
Chameleon (*Caloties versicolor*)	India	8
Birds		
Wryneck (*Jynx torquilla*)	Czech Republic, Spain	9,10
Common magpie (*Pica pica*)	Spain	1
American crow (*Corvus brachyrhynchos*)	USA	11
House crow (*Corvus splendense*)	India	8
Brahminy myna (*Sturnus pagodarum*)	India	8
Indian myna (*Acridotheres tristis*)	India	8
European starling (*Sturnus vulgaris*)	USA	12
Loggerhead shrike (*Lanius ludovicianus*)	USA	12
Common grackle (*Quiscula quisculus*)	Canada, USA	13,14
Mammals		
Opossum (*Didelphis virginana*)	USA	2
Black rat (*Rattus rattus*)	Spain	6,10,15
Norway rat (*Rattus norvegicus*)	USA	3,5
House mouse (*Mus musculus*)	Spain	10
Palm squirrel (*Fumambulus palmarum*)	India	16
Five-striped palm squirrel (*F. pennanti*)	India	15
Weasel (*Mustela nivalis*)	Spain	10
Racoon (*Procyon lotor*)	USA	2,3,5,11
Cat (*Felis domestica*)	USA, Canada	3,5,11

This table represents a compilation of the species that have been identified in various sources as nest predators of sparrows. It should be considered to be only a partial list of known nest predators of the species.

Sources: 1, Veiga (1993b); 2, Pitts (1979); 3, Murphy (1978a); 4, Anderson (1978); 5, Lowther (1985); 6, Gil Delgado et al. (1979); 7, Mirza (1972); 8, Mathew and Naik (1998); 9, Ion and Ion (1978); 10, Cordero (1991b); 11, Lowther (1996); 12, Voltura et al. (2002); 13, Gowanlock (1914); 14, A. Pieper *in litt.;* 15, Escobar and Gil-Delgado (1984); 16, Naik and Mistry (1972).

TABLE 4.4. Hatching, Fledging, and Nesting Success of the House Sparrow

Location	N	Hatching Success	Fledging Success	Nesting Success	Source
Europe (P. d. italiae)					
Italy (45°20'N)	1211	0.755	0.812	0.613*	1
Europe (P. d. domesticus)					
Spain (39°41'N)	1455	0.780	0.241*	0.188	2
Spain (39°41'N)	531	0.684*	0.427*	0.292*	3
Spain (40°38'N)	355	0.713	0.692	0.493	4
Bulgaria (42°41'N)	568	0.931*	0.728*	0.678*	5
Bulgaria (42°50'N)	1677	0.856*	0.587*	0.502*	5
Romania (46°05'N)	159	0.792*	0.683*	0.541*	6
Romania (47°10'N)	316	0.646*	0.725*	0.468*	6
Romania (47°30'N)	132	0.795*	0.819*	0.651*	6
Austria (48°12'N)	398*	0.603*	0.642*	0.387*	7
Czech Republic (49°19'N)	940	0.825	0.789	0.651	8
Czech Republic (50°13'N)	1903	0.805	0.854*	0.687	9
Poland (49°18'N)	48	0.896*	0.837*	0.750*	10
Poland (49°28'N)	57	0.842*	0.812*	0.684*	10
Poland (49°28'N)	99	0.758*	0.921*	0.698*	11
Poland (50°04'N)	171	0.749*	0.680*	0.509*	10
Poland (50°04'N)	271	0.768*	0.601*	0.462*	11
Poland (52°15'N)	1001	0.674*	0.797*	0.537*	12
Poland (52°20'N)	6239	0.684	0.805	0.551*	13
Poland (52°21'N)	238	0.782*	0.763*	0.597*	10
Poland (52°25'N)	515	0.437*	0.400*	0.175*	14
Poland (52°25'N)	1300	0.647*	0.662*	0.428*	15
Poland (54°28'N)	128	0.758*	0.776*	0.588*	11
Poland (54°37'N)	612	0.608*	0.519*	0.316*	14
Poland (54°37'N)	503	0.700*	0.662*	0.463*	15
UK (51°45'N)	3008	0.854*	0.454	0.388*	16
UK (51°45'N)	n.a.	0.801	0.523*	0.419*	17
UK (52°57'N)	404*	0.931*	0.750*	0.698*	18
Germany (52°31'N)	407*	0.877*	0.541*	0.474*	19
Germany (53°50'N)	495*	0.842*	0.849*	0.715*	19
Germany (53°50'N)	697*	n.a.	n.a.	0.696*	20
Denmark (57°12'N)	224*	n.a.	n.a.	0.765*	21
Norway (62°30;N)	150	n.a.	n.a.	0.527*	22
Norway (63°47'N)	166	n.a.	n.a.	0.542*	22
Finland (61°27'N)	952*	0.855*	0.575*	0.492*	23

(*continued*)

TABLE 4.4. (*Continued*)

Location	*N*	Hatching Success	Fledging Success	Nesting Success	Source
North America (P. d. domesticus)					
Mississippi (USA) (33°28'N)	911	0.832	0.770	0.641	24
Texas (USA) (34°10'N)	1076	0.599*	0.534*	0.320*	25
Oklahoma (USA) (34°48'N)	n.a.	0.524*	0.946	0.496	26
Oklahoma (USA) (36°07'N)	434	0.502*	0.649*	0.326*	27
Tennessee (USA) (36°21'N)	309	0.515*	0.654*	0.337*	28
Illinois (USA) (38°05'N)	1502	0.658	0.533	0.351	29
Missouri (USA) (38°55'N)	2893	0.646	0.628	0.406	30
Kansas (USA) (39°02'N)	7316	0.625	0.646	0.404*	31
Nebraska (USA) (40°50'N)	n.a.	0.51*	0.63*	0.32*	32
Illinois (USA) (41°35'N)	1027	n.a.	n.a.	0.368*	33
New York (USA) (42°27'N)	180	n.a.	n.a.	0.705	34
Wisconsin (USA) (43°00'N)	370	0.508*	0.612*	0.311*	35
Michigan (USA) (45°32'N)	1457	0.717	0.779	0.559	36
Alberta (Canada) (51°05'N)	n.a.	0.664*	0.671*	0.446*	37
Alberta (Canada) (51°05'N)	5734*	0.592*	0.542*	0.321*	38
Australasia (P. d. domesticus)					
New Zealand (43°31'S)	n.a.	0. 71	0.61*	0.43	39
New Zealand (39°25'S)	n.a.	0.82	0.61*	0.50	39
Asia (P. d. biblicus)					
Israel (32°21'N)	197	0.701	0.746*	0.523	40
Iraq (33°18'N)	608*	0.720	0.580	0.420	41
Turkey (38°24'N)	51	0.922*	0.532*	0.490*	42
Turkey (39°56'N)	351	n.a.	n.a.	0.929*	43
Turkey (39°56'N)	42	0.857*	1.000*	0.857*	44
Asia (P. d. indicus)					
India (17°19'N)	407	0.816	0.608	0.496	45
India (22°18'N)	5460	0.617	0.410*	0.253	46
India (22°18'N)					
Rural	n.a.	0.545	0.512	0.335	47
Urban	n.a.	0.686	0.550	0.360	47
India (26°18'N)	338	0.799*	0.870*	0.696	48

(continued)

TABLE 4.4. *(Continued)*

Location	N	Hatching Success	Fledging Success	Nesting Success	Source
Africa (P. d. indicus)					
Malawi (15°44'S)	509	0.644*	0.720*	0.464*	49

Hatching success is the proportion of eggs laid (or in some cases, of eggs incubated) that were known to have hatched successfully; fledging success is the proportion of hatchlings that fledged successfully; and nesting success is the product of hatching success and fledging success, or the proportion of eggs laid that fledged successfully. N is the number of eggs on which the estimates are based, with n.a. (not available) indicating that N could not be determined from the information in the source. Values designated with an asterisk (*) were calculated by the author from data contained in tables or figures in the source, and data not available in the source are indicated by n.a.

Sources: 1, Brichetti et al. (1993); 2, Gil-Delgado et al. (1979); 3, Escobar and Gil-Delgado (1984); 4, Veiga (1990a); 5, Ivanov (1987); 6, Ion and Ion (1978); 7, Vaclav and Hoi (2002b); 8, Balat (1974); 9, Novotny (1970); 10, Mackowicz et al. (1970); 11, Pinowski and Wieloch (1972); 12, Luniak et al. (1992); 13, Pinowska and Pinowski (1977); 14, Wieloch and Fryska (1975); 15, Wieloch and Strawinski (1976); 16, Seel (1968b, 1970); 17, Dawson *in litt.*; 18, Wetton et al. (1995); 19, Encke (1965b); 20, Wendtland *in litt.*; 21, Møller (1991); 22, Krogstad et al. (1996); 23, Rassi *in litt.*; 24, Sappington (1975); 25, Mitchell et al. (1973); 26, Pogue and Carter (1995); 27, North (1968); 28, Pitts (1979); 29, Will (1973); 30, Anderson (1978); 31, Lowther (1979c); 32, Pochop and Johnson (1993); 33, Lowther (1996); 34, Weaver (1942); 35, North (1972); 36, Anderson (1994); 37, Murphy (1978a); 38, McGillivray (1983); 39, Dawson (1972b); 40, Singer and Yom-Tov (1988); 41, Al-Dabbagh and Jiad (1988); 42, Siki (1992); 43, Kiziroglu et al. (1987); 44, Erdogan and Kiziroglu (1995); 45, Kumudanathan et al. (1983); 46, Naik (1974); 47, Mathew and Naik (1998); 48, Rana and Idris (1991a); 49, Nhlane (2000).

to a high of 93.1% (Gormi Bugrov, Bulgaria, and Nottingham, UK). There appeared to be regional differences in hatching success, although the reasons for this are not readily apparent. Average hatching success from 31 European studies (including those in the British Isles, and treating each study as an independent estimate of hatching success) was 76.2%, significantly higher than the average for 13 North American studies, 60.7% ($t = 4.50$, $P < .001$). One possible explanation for this difference is that most of the North American studies were from midcontinental areas where severe weather (including rapid temperature fluctuations, high winds, and heavy rains) were more common than in Europe. There did not appear to be any consistent seasonal differences in hatching success (Lowther 1983; Will 1973).

Nestling Growth and Survival

Sparrows, like other passerines, have altricial young, which means that the young hatch helpless and blind and must be fed and brooded by the parents. They have rapid growth and development, however, which leads to early fledging. The nestling period, the time from hatching to fledging, is approximately 14 d, with numerous estimates of average nestling period ranging from 13 to 16 d: 13.2 d in Italy (Brichetti et al. 1993); 13.6 d in New York (Weaver 1942); 13.9 d in Kansas (Lowther 1979c); 13.9 d in Oklahoma (North 1968); 13.9 d in England (O'Connor 1977); 14 d in Poland (Pinowska 1979); 14.1 d in Czech Republic (Novotny 1970); 14.3 d. in Michigan (Anderson 1994); 14.6 d in Tennessee (Pitts 1979); 14.8 d in Missouri (Anderson 1978); 15.4 d. in Wisconsin (North 1972); 15. d in India (Rana and Idris 1991b); and 15.8 d in Israel (Singer and Yom-Tov 1988). A few studies have reported longer average nestling periods, including 16.6 d in Turkey (Erdoan and Kizirolu 1995), 17 d in Pakistan (Mirza 1972), and 17.1 d in Mississippi(Sappington 1975). Anderson (1994) found that there was no relationship between nestling period and latitude in eight North American populations.

Begging Behavior and Parental Provisioning

As noted earlier, both parents feed the nestlings. In England, Seel (1969) observed feeding rates of broods in which the number of nestlings was known and there were two parents feeding the young. Provisioning rates increased with nestling age until the chicks were 8–11 d old, after which they remained at a stable high rate. Provisioning rate varied with brood size, increasing from 1 to 2 chicks, and from 2 to 3 chicks, but remained approximately the same for broods of 3–5 chicks (overall averages over the entire nestling period were 9.53, 15.0, 24.36, 24.20, and 25.40 visits/h for broods of 1–5, respectively. Males tended to visit the nest more frequently than females during the early part of the nestling period: they provisioned significantly more frequently for nestlings aged 0–3 d and those aged 4–7 d. For nestlings aged 8–11 d, there was no significant difference in the rate of male and female provisioning, but for nestlings aged 12–15 and 16–19 d, the female fed significantly more than the male. The drop in male provisioning rate was associated with an increase in male display near the nest.

Similar results were obtained in the United States. North (1980) reported that the overall feeding rate averaged 17.5 feeds/h in Wisconsin. Females increased their feeding rates throughout the nestling period (from an average of 7.7 feeds/h on day 1 to 21.1 feeds/h on day 13), whereas males increased their feeding rate from day 1 to day 9, after which it dropped dramatically. In New York, males contributed approximately half of the feeds during the first 10 d of the nestling period, but the percentage dropped to about 20%–25% during the last 5 d of the nestling period, and the difference was highly significant ($P < .005$) (Hegner and Wingfield 1986a). The decline in feeding by the male occurred at the same time that testosterone levels began rising rapidly to peak levels by the end of the nestling period (see Fig. 3.5 in Chapter 3).

In studies in which brood size was manipulated (see earlier discussion), provisioning rates of both parents tended to increase with increasing brood size (Forschner 1990; Hegner and Wingfield 1987a). The increases in feeding rates did not correspond to the increased number of young, but lower per-young provisioning rates may be partially offset by increased thermal efficiency in larger broods.

Recent studies have examined nestling begging behavior of sparrows, particularly to determine whether begging is an honest signal reflecting the energetic state of the nestling. One experiment examined the effect of experience on intensity of nestling begging behavior (Kedar et al. 2000). Paired nestlings were held together in the laboratory, and one was fed immediately on initiation of begging, whereas the second was fed only after it had begged intensively. Both nestlings were fed equivalent quantities of food, however, to control for physiological differences in need that might affect begging intensity. Videotapes of the begging postures of the two nestlings at subsequent feedings were used to determine their begging intensities. The mean begging intensity of chicks receiving food immediately after initiation of begging was significantly lower than that of chicks forced to beg intensively before being fed ($P = .014$). Observations in the field tended to confirm the laboratory findings: there was a significant positive correlation between the begging intensity of individual chicks in nest-boxes and the begging intensity that had resulted in their last feeding ($r = 0.44$, $P < .05$). There were no detectable differences in the growth rate of nestlings in the two laboratory groups, suggesting that there is little energetic cost of intense begging.

Nestling Growth

The most common method for monitoring nestling growth is to measure age-specific changes in total body mass, but other external measures are also frequently recorded, including tarsus length, bill length, and primary length. Some studies have also examined changes in internal organs or body components (e.g., Mathew and Naik 1994; O'Conner 1977). Ricklefs (1969) developed a model, based on the relationship of nestling growth rate to clutch size, nestling survivorship, and number of broods per season, that predicted that growth would always be maximized. "In all species, mortality is sufficient to drive the pace of development to physiological limits . . . the physiological maximum" (Ricklefs 1969:1037). Rate of development is therefore constrained by the development rate of the slowest-developing organ system in the body. Growth is dependent on provisioning by the parents, and in sparrows, as described earlier, the relative contributions of the two parents change during the course of nestling development.

Growth in mass of nestling sparrows follows a sigmoidal growth trajectory similar to that of other passerines (Mackowicz et al. 1970; Myrcha et al. 1972; Novotny 1970; Seel 1970; Weaver 1943). Maximum nestling mass is usually attained when nestlings are 10–12 d old (Barkowska et al. 1995, Ivanov 1987; Mirza 1972; Novotny 1970), followed by a recession in mass that is also typical of most passerines. In one study in England, the average mass recession amounted to 10.4% of the maximal mass achieved (O'Connor 1977).

Several studies of nestling growth have been based on finding the logistic equation that best describes the growth in mass of an individual nestling (see Ricklefs 1969); others have used a more generalized sigmoidal growth model, the Richards equation (e.g., Barkowska et al. 1995). The equations are usually determined iteratively and are found by identifying the equation that produces the least sum of squares of the deviations of the empirical values from the fitted logistic curve. Use of the logistic equation involves determination of three parameters, the asymptote of the curve (A), an inflection point constant (B), and growth rate (K) (see Ricklefs 1968). Some of the studies discussed here compared actual empirical measures of growth, and others compared parameters from fitted logistic curves. Barkowska et al. (1995) determined that mass data from a minimum of 6 d (out of the first 13 d of the nestling period) were necessary to accurately characterize the growth parameters of an individual nestling.

O'Connor (1977) examined the growth rates of different components of nestlings in England. Carcasses of nestlings were skinned, feathers were removed, and the carcasses were divided into skin (including feathers), pectoral muscles, wings, legs, head, heart, liver, gizzard, lungs, alimentary canal (including esophagus), and body shell. Wet, dry, and lean dry masses of each component were obtained. O'Connor then made allometric comparisons between the lean dry masses of different body components and total body lean dry mass. The head and body case grew relatively rapidly, whereas the legs, wings, and skin grew relatively slowly. The relative masses of the different organs were also compared, and liver, gizzard, and alimentary tract had high relative masses early in the nestling period, whereas the skin tended to increase in relative mass and the legs remained fairly constant after an initial increase during the first 3 d. The water content of all components decreased steadily with age. This decrease in relative water content has also been observed in other studies. In Poland, the percentage of water decreased from about 85% at hatching to about 71.5% in nestlings 14–15 d old (Myrcha et al. 1972), and comparable values in India were 85% and 69% (Mathew and Naik 1994). The lipid content of the Indian sparrows initially decreased (from about 3% at hatching to about 2.3% at 2–4 d), but then increased steadily (to about 7.4% at 14–16 d).

Neff (1973) examined the relative growth of various organs in both embryonic (prehatching) and nestling sparrows in Switzerland. The relative growth rates of heart, kidneys, and lungs remained fairly constant throughout embryonic and nestling life, whereas the relative growth rates of eyes and brain were initially high but dropped during embryonic life to fairly constant rates during the nestling period. The alimentary tract and liver showed an increasing relative growth rate from the initiation of embryonic development to about the fifth or sixth day of nestling life, after which they dropped slowly. In Wisconsin, Caviedes-Vidal and Karasov (2001) found that stomach mass was as great in 3-d-old nestlings as in 6- and 9-d-old nestlings, whereas small intestine mass reached its asymptote at 6 d, and both liver and pancreas masses increased until day 9.

Calorific content (Y in kilojoules per gram [kJ/g] of fresh mass) of nestlings increased linearly throughout the nestling period (X = nestling age in d) in Polish sparrows: $Y = 3.3129 + 0.2289X$ ($r = 0.935$, $P < .001$) (Myrcha et al. 1972). Calorific content per gram of dry mass did not change throughout the nestling period, however, averaging about 21.8 kJ.

Growth tends to be linear in other external characters. Tarsus length increases linearly from hatching to about 10 d of age, when it reaches a length close to that of adults (Anderson, unpublished data; O'Connor 1977). Mean tarsus length at 10 d of age for 31 nestling sparrows captured or recovered as adults during a 6-y study in Michigan was 19.2 mm (SD = 0.60), whereas their adult average was 19.3 mm (SD = 0.60) (Anderson, unpublished data). Tarsal growth is apparently typical of skeletal growth in general. Novotny (1970) measured 13 skeletal elements of nestling sparrows in the Czech Republic and found that the sum of the elements grew in an approximately linear way from hatching to 10 d of age, after which there was only a slight increase. Jones (1982), however, reported that growth in the length of the diagonal (the distance between the posterior end of the sternum and the anterior end of the coracoid) in English sparrows showed a logistic-type growth curve during nestling life. She also concluded that there was little post-fledging growth in the skeleton.

Feather growth (measured as the length of feathers after emergence from the skin) is also approximately linear but continues until fledging (Novotny 1970; O'Connor 1975a). Emergence of the remiges (primaries and secondaries) and rectrices occurred between days 4 and 6 in the Czech Republic (Novotny 1970), whereas in England the primaries first emerged in 4-d-old nestlings (O'Connor 1975a).

Nestling growth varies seasonally in some populations, but not in others. Łącki (1962) reported that sparrows in Poland reached higher masses in later broods than in early broods. Seel (1970), however, found that there was no seasonal variation in mass of 13-d-old nestlings in England, and Novotny (1970) also found no difference in growth rates with brood number in the Czech Republic. In Michigan, fitted growth curves were obtained for 522 nestlings in 152 successful broods (92 midseason and 60 late-season broods) (Anderson, unpublished data). Average hatching-day mass did not differ between middle- and late-season broods (2.95 and 2.97 g, respectively), but both average mass and tarsus length of midseason broods (26.2 g, 19.25 mm) were significantly greater than those of late-season broods (25.4 g, 19.0 mm) ($P < .02$ in both cases). An analysis of covariance of fitted logistic parameters for middle- and late-season nestlings found that both asymptote (A) and growth rate (K) were significantly lower in late-season broods ($P = .024$ and $P = .019$, respectively). Initial brood size had no effect on growth, however (Fig. 4.7).

The results of a food restriction experiment tended to support Ricklefs' hypothesis of maximal growth. Nestlings obtained in the field in Wisconsin were taken into the laboratory, maintained in constant conditions (14L:10D photoperiod, 36°C, 62% relative humidity), and hand-fed a synthetic liquid diet (Lepczyk and Karasov 2000). In two experimental groups, the quantity of food was restricted for 2 d to the amount necessary to maintain constant mass. In the early food-restriction group, food restriction began on nestling day 3, and in the late food-restriction group, it began on day 6. After the food-restriction period, the birds were provided food *ad libitum*. None of the early-restriction nestlings survived to fledge, but most of the late-restriction group survived to fledging. There was no evidence of compensatory growth in the late-restriction group compared with control nestlings maintained under the same conditions. Growth in bill length, tarsus length, and eighth primary (P8) length continued with little change during the food-restriction period but tended to reach slightly lower asymptotic values than those of control nestlings. Mass growth in the late-restricted nestlings continued along the same trajectory as that of control nestlings, reaching approximately the same peak mass 2 d later than in controls. There was therefore no evidence of compensatory growth in the food-restricted nestlings, which is consistent with the proposition that they are already growing at a maximal rate.

Brooding and Thermoregulation

Both adults participate in brooding the young for the first several days of nestling life. In a study in Wisconsin, North (1980) found that the young sparrows were brooded for 40% of the time during the first 3 d of nestling life and 35% of the time during the next 2 d. Brooding dropped considerably after that, with little brooding after day 8. The female did 66% of the brooding and also roosted in the nest at night. In France, brooding time was inversely correlated with

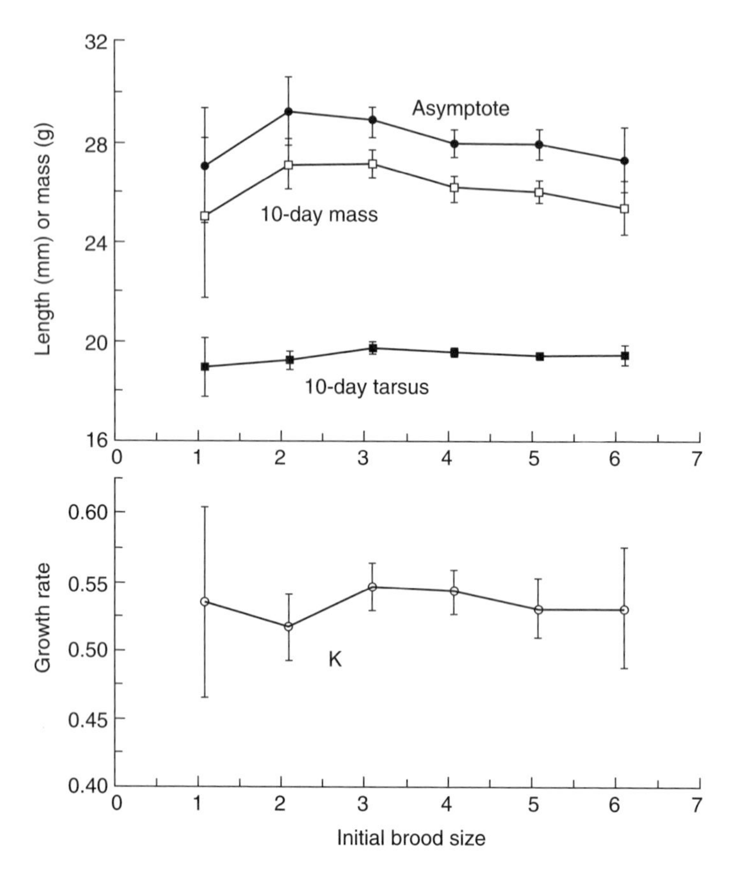

FIGURE 4.7. Effect of initial brood size on growth characteristics of nestling house sparrows in Michigan (USA). Nestlings were weighed daily from hatching day to 10 d of age, and logistic growth curves were fitted iteratively using SYSTAT to provide estimates of asymptote and growth rate (K). Length of the tarsometatarsus (tarsus) of each 10-d-old nestling was also recorded. Data are from 522 nestlings in 152 nests monitored between 1986 and 1991 (Anderson, unpublished data).

brood size, with the total time spent brooding 4- to 6-d-old chicks (Y in percent) being related to brood size (X) by the following regression equation: $Y = 70.3 - 6.2X$ ($P = .0001$) (Chastel and Kersten 2002).

The brooding schedule coincides well with the development of thermoregulatory ability by the nestlings. Seel (1969) recorded core body temperatures of nestlings in England by inserting a glass bulb containing a thermistor approximately 1 cm into the rectum. Nestlings up to 6 d old lost temperature rapidly during a 15-min period following isolation from the nest. Temperature decreases were progressively less rapid for nestlings aged 7–9 d, and 10-d-old nestlings were able to maintain a fairly high body temperature. Similar results were obtained in a study on temperature regulation in nestlings in Poland (Myrcha et al. 1972). In another experiment, Seel (1969) clipped the feathers of three nestlings capable of thermoregulation, whereas controls were only handled roughly. Featherless birds lost body temperature rapidly, but controls showed no change in body temperature. Seel concluded that the development of the feathers is probably causally related to the development of thermoregulation. Jones (1982) studied the development of the pectoralis muscles in nestlings, also in England, and found that the muscles were functional in 9-d-old nestlings, and suggested that they might be used for shivering thermogenesis to maintain body temperature.

O'Connor (1975a) attempted to identify the relative contributions of developmental changes in three factors to the ability of nestling sparrows to thermoregulate: (1) insulation due to feather development, (2) metabolic intensity, and (3) surface-to-volume ratio. The latter represents the ratio of the rate of heat loss (a function of surface area) to the volume of tissue (potential for heat production); it decreases by approximately 50% from hatching to fledging. Metabolic intensity, measured as oxygen (O_2) consumption (see Chapter 9) in the thermoneutral zone of nestling sparrows (35°C–40°C), did not change appreciably during the nestling period and averaged about 44.2 Joules/gram hour (J/gh). When nestlings were held at lower ambient temperatures, metabolic intensity decreased in young nestlings (up to about 6 d of age) but then showed a progressive increase with age. In 11–12 d old nestlings, for instance, the metabolic intensity was about 192.8 J/gh. The increase in metabolic activity coincided with the period of rapid feather development, which occurred between the sixth and ninth days of nestling life, and O'Connor concluded that all three factors play important roles in the development of thermoregulation in nestling sparrows.

O'Connor (1975c) tested the ability of nestlings to effectively dissipate heat at different stages of the nestling period, predicting that they should be better able to dissipate than generate heat at an early stage, because adults are less able to assist in heat dissipation than in heat generation. He transferred nestlings individually into respirometer chambers that were maintained at either 35°C or 40°C and monitored body temperature of the nestlings after 2–10 h in the chamber. Young of all ages were observed with open-mouth breathing and panting at 40°C. The body temperature (Y in °C) of nestling sparrows increased significantly with age (X) during the first 6 d of the nestling period at both temperatures (at 35°C, $Y = 33.2 + 0.50X$, $P < .01$; at 40°C, $Y = 36.3 + 0.49X$, $P < .01$) but did not increase significantly thereafter up to 14 d of age. The difference in the intercepts of the two regression equations was also significant ($P < .01$), and O'Connor concluded that the fact that the intercept for the

40°C regression line was lower than that for the 35°C line indicates that birds are able to dissipate heat effectively during the first few nestling days.

Fledging Success

Fledging success is usually recorded as the percentage of hatched young that leave the nest. Some of the causes of nestling mortality are the same as those for egg mortality (i.e., predation, weather, desertion), whereas others, such as starvation or accidentally falling or being pushed from a crowded nest, are unique to nestlings. Some causes of mortality may, however, act differentially on eggs and young. For instance, nestlings may be more vulnerable to predation due to their begging behavior, which could attract predators or make them more vulnerable when they go to the nest entrance to beg as they near fledging (see North 1980). Nestlings can also occasionally die due to hyperthermia during unusually warm weather (Lowther 1996). Table 4.4 includes estimates of average fledging success from numerous studies throughout much of the house sparrow's range. Fledging success ranged from a low of 24.1% in an orange grove in Spain (Gil-Delgado et al. 1979) to a high of 100% for a small sample in Turkey (Erdoğan and Kiziroğlu 1995), but the middle 90% of the values range from 42.7% to 83.7%. In Bulgaria, about 75% of nestling mortality occurred during the first few days of nestling life (Ivanov 1987).

Inspection of the data summarized in Table 4.4 fails to suggest any geographic pattern in fledging success. Indeed, unlike the situation with hatching success, there is no significant difference in average fledging success between the 31 European studies (67.5%) and the 13 North American studies (66.1%) ($t = 0.28$, $P > .80$). There is also no significant relationship between latitude and fledging success for the full data set ($r = 0.144$, $P > .10$).

Ringsby et al. (2002) examined variation in fledging success during a 6-y study on five islands off the coast of Norway. Fledging success varied significantly among the islands despite the fact that there was a strong correlation in weather patterns on the islands. However, weather, evaluated as the first principal component (PCI) of a principal components analysis including both temperature and precipitation, did have a significant effect on fledging success ($P < .001$); with good weather being associated with improved nestling survival. The island difference is apparently primarily related to differences in habitat quality and to temporal differences in nesting phenology among the islands, which resulted in different weather being experienced by nestlings on different islands.

A recent study in Austria examining the effects of male contributions at different stages of the breeding cycle found that only male provisioning rate during the first 10 d affected fledging success (Hoi et al. 2003). The residual of male provisioning rate (corrected for differences in brood size) was positively correlated with fledging success ($r = 0.57$, $P < .001$).

Intraspecific Infanticide

Intraspecific infanticide, which involves the destruction of eggs or nestlings by other adult sparrows, has been reported in several populations (e.g., Matuhin 1994; Schifferli 1978a; Veiga 1990a, 1990b, 1993c). Both males and females participate in infanticide (Veiga 1990b1993c), and losses can be considerable, with infanticide occurring in 9%–12% of nesting attempts in one population in Spain (Veiga 1990b). Individuals engaging in infanticide may be birds attempting to acquire a nesting site. Alternatively, in the case of polygynously mated females, they may be attempting to destroy the nest of the female whose nesting attempt is favored by the male (see earlier discussion) (Veiga 1990b, 1993b). Veiga (1990c) reported that cases of infanticide by a female (either directly observed or inferred from peck wounds on chicks) were more frequent in polygynous than in monogamous nests in Spain ($P < .05$). Infanticide by males is common after the disappearance of a pair male (Schifferli 1978a; Veiga 1993c).

Veiga (2003) examined two hypotheses that have been proposed to explain selection for male infanticide. The first proposes that infanticide accelerates the initiation of a new breeding attempt by the female, thereby gaining a temporal advantage for infanticidal over noninfanticidal males. The second proposes that infanticidal males obtain greater reproductive effort from females that have not completed their previous breeding cycle. Using instances of naturally occurring male replacement during a 10-y study in Spain, Veiga compared the subsequent breeding success of infanticidal and noninfanticidal males. The interclutch interval of females mated to infanticidal males was significantly shorter than that of females mated to noninfanticidal males (31.4 vs. 47.1 d, respectively, $P = .001$), and the number of subsequent breeding attempts was significantly greater (1.6 vs. 1.25, $P = .027$). The mean number of fledglings per brood did not differ between the two groups of females (2.46 vs. 2.33, $P = .84$), however. These data tend to support the temporal advantage hypothesis for male infanticide.

Number of Broods and Cessation of Breeding

The number of broods attempted by individual females in a breeding season varies geographically and also differs between resident and migratory subspecies. Determination of the actual number of attempts by individual females requires an intensive study of marked individuals throughout the breeding season, and few such studies have actually been performed. In lieu of monitoring breeding attempts by individuals, some studies have recorded the number of breeding attempts per nest site, or the length of the breeding season (usually as the number of days between initiation of the first clutch and initiation of the last clutch in the study population). Individuals in most populations are multibrooded, which raises the question of what factors are responsible for the cessation of breeding, often while conditions appear to still be favorable for reproduction.

Number of Broods

Not surprisingly, the number of broods initiated by females varies inversely with latitude in resident subspecies of the house sparrow. In India (22°18'N), a study of 16 annual nesting cycles by 11 marked females found that the average female attempted 7.0 clutches/y (Naik 1974). Two of the females successfully reared six broods (Naik and Mistry 1980). In the same population, the average number of clutches initiated per nest site was 4.13 (Naik 1974). In Spain (40°38'N), on the other hand, Veiga (1992b) reported that females had an average of 2.37 broods/y, whereas in northern Finland (65°01'N), females laid an average of 2.0 clutches/y (Alatalo 1975).

Although little is known about the breeding biology of the two migratory subspecies of the house sparrow (see Chapter 1), one study compared aspects of the breeding biology of the migratory subspecies *bactrianus* with that of the sympatric resident subspecies *domesticus* in Kazakhstan (43°16'N). Females of *bactrianus* normally lay only one clutch per year, whereas in the same area, *domesticus* females usually lay three clutches (Gavrilov and Korelov 1968). Nothing is known about the generality of this difference.

Interbrood Interval

The period between the fledging of one brood and the initiation of the next clutch is usually referred to as the interbrood interval. Some authors have referred to the period between initiations of successive clutches (if the first is successful) as the interbrood interval (e.g., Ion and Ion 1978), but this period is probably best referred to as the interclutch interval. The latter term has also been used, however, to refer to the period from the hatching of one clutch to the initiation of the next (Schwagmeyer et al. 2002). Most studies reporting interbrood intervals have used the interval between broods at the same nest site without knowing whether the same female was responsible for the successive clutches. During the interbrood interval, females not only replenish fat and protein reserves and initiate egg development (discussed earlier), but they also continue to feed fledglings from the earlier brood for several days (Summers-Smith 1963).

The average interbrood interval *sensu stricto* has generally been found to be 7–10 d (Anderson 1994; Dawson 1972b; Hegner and Wingfield 1987b; Veiga 1996a). At least three studies have reported a significant positive relationship between the length of the interbrood interval and the number of young fledging from the first brood (Dawson 1972b; Hegner and Wingfield 1987a; McGillivray 1983), which suggests that there is an intraseasonal cost of reproduction (cf. Williams 1966). In a study involving the middle- and late-season broods of 50 marked females in Michigan, however, there was no significant relationship between interbrood interval and either initial brood size or number of fledglings (Anderson 1994). This study included several females whose midseason brood sizes had been experimentally manipulated by the addition or subtraction of hatchlings. Thirty of the 50 females began their late-season clutch 4–7 d after the fledging of their midseason broods. There was a significant negative relationship between interbrood interval and the Julian date of the fledging of the midseason brood (Anderson 1994).

In a study at several breeding colonies located on farms in Kansas, Lowther (1983) found that interbrood interval was negatively correlated with the number of livestock maintained on the farm at which the breeding colony was located ($r_s = -0.75$, $P < .05$). He suggested that the number of livestock on a farm was an indicator of food availability, and he attributed the relationship to the enhanced ability of females to acquire the resources required to initiate another clutch.

In Spain, Veiga (1996a) reported that the average interbrood interval for males that changed mates (6.8 d) was significantly shorter than that of males that remained mated to the same female (11.2 d) ($P = .002$). This suggests that there may be selection pressures on males to change mates between successive broods.

Sparrows occasionally show brood overlap, in which laying of the next clutch is initiated before fledging of the preceding brood. Lowther (1979b) reported eight cases in Kansas in which egg-laying began when nestlings were 5–15 d old. Most instances occurred late in the breeding season and were presumed to involve the female that was feeding the young. Anderson (1994) observed four instances of brood overlap, each involving marked females, and all being late-season broods. In two instances, the clutch was initiated on the day before fledging, and in the other two, the first egg was laid on the day of fledging.

Cessation of Breeding

Few studies have been conducted on the factors responsible for cessation of breeding in sparrows or other multibrooded species. Both adaptive and constraint hypotheses have been proposed to explain cessation of breeding in sparrows. One adaptive explanation is that they cease breeding while conditions are still suitable for raising young because late-reared fledglings have a low probability of surviving to enter the breeding population (Murphy 1978b). Constraint hypotheses, on the other hand, suggest that sparrows cease breeding because of a declining food supply (Naik and Mistry 1980) or because females are not able to replenish their protein reserves rapidly enough to initiate another clutch (Hegner and Wingfield 1986b; Pinowska 1979). The adaptive and constraint hypotheses are not mutually exclusive, however.

Anderson (1998) attempted to test these hypotheses in a population of marked females in Michigan. One prediction of the adaptive hypothesis would be that small late-season clutches (see earlier discussion) might represent a tradeoff between clutch size and egg size, with larger late-season eggs resulting in more rapid development and larger fledglings, increasing the chances of overwinter survival of late-reared young. Females laying late-season clutches did not show any increase in average egg size, however, and there was also no increase in fledgling mass of late-reared young. Females that successfully reared enlarged midseason broods were also less likely to initiate a late-season clutch (see earlier discussion). Anderson concluded that the evidence supports the constraint hypothesis.

Reproductive Lifespan

Somewhat surprisingly, there are few long-term population studies involving large numbers of marked house sparrows. As a consequence, there is little information on the reproductive lifespan of individual sparrows. Adult mortality rates in the species are high, however (see Chapter 8), which means that few individuals live to breed for more than two breeding seasons. The oldest free-living house sparrows were two males that lived at least 13 y 4 mo (Dexter 1959; Klimkiewicz and Futcher 1987), but their breeding histories were completely unknown.

I have conducted two long-term population studies of sparrows involving many marked individuals, one in Missouri and the other in Michigan (Anderson 1978, 1994, 1998). The oldest breeding individual I recorded was a female that was at least 6 y old, and two other females were captured while breeding in five successive summers. These were the longest-lived breeding individuals in the Michigan study, among a total of 221 females captured while breeding between 1986 and 1995 (125 of which were captured as breeding adults between 1986 and 1991) (Anderson, unpublished data). Among a smaller number of individuals marked during the Missouri study, the oldest breeding female was 4 y old.

Captive Breeding

A significant impediment to the use of sparrows in many kinds of laboratory research is the relative difficulty of breeding them in captivity. There are few reports of successful captive breeding (e.g., Baker 1995; Johnston 1965; Mitchell and Hayes 1973; Moreno-Rueda and Soler 2001, 2002; Washington 1973). The conditions required for successful captive breeding have not been clearly identified, and therefore a brief review of those reported from successful studies may be useful.

Two studies have reported successful breeding by isolated pairs of sparrows (Johnston 1965; Washington 1973), but studies involving groups caged together appear to have greater success. Johnston (1965) maintained pairs in aviaries measuring 2.4 × 2.4 (3.7 m × 21.3 m^3), and Washington (1973) used two aviaries with similar dimensions (16.3 and 21.0 m^3) to house his pairs. Larger aviaries have been used in studies involving multiple birds, but the average aviary volume per bird has been much lower. Mitchell and Hayes (1973) housed 39 sparrows in two adjacent aviaries with volumes of 18.2 and 26.8 m^3, an average volume of 1.15 m^3 per bird. Baker (1995) housed about 25 sparrows in each of four outdoor aviaries with volumes of approximately 138.6 m^3, about 5.5 m^3 per bird, while Moreno-Rueda and Soler (2002) kept 51 and 34 sparrows in a 20.5 m^3 aviary in two successive years (0.4 and 0.6 m^3 per bird in the 2 y, respectively). In the successful laboratory study of hybridization between house sparrows and Spanish sparrows described in Chapter 1, Macke (1965) housed six pairs in 12 m^3 cages (1.0 m^3 per bird).

Breeding success is generally lower among captive birds than among free-living birds in the same area. Although the average clutch size of captive pairs is similar to that of free-living sparrows (Baker 1995; Mitchell and Hayes 1973; Moreno-Rueda and Soler 2002), hatching success and fledging success are often lower than in neighboring free-living birds. In Texas, Mitchell and Hayes (1973) found that hatching success in captive breeders was comparable to that of free-living birds, but fledging success was much lower. In Spain, both hatching and fledging success were lower in captive breeders, with fledging success being 0% during the more crowded year (0.4 m^3 per bird) and 32.4% in the other year (0.6 m^3 per bird) (Moreno-Rueda and Soler 2002). Interference from other sparrows, including frequent infanticide, is an important contributor to nesting failure, and the latter study may indicate that there is a threshold density above which nesting success in captive sparrows is highly unlikely.

Both Mitchell and Hayes (1973) and Moreno-Rueda and Soler (2002) provided quantitative descriptions of the diets provided to captive sparrows during the breeding period. In both cases, the diet consisted of a mixture of seeds, green vegetable material, eggs, bread, and insect larvae (mealworms or dipteran larvae). Moreno-Rueda and Soler (2002) suggested that mealworms may be toxic to nestlings when fed in large quantities and recommended the use of dipteran larvae instead.

Chapter 5

PLUMAGES AND MOLT

In such cases sexual selection must have come into action, for the males have acquired their present structure, not from being better fitted to survive in the struggle for existence, but from having gained an advantage over other males, and from having transmitted this advantage to their male offspring only. It was the importance of this distinction which led me to designate this form of selection as sexual selection.

—Darwin 1871, Vol. I: 257

The feather is the most unique feature of the class Aves. Feathers are probably the most complex derivative of the skin in any vertebrate (Stettenheim 1972), and they perform a number of functions, including flight, thermoregulation, camouflage, and display. Birds spend a considerable amount of time in feather maintenance, including activities such as preening and bathing, and in most species the feathers are completely replaced at least once annually in a complete molt. Contour feathers have tightly interlocking barbs that create the planar surface of the feather, referred to as the vane. Deposition of pigment molecules in the barbs during feather development is responsible for many of the distinctive colors that characterize the plumage of a species, as well as the color dimorphisms that are found in species such as the house sparrow.

House sparrows have 3100–3600 feathers (Summers-Smith 1963), which were found to have a total mass of approximately 1.8 g after the prebasic (postnuptial) molt in Great Britain (Schifferli 1981). Mass of the contour feathers decreases in approximately linear fashion from November to July in Great Britain, with the mass in July being about 70% of that in November (Schifferli

1981). Much of this loss is apparently due to a gradual loss of feathers that are not replaced until the next molt. Wetmore (1936), for instance, found that a sparrow captured in July near the District of Columbia (USA) had only 1359 feathers weighing 1.5 g. Most of the loss in mass is due to loss of body feathers; for birds in Illinois (USA), the mass of flight feathers (remiges and rectrices) varied only between 0.39 and 0.48 g, whereas the mass of body feathers varied between 0.90 and 1.53 g (Barnett 1970).

Newly hatched sparrows are naked, possessing no down, and the sequence of plumages and molts begins with the acquisition of the juvenal plumage in a prejuvenal molt that occurs during nestling development. Beginning a few weeks after fledging, a complete molt, the first prebasic (postnuptial) molt, results in the replacement of the juvenal plumage with the first basic plumage. Annually thereafter, a prebasic molt occurring after the breeding season results in the production of second and subsequent basic plumages.

Plumage Description

Sexual Dichromism

The adult plumage of sparrows is sexually dichromatic, with the male (Fig. 5.1a) possessing striking features on the head, throat, breast, and wings that differ from the plumage of the female (Fig. 5.1b), which has a fairly uniform gray-brown coloration. Adult males in fresh basic plumage have buffy tips on many of the feathers of the head, breast, and wings, and the striking dimorphic coloration of the breeding plumage in these regions is attained by abrasion of these tips. The juvenal plumage acquired during nestling development is similar in coloration to that of the adult female in both sexes.

The crown and center of the nape of the neck of the male in breeding plumage are gray, except in *Passer domesticus italiae* and similar groups in North Africa and on several Mediterranean islands, in which these regions are chestnut brown. The lores between the base of the bill and the eye are black, as is a short postocular stripe. These black stripes grade into broader chestnut brown stripes that proceed down the lateral parts of the nape, broadening posteriorly to form a narrow chestnut line that outlines the lower part of the cheek region, which is pale gray. There is a small white spot just behind and above the eye (postocular spot). The throat is black, as is a variable-sized region of the upper breast (the "badge"). The rest of the abdominal region is light gray. The scapulars and feathers of the upper back are streaked with black, buff, and brown, while the lower back and rump are gray-brown. On the wing, the secondary coverts are reddish brown streaked with black and have white tips pro-

FIGURE 5.1. Plumages of adult house sparrows: male in the sexual display posture (A) and female in threat position (B). See text for description of plumages. From Summers-Smith (1963: Figs. 2 and 6), used by permission of the author.

ducing a single wing bar. The rest of the wing feathers, as well as those of the tail, are brown streaked with black. Measurements of the reflectance of ultraviolet light (350–400 nm) showed no reflectance from the black throat and only slight reflectance from the gray crown in males (Radwan 1993). As noted in Chapter 3, the bill is black during the breeding season.

The crown and nape of the neck of the female are dull brown except for a buff superciliary stripe that is particularly prominent above and behind the eye. The cheek, although also buff-colored, is partially surrounded by a thin brown postocular stripe and a second thin brown line proceeding ventrally from the base of the nape. The entire ventral surface is pale grayish brown, becoming paler posteriorly. The back, wings, and tail are dull brown streaked with black. The bill is horn-colored, although it may darken somewhat during the breeding season.

Although sparrows in juvenal plumage all resemble the female in coloration, there are some differences that often permit sex identification. Keck (1934) noted that the throat of males tended to be dusky, while that of females was white, and Nichols (1935) noted that males tended to have a white postocular spot as well as a blackish throat. J. M. Harrison (1961a) examined about 500 juvenal specimens from Great Britain and noted that in general males tended to have a dusky throat, a white postocular spot, and a more colorful postocular stripe (usually cinnamon to bay or russet), whereas females tended to have whitish throats, no white postocular spot, and an ash to drab or bistre postocular stripe. There was much individual variation, however, with 5% of the males lacking the dusky throat and 6% of the females having it. Males also sometimes lacked the white postocular spot, or had it on only one side. Johnston (1967a) obtained similar results on a sample of 73 juvenal sparrows from Colorado (USA) and Kansas (USA), finding that all males tended to have darker throats and most (28 of 40) had a buffy or white postocular spot. Almost all females had whitish throats, and only 4 of 33 had postocular spots.

The plumage differences between the sexes are genetically determined and are not influenced by the sex hormones. Keck (1932a) plucked feathers from sexually dimorphic regions of the plumage of 52 castrated males, and all regenerated with male plumage coloration. Daily injection of female hormones into castrated males did not affect coloration of the replacement feathers. Keck (1932b) also performed ovarectomies on 48 females and found that regenerated feathers in dimorphic regions were invariably of the female type. He also performed subcutaneous or intraabdominal transplants of testes into 15 ovarectomized females and found on postmortem examination that at least three implants were successful. Regenerated feathers of these birds were of the female type only. Nowikow (1935) obtained similar results in castration experiments on both males and females in Russia and also concluded that the sex differences in plumage coloration are under genetic control. He proposed that the genes controlling the color dimorphism were on the sex chromosomes.

Subsequent experiments have confirmed Keck's and Nowikow's early findings. Novikov (=Nowikow) (1946) performed 121 reciprocal transplants of patches of skin from the throat region of juvenal sparrows. The grafts took in only nine cases, and in each case, the feathers that replaced the feathers in the grafted skin resembled those of the donor. This result remained the same for up to four further feather replacements.

To eliminate the possibility that early developmental differences in sex hormones could be responsible for the plumage dimorphism, Mueller (1977) performed reciprocal skin transplantations in the throat region of nestling sparrows in North Carolina (USA). Nestlings were removed from nests and cooled to torpidity by rubbing with alcohol, after which 6 × 10 mm patches of skin were cut from the throat region, rotated 180°, and transplanted to another individual. The nestlings were replaced in their nests until they were 8–11 d old, when they were taken into captivity and maintained until the first prebasic molt. Because of mortality, escapes, and graft rejection, only 12 birds survived beyond the molt. Among these were two pairs of males that served as male controls, and one pair of females plus one female whose female partner had died earlier. In all these cases, the plumage from the grafted skin, which could be unequivocally identified by the fact that the feathers were projecting forward, were appropriate for the sex of both donor and recipient. The other five birds (four males and one female) possessed intersexual grafts, and in each case the plumage on the grafted skin resembled that of the donor rather than the recipient, with the males showing white patches of forward-directed feathers and the female showing a black patch. Mueller concluded that feather coloration in sparrows is not a secondary sexual characteristic influenced by sex hormones but is genetically controlled.

Plumage Anomalies

There are numerous records of sparrows with plumage anomalies, one type of which involves intersexual plumage characteristics. The existence of these individuals has apparently even led to the naming of a species, "the enigmatic sparrow," *Passer enigmaticus* (Sarudney 1903 cited in Mayr 1949). Mayr (1949) examined a series of about 350 specimens of sparrows collected in Iran, Afghanistan, and India and found several specimens (four first-year males and six adult females) with evidence of "intersex" plumage. The males had reduced and grayish throat and breast patches and reduced or absent black and chestnut around and behind the eye, whereas the females showed grayish throats and chestnut or buffy wash on the wing coverts and chestnut on the mantle. J. M. Harrison (1961b) also found female sparrows from Great Britain that showed similar intersex characters, including a dusky throat and striations on the flank. McCanch (1992) observed a plumage gynandromorph, in which the right side had typical female plumage and the left side male plumage, in Great Britain. Cink and Lowther (1987) described a gynandromorph from Chile in which the right side had male plumage and the left side female plumage (see also Chapter 2). Gavrilov and Stephan (1980) described several instances of partial intersexes in house sparrows and Spanish sparrows in Kazakhstan. They observed four instances in house sparrows and three in Spanish sparrows from among 291,600 sparrows examined. The individuals showed partial plumage gynandromorphy; in some cases, they had an ovary on the left side and a testis on the right (with the right-side plumage tending to show evidence of male plumage patterns), and in other cases, the individuals had only a right testis.

Other plumage abnormalities that have been described in sparrows include complete albinism, partial albinism (including single white feathers), melanism, and other aberrant pigmentation. Calhoun (1947b) examined more than 1800 museum specimens from North America and reported on the kinds and frequencies of various plumage anomalies. Several birds in which the entire plumage was affected, including five albinos, five light tan, and one light gray, were presumably the result of germ-line mutations. Numerous other specimens showed various anomalies affecting one to a few feathers, presumably the result of localized somatic mutations. In a review of albinism in North American birds, Ross (1963) reported 55 cases of albinism in sparrows, 10 complete albinos and 45 partial albinos. In a similar review of plumage anomalies in Australia, Sage (1963) reported that the incidence of albinism in the Passeridae (house sparrow and tree sparrow) was 7%. Gross (1965) reported that 5.53% of North American sparrows showed partial or complete albinism, whereas Selander and Johnston (1967) reported that only 1.88% of 2877 specimens collected in North America, Bermuda, Hawaii, England, and Germany showed "conspicuous albinism" (two or more contour feathers white).

Some cases of general plumage discoloration may be due to environmental contamination. Johnston and Selander (1963) reported that the plumage of sparrows collected at a city dump in England (UK) and showing uniform dark coloration had normal coloration after being washed in white gasoline. Harrison (1963) also found that two apparently melanistic sparrows from England had normally pigmented plumage after being washed with benzene. J. M. Harrison (1961a) noted that transient plumage characteristics appear in some juvenile and immature sparrows, which he interpreted as indicating the close phylogenetic relationship of house sparrows and Spanish sparrows (see Chapter 1). These included the presence in some juvenile house sparrows of a whitish line running from the base of the bill backward and over the eye, the form of the black badge, and striations on the flanks. First-year males also showed isolated chestnut feathers in the crown (which is chestnut in Spanish sparrows).

Age Determination

Several authors have described plumage differences that distinguish first-year sparrows from 2-y-old and older individuals. J. M. Harrison (1961a) claimed that first-winter males are distinguishable from older males by a less pure gray crown and broader chestnut edges on the secondaries. He also stated that first-year females have the broad edges on the secondaries. However, in a study of 2877 specimens collected in October and November in North America, Bermuda, Hawaii, England, and Germany, Selander and Johnston (1967) concluded that there was no consistent means of distinguishing first-year birds from older birds. They specifically found no support for distinguishing on the basis of the breadth of the chestnut edges on the secondaries.

In a study of *P. d. bactrianus* in Kazakhstan, Gavrilov and Goloshchapov (1992) suggested that the pattern on the upper middle secondary coverts could be used to discriminate between first-year and older males during the spring. In adult males, the white distal portion of these feathers is larger and forms a distinct "mirror," whereas the basal portions of the feathers are black or blackish-brown. In fist-year males, the white area is smaller and the dark basal regions, which are not fully covered by the upper lesser coverts, extend into the white, resulting in a distinct dark line or partial line between the brown lesser coverts and the white tips of the middle coverts. In the fall, the feathers have buffy (ochre) tips that obscure this pattern. The yellow bill color and fleshy lateral lobes at the base of the bill have also been used to differentiate first-year from older sparrows (Kirschbaum and Ringoen 1936; Pfeiffer 1947), although Cheke (1967) reported that adult females often have yellow bases to the bill. Kirschbaum and Ringoen (1936) also stated that the feet of first-year birds are yellow, whereas those of adults are black.

Cheke (1967) described several methods for discriminating between adult females and juveniles before the first prebasic molt. Juveniles tend to have soft-textured, pliable skin on the legs, which tend to be grayish or grayish-pink, whereas adult females have brittle, hard-textured skin on the legs, which are yellow or orange to strong pink. Adult plumage is worn, and the flank feathers are coarse and have more clearcut vanes. The tenth primary of juveniles is longer and rounder than that of adult females, being more than three-fourths as long as the adjacent covert of the ninth primary; the adult female feather is only about one-half as long as the adjacent covert and is very narrow.

Pterylosis

The contour feathers of birds originate from feather follicles in restricted parts of the skin (feather tracts or pterylae), which are separated from other feather tracts by regions of skin that are bare or that give rise to down or other feathers (apteria). In a detailed study of the pterylosis of the genus *Passer* (with special emphasis on the house sparrow), Clench (1970) identified nine feather tracts: dorsal, scapulohumeral, femoral, ventral, lateral neck, capital, alar, crural, and caudal. The capital tract includes feathers originating from the head and ventral neck, and the lateral neck tract contains feathers originating on the sides of the neck. The scapulohumeral and femoral tracts include narrow bands of feathers on the dorsal surface that originate, respectively, over the scapula and humerus and over the proximal end of the femur. The crural tract is a thin scattering of feathers on the leg below the femoral tract. The dorsal tract runs down the midline of the back, and the ventral tract comprises two tracts running down the lateral surfaces of the underside. The alar tract consists of feathers that originate from the wing, including the remiges (primaries and secondaries), and the caudal tract comprises feathers that originate near the posterior end of the spine, including the rectrices. Tracts, in turn, can be subdivided into elements, portions of a tract that are usually separated by small apteria and within which there are normally uniform patterns of feather follicles.

Clench (1970) examined 176 specimens of house sparrows (*P. d. domesticus*) collected throughout the year, mostly from Connecticut (USA), two specimens of *P. d. indicus* from South Africa, and two *P. d. nilticus* from Egypt. She concentrated her study of pterylosis on the dorsal and ventral tracts (Fig. 5.2).

The dorsal tract has three elements: anterior, saddle, and posterior (see Fig. 5.2a). Each is composed of chevron-shaped rows of feathers with the apex of the chevron located medially and directed anteriorly. In the anterior element, there is an average of 11 rows, with the three anterior-most rows having nine feathers, the middle six having seven, and the final two having five. The element runs from the posterior border of the capital tract on the nape of the neck to the interscapular region where the saddle element begins. The saddle always consists of eight rows, with the average number of feathers in each row decreasing from 23 to 3, and there is little variation in number of feathers in the anterior rows. Fifteen of 42 sparrows examined in fresh basic plumage had a partial row of feathers in an anterior, lateral position on the saddle, and this partial row was seldom bilaterally symmetrical. The posterior element runs from the saddle to just anterior to the uropygial gland. The first row of the posterior element has an average of seven feathers, and the element has an average of 14 rows.

The ventral tract is sometimes described as a "wishbone"-shaped tract that is joined anteriorly at the midline. Close examination of this tract reveals that it actually comprises two tracts (see Fig. 5.2b) (Clench 1970). The anterior part, which includes the bifurcation, is composed of chevron-shaped rows with the apical feather located on the midline. The bifurcation occurs with the disappearance of central feathers in the posterior-most rows. These rows show very little internal organization, and Clench therefore considered these feathers to be part of the capital tract. One or two rows behind the scapulohumeral tract, the nature of the rows changes, with the feathers showing a different pattern and a high degree of organization. Clench (1970) considered this to be the beginning of the ventral tract.

The paired ventral tracts each have two elements, the flank element and the main element (see Fig. 5.2b). The flank element consists of chevron-shaped rows with the apex located in the middle of the element and directed anteriorly. There are seven feathers in the first row, and the number of feathers increases to 10 by the fifth row, after which it remains constant until the last few rows. The main element begins with single feathers adjacent to the medial feathers in row three of the flank element, and the number of feathers in successive rows gradually increases until, after row 17 of the flank element, the main element begins to separate from the flank element. The main ele-

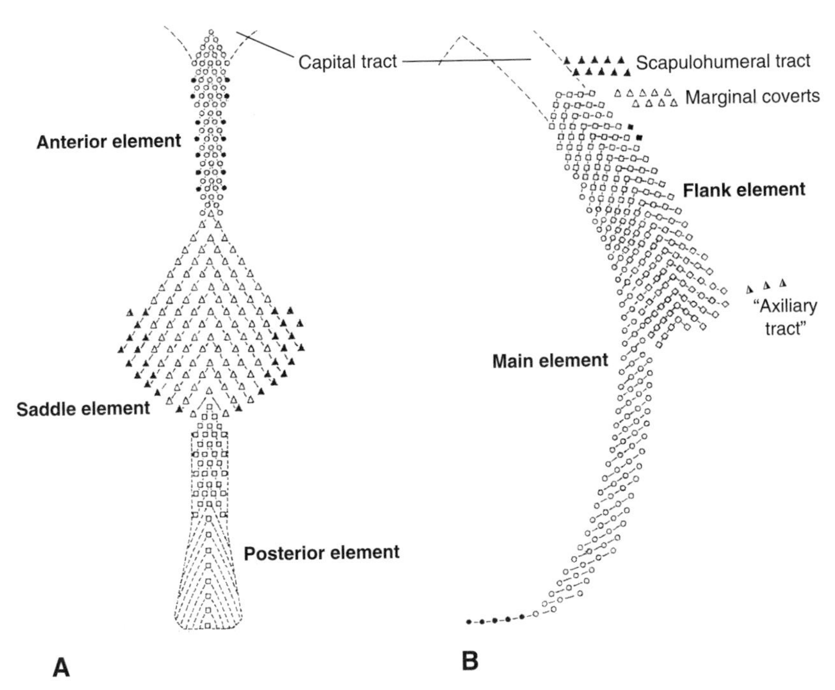

FIGURE 5.2. Schematic drawings of the feather follicles of the dorsal and ventral tracts in the house sparrow. (A) Dorsal tract: Open circles, triangles, and squares represent the feathers of the anterior, saddle, and posterior elements, respectively. Closed circles and triangles represent feathers that are not present in the 7-d-old chick. Triangles located on either side of the saddle element represent feathers in an extra lateral row that is present in some individuals. The dashed lines in the posterior-most part of the posterior element indicate irregular rows. (B) Ventral tract: Open squares and circles represent positions of feathers of the flank and main elements, respectively. Closed squares and circles represent extra feathers that are present in some individuals. Adapted from Clench (1970: Figs. 3 and 4), with permission of the American Ornithologists' Union.

ment then consists of rows of four feathers each, which continue until just anterior to the cloaca. There are typically 39 rows, with most having four feathers, but with row 38 having three and row 39 having two; some individuals have additional feathers in row 39. Variations tend to be highly bilaterally symmetrical within individuals (Clench 1970).

Variation among individuals was small across the study, with males and females having virtually identical pterylosis (Clench 1970). The first basic plumage and the second and subsequent basic plumages were also virtually identical. Not all follicles were filled with feathers in juvenal birds, but all appeared to be present. There was also no seasonal variation in pterylosis; birds did tend to replace body feathers that were lost outside the breeding season, but not those lost during the breeding season. Clench suggested that this results in reduced insulation during the warm months of the year.

The alar tract consists primarily of the remiges (primaries and secondaries) and the upper and lower wing coverts, which comprise the shorter contour feathers that cover the shafts of the remiges. Fig. 5.3a shows the dorsal aspect of a sparrow's right wing. Ten primaries originate from skin covering the carpometacarpal and phalangeal bones of the wing, and these are numbered from the most proximal to the most distal, with P10, from the second digit, being much shorter than the other nine. Nine secondaries arise from the skin overlying the radius and ulna and are numbered from distal to proximal. S7–S9 are described as "protective feathers" because they lie over the other secondaries on the folded wing (Zeidler 1966). The alula consists of three small flight feathers that arise from the skin overlying the first digit and are numbered from proximal to distal. Fig. 5.3b shows the ventral aspect of the left wing with the skeletal elements exposed, indicating the origins of the primaries, secondaries, and alula.

The upper wing coverts consist of three layers of feathers: the greater, middle, and lesser coverts (see Fig. 5.3a). There is a one-to-one correspondence between each greater, middle, and lesser covert and a primary or secondary feather, with the origins of the corresponding feathers being arranged as illustrated in Fig. 5.4. The lower (or under) wing coverts also comprise three layers of feathers termed greater, middle, and lesser coverts. Lower wing coverts are often missing in sparrows in juvenal plumage (Zeidler 1966).

The caudal tract includes the rectrices (or tail feathers) and the upper and lower tail coverts that cover the shafts of the rectrices. Sparrows have six pairs of rectrices, and these are numbered on each side from the innermost to the outermost.

Molt

Molt is the periodic replacement of all or part of the plumage. In both temperate resident and temperate migratory populations of sparrows, there is a complete molt that follows the breeding season. This molt is known as the prebasic (postnuptial) molt, and it results in the acquisition of the basic plumage. In tropical resident populations, this molt may begin during a hiatus in the middle of the breeding season, be interrupted during a resumption of breeding, and then be completed after the end of breeding (Mathew and Naik 1986). Although most authors have not reported a prealternate (prenuptial) molt in

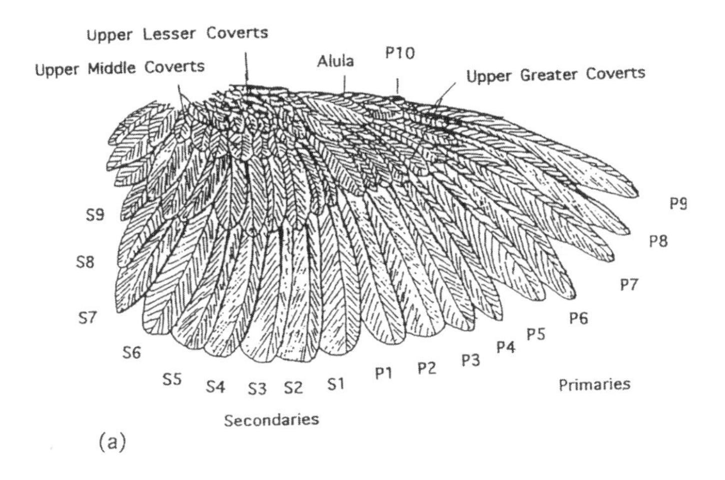

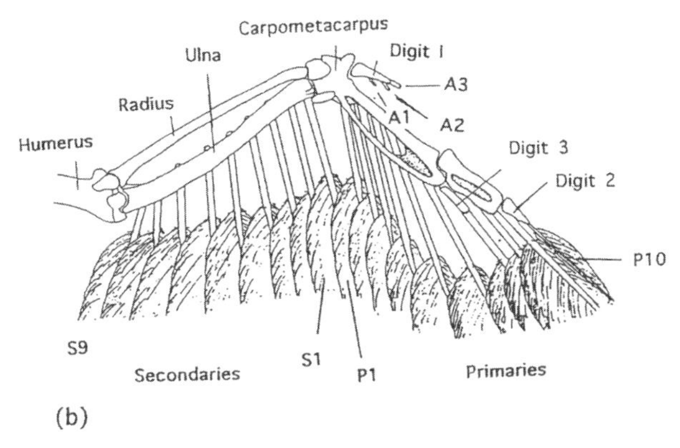

FIGURE 5.3. The wing of the house sparrow. (a) Dorsal aspect of the right wing, with the primaries, secondaries, alula, upper greater coverts, upper middle coverts, and upper lesser coverts identified. S7–S9 are sometimes referred to as tertials, and they cover the other secondaries when the wing is folded. Adapted from Zeidler (1966: Fig. 1). (b) Ventral aspect of the left wing, with the major skeletal elements of the wing exposed and identified; the origins of the primary, secondary, and alula feathers are shown relative to the bones of the wing. P1–P10, primaries; S1–S9, secondaries; A1–A3, alula. Adapted from Zeidler (1966: Fig. 4), with permission of Deutsche Ornithologen-Gesellschaft.

sparrows, Fulgione et al. (1998) reported that Italian sparrows (*P. d. italiae*) have a partial prealternate molt in February that involves primarily head and body plumage.

The molt of a temperate resident population was intensively studied by Zeidler (1966) in Germany. He described molt based both on observations of more than 200 sparrows that were captured and released and on another 78

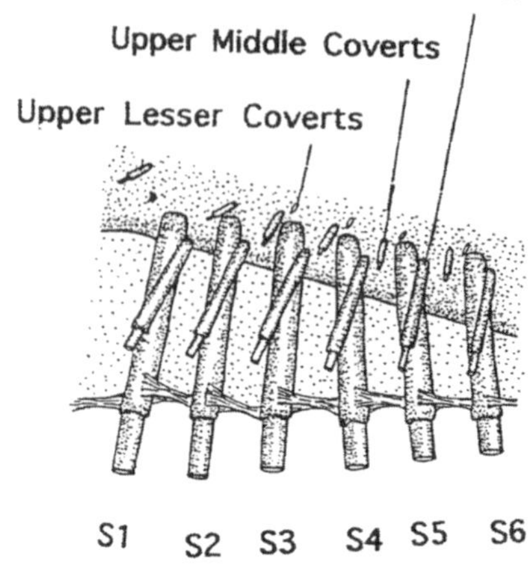

FIGURE 5.4. Schematic drawing of the location of the origins of the upper lesser, middle, and greater secondary coverts in relation to the origins of S1–S6 of the left wing of the house sparrow. Adapted from Zeidler (1966: Fig. 8), with permission of Deutsche Ornithologen-Gesellschaft.

birds that were captured, kept in the laboratory, and examined every 1–2 days for the progression of molt. Old, growing, and new feathers were determined by coloration. Birds just beginning to molt were marked and released, and information from subsequent captures was used to compare feather growth rate and molting schedule in the free-living birds with those in the captives. This resulted in the conclusion that molt continued on schedule (compared with the molt in free-living birds) for only about 14 d in captivity, and then slowed or stopped entirely. Mathew and Padmavat (1985) also reported that molt could be interrupted in sparrows taken into captivity.

Fig. 5.5 illustrates the temporal relationships of the molting of the primaries, secondaries, and rectrices. The primaries normally molt in order from P1 to P10, with the average interval from the shedding of P1 until that of P10 being 63 d. Growth of the primaries is as rapid for those replaced in June and July as for those lost later. Secondaries begin to molt at the time of the shedding of P5 and tend to molt in the following order: S8, S1, S9, S7, S2, S3, S4, S5, S6. Zeidler (1966) recorded only three deviations from this order. In a study of molt in both house and Spanish sparrows in Spain, Alonso (1984b) also reported that the "tertials" (S7–S9) usually molt in the order S8, S9, S7, but occasionally S8, S7, S9 or S7, S8, S9. He also found that the remaining secondaries molt in order from S1 to S6 but did not specify the relationship between S1–S6 and the tertials. Molt of the rectrices also begins, with R1, at about the time of the shedding of P5. Rectrices are then lost in order at intervals of about 5 d, with all being lost before the shedding of P8. Growth of tail feathers is similar to that of wing feathers, with each completing growth in about 22 d. The alula feathers molt in order, A1–A3, beginning at the time that P7 is growing. A

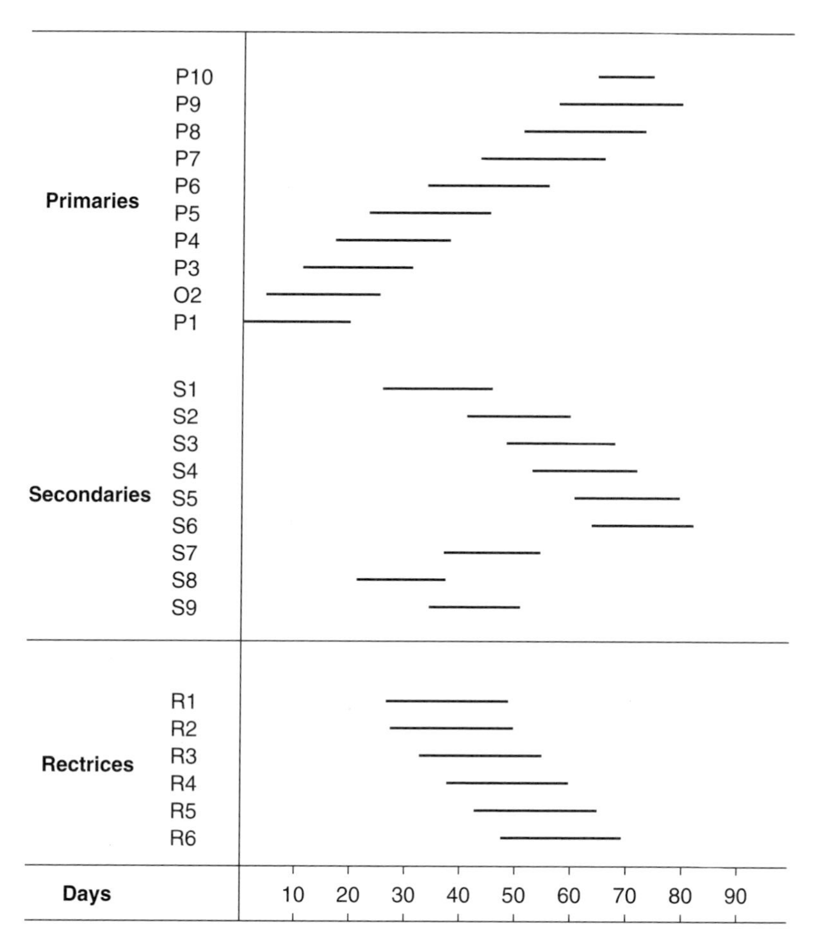

FIGURE 5.5. Timing and sequence of molt of flight feathers (primaries, secondaries, and rectrices) of the house sparrow in Germany. The beginning of each horizontal line represents the timing of the dropping of the old feather, and the length of the line indicates the period of growth of the replacement feather. Adapted from Zeidler (1966: Fig. 14), with permission of Deutsche Ornithologen-Gesellschaft.

similar sequence of molting in the remiges and rectrices was observed in subtropical resident *P. d. indicus* in India (Mathew 1985; Mathew and Padmavat 1985).

Dorsal greater primary coverts are lost at the same time as the corresponding primary but are replaced more quickly (due to their smaller size) (Zeidler 1966). Thus, the new covert lies over and provides some protection for the feather bud of the developing primary. The dorsal greater secondary coverts are all lost simultaneously, at about the time of the shedding of P4 and shortly before the molting of the first secondaries, and growth of new coverts requires about 14 d. The middle coverts are lost after the greater coverts have completed growth, protecting the feather buds before that time. The carpal covert does not lie over a secondary but over P1; however, it molts at the same time as the other secondary coverts. The middle primary coverts molt in order from proximal to distal, beginning after the shedding of P4, and the new feathers are hardened before the growth of P8 is complete. The middle secondary coverts are shed from distal to proximal after the renewal of P4. The underwing coverts are the last wing feathers to molt, the greater coverts beginning after the loss of P7, and the middle coverts between the losses of P8 and P10. Lesser underwing coverts are molted more or less simultaneously.

Body molt begins after the loss of P3 (Zeidler 1966). The first body feathers molted are the central feathers of the back and crown, followed by the underside of the breast. Feathers molt sequentially outward (peripherally) from these locations, with the most rapid molt during the growth of primaries P4, P5, and P6. The rump feathers are renewed by the time P8 is shed, and the belly feathers are renewed last. Fig. 5.6 shows the timing of the molt of different regions of the body plumage of Spanish sparrows in Spain, which Alonso (1984b) described as being identical to that of the house sparrows that he studied concurrently. The general pattern of body molt is very similar to that described by Zeidler (1966) for sparrows in Germany, except that Alonso (1984b) reported that molt in three regions—the crown, throat, and upper back—began up to 3 d before the shedding of P1. Alonso also found that the completion of development of S5 and S6 required 3 d after the completion of development of the last primary, P10. Zeidler (1966) suggested that the molting sequence is evolutionarily old and relatively stable.

Zeidler (1966) estimated that molt lasted 82 d in Germany, whereas Alonso (1984b) found that the duration of molt for adult house sparrows was about 75 d in Spain. Both authors reported that early-molting juveniles (which initiated molt as early as May in Germany) required longer to molt than did adults. Late-reared young molted at a younger age and also molted faster than early-reared young in both populations. In Germany, for instance, juveniles captured in June and July were growing one or two primaries, while juveniles molting in late August or September were growing three, and rarely four, primaries (Zeidler 1966). Based on more than 1000 molt cards submitted to the British Trust for Ornithology, Ginn and Melville (1983) estimated the duration of molt in adult sparrows in Great Britain to be 60 d, and that of juveniles to be 60–80 d, with late-molting juveniles molting more quickly. Bährmann (1967) found in Germany that first-brood juveniles initiated prebasic molt earlier than adults did. Early molt of juveniles, as early as May, has also been observed in Texas (USA) (Casto 1974).

Region of Plumage	Before P1	P1	P2	P3	P4	P5	P6	P7	P8	P9	P10	S5 S6
Forehead										•	•	•
Crown	•	•	•	•	•	•	•	•	•	•	•	
Neck		•	•	•								
Nape					•	•	•	•	•			
Shoulder				•	•	•	•	•	•	•		
Upper back	•	•	•	•	•	•	•					
Lower back				•	•	•	•	•	•	•	•	•
Rump				•	•	•	•	•	•	•	•	
Upper tail coverts					•	•	•	•	•			
Chin											•	•
Ear coverts and cheek		•	•							•	•	•
Throat	•	•	•				•	•	•	•	•	•
Breast			•	•	•	•	•	•	•	•		
Belly							•	•	•	•	•	
Flank		•	•	•	•	•	•	•	•	•	•	•
Leg						•						
Under tail coverts						•	•	•				

FIGURE 5.6. Sequence and timing of molt of the body plumage of Spanish sparrows in Coria, Spain, 1979–1980. The timing of the molt is presented relative to that of the 10 primaries, which molt sequentially from P1 to P10, and includes short periods preceding the beginning and after the completion of primary molt. The time required for the completion of primary molt in the Spanish sparrow is 66 d. The sequence of molt in the house sparrow at Coria is identical to that of the Spanish sparrow, except that the duration of primary molt is 69 d. From Alonso (1984b: Table 4), with permission of Deutsche Ornithologen-Gesellschaft.

In a comparative study of molt in captive sparrows from a sedentary population and from a migratory population of *P. d. bactrianus*, Dolnik and Gavrilov (1975) found that molt began in both groups at the same time, but that *bactrianus* completed the molt 1 mo earlier than the sedentary birds, suggesting that molt is more rapid in migratory than in sedentary populations. These conclusions were generally confirmed in comparative studies of free-living birds in Kazakhstan, where two studies found that molt in *bactrianus* reaches its peak in mid-August, whereas that in the sedentary *domesticus* peaks in mid-September (Gavrilov 1979; Stephan 1982).

Interruption of Molt

Harper (1984) defined two types of molt interruption: suspended molt, in which the resumption of molt begins at the point where molt was interrupted, and arrested molt, in which molt begins again at the normal starting point. He reported interrupted molt in seven juvenal male sparrows in England and concluded that it might be caused by molt suspension during the period of autumnal defense of nesting sites. Casto (1974) captured six adult sparrows in which molt had been interrupted after the replacement of P1 and P2. Molt resumed in all four birds that he subsequently maintained in the laboratory, indicating that this also was a case of suspended molt.

Mathew and Naik (1986) captured 189 sparrows during 2 y at Rajkot, India (22°18'N) and determined the stage of molt of the primaries. Virtually all birds had begun to molt by July, and approximately half of the birds had completed their molt by September. Many birds interrupted their molt when they resumed breeding during the late-monsoon breeding season. (The breeding season is bimodal in this population, with peaks during the summer months of February–May and during the late monsoon, September–October; also see Chapter 3). Molt interruption was observed in both sexes in birds captured during the late-monsoon breeding season but was more common in females (18 of 35 birds) than males (6 of 26 birds) ($\chi^2 = 3.91$, $P < .05$). In 21 birds from known active nesting sites, 15 had interrupted molt, but of 27 marked birds that were not breeding when captured, only 6 had interrupted molt and all others were molting. In birds in which molt was interrupted, the dropped feathers continued to grow to full size even though adjacent primaries were not shed. Mathew and Padmavat (1985) captured 19 sparrows during September, many in interrupted molt, and maintained them in individual cages in an outdoor aviary. Some resumed molting at various intervals after capture, indicating that the interrupted molt was suspended rather than arrested. Mathew and Naik (1986) concluded that the overlap in molting and breeding in the population indicates that the initiation of molt is controlled by a mechanism that is independent of the mechanism controlling breeding.

Control of Molt

The prebasic molt usually follows the termination of the breeding season in temperate resident and temperate migratory populations of sparrows. As has just been described, however, molt in subtropical resident populations in India typically begins during a hiatus in the middle of the protracted breeding season. It is then often, but not always, suspended when breeding resumes in late summer, and it is completed after this second period of breeding activity. Most of the work on the control of molt has been done on birds from temperate resident populations, and the relevance of these findings for subtropical resident birds is not known.

Changes in photoperiod clearly play a role in the onset and speed of molt in temperate resident populations (see Chapter 3). In an early experiment, Lesher and Kendeigh (1941) showed that sparrows that had been maintained on a 15L:9D photoperiod in the laboratory initiated molt when they were exposed to either an abruptly shortened or a gradually shortening photoperiod. Blackmore (1969) also found that sparrows transferred to 10L:14D after being maintained for at least 2 mo on 15L:9D were induced to molt. On the other hand, Murton, Lofts, and Westwood (1970) exposed sparrows captured in late July in England to one of three skeleton light treatments: 6L:1D:1L:16D, 6L:5D:1L:12D, or 6L:9D:1L:8D. Molt had just begun in all individuals at the time of capture. After 35 d, birds on 6L:1D:1L:16D showed considerable molt of both body and wing feathers, but the birds on the other two photoperiods remained essentially unchanged. At the termination of the experiment, the 6L:1D:1L:16D birds had completed molt, and molt was proceeding slowly in birds from the other two groups, with only about half of the primaries replaced. Dawson (1991) demonstrated that maintaining male sparrows on long photoperiods beginning at the summer solstice retarded the onset of molt. For individuals maintained on natural photoperiods, molt began by late July in 20% of individuals, and in all individuals by early August, and was complete by early November. None of the birds placed on 18L:6D on 22 June (the natural photoperiod at the summer solstice at the latitude of the experiment) had begun to molt by mid-August; however, although 40% began by late August, and all began by mid-September. Sparrows can therefore be induced to initiate molt by exposure to short or shortening photoperiods, while exposure to long photoperiods (or skeletons of long photoperiods) at the time that molt is about to begin retards the onset of molt.

In an intriguing follow-up study (elements of which were also described in Chapter 3), Dawson (1998a) captured males during the winter in England and placed them on one of three photoperiods in February: 18L:6D, 16L:8D, or 13L:11D. Additional males were placed on 18L:6D for either 6 or 12 wk, after which they were transferred to either 16L:8D or 13L:11D for the duration of the experiment. Dawson monitored left testis size by periodic laparoscopy and stage of primary molt. Molt began in the 18L:6D and 16L:8D groups about midway through the period of testicular regression, but it began at the beginning of regression in the 13L:11D group (which initiated regression several weeks later than the other two groups). Birds transferred to 13L:11D after 6 wk began the molt sooner than those maintained on 18L:6D, as did birds transferred to 16L:8D after 6 wk. The timing of the onset of molt, like the timing of testicular regression, is apparently determined early in the exposure to long days (Dawson 1998a). Exposure to short days before molt begins serves to advance the onset of molt but has little effect on its rate, whereas exposure to short days near the beginning of molt increases the rate of molting.

Although it is widely believed that the endogenous control of molt is hormonal, there is no consensus on the mechanism of endocrine control. The most frequently discussed candidate is thyroxine (T_4), which has a broad effect on the organism by increasing the metabolic activity of many tissues (see Chapter 9). Such an increase in metabolic activity could certainly be important for the rapid biosynthesis of proteins that is integral to feather replacement, and also for additional heat generation during a period when feather loss reduces the insulative quality of the plumage.

Early studies that attempted to assess thyroid activity by histological examination of the secretory epithelium of the thyroid follicles obtained equivocal results. Miller (1939) found that the thyroid glands of several molting sparrows taken in September in Iowa (USA) showed no activation. Davis and Davis (1954) examined the thyroid glands of sparrows taken throughout the year in California (USA). They found increased thyroid activity in males during June and July, which they postulated was in preparation for molt. Kendeigh and Wallin (1966) found increased thyroid activity only in the winter in Ohio (USA).

More recently radioimmunoassay techniques have been used to estimate the levels of circulating thyroid hormones. Smith (1982) maintained sparrows for 1 y in outdoor aviaries in Washington (USA) (48°N) and collected blood samples at biweekly intervals for radioimmunoassay of plasma levels of T_4 and triiodothyroxine (T_3) (see Fig. 9.6 in Chapter 9). Plasma levels of T_4 and T_3 did not differ at any time between the sexes, and levels of T_3 did not vary significantly throughout the year. Significant seasonal changes did occur in plasma levels of T_4, however, with higher levels during the molt than at other times of the year. Levels increased before the start of molt but after the regression of the gonads. Although this pattern suggests the possibility of a role for T_4 in the control of molt, it is also possible to explain the relationship as the result of increased metabolic demands for synthesis of feather proteins and thermoregulation during molt causing increased thyroid activity.

Other hormones that may be involved in the control of avian molt are progestins and prolactin (Payne 1972). Although both progesterone and prolactin may play some role in defeathering during the formation of an incubation patch in female sparrows (see later discussion), there are currently no studies linking these hormones to the control of molt in sparrows.

Energetics of Molt

Blackmore (1969) estimated the energetic cost of molt at three temperatures (3°C, 22°C, and 32°C) for sparrows being maintained in the laboratory on *ad libitum* food. The maximum increase in daily energy cost during molt was 23.0 kJ at 3°C, 17.6 kJ at 22°C, and 19.6 kJ at 32°C, but the mean increases were only 0.8, 2.5, and 4.2 kJ/d at the three temperatures, respectively. This represented increases in daily metabolic rate of only 0.9%, 3.9%, and 8.6%, respectively. Much of the energetic cost of molt was apparently compensated for by the increased insulative quality of the new feathers, particularly at the lowest temperature.

Ptilochronology

Ptilochronology is the study of the growth rates of feathers (Grubb 1989). The shafts of mature flight and tail feathers have "growth bars" consisting of alternating bands of lightly and darkly pigmented regions that are oriented approximately perpendicular to the axis of the shaft. Each growth bar (consisting of a dark region and an adjacent light region) represents the growth in length of the feather during one 24-h period of its development. Grubb (1989) proposed that the width of these bars provides a chronology of the nutritional state of the bird during the period of feather development. Comparisons of the growth bars of plucked feathers and the induced replacement feathers of the same individual can then be used to draw inferences about the nutritional state of the individual at the time of feather replacement (but see Murphy and King 1991).

Grubb and Pravosudov (1994) captured 12 male sparrows in October (after the prebasic molt) and maintained them in the laboratory under constant photoperiod and temperature (8L:16D and 22°C). Birds were also maintained on *ad libitum* food throughout the experiment, and therefore were presumably in optimal condition. Either one or both outer rectrices (R6, or R6 and L6) were removed at the beginning of the experiment, and replacement feathers were removed at intervals thereafter. Growth bar width did not vary either between induced R6 and L6 rectrices or between repeated induced R6 rectrices. Total length of induced feathers did vary, however, in parallel with body mass, which peaked in December and was lower in November and February. Although these results suggest that individual growth bar width does not vary under optimal conditions, it leaves open the question of whether growth bar width is an accurate indicator of a suboptimal nutritional state.

Aparicio (1998) investigated the replacement growth rates of plucked seventh primaries (P7) in a study of fluctuating asymmetry in feather length.

Sparrows were captured in January in Spain and maintained in individual cages at 20°C on natural photoperiods and *ad libitum* food. P7 was removed on both sides from each individual, and the replacement feather was measured every 2–3 days throughout its growth. Three treatment groups were established: a control group in which the two feathers were removed on the same day, and two experimental groups in which either the right P7 or the left P7 was removed 2 d after the other feather. Signed asymmetry (left – right) exhibited the characteristics of fluctuating asymmetry, with a mean of 0 and no skewness or kurtosis. Mean asymmetry in the control birds increased from the beginning of growth until midgrowth (peaking at about 1.3 mm) and then decreased as the feathers reached full size (to a mean asymmetry of about 0.5 mm). In the experimental birds, the mean asymmetry increased until the feathers were about 40% grown (peaking at about 7 mm difference) and then decreased until full growth was achieved (with only a slightly higher mean asymmetry than in control birds). Aparicio found no evidence of compensational growth in the experimental groups.

Plumage Maintenance

Sparrows devote a considerable amount of time to performing several different activities whose primary function is plumage maintenance. These include preening, bathing, head-scratching, and anting. Many of these activities are performed socially, and often in sequence (i.e., bathing, drying, oiling, preening).

Preening is a highly ritualized activity in which the bird wipes its bill over the uropygial gland at the base of the tail and then systematically grooms the plumage by drawing the feathers through the bill. Sparrows preen frequently throughout the day. Møller and Erritzoe (1992) reported on the preening rate of sparrows from October to March in Denmark. They defined a preening bout as periods of preening that were separated from other such periods by 2 min or more of another activity. Monthly preening rates varied from 1.3 to 3.8 bouts/h for males and 0.73–3.67 bouts/h for females, with lower rates tending to be in the months December–February in both sexes. There was no significant difference between the preening rate of males and that of females.

Head-scratching involves using the foot to scratch areas around the head that cannot be reached with the bill. Birds perform the head-scratching activity in one of two species-specific ways, by bringing the foot up past the folded wing (direct head-scratching) or by lowering the wing and bringing the foot up over the wing (indirect head-scratching) (Simmons 1957). Sparrows, like all other passerines except babblers (Timaliidae), use the indirect method (Simmons 1957).

Sparrows often engage in social water-, dust- (or sand-), and sun-bathing (Daanje 1941). During sun- and dust-bathing, the birds typically lean forward with wings lowered and throw the medium into the fluffed-out feathers with the feet or with flicking movements of the wings. Although water-bathing typically takes place with the birds standing in shallow water, sparrows have been observed bathing in deep water (minimum depth, 30 mm), either by flying slowly over the water, dragging their legs and tail in the water, or by using a submerged water lily stem as partial support while "hovering" at the water's surface (Washington 1990).

Water-bathing occurs throughout the year in Great Britain, where it has even been observed in melting snow (Summers-Smith 1963). Stainton (1982) reported that bathing was approximately equally distributed throughout the year among 40,412 observations of sparrows bathing in London (UK). The diurnal pattern of bathing showed an approximately normal distribution, peaking at midday (1100–1400) in both summer and winter. Two sparrows were observed snow-bathing in 10 cm of fresh powder snow in Germany (Berndt 1961).

Dust-bathing frequently follows water-bathing and may accelerate the drying of the feathers (Daanje 1941). In Great Britain, dust-bathing occurs throughout the year but is much more common during middle and late-summer (Stainton 1982). Davis (1945) reported a case of a newly fledged sparrow dust-bathing on an active anthill (see later discussion).

When sun-bathing, birds normally lie motionless on their bellies with their feathers fluffed out and their wings held away from the body. Summers-Smith (1963) reported seeing birds remain in this position for longer than 30 min and stated that sun-bathing can occur at any time of the year in Great Britain, whereas Potter and Hauser (1974) found that dust-bathing and sun-bathing occurred commonly during the warm months in North Carolina. Stainton (1982) stated that sun-bathing was sometimes difficult to identify, often occurring when birds were also dust-bathing. Daanje (1941) described an alternative sun-bathing position in which the bird lay on its right side with its left leg extended and feathers fluffed out. Summers-Smith (1963) reported that sun-bathing is usually done in a protected area (e.g., on a rooftop), whereas Rose (1983) observed six sparrows entering clear glass jars lying on the ground to sun-bathe. The latter observation, which took place in July in England, also occurred during dust-bathing.

Anting is a curious behavior that has been observed in a large number of bird species, including several times in the house sparrow (Common 1956; Hauser 1973; Potter 1970; Potter and Hauser 1974). Birds either place ants among their feathers or perform preening-like movements holding an ant in their bills. The function of the behavior is as yet unknown, although a number of possible functions have been suggested. Perhaps the most plausible explanations involve relief of skin irritation during molt and the removal of ectoparasites using formic acid secreted by the ants. In a review of the timing of many records of anting in numerous bird species, Potter (1970) concluded that the relief of skin irritation during molt is the most likely function.

The effectiveness of feather maintenance in the removal of ectoparasites has recently been demonstrated experimentally. Polani et al. (2000) observed a significant increase in grooming rate (preening and scratching) after the experimental introduction of 26 feather mites (*Dermonyssis gallinae*) into the feathers of each of 28 sparrows that had been previously deparasitized. An average of 0.5 mites were removed from the birds by the same deparasitizing method 13 d after the experimental infestation, and there was a significant negative correlation between grooming rate and parasite load at the end of the experiment.

Incubation Patch

An incubation patch, or brood patch, develops in female sparrows during nest-building and egg-laying of the first brood and persists throughout the breeding season. It is a large area of highly vascularized skin in the breast and belly region that is devoid of down feathers. Jani et al. (1984) described five phases in the incubation patch cycle in India: (1) the initiation phase, during which partial defeathering takes place; (2) the vascularization phase, during which defeathering is completed and vascularization of the exposed ventral skin occurs; (3) the edematous phase (the fully functional incubation patch); (4) the regression phase, in which there is a decrease in edema and vascularization; and (5) the refeathering phase.

Selander and Yang (1966) collected 64 females during the breeding season in Texas and cut two samples of skin from the breast area, using a template to ensure uniform size (2.0 cm^2). One sample was weighed for wet mass, dried for 24 hours at 50°C and reweighed for dry mass, and extracted with petroleum ether to obtain lean dry mass and lipid mass. The second sample was prepared for histological examination to obtain information on composition, thickness, and vascularity. The fully developed incubation patch of incubating sparrows involves loss of down feathers (including the follicles and papillae) from the abdomen and breast, a thickening of the epidermis, increased vascularity of the dermis, and increased edema. Selander and Yang also found that defeathering, which requires at least 1 wk, is virtually complete by the end of egg-laying, whereas integument mass does not peak until late incubation (going from a mean of 10.19 mg for nonbreeding females to 93.92 mg for females during late incubation). Although nonfat dry mass of the integument increases dramatically (from a mean of 3.69 mg to 13.94 mg), most of the gain is in water (6.42 mg to 79.75 mg); the fat content actually decreases during late incubation (from 0.43 mg to 0.28 mg). The increase in skin mass, and particularly its water content (edema), is moderate until egg-laying, after which skin mass (particularly edema) increases rapidly. The epidermis of the incubation patch increases in thickness by 5.2 times, and dermal blood vessel number increases 7.2 times, while vessel size increases 5.9 times. Between broods, the brood patch returns to about the same condition as in females preparing to lay their first clutches.

Menon et al. (1978) examined the dermis of the incubation patch of female sparrows from India for evidence of feather papillae. Although the dermis appears to lack papillae on macroscopic inspection, microscopic examination shows that the papillae are still present and stain readily with hematoxylin.

The development of the incubation patch is apparently controlled by a complex interplay of female sex hormones. Selander and Yang (1966) captured both male and female sparrows in December in Texas and maintained them on 9L:15D photoperiods beginning in January. Five treatment groups were established: (1) estradiol implanted subcutaneously, (2) estradiol implant plus progesterone injections on alternate days, (3) estradiol implant plus prolactin injections on alternate days, (4) estradiol implant plus both progesterone and prolactin injections, and (5) control groups receiving either saline or oil. Both sexes showed the same responses, and data were pooled for each of the five groups. Prolactin augmented the effect of estradiol in promoting thickening of the epidermis and increased vascularity of the dermis. Defeathering appeared to be inhibited by progesterone but enhanced by prolactin, whereas both progesterone and prolactin increased the rate of development of edema. Subcutaneous fat in controls was double that in experimental birds, which Selander and Yang attributed to the normal tendency for an increase in caged sparrows.

Heat for transfer to the eggs during incubation and to the young during brooding comes not only from the blood circulating through the highly vascularized dermis of the incubation patch but also from *in situ* thermogenesis. Shah et al. (1979) reported that there was an increase in concentration of glycosaminoglycans in the skin of the incubation patch during its development. Jani et al. (1984, 1985a, 1985b) examined the metabolic activity of cut sections of skin from the incubation patch of breeding females in India by monitoring the activity of several enzymes: acid phosphatase (ACP), alkaline phosphatase (ALP), succinate (SDH), lactate (LDH), and malate (MDH) dehydrogenases; glucose-6-phosphate dehydrogenase (G-6-PDH); α-glycerophosphate dehydrogenase (α-GPDH); and β-hydroxybutyrate dehydrogenase (βDH). Activities of both ACP and ALP decrease during the initiation and vascularization phases, whereas the activities of SDH, LDH, MDH, G-6-PDH, α-GPDH, and βDH all show increases during the development of the patch. Activities of the latter enzymes peak during the edematous phase, and activities of ACP and ALP also increase during this functional phase. Activities of all of the enzymes decrease during the regression phase and then increase during the refeathering phase, particularly in the feather germs. The increased activity of βDH indicates that fats are being mobilized during the developmental and functional phases of the incubation patch. This conclusion is further supported by the fact that lipids are particularly prominent in the epidermis during defeathering and the edematous phase (Jani et al. 1985a).

Badge Size in Males

As noted previously, adult males differ from females in having a black throat and breast patch (sometimes referred to as a bib, gorget, or badge). This type of sexually dimorphic characteristic is found in a number of avian species, and its function has been the subject of considerable speculation. Rohwer (1975) proposed a novel hypothesis (the Badge of Status Hypothesis) for the function of a similar badge in male Harris's sparrows (*Zonotrichia querula*), suggesting that the size and blackness of the badge serve as a signal of status in intraspecific interactions among males competing for access to food in winter foraging flocks. Rohwer suggested that males with larger, blacker badges are "studlier" than their smaller-badged counterparts and are socially dominant to them. In the same year, Zahavi (1975) proposed that plumage dimorphisms in birds may represent one example of sexual selection that operates on the "handicap principle." According to this hypothesis, sexually selected traits involve mate-choice decisions by females in which the choice of the selected character is based on a cost associated with the development or maintenance of the character that makes it a reliable indicator of male quality. Males that possess the trait, or that have more intense manifestations of a variable trait, demonstrate their greater fitness by having overcome the costs of producing and maintaining the trait. In a critique of Rohwer's hypothesis, Shields (1977) suggested that variable badges may facilitate individual recognition and that correlations between badge size and status may be explained by sex- or age-related differences in plumage. Hamilton and Zuk (1982) proposed a specific mechanism for the operation of Zahavi's handicap principle by suggesting that parasites may play a central role in sexual selection for strongly dimorphic plumage. They suggested that males communicate their resistance to or freedom from parasitic infection by the size or intensity of the dimorphic character. These hypotheses and numerous extensions and modifications have stimulated a veritable "cottage industry" of studies on the significance of such plumage characteristics in birds, including many on the house sparrow (see Jawor and Breitwisch 2003). Recently, the potential value of asymmetry in the size and shape of the badge has also been explored as a means of signaling important information among individuals.

Size of the Badge

One of the major problems in attempting to critically assess the results of the numerous studies on badge size in sparrows is that there is no standard method for measuring badge size. Some investigators have used linear measurements to estimate area, and others have attempted to measure area directly. Møller (1987b, 1989) measured the maximum length (L) and maximum breadth (B) of the badge by flattening the badge against a rule with the bird's beak kept at a right angle to the body, and then calculated the area (A) of the badge as $A = 166.7 + 0.45LB$ (a formula derived from the badge sizes of 29 study specimens) (Møller 1989). This method has been used in several other studies as well (e.g., Gonzalez, Sorci and de Lope 1999; Gonzalez et al. 2001, 2002; Hein et al. 2003; Strasser and Schwabel 2004; Vaclav and Hoi 2002a). Solberg and Ringsby (1997) also used this method for estimating total badge size, but they used the product of length and breadth to estimate the area of the visible badge (i.e., the portion of the badge lacking gray or buffy feather tips). Veiga (1993a) measured height (H) and width (W) of the "throat patch" while maintaining the bird in a "natural position" and estimated the area of the badge as the area of a circle with radius H and chord W.

Cordero et al. (1999a) determined badge size by photographing the bird in a standard position, projecting the picture onto a screen, and measuring the area of the badge on the screen. They also measured the length and breadth of the badge with a rule. Stepwise multiple regression of length (L) and breadth (B) on badge area (A) (from the photograph) indicated that only breadth contributed significantly to badge area: $A = 1.28B + 287$ ($P = .04$). Westneat et al. (2002) used the first principal component (PCI) of four linear measurements of the badge to represent badge size. Ritchison (1985) measured badge size directly by tracing each bird's throat patch, while Evans et al. (2000) and Buchanan et al. (2001) determined badge size by tracing the badge on acetate paper, cutting out the outline of the badge, weighing it, and comparing it to the weight of a known area of acetate paper. Andersson and Ahlund (1991) used an image-analyzer to determine badge area from a photograph of the male, a method that was also used in several other studies (Kimball 1997; McGraw et al. 2003; Voltura et al. 2002). In other studies, however, investigators simply assigned individuals to one of three to five badge size classes (Cordero et al. 1999a; Hein et al. 2003; Møller 1987b, 1988; Reyer et al. 1998) or based their studies on one linear measurement of the badge (Griffith 2000; Griffith, Owens and Burke 1999a, 1999b).

Table 5.1 presents a compilation of data on badge size of sparrows from several sites in Europe and North America. A quick perusal of the values for average badge size and range in badge size raises some obvious concerns. For example, badge size is reported to range from 37–107 mm² in a study in Kentucky (USA) to 381–1058 mm² in a study in Denmark, and the average badge size varies from 294.5 to 677.7 mm² in two locales in Denmark, both measured by the same investigator (Anders Pape Møller). Variation such as this would suggest that the trait in question is highly phenotypically plastic and subject to local environmental conditions, is subject to intense local selection pressures if it is primarily genetically determined, or is not being accurately and/or consistently measured. The reported locations span a range of latitudes from 35–67°N and include both xeric (New Mexico [USA] and Spain) and relatively mesic (Norway, Kentucky and New York [USA], Scotland, and Denmark) environments. No consistent pattern in badge size emerges along either of these gradients. The great variability both within and among sites suggests that a standard method for measurement of badge size needs to be adopted, and that the repeatability of this method needs to be established.

A few recent studies have reported on the repeatability of the measurement methods employed, although it is not clear that all used the appropriate estimation of repeatability (see Lessells and Boag 1987). Evans et al. (2000) reported a repeatability of 68% ($P < .025$), and Buchanan et al. (2001) reported a value of 0.87 using the same measurement method (weighing acetate paper cutouts of the badge). McGraw et al. (2003) reported a repeatability of 0.91 ($P < .0001$) based on digitizing of photographs of the badge. Solberg and Ringsby (1997) reported that they found no significant change in total badge area of 11 males that were captured in December and in March during the same winter using the Møller (1987b, 1989) method. It seems clear that methods utilizing photographs, although more time-consuming and cumbersome, are much more reliable and accurate than the estimates based on linear measurements.

Badge Size and Age

Most studies have found that badge size increases with age, at least between first-year and older males. Badge size in first-year males may also be related to age, with males hatched earlier in the season having larger badges (Griffith, Owens, and Burke 1999a; Veiga 1993a). Ritchison (1985) assigned 35 male sparrows used in dominance trials to one of 15 badge size categories based on measured badge area (1 = 105–109 mm²; 2 = 100–104 mm²; . . . 15 = 35–39 mm²), and the average badge category of 23 adults was 4.2 (SD = 2.9), whereas that

TABLE 5.1. Badge Size of Male House Sparrows

Badge Size (mm²)				
Mean (SD)	Range	*N* (Age Group)	Location (Latitude)	Source
321.7* (64.8*)	211–446	37 (All)	Denmark (57°N)	1
352 (141*)	—	32 (First-year)	Denmark (57°N)	1
354 (98*)	—	22 (Second + year)	Denmark (57°N)	1
335.5* (72.5*)	—	84 (Autumn, first-year)	Denmark (57°N)	2
326.5* (73.7*)	—	34 (Autumn, adult)	Denmark (57°N)	2
325.1* (79.9*)	—	61 (Winter, first-year)	Denmark (57°N)	2
294.5* (73.7*)	—	28 (Winter, adult)	Denmark (57°N)	2
322.3* (83.4*)	207*–491*	33 (All)	Denmark (57°N)	3
— (—)	400*–970*	148* (All)	Denmark (56°N)	4
537.0 (137.0*)	216–1058	418 (All, visible badge)	Denmark (56°N)	5
677.7 (116.5*)	381–1058	418 (All, total badge)	Denmark (56°N)	5
252.6* (45.4*)	148*–368*	91 (First year)	Spain (41°N)	6
288.5* (54.6*)	157*–395*	39 (Second-year +)	Spain (41°N)	6
252.8* (51.1*)	172*–354*	18 (First-year)	Spain (41°N)	7
298.1* (51.4*)	156*–405*	59 (Second-year +)	Spain (41°N)	7
113.2* (18.6*)	80–140	9 (December, all)	Norway (67°N)	8
177.6* (116.9*)	84–462	18 (March, all)	Norway (67°N)	8
83.3* (15.9*)	37*–107*	35 (All)	Kentucky (USA) (38°N)	9
417.4 (61.9)	—	3 (All)	New Mexico (USA) (35°N)	10
387.1 (139.3)	175.3–589.7	—(All)	New Mexico (USA) (35°N)	11
365.6* (41.5*)	—	74 (March, all)	France (46°N)	12
654 (90*)	—	42 (Feb.–May, all)	New York (USA) (42°N)	13
364.4* (119.1*)	177–693	27 (May–July, all)	Oklahoma (USA) (35°N)	14
368.2* (43.2*)	296*–513*	63 (November, all)	Spain (39°N)	15
383.2* (54.5*)	264*–512*	49 (April, all)	Spain (39°N)	15
328.2* (116.5*)	114*–621*	66 (Summer, all)	Scotland (UK) (56°N)	16
338.6* (53.4*)	270*–471*	20 (Summer, all)	Wisconsin (USA) (43°N)	17
303.3* (49.6*)	244*–384*	10 (Winter, all)	Hungary (47°N)	18

This table contains a compilation of estimates of badge size in male house sparrows from various locations in Europe and North America. Method of measuring badge size differed among studies (see text), and in many cases (designated with an asterisk [*]) the values presented in the table were computed from data presented in figures or tables in the sources.

Sources: 1, Møller (1987b:Tables I, II, III): 2, Møller (1989); 3, Møller (1990:Fig. 1b); 4, Møller and Erritzoe (1988:Fig. 1); 5, Møller and Erritzoe (1992); 6, Veiga (1993a:Fig. 3); 7, Veiga (1996b:Fig 2), Veiga and Puerta (1995:Fig. 3); 8, Solberg and Ringsby (1997:Tables 1–3); 9, Ritchison (1985); 10, Kimball (1996); 11, Kimball (1997); 12, Marzuc, Bonneaud, et al. (2003); 13, McGraw et al. (2003); 14, Voltura et al (2002:Fig. 2); 15, Gonzalez, Sorci, and de Lope (1999:Fig. 2); 16, Buchanan et al. (2001:Fig. 3); 17, Riters et al. (2004); 18, Liker and Barta (2001).

of 12 first-year males was 8.7 (SD = 3.7). In two samples from Spain, Veiga (1993a, 1996b) found that first-year males had significantly smaller badges than second-year and older males. Solberg and Ringsby (1997) found that both visible badge size and total badge size were significantly larger in 47 adult males compared with 44 first-year males on two Norwegian islands. Similarly, Griffith, Owens, and Burke (1999a) found that badge length was significantly greater in older males than in first-year males on Lundy Island (UK), and Cordero et al. (1999a) reported that first-year males in both England and Spain had significantly smaller badges than older males.

Møller (1987b), however, reported that badge size did not increase with age in Denmark. Hein et al. (2003) also reported that badge size did not correlate significantly with minimum male age in 1 y of a 2–y study in Kentucky, although there was a significant positive correlation in the other year. Although differential survivorship based on badge size could account for a pattern of larger badge size with age, Veiga (1993a) found that 13 first-year males captured in subsequent years had significantly larger badges (mean increase = 59.8 mm^2, $t = 6.2$, $P < .0001$). Møller (1987b), however, reported that individuals did not show unidirectional change in badge size with age, although it is unclear whether he was comparing individuals only between their first and subsequent years. Collectively, the data suggest that badge size tends to increase between first- and second-year males but shows no consistent directional change after the second year.

Heritability of Badge Size

Little is known about the genetic basis of badge size. Møller (1989) reported a heritability of badge size for sparrows in Denmark of 0.60 (SE = 0.23), based on 11 father-son combinations. In cross-fostering experiments on Lundy Island, however, offspring badge length was significantly correlated with the badge length of the foster male but was not correlated with that of the genetic father (Griffith, Ownes, and Burke 1999a). Obviously, much more work must be done to determine the extent to which this trait is heritable.

Badge Size and Body Size or Condition

No relationship has been found between badge size and body size in sparrows. Møller and Erritzoe (1988) found no correlation with either tarsus length (r = 0.04) or body mass (r = 0.07) in males taken in Denmark during the summer. Veiga (1996b) performed a principal components analysis on a sample of 178 male sparrows in Spain using 10 morphological variables (including body condition as the residual of a regression of body mass on tarsus length). PCI had high positive loadings from badge size, badge feather length, and size of pigmented area on badge feathers and was clearly associated with badge size. PCI also had the highest positive loading for wing length, which is sometimes used as an indication of body size in birds, although it is primarily a plumage trait. PCII and PCIII represented body size and body condition, respectively, and the fact that these components were orthogonal to PCI means that badge size was unrelated to either body size or body condition in this sample. In Norway, no significant relationship was found between either visible badge size or total badge size and PCI (the principal component indicating overall body size) (Solberg and Ringsby 1997). Likewise, badge size was unrelated to PCI (body size based on principal components analysis of wing and tarsus lengths) in Kentucky (Hein et al. 2003). In France, badge size was not related to either body size (PCI based on tarsus, wing, head-bill, and sternum lengths and body mass), nor was it related to body condition (estimated as the residuals of body mass on PCI) (Mazuc, Bonneaud et al. 2003).

Other studies that have examined the relationship between badge size and body condition have obtained conflicting results. In samples of males from four different seasons in Denmark, Møller (1989) found no significant correlation between badge size and body condition (body mass/tarsus length^3) during the autumn or in two breeding-season samples. Badge size was significantly correlated with condition in the winter, however ($r = 0.50$, $P < .001$). Veiga (1993a) also found a significant positive correlation between badge size and body condition in a sample from Spain ($r = 0.209$, $P = .01$). When the sample was subdivided by age, the first-year males did not show a significant correlation ($r = 0.12$, $P = .12$), but the older males did ($r = 0.329$, $P = .02$). Similar results were obtained by Cordero et al. (1999a), who found that body condition (estimated as the residual of body mass regressed against tarsus length) was correlated with badge size for older males ($r = 0.75$, $P = .008$) but not first-year males ($r = 0.167$, $P = .75$) at another site in Spain. In England, however, body condition was unrelated to badge size in either first-year or older males (Cordero et al. 1999a). No relationship was found between body condition (mass/tarsus length) and either visible or total badge size in Norway (Solberg and Ringsby 1997).

Møller and Erritzoe (1988) found that badge size was positively correlated with estimated testes volume in a sample of 148 breeding-season males ($r = 0.23$, $P < .01$, both variables $\log_{10}$-transformed). Working with the same sample, Møller (1994) reported that the difference in testis volume (left – right) actually explained more of the variance in badge size than did testes volume itself ($R^2 = 0.09$, $P < .0001$ vs. $R^2 = 0.05$, $P = .0043$, respectively). A positive but nonsignificant relationship was also found between testes mass and badge size in a sample of 12 males collected during the egg-laying period in Spain ($r = 0.484$, $P > .05$) (Birkhead et al. 1994).

In summary, estimates of average badge size vary by more than 8-fold among different studies, and although some of this variation is undoubtedly due to a lack of standardization in the method of badge measurement, there is also little consensus among studies on attributes that are related to badge size. Badge size does not appear to be consistently related to body size or condition, although condition may be related to badge size at certain times of the year. Almost all studies have found that first-year males have smaller badges than older males, and badge size also appears to be positively correlated with testes size (see later discussion).

Sexual Selection and the Badge of Status

Sexual selection can act in basically two ways in monogamous species, such as the house sparrow, that have a typical division of labor. Males acquire and defend nest sites and mates and assist in provisioning and protecting the young, whereas females build the nest, lay and incubate the eggs, and brood and assist in provisioning and protecting the young (see Chapter 4). In systems such as this, sexual selection can operate either by male-male contest or by female mate-choice, and in some cases, by both. Females should preferentially choose males with large badges if badge size is indicative of either genetic or phenotypic qualities that will enhance the female's fitness. Numerous studies in recent years have attempted to determine whether the so-called badge of status of male sparrows plays a role in either male-male contest or female mate-choice.

Badge Size and Dominance

Ritchison (1985) established three flocks of 10–15 wild-caught male sparrows in a 1.9 m^3 flight cage for about 2 wk. He then determined dominance relationships among the individuals during a 6-h period of observation after the withholding of food for 12 h. Only encounters involving active displacements were counted, and the winner and loser of each encounter were identified. Of 185 encounters, the male with the larger badge won in 118 (64%, $P < .05$), and of 47 encounters involving an adult and a juvenile, the adult won in 33 (70%). Further analysis indicated that age was the primary predictor of success in encounters and that the badge size relationship was therefore only due to the fact that juvenile males generally had smaller badges (see earlier discussion). In another study using dyads to investigate the effects of hunger-induced motivation on the dominance relationships of captive sparrows, Andersson and Ahlund

(1991) found that badge size had no effect on the outcome. They suggested that the data supported the individual recognition hypothesis for badges, as opposed to the status-signaling hypothesis.

Møller (1987b) observed one field flock (13 males) and two laboratory flocks (10 and 14 males) during the winter in Denmark. All individuals were color-banded, and each male was assigned to one of five badge size classes. Based on observations of direct encounters between individuals in which one male attacked and displaced another, a dominance hierarchy for each flock was constructed. Males with large badges tended to have higher ranks than those with small badges, and there was a significant positive relationship between badge size and dominance rank in the field flock and in one of the laboratory flocks. Because Møller (1987b) found no significant relationship between age and badge size (discussed earlier), he concluded that the results support the status-signaling function of badge size and fail to support an individual recognition function.

In a study of small populations of sparrows living on islands in Norway, Solberg and Ringsby (1997) examined the effect of badge size on dominance relationships of captive males during the winter. Captive flocks of 6–11 males were established in 2 × 2 × 2 m cages during both December and March. Dominance relationships in the flocks were then determined based on agonistic interactions at both feeding stations and perches (with each interaction having a winner and a loser, the winner being the individual that occupied the feeding station or perch after the interaction). In none of the groups individually was either visible badge size or total badge size significantly correlated with dominance rank, but in the groups in each month collectively, visible badge size was positively correlated with dominance rank when both values were converted to standard scores.

Four other recent studies have also reported on the relationship between badge size and dominance status among male sparrows. In a mixed laboratory flock consisting of 10 males and 10 females in Hungary, Liker and Barta (2001) found that dominance rank among the males (1 = highest-ranking male) was negatively correlated with badge size ($r = -0.83$, $P < .05$) but was unrelated to body mass or wing, beak, or tarsus length. Gonzalez et al. (2002) reported on aggressive interactions in aviary flocks of 6–7 males in Spain. Large-badged males won a significantly greater proportion of aggressive encounters than did small-badged males. Gonzalez et al. also tested the effect of badge-size manipulations, increasing the badge size of the two lowest-ranking males in each aviary flock to approximately 500 mm^2 (the 95th percentile of badge sizes in the population). Although the males with experimentally enlarged badges still won fewer aggressive encounters than control individuals, there was a significant trend toward an increase in proportion of encounters won ($P = .025$). Hein et al. (2003) observed aggressive interactions at a winter feeding station in Kentucky that permitted only one individual to feed at a time. In the 2 y of the study, dominance rank (based on proportion of interactions won) increased significantly with badge size in one year, but there was no such relationship in the second year. In a study of laboratory flocks of five males in Wisconsin (USA), however, Riters et al. (2004) found no relationship between badge size and dominance rank.

In the mixed flock of 10 males and 10 females in Hungary, large-badged males also tended to be dominant to females (Liker and Barta 2001). Females initiated fewer aggressive encounters with increasing badge size of males ($r_s = -0.65$, $P < .05$), and the proportion of aggressive encounters won decreased with increasing badge size ($r_s = -0.84$, $P < .01$).

Badge Size and Nest Site Acquisition

Møller (1988) determined mating status during the first half of the breeding season in Denmark by walking through colonies and inferring mating status from song type of each male (see Chapter 7). He estimated badge size of each male from a distance of 10 m or less and assigned each to one of five badge size categories (see Møller 1987b). He also counted the number of nest sites and nest-site types (cavities vs. open sites) in the area occupied by each male. Males with small to medium-sized badges (size categories 1–3) had significantly fewer nest sites in their defended area than those with larger badges (categories 4 and 5), and they had cavity nest sites significantly less often than did the larger-badged males.

Veiga (1993a) examined the effect of badge size on the acquisition of a nest site by experimentally manipulating badge size during 2 y at a breeding colony in Spain. He enlarged the badge with black dye to twice its size in some males, and decreased badge size by one-half in other males by applying hair decolorizer to the margins of the badge, thereby establishing four groups: (1) badge-enlarged, (2) badge-reduced, (3) control-1 (normal badge size maintained but dye added), and (4) control-2 (untreated). There were no differences between the two control groups, so they were pooled. Badge-enlarged males occupied more nest-boxes than either badge-reduced or control males (Mann Whitney U test, $P = .04$ and $P = .046$, respectively). Successful males (those that occupied at least one nest-box) were significantly older than unsuccessful males. Fig. 5.7 reproduces Fig. 4 from Veiga (1993a), and close scrutiny of the data suggests that age and original badge size may be confounding variables to a straightforward interpretation of the results. Furthermore, a test of independence of the three treatment groups and the two outcomes ("nest owners" and "without nest") fails to find a significant difference in success in nest-site acquisition among the treatment groups ($\chi^2 = 4.03$, df $= 2$, $P > .10$).

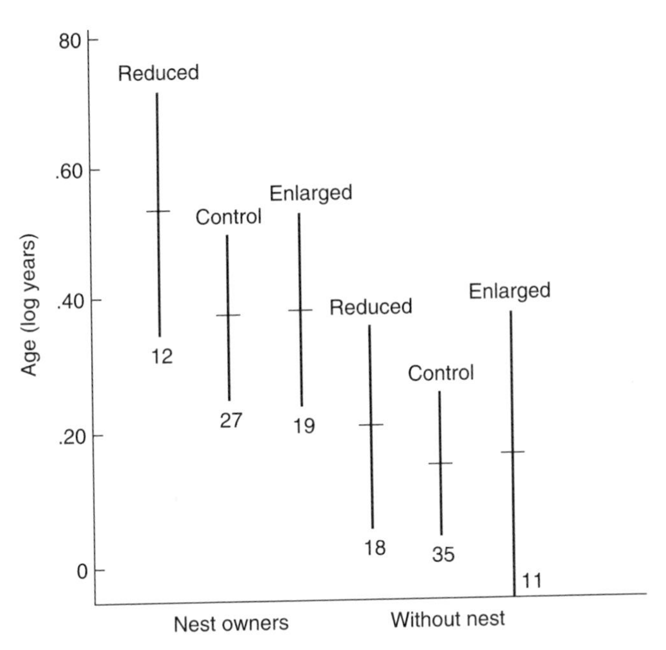

FIGURE 5.7. Nest occupancy by male house sparrows with experimentally altered badge sizes at Collado Villalba, Spain, in 1990 and 1991 (Veiga 1993a). Badges of some males were experimentally modified before the breeding season by either enlarging the badge with black dye or decreasing it by applying hair decolorizer to the margins of the badge. These treatments resulted in four treatment groups: (1) badge-enlarged (2× normal badge size), (2) badge-reduced (1/2 normal badge size), (3) control-1 (normal badge size maintained but black dye added to existing badge), and (4) control-2 (untreated). Because control-1 and control-2 males did not differ in any comparison, their data were lumped for analysis. Average age of males occupying nest sites ("nest owners") and males failing to occupy nest sites ("without nest") in the three treatment categories is depicted in the graph (vertical lines represent 95% confidence intervals). Numbers at the bottom of each confidence interval line represent the number of males in each treatment group in the 2 y combined. From Veiga (1993a: Fig. 4), reprinted with permission of *Evolution*.

In an effort to test the hypothesis that large badge size contributes to success in acquiring high-quality nest sites, Kimball (1997) placed nest-boxes on trees either singly (single-box nest site = SBNS) or in pairs, with the entrance holes of the two nest-boxes only 23–25 cm apart (double-box nest site = DBNS). Badge size of males was measured by digitizing a photograph of the male containing a ruler, and bill depth, length, and width; tarsus length; and wing chord were measured. There was no difference in badge size, wing chord, tarsus length, or bill length, width, or depth between males holding DBNSs and those holding SBNSs (only bill length approached significance; $P = .07$, one-tailed).

Badge Size and Mating Success

As noted earlier, females should selectively choose males with larger badges if badge size is a reliable indicator of genetic and/or phenotypic qualities of the male that would enhance the female's fitness. Qualities that have been identified as possible candidates for such selection include disease resistance or freedom from parasitism, nest defense, and parental quality (provisioning ability). Some studies have used the initiation date of the first clutch of a female as an indicator of mating preference, and other studies have examined other aspects of female reproductive behavior to attempt to identify such female mate-choice.

In the same study in which he examined the number of nest sites in defended areas in relation to badge size (described earlier), Møller (1988) also attempted to determine whether female sparrows used badge size and/or territory quality in selecting a mate. He reported that mated males consistently had larger badge sizes than unmated males. He also performed an experiment in which he attempted to determine whether females were responding to badge size. He captured 33 females in late February 1987 and maintained them in an indoor aviary on a 16L:8D photoperiod. In mid-March, all females were implanted subcutaneously with a Silastic tube containing 17-β-estradiol. On 30 March, females were tested with either male song ($n = 6$) or male song plus a mounted male with one of three badge sizes (170, 406, or 646 mm^2). Latency, number of solicitation displays, time spent displaying (during a 2-min exposure), frequency and intensity of displays (scored as 0–2 for five measures of intensity, with scores ranging from 0 to 18 [sic]) were recorded for each female. Latency was significantly shorter for male model plus song, and display intensity was also greater with the model. The number of displays and display intensity were greater with the model with the largest badge size. Møller (1988) concluded that females may use both badge size and territory quality in choosing a mate.

Møller (1990) also attempted to determine whether badge size is sexually selected by examining the relationship between badge size and mating activity in a population in Denmark in which the badge sizes of the color-marked males were known. The number of copulation bouts, the number of multimale communal displays (see Chapter 4), and the number of sexual displays performed toward nonmate females were recorded for each male during 1079 h of focal male observations during three breeding seasons, 1984–1986 (including an average of 9.1 h/d for the 81 d of observation in 1985). He reported that males with larger badges directed sexual displays more frequently both toward their mates and toward nonmates, and they copulated more frequently with their mates. The behavior of the males in response to the flight of an unaccompanied female was also investigated. In 1985, three females from the breeding colony were released each day near a dense bush in which males often congregated. The females were released during three periods in the breeding cycle, the prelaying period (11–20 d before egg-laying began), egg formation and laying, and incubation, with eight trials being run during each period. In a second year, females with Silastic implants containing 17-β-estradiol were released between 0700 and 1000 near the same bush (six females implanted with empty Silastic tubes were also released). Møller concluded that males were able to identify females that were impregnable, because they directed almost no communal displays toward females during the prelaying or incubation periods, and none toward the six control females. Males with larger badges participated in more communal displays directed toward the females than did smaller-badged birds. Males that engaged in either forced or solicited extra-pair copulations had significantly larger badges than males that did not participate in extra-pair copulations (Fig. 5.8). Møller concluded that badge size was strongly directionally selected by increased mating success of large-badged males.

Several other studies have either failed to confirm these conclusions or have reported contradictory results regarding female mate-choice and badge size. Kimball (1996) tested mate-choice by 40 captive female sparrows using a cross-shaped arena in which one male was visible through a Plexiglas wall at the end of each of the four arms. Badge size of males was altered experimentally, either by cutting off the black tips of badge feathers to expose the gray bases of these feathers (which are similar in color to the surrounding plumage) or by coloring feathers with a black marker (the central badge feathers of all males were colored with the marker for uniformity of color). Four badge sizes were created with templates made from museum specimens: small, 117 mm^2; medium, 230 mm^2; large, 389 mm^2; and extra-large, 619 mm^2. Females were then placed in a central chamber in the test apparatus, with a male of each badge size at the ends of the four arms; the males were visually but not acoustically isolated from each other. Females were permitted to move freely among the four arms for 60 min, and only females that visited all four males during this period were considered. The time spent by the female with each of the four males was then determined during the following 30 min. Females did not favor one of the arms, or the first male that they visited. There was no significant relationship between any measure of badge size (area, chin width, maximum length, or maximum width) and the amount of time spent by the female with the male. This applied to all females (n = 40), as well as to three subsets of females that showed potentially higher motivation. The only correlates of time

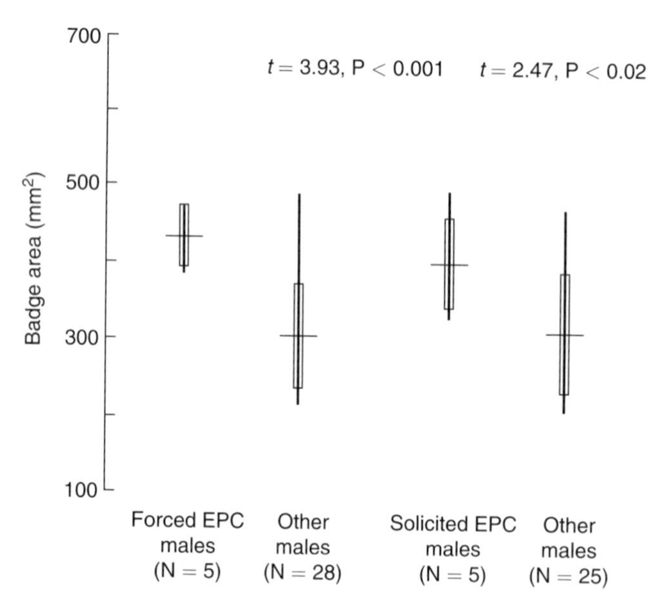

FIGURE 5.8. Badge size and incidence of forced and solicited extra-pair copulations (EPC) in house sparrows in Denmark, 1984–1986. Communal displays involving color-marked sparrows were observed, and incidents of both forced and solicited EPCs were recorded. The badge sizes of males participating in EPCs were compared with those of males that were not observed performing EPCs (horizontal lines represent means, vertical boxes represent standard deviations, and vertical lines represent ranges). Adapted from Møller (1992: Fig. 2), with permission of University of Chicago Press.

spent with a male were a positive relationship with bill depth and a negative relationship with male activity during the trial (males that showed courtship behavior and singing tended to move less). Kimball concluded that there was no indication of mate-choice favoring large badge size in this population. The fact that females tended to favor displaying males (reduced locomotor activity) confounds the interpretation of these results. If male sexual display was independent of badge size, which it presumably was, and if there was considerable variation among males in intensity of display, then one would not expect to see a relationship between badge size and mate-choice even if there were a tendency for females to choose a large-badged male over a small-badged male given equal intensities of sexual display.

Reyer et al. (1998) attempted to assess the effect of male badge size on the nest defense behavior of male and female sparrows in Switzerland. Each male was assigned to one of three badge size categories (small, medium, and large).

Each pair was tested once, and a test consisted of placing a mounted mustelid onto a ledge within 2 m of the nest and observing the behavior of both the male and the female. Data recorded included latency (time to discover the presence of the "predator"), number of attacks in the next 20 min (approaches to 30–200 cm), alarm calls (recorded every 15 s), distance between perched bird and predator (noted every 15 s as 0–2 m, 2–5 m, >5 m but in sight, or out of sight). A principal components analysis was performed on the aggression variables, and two principal components explaining 65.4% of the variation were identified. PCI, "Approach," had high positive loadings from number of attacks and time spent 0–2 m and 2–5 m from predator, whereas PCII, "Distant Warning," had high positive loadings from time spent >5 m from the predator and time spent calling. A MANOVA was then performed with PCI and PCII as the dependent variables and seven independent variables (distance of observer from nest, date, number of young, age of young, year, sex, and badge size). The only significant relationships were with sex ($P = .027$) and badge size × sex ($P = .006$). Overall, females scored higher on approach and lower on distant warning than males. In the badge size × sex relationship, females with large-badged mates tended to approach less and large-badged males tended to approach more, whereas females of small-badged males tended to approach more and small-badged males approached less. Reyer et al. concluded that females could predict that large-badged males would provide higher levels of nest defense.

In a study attempting to test the effect of experimentally handicapping females on the rate of mate replacement between first and second broods, Veiga (1996a) watched nests for two 1-h periods (before and after noon) when nestlings were 5 and 12 d old, counted the number of feeding trips by each parent, and scored male courtship behavior from 0 (no courtship behavior) to 4 (full courtship behavior). Badge size and courtship score were positively correlated ($r_s = .43$, $P = .048$), but courtship score was not correlated with other morphological attributes of males or their mates. Veiga (1993a, 1996b) also found, however, that first-year males had significantly smaller badges than second-year and older males in this population, so the increased courtship behavior may have been the result of age rather than badge size.

Griffith, Owens, and Burke (1999b) examined female mate-choice and badge size (= length) on Lundy Island. There was no relationship between male badge length and either laying date of the first clutch or clutch size. Polygynously mated males had significantly smaller badges than those of unpaired males in the population, and there was a significant negative relationship between number of young fledging and badge length. The authors concluded that Lundy Island females tended to favor small-badged males.

In a similar study in Austria, however, Vaclav and Hoi (2002a) reported that females mated to males with average-sized badges initiated their first clutches significantly earlier than did females mated to smaller- or larger-badged males. They also found that females mated to males with average-sized badges laid significantly larger clutches than did those mated to other males, and they concluded that females in this population favored males with average-sized badges.

In a study at two breeding colonies, one in England and the other in Spain, Cordero et al. (1999a) used DNA fingerprinting to identify offspring not fathered by the pair male (extra-pair offspring). The badge size of cuckolded and uncuckolded males did not differ in England, but in Spain, the proportion of extra-pair young in nests of large-badged males (15%) was marginally greater than in nests of small-badged males (4%) ($P = .07$).

A recent study suggested that badge size may be an indicator of the parental quality of the male. Voltura et al. (2002) examined the provisioning rates of male sparrows in Oklahoma (USA). The proportion of male feeds was marginally correlated with badge size ($r = 0.36$, $P = .06$), and nestling survivorship also increased with badge size (logistic regression, $P = .007$).

In summary, the results of the numerous studies of sexual selection on badge size by either male-male contest or female mate choice are inconclusive. Some studies have reported significant relationships between badge size and dominance or nest site acquisition, whereas others have found that any differences in these measures of success in male-male contest were attributable to age rather than to differences in badge size. Results of studies on female mate choice have also been inconclusive, with some indicating that females tend to select large-badged males and others finding no such relationship. Clearly, much more research is needed before firm conclusions can be drawn regarding the selective basis for badge size in males.

Badge Development

Although the various hypotheses for the function of badge size all presume that badge size is related to male quality in some way, virtually nothing is known about the actual mechanism of badge size development. As noted earlier, there is some evidence of a significant genetic component to badge size (cf. Møller 1989). On the other hand, Veiga (1993a) found that badge size in first-year males in Spain was inversely correlated with fledging date ($r = -0.495$, $P = .05$), and Veiga and Puerta (1995) argued that badge size acts as a phenotypically plastic signal that is related to the nutritional state of the male at the time of molt. The results of cross-fostering experiments also suggest that nutritional state may play a role in badge size determination in first-year males (see earlier discussion) (Griffith, Owens, and Burke 1999a). Curiously, despite the intense interest in the adaptive significance of badge size differences in male sparrows, little is known about the actual mechanism of badge size determination. Is the variation due to differences in feather size, as suggested by the positive relationship observed between badge size, badge feather length, and wing length in Spain (Veiga 1996b), or is it due to variation in the number of pigmented feathers originating presumably from the ventral portion of the capital tract, or possibly from the scapulohumeral tract (see Fig. 5.2b), or do feather number and feather size both contribute to increased badge size?

Several recent laboratory studies have examined the effects of various factors on badge size determination in molting sparrows. Factors that have been examined include nutritional state (including body condition), hormone levels (particularly testosterone levels), dominance status, and parasitism. One field study examined the effect of reproductive effort on badge size determination (Griffith 2000).

Veiga and Puerta (1995) captured male sparrows in late June and early July in Spain and housed them in outdoor aviaries, with 10–15 birds per aviary. Birds were maintained on an *ad libitum* diet of mixed seeds supplemented with vitamins. Various morphological and physiological measurements were taken before and after the molt, including badge size, bill measurements, wing and tarsus lengths, fat content (using electrical conductivity), and protein content of the plasma. First-year sparrows in the aviary ($n = 22$) developed badges that were not significantly different in size from those of adults molting either in the aviary ($n = 17$; $t = 0.3$, $P = .76$) or in the wild ($n = 59$; $t = 0.83$, $P = .4$) but were significantly larger than those of first-year males molting in the wild ($n = 18$; $t = 3.35$, $P = .002$). Change in plasma protein level during molt was negatively correlated with badge size in first-year males ($r = -0.66$, $P = .011$) but not in adults ($r = -0.014$, $P = .89$). Stepwise multiple regression indicated that reduction in blood proteins had significant negative effects on badge size ($P = .01$) but fat reserves did not. Veiga and Puerta concluded that, because badge coloration is caused by melanin deposition during feather development, if protein levels become low, they will be allocated toward higher-priority attributes, and badge size will be concomitantly smaller.

In a subsequent study, also in Spain, Gonzalez, Sorci et al. (1999) maintained 96 juvenile males in 10 cages in outdoor aviaries on one of two diets, *ad libitum* canary seed or the same diet plus a protein supplement. Neither badge size nor badge color (measured photometrically) was affected by food treatment or by body mass. Two measures of the immunocompetence of the birds were also recorded, T-cell response and humoral response. The former was determined by measuring the degree of wing web (propatagium) swelling in response to an injection of phytohemagglutinin, and the latter was measured as the concentration of circulating immunoglobulins in blood samples from the birds. Badge size showed no significant relationship with either measure of immunocompetence. Navarro et al. (2003) also found no relationship between badge size and wing web swelling (T-cell response) in 24 males captured in Spain in December.

The results of two other studies suggest that badge size may be influenced by levels of either endoparasitic or ectoparasitic infestation (cf. the Hamilton-Zuk hypothesis). Kruszewicz (1994) studied the progression of molt in first-year house sparrows in Poland. Although badge size was not examined, molt score was determined, body mass was measured, and a blood sample was taken for determination of hematocrit and level of *Plasmodium* infection of the erythrocytes. Significant positive correlations were found between molt score and body mass ($r_s = 0.602$. $P < .0002$) and between molt score and hematocrit ($r_s = 0.446$, $P < .006$). Hematocrit was inversely correlated with number of protozoans ($r_s = -0.452$, $P < .004$). These results suggest that the pace of molt may also be controlled, or at least influenced, by the physiological condition and disease state of the organism. If the development of badge size is similarly influenced, then badge size may reflect either general body condition or state of health of the bird at the time of the molt.

Møller et al. (1996) examined the relationship between the size of the bursa of Fabricius and badge size in 70 first-year males collected between late September and late March in Denmark. The bursa of Fabricius, a dorsal diverticulum of the proctodeal region of the cloaca, is an organ of immune defense that disappears after sexual maturity (see Chapter 9). Badge area was estimated by the formula of Møller (1987b, 1989), and badge asymmetry was determined using a digitized photograph of the badge. The amount of asymmetry was negatively correlated with badge size ($r = -0.32$, $P < .001$). There was also a significant negative relationship between volume of the bursa and badge size ($R^2 = 0.11$, $P < .005$). The volume of the bursa was also significantly correlated with date in days following 1 May ($R^2 = 0.52$, $P < .0001$), and there was no significant sex difference in size of bursa (Kendall $\tau = 0.22$, NS). Because badge size may be inversely related to fledging date (see Veiga 1993a), the authors mentioned that differences in fledging time might account for some of the observed relationship between badge size and bursa volume. The incidence and severity of Mallophaga infestation was estimated by counting the number of holes in the primaries and secondaries of both wings of males. The incidence of Mallophaga damage was 7%, with the average number of holes in infested individuals being 22 (range, 5–42). There was a positive correlation between intensity of mallophagan infestation and volume of the bursa of Fabricius (Kendall $\tau = 0.25$, $P < .05$).

Recent studies have also suggested that hormone levels, particularly those of testosterone (T) and corticosterone, may influence badge size in male sparrows. Evans et al. (2000) captured male sparrows in Scotland in late winter and maintained them in the laboratory through the subsequent prebasic molt. Individuals were assigned to one of four treatment groups: (1) castrated, high T; (2) castrated, low T; (3) castrated, no T; and (4) sham-operated, no exogenous T. Silastic tubes containing T were implanted subcutaneously at the time of castration, with high-T birds receiving a 6-mm tube, low-T birds receiving a 1-mm tube, and no T and sham-operated birds receiving empty tubes. At the beginning of the molt, the Silastic tubes were replaced with tubes containing 1 mm of T in the high-T birds, or 0.5 mm T in the low-T birds. Levels of both T and corticosterone varied significantly among the treatment groups and tended to vary in concert. Change in badge size varied significantly among the treatment groups, with the high-T group having the largest change in badge area (an average increase of about 170 mm^2), while the other three groups showed about the same increases in area (about 70–95 mm^2). In the latter three groups, badge size increases were positively related with T levels in the individuals, whereas in the high-T group, change in badge size decreased slightly with T level. Evans et al. concluded that T level, either during the breeding season (if badge size is determined during that time) or during molt, has a positive effect on badge size. In the same experiment, assignment of the birds to two dietary regimes, either high-quality (consisting of *ad libitum* mixed seeds plus mealworms) or low-quality (consisting of *ad libitum* mixed seeds but mealworms only once weekly), had no effect on the change in badge size (Buchanan et al. 2001).

Gonzalez et al. (2001) found that there was a significant positive correlation between badge size and the log of plasma T levels (measured using radioimmunoassay) in 42 male sparrows captured in April in Spain ($r = 0.41$, $P = .007$). Badge size of the males was again measured in October, after the molt, and was significantly correlated with April plasma T levels after controlling for the positive relationship between initial badge size and October badge size (partial $r^2 = 0.12$, $P = .007$).

Badge size in first-year males may be influenced by the amount of T deposited in the yolk by the female. Strasser and Schwabel (2004) injected 200 ng of T into yolks of eggs in Washington and found that the average badge size after the first prebasic molt of males hatched from eggs with experimentally elevated T levels was significantly greater than that of control males.

Another possible mechanism for the determination of badge size involves the levels of circulating thyroxine (T_4). Miller (1935) studied the effect of exogenous T_4 on the replacement of feathers plucked from sexually dimorphic regions of male and female sparrows, including the throat badge. The replacement feathers from the badge area of treated males were gray due to a great reduction in melanin deposition in the barbs (Fig. 5.9). This suggests the possibility that feathers near the margins of the male badge, if they are labile with respect to melanin deposition, may respond to differences in endogenous T_4 levels, thereby affecting badge size. Thyroxine may thus act as a mediator of body condition, disease state, or other factors that might influence badge size.

The results of a recent study suggest that dominance status may influence badge size determination. McGraw et al. (2003) maintained 14 groups of three males each in the laboratory from spring through the prebasic molt in New York. Aggressive interactions between the birds were monitored, and, although dominance relationships were quite fluid, males were identified as α, β, or χ based on their overall dominance rank. During the subsequent molt, only β males increased their badge size significantly ($P = .04$), and β males had significantly larger postmolt badges than either α or χ males ($P = .02$).

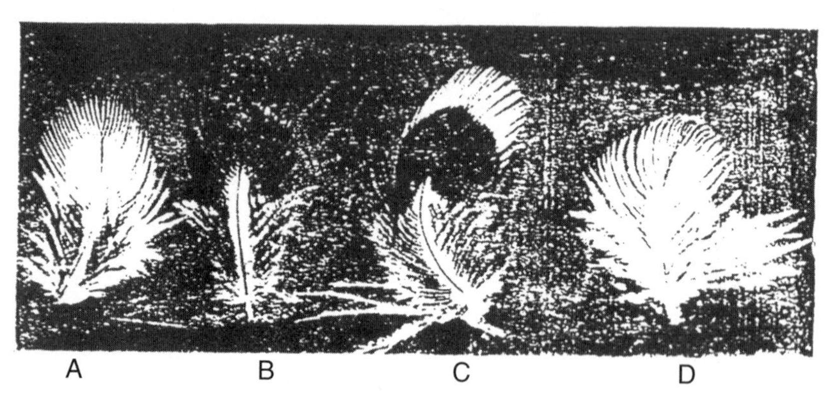

FIGURE 5.9. The effect of exogenous thyroxine (T_4) treatment on the pigmentation of feathers from the throat badge of male house sparrows. Birds were captured in Iowa (USA) and given two doses of 1 mg of thyroxine at 4-d intervals, beginning 5 d after plucking of feathers from the breast (and other sexually dimorphic regions of the plumage). The feathers illustrated (2×) include (A) a typical breast feather from an adult female, (B) a breast feather from an adult male in breeding plumage, (C) a newly molted feather from the badge area of an adult male, and (D) a replacement feather from the badge area of a T_4-treated male. The replacement feather in the T_4-treated male closely resembles the normal female breast feather and also resembles the feathers surrounding the badge in males. Note that newly molted feathers in the badge region of male house sparrows (C) have three distinct regions: a light buffy tip, a black bar across the middle of the feather, and a downy basal region. From Miller (1935: Fig. 5), reproduced with permission of Journal of Experimental Zoology.

In a field study, Griffith (2000) experimentally altered the reproductive effort of male sparrows on Lundy Island by manipulating brood size shortly after hatch. The badge sizes of 31 males after the subsequent prebasic molt were compared with their previous badge size. Males with reduced brood sizes had proportionately larger badges after the molt, whereas males rearing enlarged broods had proportionately smaller badges. Males rearing the number of chicks that hatched in their nests showed no change in badge size.

Many of the new feathers in the badge area are tipped with whitish or buffy ends after the prebasic molt, particularly around the margins of the badge. These buffy tips partially conceal true badge size, which is fully exposed only after the tips are abraded (see Fig. 5.9). The presence of the buffy tips in the newly acquired plumage means that the apparent badge size can be quite different from the actual badge size. Bogliani and Brangi (1990) measured the width of the buffy tips of newly molted feathers on the throat badge and on the crown at three distances from the bill (T1–T3 and C1–C3, respectively) on male Italian sparrows (*P. d. italiae*). Buffy tips were narrower close to the bill and became wider the further from the bill, particularly on the throat, with T3 averaging more than six times wider than T1. The pattern of abrasion of the buffy tips was similar for crown and throat, with the feathers closest to the bill (T1 and C1) abrading between November and February, whereas the other feathers show little abrasion during that period. Abrasion of feather tips at T2 and T3 resulted in the acquisition of full badge size fairly quickly thereafter, however, coinciding with the timing of the breakup of winter flocks.

Veiga (1996b) experimentally removed the buffy tips of male sparrows during the winters preceding two breeding seasons in Spain. Control males had an equivalent number of feathers clipped from the breast region outside the badge or were left unclipped. Nesting success of experimental and control birds was monitored in both years. In one year, badge size had no effect on nesting success in the control group, but birds with large badges in the experimental group had significantly lower nesting success. In the other year, only badge size had a significant effect on nesting success (20 of 51 males with badge sizes above the median were successful, whereas only 9 of 50 males with badge sizes below the median were successful) ($\chi^2 = 5.6$, df $= 1$, $P = .02$). Veiga concluded that the ability to facultatively conceal badge size contributes to male fitness.

The rate of abrasion of the buffy tips may be affected by T levels. Gonzalez et al. (2001) captured males in April in Spain and implanted either empty Silastic tubes or tubes containing cyproterone acetate (CPA), an antiandrogen. Males implanted with CPA had significantly less abrasion of the buffy feather tips on the bib than did the control males.

Evolution of Badge Size

Owens and Hartley (1991) developed a theoretical, game-theory model to explore the evolutionary implications of the Badge of Status Hypothesis. The model explores the question of whether populations can develop in which there are two evolutionarily stable strategies involving "honest" status signalers—large-badge aggressive (LA) and small-badge submissive (SS) individuals—and whether these populations remain stable in the presence of two deceptive groups—large-badge submissive (LS) birds and small-badge individuals that are aggressive when food resources are intermediate but submissive when food resources are abundant (SA = "Trojan sparrows"). In asexually reproducing populations with honest signalers, a stable equilibrium can develop between LA and SS individuals, with proportions of the two types depending on the relationship between the value of the food items and the costs of fighting. Such an equilibrium can be invulnerable to invasion by LS individuals if the punishment costs for "cheating" are high. The equilibrium is, however, almost always vulnerable to invasion by Trojan sparrows, with Trojan sparrows being the only survivors in some cases.

Although it is premature to conclude that the throat badge of male sparrows is in fact a badge of status (see later discussion), it is appropriate to ask whether there is any evidence for "punishment" of badge size cheaters in sparrows. Veiga (1995) experimentally manipulated badge sizes of males captured during February and March in an isolated breeding colony in central Spain, creating four badge size categories: enlarged, reduced, dyed control, and unmanipulated control. The males were also categorized into three age groups: first-year, 1-year-old, and >1-year-old. Survivorship to the following February was monitored by recapture and sighting of marked birds. When badge-enlarged birds were compared to all other treatment groups, there was a significant three-way interaction in the three-way contingency analysis ($G = 8.41$, $P = .014$), with more first-year males with enlarged badges dying than would be expected. This suggests that there is a fitness penalty for first-year males with "deceptively" large badges, although the mechanism of the "punishment" for the deception is unknown.

Conclusions

Although a number of highly original and intriguing suggestions have emerged from the studies of badge size in male sparrows during the last 15 y, the evidence from these studies remains equivocal and, as yet, unconvincing. As mentioned earlier, one of the reasons for this is undoubtedly the lack of a standard protocol for measuring badge size. This deficiency obviates possible comparisons among studies, but presumably the various methods of measuring badge size at least accurately reflect relative badge size differences within the populations studied. There is, however, little agreement in the findings of the studies. Badge size is not consistently related to dominance or to command of resources in male-male interactions, nor is it consistently related to female mate choice. It also is not consistently related to either body size or condition.

The most consistent finding is that badge size increases with age, at least between first-year and older males (but see Møller 1987b). This relationship has been reported both for different age cohorts in the same population and for individuals over time. This suggests that badge size may be a fairly consistent predictor of age in a population, a fact that may alone account for most of the patterns reported in the studies described. The greater experience of older males may account for their success in dominance interactions and in competition for limited resources such as nest sites. In mate choice by females, the fact that large-badged males have usually survived at least two winters rather than just one may constitute the best "rule of thumb" for selecting a mate with the "best genes" (see Riters et al. 2004). To the extent that a female also uses other factors, such as nest site quality or availability of alternative nest sites, in making her mate choice, these factors may become confounded with male badge size in the mate-choice decision. Teasing apart the roles of these and other factors in female mate choice will require much careful and ingenious empirical work if the role of male badge size in female mate choice is to finally be understood.

Several questions need to be answered: Is badge size primarily genetically determined, or is there a major environmental contribution to badge size? Does badge size show unidirectional change with age? What is the proximate mechanism of badge size determination? Although phenomenological studies of intrapopulation correlates of badge size may continue to generate interesting suggestions about the function of male badge size in the species, it is unlikely that any consensus will emerge until these questions are answered. The relationship between badge size and age, at least for first-year versus older males, also suggests that future studies should always control for age effects, perhaps by using only 2-y-old or older males for experimental studies.

Chapter 6

FORAGING BEHAVIOR AND FOOD

No bars are set too close, no mesh too fine
To keep me from the eagle and the lion,
Whom keepers feed that I may freely dine,
This goes to show that if you have the wit
To be small, common, cute, and live on shit,
Though the cage fret kings, you may make free with it.
—Howard Nemerov, *The Sparrow in the Zoo*
(by permission of Margaret Nemerov)

The word most often used to describe the foraging behavior of the house sparrow is "opportunistic." Sparrows have been observed fluttering in front of the electronic sensor to open automatic doors in a grocery store in Australia, and in a bus station in New Zealand, to enter to feed (Breitwisch and Breitwisch 1991; Hubregste 1992). In New Zealand, they have also been observed to peck holes in the base of flowers of the Kowhai tree to rob nectar (Sitdolphi 1974b), and in Spain they have learned to pull the tray from the bottom of an outdoor birdcage to obtain uneaten seeds (Garcia 1994). They remove dead insects from spider webs and from the radiators of automobiles (Flux and Thompson 1986; Rossetti 1983), beat bushes with their wings to flush moths and beetles which they then catch on the wing (Guillory and Deshotels 1981), and sally from a perch to capture flying periodical cicadas (*Magicicada* spp.) during the peak of their emergence (Anderson 1977; Howard 1937). They have also been observed to cling to the walls of beachfront hotels in Hawaii (USA) to monitor feeding opportunities provided by vacationers eating on their balconies (Kalmus 1984). This latter activity, which is timed to coincide with breakfast and lunch times of the vacationers, is a complex suite of behaviors that significantly reduces the energetic costs of monitoring the feeding opportunities that would be expended by flying. Opportunism is not limited to adult sparrows, because recent fledglings have been observed entering nests containing nestling sparrows to obtain food from the attending adults (Schifferli 1980c).

This remarkable opportunism, or adaptability, has undoubtedly played a major role in the house sparrow's success as a commensal of urban and agricultural humans. Sparrows, although primarily granivorous, feed on a wide variety of items, utilize a variety of foraging techniques, and even feed nocturnally under certain conditions. This opportunistic foraging strategy raises questions about how sparrows learn to exploit feeding opportunities and how they deal with neophobia, the fear of new things, which is common in many animals.

The house sparrow is also a very social species. It breeds in colonies, roosts communally, feeds in flocks, and even engages in communal sexual displays (see Chapter 4). It has a clear preference for feeding in flocks as opposed to feeding alone (Popp 1988), and it frequently feeds in heterospecific as well as homospecific flocks (Clergeau 1990; Kalinoski 1975). Clergeau (1990), for instance, reported that sparrows landed among stuffed models of foraging European starlings (*Sturnus vulgaris*) placed in the middle of urban lawns. Why do house sparrows forage in flocks? What are the selective advantages of flock foraging that outweigh the obvious disadvantage of increased competition among individuals for potentially limiting food resources? What determines flock size in foraging sparrows? How do foraging flocks form? Although many answers to these questions have been proposed, one of the most cogent and persuasive arguments has focused on the increased foraging efficiency gained by flock foragers due to reductions in individual vigilance behaviors required to avoid predation during foraging.

Sparrows spend a great deal of time foraging on the ground for seeds, usually in open, exposed areas. Open areas are preferred over areas with concealing vegetation, presumably because of the danger of undetected approach by terrestrial predators (Popp 1988). Foraging in open areas increases the risk of detection and successful attack by predators approaching in the air, however. A theoretical model proposed by Pulliam (1973) for ground-foraging, seed-eating birds that are vulnerable to avian predators has helped to stimulate numerous studies of foraging house sparrows to test both the assumptions and predictions of the model. The major variables in the model are scanning frequency (λ),the time required for an exposed predator to reach its potential prey (τ), and flock size (N). Scanning rate is defined as the number of scans per minute of foraging time, and the interscan interval of an individual is assumed to be randomly distributed around its average scanning rate and independent of the scanning behavior of other individuals in the flock. The model assumes that each member of a flock has an equal chance of being captured by the predator.

One of the major predictions of the model is that birds foraging in flocks can reduce their individual scanning frequencies and still enjoy the same degree of protection from predation. This means that a higher percentage of foraging time can be devoted to feeding, resulting not only in greater foraging efficiency, but also in a reduction in the amount of time that the birds are exposed to predation. The model also shows that the probability of detecting an approaching predator levels off quickly as flock size increases, with asymptotic detection rates being reached at flock sizes of five or six for realistic estimates of λ and τ.

Pulliam et al. (1982) extended this model, by including additional variables, and used game-theory modeling to attempt to determine under what conditions flock foraging behavior would be an evolutionarily stable strategy (ESS). An ESS is an evolved behavior that cannot be successfully exploited by "cheaters," individuals that derive the benefits of the system without sharing the costs. In this case, cheaters would be individuals feeding in flocks that do not participate in scanning behaviors to detect approaching predators. The extended model assumes that the length of a scanning event is constant and that the time required to change from feeding to scanning is negligible (instantaneous). It also takes into account the probability of the predator's being successful (b), and it assumes that $b < 1$ if the predator has been detected by the potential prey. The distance to cover (thick bush, thicket, or hedgerow) for the foraging birds will have an effect on b, with the probability of successful attack increasing with increasing distance from cover. Three types of birds were treated in the analysis of the model, "cooperative," "selfish," and "judge." Cooperative birds always engage in scanning for predators while foraging, selfish birds feed in flocks without participating in scanning for predators, and judges adjust their scanning behavior in response to that of other birds in the flock, scanning if others scan and not scanning if others do not scan. The model shows that the judge strategy is evolutionarily stable, while the cooperative strategy is unstable because it can be successfully exploited by "selfish" birds.

Both of the models assume that the probability of an individual's being the victim of a successful predator is proportional to $1/N$, which means that the individual probability of being the victim of predation would continue to decrease with increasing flock size as long as detection by the predator is independent of flock size. This suggests that optimal flock size, beyond the asymptotic size for predator detection, must be determined by some other factor, such as food density, increased competition among flock members, or increased rate of detection by the predator.

Increased competition for food encountered by socially foraging birds may well lead to aggression among flock members. Increased aggression among flock members may in turn lead to reduced efficiency in food acquisition, or it may even increase the risk of detection by predators. Other factors that may influence house sparrow foraging strategy include temporal pattern of foraging, environmental conditions (e.g., ambient temperature), and individual variation in foraging pattern.

Numerous observational and experimental studies of sparrows have been done in recent years (e.g., Barnard 1980c; Caraco and Bayham 1982; Elgar 1986b; Lima 1987). Most of these studies have been performed during the late autumn and winter (October–March in the Northern Hemisphere), when flocks are easier to attract and observe. Because sparrows regularly use bird feeders, experimenters can readily control such factors as arena size, food density, and distance to cover. In some cases, this has permitted direct tests of either the assumptions or the predictions of Pulliam's model.

This chapter first treats the flock foraging strategy of house sparrows and then discusses their diet throughout the year.

Foraging Strategy of the House Sparrow

Temporal Pattern

Beer (1961) studied the diurnal pattern of feeding by sparrows at a feeding station in Minnesota (USA) between November and March. The typical pattern included a peak of foraging activity immediately after sunrise, followed by a decline in feeding activity until shortly before sunset, when a second peak occurred (Fig. 6.1). Little feeding was observed before sunrise or after sunset except during the coldest weather. Beer also examined the effect of temperature on foraging pattern by comparing patterns on days with mean temperatures falling in three different ranges: –25°C to –20°C, –14°C to –9°C, and –3°C to 2°C. On the coldest days, sparrows often arrived before sunrise and fed constantly while at the feeding station with very little conflict. Feeding was less intense on warmer days. Two birds were observed to freeze to death on the feeding tray on very cold days. In a study conducted during the summer months of May to July in New York (USA), sparrows usually arrived at the feeding stations before dawn (Lima 1987).

Nocturnal foraging has been reported in the house sparrow. Broun (1971) reported sparrows feeding on insects attracted to floodlights at the Bangkok

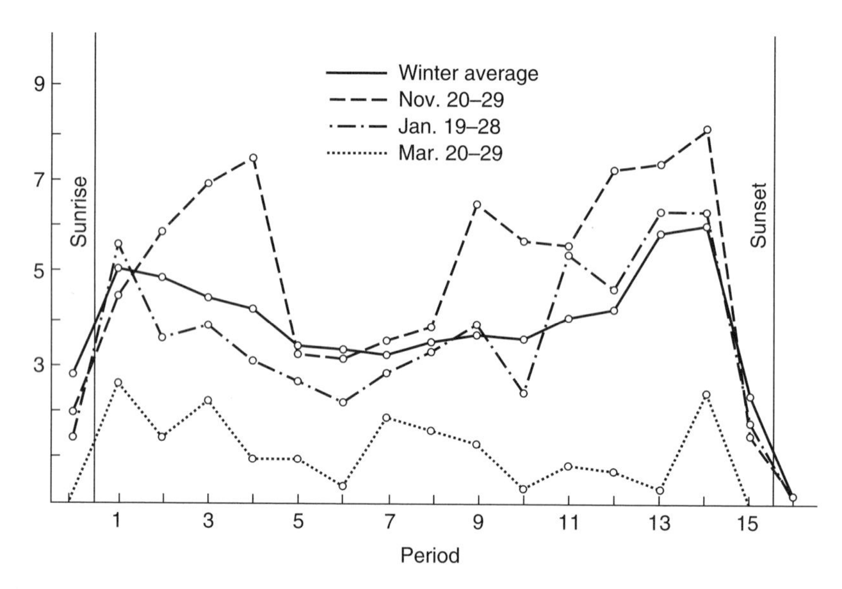

FIGURE 6.1. Diurnal pattern of winter foraging in house sparrows in Minnesota (USA). Due to variation in daylength throughout the observation period (November–March), each day was divided into 15 intervals of equal length from sunrise to sunset, and these are plotted on the abscissa. The average number of sparrows feeding during each interval throughout the observation period is plotted on the ordinate. From Beer (1961: Fig. 1), with permission of the American Ornithologists' Union.

Airport (Thailand) between 2250 and 2300. A similar observation was made at 2330 on the observation deck of the Empire State Building (New York) (Brooke 1973). Haftorn (1994) observed sparrows feeding at a feeding station during the continuous darkness of the arctic winter above the Arctic Circle in Norway (69°27'N) (see also Chapter 3).

Flock Size

Flock size in sparrows varies from a few birds (4–8) to several hundred birds (Heij and Moeliker 1990; Summers-Smith 1954a), with one flock of more than 5000 birds being reported (Summers-Smith 1956). The larger flocks are most often reported from grain fields near the time of harvest and usually include many juveniles. In winter, flock sizes tend to be smaller. Beer (1961), for instance, stated that several flocks of 5–15 sparrows visited his winter feeding station. Other studies done at feeding stations have also reported small flock sizes. Breitwisch and Hudak (1989) recorded an average flock size of 6.6 for 178 flocks visiting a winter feeding station in Pennsylvania (USA). The largest flocks observed during each of 18 observation periods averaged 16.9, with the largest comprising 27 sparrows. Flock size (N) at a winter feeding station in New York had a weak but highly significant inverse correlation with ambient temperature (T in °C): $N = 2.3 - 0.06T$ ($df = 359$, $P < .0005$) (Caraco and Bayham 1982). Barnard (1980b), however, reported that flock size was positively correlated with ambient temperature in an English field. The difference between these two studies may be due to differences in ambient winter temperatures at the two sites, which are usually below freezing in New York but usually above freezing in England. Barnard (1980d) reported that flock size increased with increasing distance to cover in an English field.

Barnard (1980a) attempted to identify what factors contribute to flock size in foraging sparrows by examining equilibrium-sized flocks feeding at cattlesheds in England. An equilibrium-sized flock was defined as a flock in which the number of departing sparrows matched the number of arriving sparrows (flocks of static size were not included). Flock size at equilibrium for 24 such flocks was compared with several environmental factors: size of the high-density food patch, total area of the foraging site, seed density in the high-density patch, seed density in the area surrounding the high-density patch, time of day, daylength, ambient temperature, and average temperature during the preceding 24 h. The two factors that contributed significantly to equilibrium flock size were seed density in the area surrounding the high-density patch ($P < .02$) and area of the high-density patch ($P < .05$). The density of seeds in the high-density patch had only a negligible effect on flock size. Individuals tended to leave the flock if their individual pecking rate dropped significantly below the average pecking rate in the flock, and Barnard concluded that the seed density in the area surrounding the high-density patch influenced flock size by determining the "giving-up time" of individuals that were not in the high-density patch.

A study by Johnson et al. (2001) suggested that increasing flock size may have negative effects on foraging efficiency due to increased interference among flock members. They studied two aspects of foraging efficiency, search time and handling time, in foraging sparrow flocks in Quebec (Canada). Foraging efficiency increased with flock size up to three birds, but decreased with further increases in flock size (Fig. 6.2). The decrease was caused by an increase in handling time rather than increased search time, and the authors attributed the decrease in foraging efficiency to increased social vigilance by the foraging sparrows. They suggested that this social vigilance enabled the birds to detect and avoid particularly aggressive individuals.

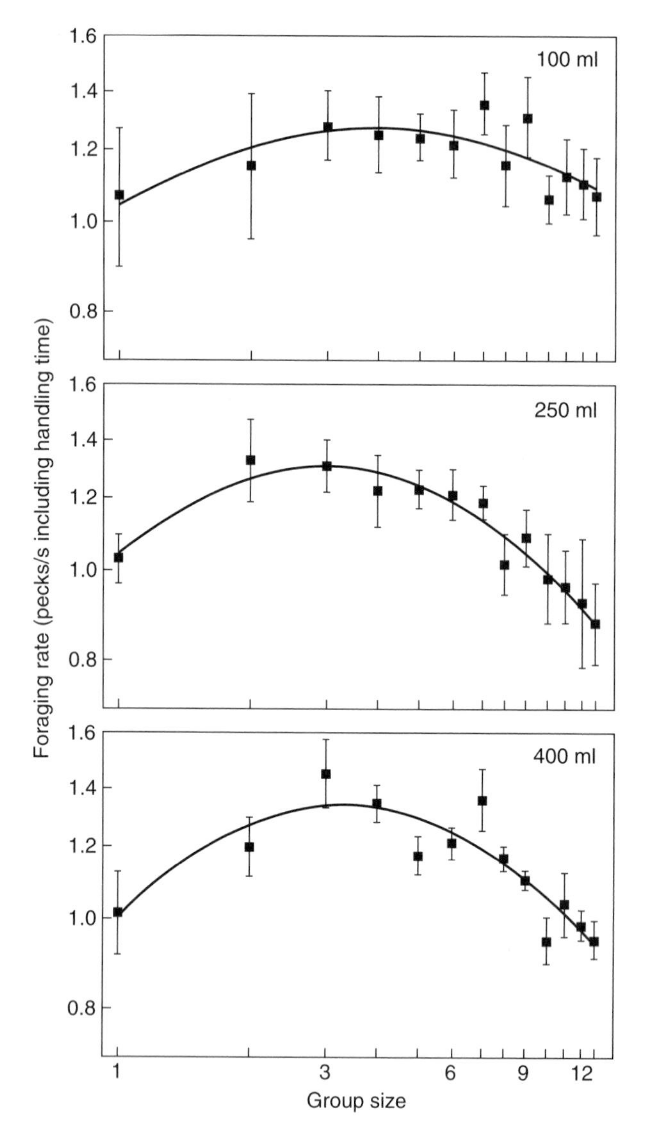

FIGURE 6.2. Foraging rate of house sparrows plotted against flock size. The data were obtained from observations of six individual sparrows feeding on a 0.59 × 0.61 m feeding platform at three different food densities (100, 250, and 400 ml of white millet seed) and at flock sizes ranging from 1 to 13 individuals. The foraging rate was the pecking rate of the individuals, including the handling time for seeds ingested. The quadratic relationships at the 100- and 400-ml seed densities were both highly significant ($P < .0005$). From Johnson et al. (2001: Fig. 3).

Vigilance

Foraging sparrows frequently raise their heads until the bill is horizontal and turn their heads from side to side about the vertical axis (Barnard 1980c). This behavior, which Barnard called "looking," has also been called "scanning" (Elcavage and Caraco 1983) or "head-jerk" (Elgar and Catterall 1981). It is usually interpreted as vigilance behavior, with its primary function being the detection of potential predators. Elgar et al. (1986) suggested that it might also function to monitor the behavior of other foraging sparrows.

One of the principal predictions of the Pulliam model is that participation in flock foraging should increase foraging efficiency without sacrificing predator detection. There are several ways in which sparrows could achieve this objective (see McVean and Haddlesey 1980), but two of the simplest would be decreasing the time of individual scanning events and decreasing the scanning rate (with individual scanning events remaining constant). The Pulliam model postulates that the increase in efficiency is achieved by the latter method. The model shows that, if scanning by an individual is independent of scanning by other individuals in the flock, the overall scanning frequency of the flock, and hence the chances of detecting an approaching predator, does not change even if individual scanning rates decrease as flock size increases. Both observational and experimental studies have attempted to test this prediction for house sparrow flocks.

Barnard (1980c, 1980d) studied foraging flocks of sparrows in two habitats in England. Although both habitats had comparable densities of food, primarily barley and other grains, one (open field adjacent to a hedgerow) had many more potential predators than the other (cattlesheds). Sparrows foraging in the field scanned at a significantly higher rate than did those foraging in cattlesheds (29 and 14 scans/min, respectively, $P < .01$) (Barnard 1980c). Individual scanning rate in the field decreased with increasing flock size up to a flock size of about 17 birds ($r = -0.849$, $P < .001$) but was not correlated with flock size in the cattlesheds (Barnard 1980c).

Elgar and Catterall (1981) observed individual scanning rates in sparrow flocks foraging on lawns and along roadsides in the winter in Australia. The average duration of a scanning event was 0.63 s, and it did not differ in any systematic way with flock size. Elcavage and Caraco (1983) monitored scanning rates in foraging sparrows at a winter feeding site in New York. They found that the average duration of a scanning event was 0.74 s, and that duration was not significantly correlated with flock size for flocks of one to six birds. Although these studies tend to support the assumption of the Pulliam model that scanning events do not vary in length, other studies suggest that the duration of individual scanning events can vary considerably in sparrows. McVean and Haddlesey (1980) observed sparrows engaged in several activities, including foraging, preening, and perching (either on posts or in trees), and found that the duration of a scanning event varied from just under 1 s to about 6 s, with the large majority of scanning events lasting less than 3 s. Studd et al. (1983) observed sparrows feeding at a four-sided feeder that permitted only four birds to feed simultaneously but also permitted each feeding bird to see all other feeding birds. The length of individual scanning events decreased with the number of birds feeding simultaneously, from almost 10 s for lone birds to about 4 s for birds feeding with three other sparrows ($r_s = -0.37$, $P < .001$). The assumption that individual scanning times are constant and independent of flock size therefore must apply only in certain situations for house sparrows.

Several studies have found that scanning frequency decreases with increasing flock size in sparrows. In Australia, scanning rates of individuals decreased significantly with flock size up to a flock size of five, and then leveled off at about 10 scans/min (Fig. 6.3) (Elgar and Catterall 1981). Elcavage and Caraco (1983) found that time between scans increased with increasing flock size up to six sparrows ($r = 0.98$, $P < .01$), and the regression equation that they calculated suggested that the lowest scanning frequency would be about 25 scans/min. Elgar et al. (1984) also found a significant positive correlation between flock size and interval between scans ($r = 0.267$, $P < .0001$) and suggested that the minimum scanning rate for a flock of five or more birds would be about 16/min. Both Elgar (1986b) and Lima (1987) also reported that time spent scanning decreased significantly with increasing flock size in foraging sparrows.

Elgar et al. (1984) performed two experiments to test how sparrows determine flock size. In the first, they placed a 12-cm high barrier across a 2.4 × 2.4 m feeding tray. Interscan interval was significantly correlated only with the number of birds on the same side of this visual barrier, not with the total number on the feeding tray. In the second experiment, two feeding trays were placed either adjacent to each other or 1.2 or 2.4 m apart. The interscan time of sparrows on one of the trays (the focal tray) was significantly correlated with the total number of birds on both trays only when the trays were adjacent. If the two trays were 1.2 or 2.4 m apart, the interscan time of birds on the focal tray was strongly influenced only by the number of birds feeding on the focal tray. Elgar et al. concluded that sparrows use information only on the number of visible sparrows within about 1.2 m to influence their scanning rate while feeding.

The variation in minimum scanning rates identified previously (from 10 to almost 30 scans/min) suggests that factors other than flock size affect scanning rate in the house sparrow. These factors may include distance to cover, frequency of predator attacks in the area, or other environmental variables.

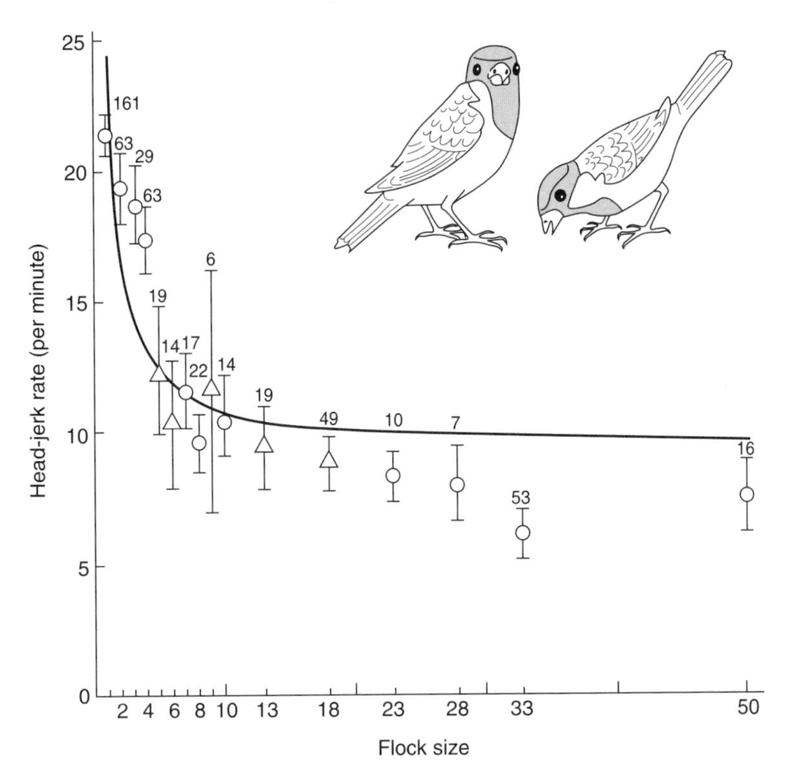

FIGURE 6.3. Scanning rate (λ) plotted against flock size (N) in house sparrows foraging in open fields and lawns in Australia (the sparrow on the left is in the scanning position). Points represent mean scanning rates ± SE. Open circles indicate points at which the interscan intervals did not differ significantly from random (as assumed by the Pulliam model—see text), whereas triangles indicate points at which the interscan interval deviated significantly from random. Numbers represent the number of flocks of each size, and flock sizes greater than 10 are represented by the midpoint of intervals of five (e.g., 13 represents flock sizes of 11–15). The curve is fitted by regression of λ on 1/N for all flock sizes and is represented by the equation $\lambda = 9.35 + 15.13/N$. From Elgar and Catterall (1981: Fig. 2), with permission from Elsevier.

Foraging Efficiency

The Pulliam model predicts that foraging efficiency should increase with increasing flock size due to a reduction in vigilance time on the part of individuals. Measures of foraging efficiency are usually indirect, such as pecking rate or time spent feeding. The hypothesis that individuals should select foraging patches according to an ideal free distribution, on the other hand, suggests that individuals should distribute themselves in such a way that intake rates are equal across a range of flock sizes and food densities (see Fretwell and Lucas 1970). This hypothesis assumes that individuals are able to accurately assess the foraging potential (and possibly predation risk) of feeding sites, that they choose foraging sites that maximize their individual intake rates, and that they act freely (that is, make decisions independently of other individuals).

Some evidence from sparrows tends to support ideal free distribution theory. Barnard (1980c), for instance, found that average pecking rates were very similar in open fields and cattlesheds in England (0.643 vs. 0.623 pecks/s, respectively). In the fields, which had a greater risk of predation, pecking rate was significantly correlated with both flock size ($r_p = 0.624$, $P < .001$) and seed density ($r_p = 0.260$, $P < .01$). In the cattlesheds, however, pecking rate was significantly correlated with seed density ($r_p = 0.715$, $P < .0001$) and minimum ambient temperature ($r_p = 0.457$, $P < .0001$). Pecking rate in both areas tended to show a Holling type 2 functional response to increasing food density (Fig. 6.4). Seed densities at the field and in the cattleshed sites were comparable (Barnard 1980d).

Barnard (1980b) also conducted an experiment to determine the effect of food density on flock size in the open field habitat. He baited strips 0.25 m

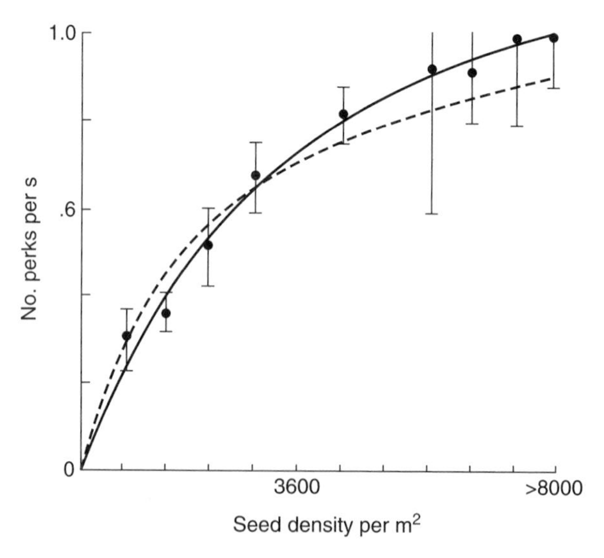

FIGURE 6.4. Holling type 2 functional response of foraging house sparrows (*solid line*). The pecking rate of foraging sparrows is plotted against seed density for 54 feeding bouts in cattlesheds in England (UK). Points are mean pecking rates (± SE). The dotted line is the disc equation (Holling type 2) for an empirically determined handling time of 1.02 s. From Barnard (1980c: Fig. 4), with permission from Elsevier.

wide × 20 m long located at the edge of a field with seeds at densities of 300, 600, or 1200 seeds/m^2. The number of sparrows feeding at each site was determined before the placement of the seeds, and again 3, 4, and 5 d after the manipulation. Although the number of birds foraging on the patches did not differ in any consistent way before the baiting, the number of birds feeding at the three sites showed a very rapid response to the changes in food density. The average number of birds feeding at the sites with seed density of 1200 seeds/m^2 increased by 37.8, whereas numbers decreased by 19.5 and 12.9 at sites with seed densities of 300 and 600 seeds/m^2, respectively.

Two explicit tests of the theory of an ideal free distribution failed to support the theory for the house sparrow. One involved free-living sparrows (Elgar 1987), and the other involved a laboratory flock of six females (Gray 1994). Elgar (1987) attempted to test whether sparrows make the decision to forage in flocks based on the expected return rate. The study was conducted on the roof of the Zoology Building at Cambridge University (UK) and involved the use of feeding trays of two different sizes (1 × 1 m and 0.15 × 0.15 m) along with differences in food density. The larger trays were provisioned with 100 ml of seed, and the smaller trays received 50 ml of seed, so that the food density on the smaller tray exceeded that on the larger tray by 22 times. Flock size, pecking rate, scanning rate, and rate of aggressive encounters were monitored at single feeders of one size or the other on randomly determined days. For 151 flocks at the larger tray, pecking rate increased with increasing flock size (to nine birds) ($r_s = 0.344$, $P < .001$), and the scanning rate decreased ($r_s = -0.405$, $P < .001$). On the other hand, pecking rate decreased significantly with increasing flock size on the smaller tray ($r_s = -0.270$, $P < .001$). The average pecking rate of solitary sparrows did not differ with tray size (0.93 pecks/s on large trays vs. 0.91 pecks/s on small trays). Aggression rate increased with flock size on both feeding trays, but the rate of aggression with two birds on the small tray was 15 times greater than that with seven or more birds on the large tray.

Based on these results and on the previously described observation that sparrows adjust their scanning rate on the basis of the number of sparrows feeding within about 1.2 m, Elgar (1987) attempted to determine whether sparrows make decisions about joining a flock based on optimizing their individual rates of return. He placed two trays of the same size, large or small, 3–4 m apart and supplied each with the same seed amounts as described earlier. If birds select a feeding site based on expected average return rate, they should always select the tray with the larger number of foraging sparrows when there are two large trays, but they should select the tray with the smaller number of sparrows when there are two small trays. The results, however, did not conform to these predictions. When birds had the choice between feeding at an empty tray or joining a flock feeding at the other tray, they consistently chose the latter regardless of tray size. The number choosing an already occupied feeding tray did increase with increasing flock size at large trays ($r_s = 0.928$, $P < .01$), but it decreased significantly with increasing flock size at small trays ($r_s = -0.943$, $P < .005$). These results suggest that factors other than expected individual intake rate affect the choice of whether or not to join a flock. Elgar suggested that this might be due to an inability to gather accurate information about expected intake rate at a site, in part because of the short duration of feeding bouts (mean = 23.1 s). He also suggested that sparrows may operate under a simple rule of thumb to join a flock of any size. Grubb and Greenwald (1982) also observed a tendency for sparrows to join existing foraging flocks in a choice experiment in the field.

In the second experiment, Gray (1994) studied the foraging choices of six female sparrows in an indoor aviary in which constant conditions were maintained (23°C and 8L:16D). Food delivery rate at two feeding stations was controlled at one of five ratios (8:1, 1:4, 4:1, 1:8, or 1:1) for 15-d periods, and the individual and group use of the two feeders was monitored for the last 5 d. The feeding time allocation of the birds undermatched the food output ratio in all cases (the slopes for the regression equations were 0.70 for time spent on the feeding perches and 0.71 for time spent on either the feeding tables or the perches, both significantly different from the predicted 1). The 10-d period of prior exposure to the difference in food delivery rates at the two feeders suggests that the failure of the sparrows to conform to an ideal free distribution in their foraging is not due to an inability to gain accurate information on potential food intake. Gray noted that the sparrows tended to feed socially and suggested that this tendency to feed socially violates the "free" assumption of ideal free distribution theory.

These results indicate that, to the extent that foraging sparrows appear to conform to an ideal free distribution (cf. Barnard 1980c), the decisions responsible for this distribution are made at the level of groups, not at the level of individuals. These groups may, however, distribute themselves over the potential foraging area in a way that conforms to the expectations of the ideal free distribution. The large congregations of sparrows that gather periodically throughout the day at communal roosting sites may play a role in information exchange that facilitates the achievement of this foraging distribution. Although such information transfer has been demonstrated in other social species (i.e., Brown 1986; Ward 1965; Ward and Zahavi 1973), I know of no study demonstrating such information transfer in the house sparrow. Information transfer does occur within foraging flocks, however (see later discussion).

Distance to Cover

If the threat of predation is a major factor in shaping foraging behavior in sparrows, the distance from protective cover of a foraging location should influence sparrow behavior. All other things being equal, sparrows should select foraging sties closer to protective cover. This in turn might lead to increases in foraging efficiency if sparrows adjust their scanning rate based on distance to cover.

Sparrows tend to forage near protective cover when possible. Cowie and Simons (1991) placed feeders at four distances from a hedgerow (0, 2.5, 5.0, and 7.5 m) and found that sparrows fed only at the feeder adjacent to the hedgerow. Giesbrech and Ankney (1998) placed feeding stations 1.7, 2.7, and 3.7 m from protective cover and found that sparrows fed significantly more often at the closest feeding stations; they used the more distant stations more frequently later in the observation periods, when food depletion had occurred in the closer stations. Sodhi (1992b) observed the distance to protective cover of foraging sparrows in Saskatoon, Saskatchewan (Canada) during May–July. House sparrows make up about 64.5%–69% of the prey items of breeding merlins (*Falco columbairus*) in Saskatoon (Oliphant and McTaggart 1977; Sodhi and Oliphant 1993) (see also Chapter 8). In each of the three months of the study, 67% or more of the sparrows observed were feeding within 1 m of protective coverage.

Barnard (1980d) reported that flock size increased significantly with distance to cover in the field habitat in England. He also found that the duration of foraging bouts decreased significantly with distance to cover, suggesting that sparrows flush more easily when feeding at a greater distance from safety.

Two experimental studies have reported conflicting results with respect to vigilance behavior and distance to cover. In the first, Lazarus and Symonds (1992) used a frame (2 m long, 1 m high, and 0.5 m thick) to establish either "protective" or "obstructive" cover for winter foraging flocks in England. For protective cover, the frame was filled with cuttings of *Spiraea*, and for obstructive cover, it was covered by a solid wooden board. Food was placed either 0.5 or 5 m from the frame, resulting in four treatments. Individual birds feeding in mixed flocks (mostly house sparrows and European starlings) were observed for up to 200 s and scored every 5 s for vigilance. Vigilance increased with distance to protective cover but decreased with distance from obstructive cover. In another study, Lima (1987) investigated the effects of distance to cover in free-living sparrows in an old field in New York. Feeding patches (2.5 × 0.4 m) were located 0.5 and 5 m from a hedgerow with the long axis of the patch parallel to the hedgerow. In some treatments, walls (7 cm high) were placed entirely around the feeding patch; in others, the patch was left open. In subsequent choice experiments, food was placed in both patches, with both patches either surrounded by walls or open. Sparrows showed a strong preference for the patch closest to cover, with 82% of individuals feeding in that patch when the walls were present and 73% feeding there when there were no walls (both significant at $P < .05$). Contrary to expectation, however, scanning rates decreased significantly with increased distance from cover, both with and without walls. Lima interpreted this somewhat surprising result as sparrows' minimizing the amount of time they were exposed to predation while obtaining a certain amount of food.

Aggression and Dominance Hierarchies

Aggressive interactions ranging from various types of threat to overt fighting are common in flocks of foraging sparrows. Fighting is apparently uncommon in winter foraging flocks (Beer 1961; Breitwisch and Hudak 1989), but aggressive encounters such as attempts to displace another individual at a feeding station are frequent. Although many flock foraging species exhibit dominance hierarchies that serve to limit overt fighting and lesson the risks of injury, it is not clear whether sparrow flocks regularly show such dominance hierarchies.

Dominance hierarchies have been reported in captive flocks of house sparrows (e. g., Andersson and Ahlund 1991; Gray 1994; Hegner and Wingfield 1987c), although Watson (1970) found no evidence for stable linear hierarchies in groups of 4, 6, or 8 caged sparrows. Hegner and Wingfield (1987c) observed aggressive interactions for 2 wk in 18 caged flocks each consisting of six sparrows in New York. They were able to identify unambiguous hierarchies in 62% of the flocks, with the remaining flocks having at least one ambiguous or intransitive relationship. Gray (1994) determined the dominance relationships in a group of six female sparrows over a 75-d period in New Zealand. He scored all agonistic interactions at feeding locations on day 10 of each of five 15-d feeding trials and constructed a matrix of dominance relationships among the six birds. The dominance hierarchy based on these data changed during the five trials, being A>B>C>D>E>F in the first two, B>C>A>D>E>F in the third, and B>C>D>A>E>F in the final two). Overall, two equally parsimonious hierarchies were found, but A>B>C>D>E>F was selected because it also minimized the total proportions of reversals (intransitive relationships). The hierarchy was strongly nonlinear, with two pronounced violations (E>B and C>A). Solberg and Ringsby (1997) used both direct and videotaped observations to study the dominance relationships of five caged groups consisting of 6–11 male sparrows in Norway. There were significant linear hierarchies in only two of the five groups, although a third was close to significance. Riters et al. (2004) found little evidence for stable linear hierarchies in four caged flocks each consisting of five males. These results suggest that even in stable groups of caged sparrows, dominance hierarchies are not fixed or rigidly linear.

Andersson and Ahlund (1991) studied the effect of food deprivation on dominance relationships in groups consisting of three caged sparrows. Subordinate birds won significantly more aggressive encounters with dominants if they had been deprived of food for 20 h before the trial than if they had been fed immediately before the trial. In a second experiment, Andersson and Ahlund paired two sparrows that were strangers (no established dominance relationship) and tested them when one bird had been food-deprived for 20 h while the other had been deprived for only 3 h. In 16 of the 17 trials in which there were aggressive interactions, the 20–h bird won a majority of the conflicts ($P < .001$). Andersson and Ahlund concluded that hunger-induced motivation had an important effect on the outcome of aggressive encounters among sparrows. Cink (1977) reported an increase in aggressiveness in caged sparrows after a 3-h period of food deprivation.

In a recent experiment, Lendvai et al. (2004) tested the effects of loss of energy reserves on foraging tactics of captive sparrows in Hungary. Birds that were ranked in the middle on degree of aggressiveness during a pretrial period were separated into experimental and control groups, with the former being exposed overnight to an electric fan. Birds in the experimental group lost about twice as much mass overnight as the control birds ($P = .008$). They also were more likely than controls to feed by aggressive replacement on the morning after the treatment ($P = .001$).

It is not known whether dominance hierarchies play a role in the foraging interactions of free-living sparrows. Møller (1987b) mentioned a dominance hierarchy among free-living male sparrows in Denmark but provided no details on how the dominance relationships were determined. At least two elements would be necessary for dominance hierarchies to affect the foraging behavior of free-living sparrows: temporally stable flocks and individual recognition among flock members. The evidence of dominance hierarchies in some caged flocks of sparrows (described previously) suggests that sparrows are capable of individual recognition, presumably using visual and/or auditory cues. McGillivray (1980a) studied sparrows nesting in close proximity to each other in conifers in Alberta (Canada) and reported that strange birds entering nest trees were vigorously chased by the nesting birds, but neighbors were tolerated, implying individual recognition among the nesting birds. Badge size or shape in males is one possible visual cue for individual recognition (see Chapter 5). Whether the composition of a foraging flock is temporally stable is not known, although Beimborn (1967) reported that individuals tend to be found together at different times. Although the size and composition of foraging and roosting flocks are obviously very dynamic, it is possible that small groups of individuals maintain a high degree of cohesiveness.

Some studies of free-living sparrows have examined the outcome of aggressive encounters by monitoring attempted displacements at feeding sites in which either the challenger is successful in displacing a feeding individual or the latter is able to retain its feeding position. Johnston (1969a) reported that females were more aggressive and generally dominant to males during a three-month (March–May) period in Kansas (USA). Kalinoski (1975) also reported that females were dominant to males at a winter feeding station (October–March) in New Mexico (USA). Cink (1977), however, reported that males were generally dominant to females from early September to early February in Kansas, but from late February through August, females tended to be dominant. There is therefore apparently a seasonal shift in dominance relationships between the sexes, with males tending to be dominant to females during the late fall and early winter months, but with females tending to become dominant in late winter and during the breeding season. Watson (1970) reported that females were generally subordinate to males in his captive flocks, but both Hegner and Wingfield (1987c) and Solberg and Ringsby (1997) reported that dominance status was not sex-related in their captive flocks.

Dominance relationships may also be site dependent. Cink (1977) studied dominance relationships in winter flocks containing many color-banded individuals in Kansas. Dominance matrices were constructed from data on aggressive interactions at five different feeding sites. The one or two individuals that were consistently dominant at one feeding station had usually bred near that feeding station. Among eight males scored at all five sites, Cink found that an individual that was dominant at one feeding station was always subordinate at the other sites. In fact, it was not uncommon for the most dominant individual at one site to be the lowest-ranking individual at another site.

Barnard (1980b) found that flock size and fighting rate (fights/s per bird) were positively correlated ($r = 0.549$, $P < .001$) for sparrows foraging at his experimental sites in England. He also reported that ambient temperature had a significant negative effect ($P < .001$) and time of day a significant positive effect ($P < .0001$) on the fighting rate and suggested that fighting rate tended to increase when feeding priority was high (late in the day and at lower ambient temperatures). Beer (1961), however, observed that on the coldest winter days in Minnesota (−25°C to −20°C), sparrows tended to feed constantly and with very little conflict.

Social Facilitation and Information Transfer

Protection from predators may not be the only function of flock foraging in house sparrows. Caraco and Bayham (1982) noted that foraging sparrows tend to orient themselves so as to keep neighboring birds in their visual field. Scanning may therefore serve another function in addition to predator detection, that of monitoring the behavior of other flock members. Monitoring the feeding behavior of other individuals may enhance the feeding efficiency of the monitoring individual, and the collective activities of the flock may improve the chances of finding and exploiting hidden or new food resources.

Elgar and Catterall (1982) compared the feeding efficiencies of single sparrows and flocks of four and eight sparrows in the laboratory. They employed a test arena in which grain was placed in only 1 of 36 feeding cups hidden by shredded newspaper. The sparrows had been maintained in the laboratory for 5 months and had learned to search for hidden food in the cups. In the test trials, birds in flocks were much more likely to obtain food than were single birds ($P < .005$). The proportions of birds obtaining food were 0.125 for single birds and 0.875 and 0.792 for birds in flocks of four and eight, respectively. Individuals in flocks tended to search in cups near the cup containing food after observing another bird feeding from it.

In a carefully designed series of laboratory experiments, Turner (1965) showed that sparrows are stimulated to feed by observing another sparrow feeding. The experiments also showed that sparrows learn to eat a novel artificial food item by observing other sparrows eating the item. Juvenile sparrows were particularly prone to learn by observation to exploit the novel food. The novel food used in the experiments was bright green, and Turner reported that it was difficult to train sparrows to take the food. Once one individual had learned to use the food, however, it was easy to train other individuals to eat it by using the first sparrow as a demonstrator.

Fryday and Greig-Smith (1994) conducted three experiments to test for social learning in the selection of novel foods by sparrows. The novel foods consisted of red- or yellow-colored foods that had not been previously encountered by the test individuals. In the first experiment, naive individuals were paired for 4 d in an apparatus that permitted them to observe demonstrator individuals eating either red- or yellow-colored food. On the fifth day, the test birds were offered a choice between red- and yellow-colored food. The test birds ate a significantly greater proportion of the observed color than did controls that had not observed a demonstrator feeding. Demonstrators ate significantly more than either naive test or control birds, however, and naive birds had a preference for yellow food over red. In the second experiment, naive birds were placed with demonstrators that were fed either untreated red food or quinine-treated red food (to which they have an aversion). The test birds were then offered a choice between red and yellow foods, neither of which they had previously experienced. There was a significant tendency for test birds to take less of the novel foods if the demonstrator was given quinine-treated food. Fryday and Greig-Smith suggest that one explanation for this result might be that the food consumption rate of the observer was tied to the consumption rate of the demonstrator. To test this possibility, they conducted a third experiment in which test birds that had previous experience of either untreated or quinine-treated food were paired with demonstrators that were fed either treated or untreated red food. Only the pretrial treatment of the test birds had a significant effect on their consumption rate during the trial, and there were no significant effects of demonstrators on test birds. Fryday and Greig-Smith concluded that the observation of feeding by conspecifics is important only in the presence of a novel food.

Barnard and Sibly (1981) studied the individual feeding strategies of sparrows feeding in a laboratory arena and identified six individuals, termed "searchers," that obtained only 26.0% of their food by interactions with other birds and six individuals, termed "copiers," that obtained 58.1% of their food by interacting with other sparrows. Barnard and Sibly used an experimental setup in which 30 decapitated mealworms were placed individually in 0.5-inch diameter wells placed at 4–inch intervals in an arena (five worms were clumped in each case to simulate a clumped food resource). They manipulated the ratios of searchers to copiers in a series of trials and found that the copiers tended to consistently forage by interacting with other sparrows and did not adjust their foraging strategy based on the ratio of searchers to copiers. Liker and Barta (2002) used a similar experimental design, except that they provided 30 millet seeds per well in two clumps of six wells on a feeding grid and observed the foraging behavior of groups of sparrows from Hungary that had been housed together for at least 2 wk before the trials. More aggressive birds tended to forage more often by joining birds at food clumps than did less aggressive individuals.

Collectively, these findings indicate that flock-foraging sparrows are responding to the behavior of other sparrows in the flock.

Flock Formation

Sparrows that arrive at a potential foraging site usually attempt to attract other sparrows to the area before initiating foraging. The intensity of this effort apparently depends both on the type of food and on the perceived predation risk at the site. A game-theory model developed by Newman and Caraco (1989) suggests that attracting other sparrows to a food source is selectively advantageous if predation risk is high, even if the behavior results in reduced food availability to the individual.

Elgar (1986a) studied the behavior of the first sparrow (the "pioneer") to arrive on the parapet above a feeding station on the roof of the Zoology Building at Cambridge. The pioneer, which was equally likely to be a male or female, usually gave numerous "chirrup" calls (see Chapter 7) before either flying down to the feeding tray or flying away. The rate at which "chirrup" calls were given by the pioneer was inversely correlated with the time until the pioneer was joined by other sparrows ($R^2 = 0.36$, $P = .006$). Elgar attempted to test whether "chirrup" calls serve to attract other sparrows. He counted the number of sparrows that arrived on the parapet (above a feeder that had been empty for 2 d) in the 5 min after either playback of taped "chirrup" calls at the rate of 52 calls/min, playback of human whistles at the same rate, or no playback. Sparrows arrived on the parapet significantly more often after the playback of "chirrup" calls than after the other two treatments ($P < .01$), results that strongly suggest that the "chirrup" call is used to attract other sparrows.

Elgar (1986a) also tried to identify under what conditions sparrows attempt to attract other sparrows to a food resource. He placed one of five food treatments (no food, 1/8 slice of bread whole, ½ slice of bread whole, 1/8 slice of bread crumbled, or birdseed) on a feeding tray and observed the behavior of the pioneer sparrow. The chirrup rate of pioneers varied significantly among the various food trials, being very low for no food, intermediate for whole bread pieces (<10 calls/min), higher for bread crumbs, and highest for birdseed. The chirrup rate of the pioneer was therefore determined by both the presence and the divisibility of the food. Pioneers were more likely to feed on a divisible food source (bread crumbs or birdseed) if they were joined by other sparrows, but they were as likely to feed on the 1/8 slice (whole) if they were not joined as if they were joined.

The position of the feeding tray relative to the parapet (safety) and the position of the observer (potential "predator") also influenced the calling frequency of pioneer sparrows. In another experiment, Elgar (1986b) placed the feeding tray (containing birdseed) either adjacent to or 2 m from the parapet and positioned the observer either 15 or 25 m from the feeding tray. Pioneer birds gave significantly fewer chirrup calls when the feeder was adjacent to the parapet and the observer 25 m away (approximately 7.5 calls/min), compared with pioneers in the other three experimental configurations (approximately 15 calls/min). Pioneers were as likely to feed at the tray when it was adjacent to the parapet and the observer was 25 m away whether or not they were joined by other sparrows. In all other configurations, pioneers were more likely to feed if they were joined by other sparrows than if they were not. These results suggest that house sparrows use information both about the type of food source (whether or not it is readily divisible) and predation risk at the foraging site in deciding whether to feed socially.

Breitwisch and Hudak (1989) used a similar situation to that used by Elgar (feeding tray on a flat roof surrounded by a parapet) to attempt to determine whether one sex was more likely to initiate flock foraging, and which sex was more likely to remain on the foraging patch when all but one feeding bird flushed from the patch. There was no difference in sex in sparrows initiating feeding (the initiators were 53.9% male and the groups of sparrows on the parapet at the time of initiation were 53.3% male, $G = 0.06$, $P > .05$). There was, however, a significant difference in which sex remained behind when all but one feeding sparrow flushed. Males were the only bird remaining in 48 (71.6%) of 67 instances in which only one bird remained (the sex ratio of these 67 flocks was 54.7% male, $G = 7.04$, $P < .01$).

Barnard (1980a) also looked at the individuals that remained behind when other flock foragers left and found that the pecking rate of the remaining bird just before the flock departed apparently affected its decision to remain. He compared the pecking rate of the 10 pecks prior to the flock's departing for the bird that remained to those of five randomly selected birds in the flock. The average pecking rate of the remaining bird was significantly lower than that of the other birds in the flock (0.51 vs. 1.11 pecks/s, respectively, $P < .001$). This rate was also significantly lower than that of the first 10 pecks of the remaining bird after the departure of the other birds (1.27, $P < .001$).

Summary

The foraging strategy of the house sparrow is a complex suite of behaviors that appears to require the integration of several kinds of information to determine the particular behaviors used in a specific situation. This information apparently includes factors such as number of simultaneously foraging sparrows, food density, pecking rate of other foraging sparrows, distance to cover, and relative predation risk of the site. Sparrows appear to generally obey the rule of thumb, "join a foraging group rather than feed alone." They may, however, gauge the divisibility of the food source before attempting to attract other sparrows to a new site. In general, the flock foraging behavior of sparrows conforms well to the Pulliam model, suggesting that it has evolved to maximize food-gathering efficiency while minimizing predation risk. The scanning frequency of foraging sparrows decreases significantly with flock size (i.e., Elcavage and Caraco 1983; Elgar and Catterall 1981), whereas the pecking rate (= feeding rate?) increases significantly with flock size (e.g., Barnard 1980c).

The probability of early detection of an approaching predator may actually be lower for a sparrow feeding alone than for small flocks. Several other factors appear to influence the foraging strategy of sparrows, however, including food density, distance to cover, and possibly site-specific predation risk.

Sparrows appear to form small foraging groups outside of the breeding season that have some temporal stability. This temporal stability provides the opportunity for fairly stable dominance relationships (dominance hierarchies) to develop in such groups. The evidence for the actual existence of such hierarchies in free-living sparrows is equivocal at best, and even some studies of laboratory flocks have failed to find stable hierarchies. Stable dominance relationships, if they occur in sparrows, presumably benefit both dominant and subordinate individuals by reducing the costs of aggression for both individuals. Dominance relationships in house sparrows have been reported to vary with sex and season (males tending to dominate females during late fall and early winter, but females tending to be dominant to males during late winter and the breeding season), with location (Cink 1977), and with hunger-induced motivation (Andersson and Ahlund 1991).

Diet of the House Sparrow

Although specialized as a granivore, the house sparrow has remarkably catholic tastes. It is primarily its proclivities for certain foods that have resulted in its being regarded as a pest, with the consequence that humans have made repeated efforts to control sparrow populations (see Chapter 10). Why does it choose the particular foods that it utilizes at different times of the year? What determines how much it eats? How does it overcome neophobia to exploit novel food sources?

The principal evolutionary paradigm that has been developed to explain food choice and the foraging strategy of animals is optimal foraging theory (Pyke et al. 1977). This theory has focused primarily on the energy content of the food and postulates that an animal should attempt to maximize its net rate of energy acquisition. In addition to the predation risks incurred while foraging (described earlier), there are other costs associated with foraging. Two costs that are frequently identified are the cost of searching for a particular item and the cost of handling the item once it is found. This latter cost includes the cost of capturing the item as well as the actual costs of handling and processing the item during ingestion, digestion, and absorption. Optimal foraging theory typically attempts to convert all of the costs of searching for and handling a particular prey type into energetic units, and then to compare the total energetic cost to the energetic benefit derived from eating that prey type.

Potential prey (or food) types can then be ranked according to their net benefit to the organism, and a food type should be included in the diet only if its net energetic benefit is greater than 0. For seed-eating species such as the house sparrow, the great variety in size and hardness of seeds means that handling time for a particular seed may be largely based on the size and structure of the bill (see Pulliam 1985; Ziswiler 1965).

Pyke et al. (1977) identified four categories of foraging decisions that animals must make: (1) optimal diet (what to eat), (2) patch choice (where to forage), (3) time allocation to patches (when to move), and (4) foraging path in a patch (how to search). Most of the empirical work has attempted to address the first of these decisions, the optimal diet. Pyke et al. identified three major predictions of optimal foraging theory with respect to optimizing the diet. First, increasing food abundance leads to greater food specialization, because foragers should concentrate on the foods with the highest benefit-to-cost ratios. Second, if one food item becomes very abundant, all items of lesser net return should be dropped from the diet. Third, when a forager is changing its diet, food items should be added or dropped according to the rank order of their benefit-to-cost ratio.

I know of no study that has attempted to test these predictions for sparrows. Some studies have examined seed selection in sparrows (e.g., Diaz 1990, 1994), but these studies did not attempt to measure the handling costs associated with each seed type. Other studies have attempted to identify other elements of seed selection in sparrows, such as taste, sometimes as elements of pest management of the species (e.g., Rana 1991).

Diaz (1990) used 13 species of seeds to test the seed preferences of 10 species of granivorous passerines, including the house sparrow and the congeneric tree sparrow. After being exposed for at least 1 wk to all 13 seed species, birds were tested individually for their seed preferences during a 2-h trial in which all seeds were provided in quantities that would not allow complete depletion of any seed type. A principal components analysis on various nutritional and size attributes of the seeds resulted in the identification of two principal components. PCI clearly represented nutritional value (with protein, fat, minerals, and energy loading positively and water and carbohydrate loading negatively), and PCII represented seed size (length, width, and weight all loading positively). The seed selection preferences of the 10 granivores were then plotted on the two principal component axes. The house sparrow and the tree sparrow were located very close to each other, based both on the number of seeds eaten and the weight of seeds consumed, and both were located in the quadrant of small seed size and low nutritive value. Specific seed preferences of the house sparrow were not presented, but the seed species that clustered close to the coordinates of the house sparrow in principal component space included millet (*Panicum miliaceum*), canary grass (*Phalaris canariensis*), oats (*Avena sativa*), rye (*Secale cereale*), common vetch (*Vicea sativa*), wheat (*Triticum aestivum*), and barley (*Hordeum vulgare*). For all 10 granivorous species, seed size preferences did not correlate with either body mass or bill size.

In a second experiment, Diaz (1994) used 7 of the 10 species (again including the house sparrow and tree sparrow) to test the Niche Variation Hypothesis, which states that greater intraspecific variance in bill size or body size should result in greater variance in seed size utilization (see also Chapter 2). Birds were tested on the same 13 species of seeds as before, with a similar protocol. Variability in seed size consumption tended to vary only with mean bill size (particularly bill depth and bill width), but not with variability in bill size, and therefore failed to support the Niche Variation Hypothesis.

Sparrows apparently use taste to make foraging decisions. Madej and Clay (1991) performed seed preference tests on house sparrows and four other seed-eating species on fungi-infected tall fescue (*Festuca arundinacea*) seeds. Birds were offered a choice of either uninfected fescue seeds (less than 10% infected with the endophytic fungus *Acremonium coenophialum*) or millet seeds, or a choice of either uninfected fescue or infected fescue seeds (more than 90% of seeds infected). Sparrows had a strong preference for millet over fescue ($P < .01$), but also showed a preference for uninfected over infected fescue seeds ($P < .02$). The fungi produce a high alkaloid content that can apparently be tasted (as bitter) by the birds (Madej and Clay 1991). Sparrows are therefore capable of discriminating between infected and uninfected seeds, which suggests that seed selection may not only be a function of seed size, caloric and/or nutrient content, and handling time, but may also be affected by taste. Rana (1991) also reported that treatment of millet with various concentrations of five different fungicides significantly reduced consumption rates by sparrows.

Moulton and Ferris (1991) examined the summer feeding preferences of 10 species of introduced finches from four families in urban parks in Hawaii. They recognized six food categories: grass seeds, forb seeds, sedge seeds, insects, fallen fruit, and garbage (human refuse). The house sparrow regularly fed on four of the six categories (all except sedge seeds and fallen fruit) and had the broadest foraging niche among the eight species for which there were sufficient observations for comparison.

At least one study has attempted to identify the mechanism determining how much food sparrows consume. Kendeigh et al. (1969) maintained captive sparrows on nine different ambient temperature regimes or under enforced nighttime exercise to determine whether nocturnal events were responsible for the amount of food consumed during the daytime. In the temperature experiments, birds were maintained on 12L:12D photoperiods and at one of nine night:day temperature regimes: −20°C:20°C, 20°C:−19°C, −15°C:23°C,

23°C:−15°C, −5°C:23°C, 23°C:−5°C, 7°C:23°C, 23°C:7°C, or 23°C:23°C (control). After the birds were fully adjusted to the experimental regime, the daytime gain in body mass matched the amount of mass lost at night. Birds maintained on cold night treatments had greater daily fluctuations in mass than did birds maintained on cold days, and the amount of the fluctuation increased with decreasing nocturnal temperature, except that there was no significant difference between −15°C and −20°C. In all cases, the birds maintained on cold nocturnal temperatures had a significantly greater body mass in the evening than did birds maintained on cold daytime temperatures. In the enforced nighttime exercise experiment, birds were maintained at constant conditions (24°C, 10L:14D) but were forced to exercise for 12 h during the darkness by means of a slowly rotating cage. During the early part of the experiment, the mean morning body mass dropped and the mean evening body mass increased, but after 1 mo the mean morning body mass had reached approximately the same level as in the pretreatment birds, and the mean evening mass was about 1 g higher than that in pretreatment birds. Kendeigh et al. concluded that sparrows regulate their energy intake (and body mass) based both on the amount of energy expended the previous night and on daytime temperature. They also adjust to higher nocturnal mass losses by increasing their evening body mass rather than by continuing to permit a greater overnight mass loss.

Diet of Adult House Sparrows

Several techniques have been used to attempt to identify and quantify the diet of adult sparrows. Most studies have not differentiated between adult and juvenile individuals, so this analysis actually refers to the diet of free-living sparrows. One of the more commonly used methods is the analysis of stomach contents (usually including crop, proventriculus, and ventriculus) of sacrificed specimens that have been mist-netted, trapped, poisoned, or shot. The fact that sparrows are considered serious pests over significant portions of their range (see Chapter 10) has made this technique acceptable. One of the difficulties with this technique in quantitative studies of the diet is the difference in digestive and clearance rates of different types of food, which results in a bias toward items with longer digestion and clearance times. Hammer (1948), for instance, noted that samples containing insects were biased toward heavily chitinized groups, apparently because soft-bodied insects had rapid clearance rates that resulted in little or no evidence remaining in the stomach of their occurrence in the diet.

Some dietary studies are made in the context of other types of studies of sparrow biology, however, and these require methods that minimize the impacts on the birds being studied. Several studies have been done to determine the efficacy of various emetics (substances that induce vomiting) in providing useful samples for dietary analysis. Effective emetics should have a high rate of success in inducing emesis, have a short latency (the time between administration of the emetic and regurgitation), have a short recovery time, result in minimal or no mortality, and result in complete or nearly complete regurgitation of the contents of the esophagus, crop, and proventriculus. Table 6.1 summarizes results of studies examining the effectiveness of various emetics. All have one or more drawbacks. Saline flushing appears to be the most effective in completely emptying the stomach contents, but it requires anesthetizing the birds and has a rather high mortality rate (8.2%), which included 22.6% mortality in a sample of field-captured sparrows (Gionfriddo et al. 1995). The other emetics tested had less than full recovery of stomach contents. In addition to the emetics listed in Table 6.1, Radke and Frydendall (1974) reported on several other possible emetics that they concluded were unsuitable for use in sparrows, usually because of unacceptably high mortality rates or long latencies. These included Ipocoac (administered orally) and apomorphine (oral) at any dosage, copper sulfate (oral) at doses greater than 0.2 ml, and apomorphine (intraperitoneal) at doses greater than 0.1 ml.

One other problem that arises when attempting to compare results across dietary studies is that some authors summarize their data based on numbers of items in the diet, whereas others report proportions by dry mass or volume found in each category. These methods yield comparable results only if all items in the diet have equal mass or volume, a condition that is never met. Although there are theoretical reasons for considering both measures, I consider dry mass or volume to be better descriptors of the diet because they more closely approximate the proportions of energy and nutrients supplied by each dietary category.

Some of the early studies of the diet of adult sparrows were based on stomach analyses of large numbers of specimens collected during several years and over large geographic areas. Kalmbach (1940) reported on the stomach contents of 4848 adult (= free-living) sparrows collected from throughout the United States. Vegetable material comprised almost 97% of the annual diet. Cereal grains and grain-based animal feeds constituted the bulk of the diet throughout the year, with the percentage of items in this category falling below 60% only in the month of October. Weed and grass seeds were prominent in the autumn and early winter (September–December), peaking in October with 41% of the identified items. Animal food, mostly beetles (Coleoptera), occurred in the diet primarily during the breeding season (March–August), peaking in May at about 10%. Very similar results were obtained by Hammer (1948) based on 2657 sparrows collected from various parts of Denmark from 1941 to 1944. Grains, principally oats, barley, and wheat (and, to a lesser extent, rye), were

TABLE 6.1. Effectiveness of Emetics Used to Obtain Contents of Upper Digestive Tract of House Sparrows

Emetic	Dosage (ml)	Latency (min)	Recovery (min)	Mortality (%)	Effectiveness (%)	Source
Flamed tincture of digitalis	0.05, 0.2 (oral)	4.7	0	0	82–83	1
Apomorphine (0.04 g/mL)	0.04 (ocular)	3.2	2.7	0	74	1
Apomorphine (0.02 g/mL)	0.1 (intraperitoneal)	6.8	2.0	0	85	1
Copper sulfate (1.5% in water)	0.2 (oral)	14.7	n.a.	0	85	1
Antimony potassium tartrate (1%)	0.5 (oral)	n.a.	n.a.	0	<100	2
Antimony potassium tartrate (5%)	0.04 (oral)	<15	n.a.	n.a.	58	3
Saline solution (0.9% Na CL)	30 (oral-anesthetized)	n.a.	<60	8.2	≈ 100	4

Data not reported in the studies are indicated by n.a. (not available).

Sources: 1, Radke and Frydendall (1974); 2, Prys-Jones et al. (1974); 3, Gavett and Wakeley (1986a); 4, Gionfriddo et al. (1995).

again the most common food items and were taken throughout the year. Only in October 1941 did the percentage of specimens with grain in their stomachs fall below 80% (to 78%), and in most months of the study the percentage containing grain was close to 100%. Weed seeds were again most prominent in the fall and early winter diet (September–December). Animal items—primarily beetles; caterpillars (Lepidoptera); larval, pupal, and newly emerged adult flies (Diptera); and aphids (Homoptera)—were most prominent during the breeding season (April–August). Hammer reported that there were no differences in the diet based on either sex or age of the sparrows. She did note that there were dietary differences among the sites from which the specimens were collected, including the presence of such items as molluscs and crustaceans at locations where sparrows could feed on beaches (e.g., Vorso Island).

Dietary studies at more restricted agricultural sites in other temperate-zone parts of the house sparrow's range have also found that grains and weed and grass seeds comprise the bulk of the diet, particularly during fall and winter. Simeonov (1964) examined the contents of 785 stomachs of sparrows from Bulgaria and reported that grains and weed and grass seeds made up 100% of the diet during the winter (December–February), 91.0% during spring (March–May), 51.3% during summer (June–August), and 96.5% during autumn (September–November). The remainder of the diet was composed primarily of insects. Rekasi (1968) reported that grains (primarily maize, sorghum, and wheat) and weed and grass seeds were the principal food items found in 334 stomachs of birds taken in late summer and early fall in Hungary. Keil (1972) analyzed the stomach contents of 751 sparrows collected from October to March in Germany. Cereal grains made up 48% of the diet by dry mass; weed seeds, 38%; other plant parts, 15%; and insects, 1%. Ion (1992) examined stomach contents of 394 sparrows collected in all seasons of the year at three rural agricultural locations in Romania. Overall, the diet consisted of 72% vegetable items and 28% animal material, the latter consisting primarily of beetles, flies, and lepidopteran larvae. Animal material was prominent during the spring (39% of items) and summer (68%). The principal vegetable materials included wheat, sunflower, maize, *Panicum*, and weed seeds. In New Zealand, stomach analyses of 401 adult sparrows collected throughout the year in a mixed farm region near Hawke's Bay showed that they fed primarily on cereal grains and weed seeds, with the highest proportion of animal material being used during the breeding season (November–February) (MacMillan 1981). The sparrows fed primarily on wheat during the summer and maize during the winter, and there was no significant difference in the diet based on sex. In Spain, weed seeds and cereals comprised 95.4% and 97.1% by mass of the diet of adult sparrows in winter and spring, respectively, with wheat (82.2% in winter and 95.7% in spring) being the single most important item (Sanchez-Aguado

1986). Insects, primarily Coleoptera in winter and Homoptera, Coleoptera, and Diptera in spring, made up the remainder of the diet.

Studies in subtopical regions of India (on *Passer domesticus indicus*) have shown similar patterns. Beri et al. (1972) collected 102 sparrows throughout the year and reported on the percentage of stomachs containing the following food items: wheat, 35.2%; pearl millet, 32.9%; jowar (= barley), 31.8%; green vegetables, 26.7%; weed and grass seeds, 22.3%; and insects, 1.7%. Insects occurred most commonly between May and October (during the breeding season). Simwat (1977) analyzed the gut contents of 779 adult sparrows collected at weekly intervals throughout the year in the region of Ludhiana, Punjab. Wheat was the most common food item, comprising 30.8% of the diet. Weed seeds (16.2%), pearl millet (12.0%), rice (11.1%), and insects (8.1%) were also common food sources, while other categories (maize, sorghum, oats, groundnut, leguminous seeds, and green leaves) made up 0.9%–2.6% of the diet. There were marked seasonal changes in the proportions of the various food items, however. Arthropods, which occurred most frequently in the diet from March to July, included beetles, caterpillars, flies, aphids, earhead bugs (Hemiptera), ants and bees (Hymenoptera), leaf folders (Lepidoptera), wireworms (Coleoptera), grasshoppers (Orthoptera), and spiders (Arachnida). In a second study in Punjab, cereal grains made up 83.6% by mass of the diet of 96 adult sparrows collected throughout the year at farms of Punjab Agricultural University (Saini and Dhindsa 1991). Wheat (38.8%) and pearl millet (29.9%) were the predominant cereals, and rice (7.7%) and maize (7.2%) made up the remainder. Weed seeds comprised 6.2% of the diet, whereas insects and other animal material made up only 6.1%. Rana and Idris (1987) examined the stomach contents of 510 sparrows collected throughout the year in an arid region near Jodhpur. Cereal grains (particularly wheat, pearl millet, and barley) were again prominent in the diet, with insects and leguminous seeds being most numerous in the period from July to November.

The diet of sparrows in urban environments is quite different from that in rural areas, primarily due the lack of availability of cereal grains in urban areas. Gavett and Wakeley (1986b) compared the diets of urban and rural populations in Pennsylvania during late spring. Food items obtained by emesis were categorized into five broad groups: (1) commercial cereal grains, (2) commercial birdseed, (3) other seeds, (4) plant fragments, and (5) insects and arachnids. In the rural sparrows, grains (97.3%) and insects and arachnids (94.3%, mostly coleopterans) were the most frequently occurring items in the diets, whereas in the urban sparrows, insects (93.1%), other seeds (81.2%, including many elm seeds), cereal grains (66.5%), and birdseed (61.9%) were most frequently represented. In Hamburg (Germany), 54% of the food of adult sparrows was provided directly or indirectly by humans; it consisted primarily of wild bird seed and human refuse (Bower 1999). As described earlier, the diet of sparrows in seven urban parks in Hawaii included large quantities of grass seeds, forb seeds, insects and human refuse (Moulton and Ferris 1991).

In addition to the items already identified in the diet of free-living sparrows, there are infrequent appearances of many other types of food. Arthropods, in addition to the spiders and insect orders already mentioned, include Collembola, Odonata, Dermaptera, Thysanoptera, and Neuroptera (Hammer 1948; Macmillan 1981). Other invertebrate prey besides the arthropods and molluscs include earthworms (Annelida) (Hammer 1948). Vertebrate prey are taken very rarely, with some of the species taken being lizards, such as the wall lizard (*Podarcis muralis*) (Gallelli 1948; Moltoni 1954), Turkish gecko (*Hemidactylus turcicus*) (Angelici 1993), yellow-headed gecko (*Gonatodes albogularis*), and *Anolis* sp. (Bello 2000), as well as juvenile house mice (*Mus musculus*) (Buges 1991).

One sex difference in the diet may occur during the egg-laying period. Pinowska (1975) analyzed the stomach contents of 1347 adult females and 34 adult males collected around cattlesheds in northern Poland. For 169 females captured during the winter, wheat (68.4% of dry mass) was the principal food, with barley (22.7%), oats (3.4%), canary grass (3.9%), and small amounts of rye, weed seeds, and green plant parts making up the remainder. Females collected during the breeding season were identified as belonging to one of the following categories: (1) prebreeding, (2) egg-laying, (3) early incubation, (4) late incubation, (5) feeding nestlings, and (6) nonbreeding. Females continued to feed on grains during the breeding season, with wheat found in about 65% of individuals, barley in about 25%, and oats in about 20%. The amount of animal food in the diet changed with the stage in the breeding cycle, with egg-laying females having more animal food than those in other parts of the cycle. The animal portion of the diet of breeding females consisted entirely of insects, primarily beetles and flies. The proportion of females captured in mist nets during nestling feeding that contained insects in their stomachs was higher than the proportion in individuals that were shot, and Pinowska concluded that the proportion containing animals was artificially high because females captured in mist nets apparently swallow the food they are bringing to nestlings. The high protein and nutrient demands of egg formation are presumably responsible for the shift toward insects during egg-laying (see Chapter 4).

Murphy and King (1982) developed a semisynthetic diet for house sparrows being maintained in the laboratory, using a modification of the Illinois crystalline amino acid diet. They placed 12 sparrows on the diet for 28 d during the annual molt and held them in the laboratory under constant conditions (16L:8D, 21°C). The birds maintained their body weight, and their molt schedule was unaffected compared with control birds being fed chick starter.

Nestling Diet

The two principal methods of determining the nestling diet of sparrows have been stomach content analysis of sacrificed individuals (e.g., Collinge 1914; Simwat 1977) and the collar method (e.g., Łącki 1962; Mathew and Naik 1993). In the latter method, a ligature is placed around the neck of nestlings that is tight enough to prevent swallowing, yet loose enough to permit unimpaired breathing. After a length of time, the food that collects in the distensible esophagus above the ligature is collected and identified. Both methods have potential biases in characterizing the nestling diet; in the former method, unequal clearance times lead to underrepresentation of certain types of food, and the latter method may affect the provisioning behavior of the adults and/or the begging behavior of the young, resulting in inaccuracies in determining the quantity of food provided by adults under normal circumstances. The relative proportions of items fed by the adults should be unaffected, however. Another method, used by Seel (1969), involved the removal of crop contents from nestlings during periodic nest checks. This method may also provide reasonable estimates of the proportions of various food items in the nestling diet, but it would not provide accurate quantitative data on the rate of food delivery.

Most studies of the nestling diet of sparrows have been conducted in rural, agricultural areas. The nestling diet consists mostly of animal material, primarily arthropods. In a study based on an analysis of stomach contents of 2819 nestlings collected throughout the United States, Kalmbach (1940) reported that 68.1% of the diet by volume consisted of animal material, 99.7% of which were arthropods. Similar high percentages of animal material have been reported in numerous other studies from different parts of the house sparrow's range. Examples include 96.1% from Bulgaria (Simeonov 1964); 87.7% and 98.9% for rural and urban sites, respectively, in Germany (Encke 1965a); 76% in Romania (Ion 1971); 80.3% and 84.3% at two locations in Poland (Wieloch 1975); and 78.4% in Missouri (USA) (Anderson 1978). The plant material consists primarily of grains, particularly wheat, and weed seeds. Small amounts of grit, including snail shells, pieces of eggshells, and small stones, are also present in the nestling diet.

The bulk of the animal material consists of insects, although spiders (Arachnida) are also usually present. The principal insect groups are remarkably similar in many parts of the sparrow's range and include beetles, caterpillars, flies, and grasshoppers and crickets (Orthoptera). Other groups that sometimes contribute significantly to the diet include bugs (Homoptera) and ants and sawflies (Hymenoptera). The beetle families most often encountered in the diet include Carabidae, Curculionidae, Coccinelidae, Chrysomelidae, and Scarabaeidae, and the principal lepidopteran larvae belong to the families Geometridae and Noctuidae. The most frequently mentioned dipteran family is Syrphidae, and the orthopterans typically belong to the Acrididae or Gryllidae. When bugs have been prominent in the nestling diet, they typically belong to the Aphididae, Cicadellidae, or Cicadidae, and when hymenopterans are common, they belong to the Formicidae or Cimbicidae. Wieloch (1975) analyzed the trophic position of the arthropods in the nestling diet at two sites in Poland and concluded that approximately 38% were phytophagous, 46% carnivorous ("zoophagous"), 6% detritivorous ("coprophagous," "necrophagous," or "saprophagous"), and 10% undetermined.

The proportion of animal material tends to decrease with nestling age, with the nestling period typically lasting 14–16 d (see Chapter 4). Fig. 6.5 shows the percentage of animal material at various ages. The heavy reliance on animal material during the first few days of nestling development is apparent, and the steady decline in the percentage of animal material with increasing age is also consistent.

Mueller (1986) attempted to determine when young sparrows switch from an insect diet to a seed diet by taking 8–10 d old nestlings into captivity. She housed the nestlings in "broods" of 3–5 in paper bags and fed them either moistened puppy food or adult dipterans with forceps until they "fledged." Eleven-day-old nestlings preferred the insects to the puppy food, but by 14 d they would pick up puppy food from the floor of the "nest." When they were 16–17 d old, they suddenly showed no interest in insects but continued to beg and to accept puppy food. At 22–24 d, they began to feed on millet seeds, although they continued to beg; at 25–30 d, they no longer accepted food from the forceps. Mueller suggested that the sudden aversion to insects, followed shortly thereafter by beginning to feed on seeds, represents an innate behavioral change in developing sparrows.

Fig. 6.5 also illustrates the fact that the composition of the nestling diet changes seasonally. The percentage of animal material in the diet decreases much more rapidly with nestling age during the summer breeding season in India than during the late monsoon breeding season (Mathew and Naik 1993). The total amount of food delivered during the late monsoon is also greater than that delivered during the summer, suggesting that animal material may be the preferred food throughout the nestling period but that lack of readily available animal material may result in a shift to more vegetable material as the energetic and nutritive requirements of the growing nestlings increase.

Other studies have also shown seasonal changes in the nestling diet. In general, the percentages of dipterans and lepidopteran caterpillars tend to be high early in the breeding season and to decrease later in the season, whereas the percentages of orthopterans and homopterans tend to increase later in the breeding season (Anderson 1984; Encke 1965a; Ion 1971; Macmillan and Pollock

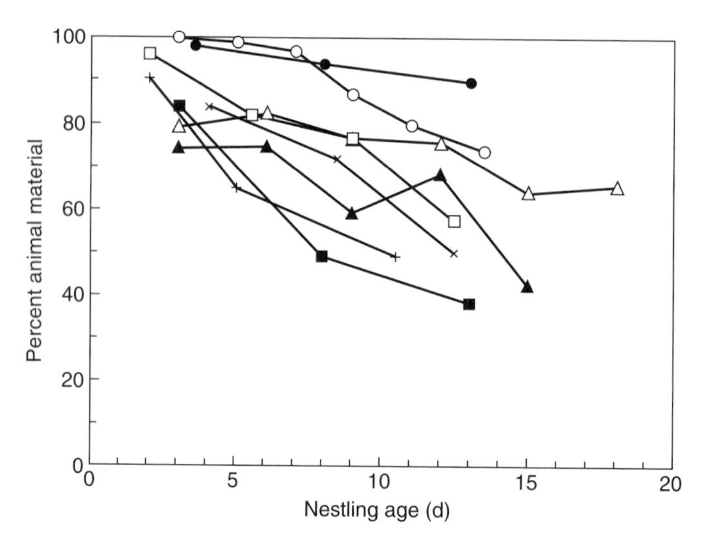

FIGURE 6.5. Changes in the percentage of animal food with age in the diet of nestling house sparrows. For each study, nestling age represents the midpoint of an age interval. All percentages indicate the percentage by dry mass or volume, except those of Encke (1965a) and Wieloch (1975), which are percentages of numbers of items. Wieloch (1975) reported that the overall percentage of animal items in the diets for all ages combined at Turew, Poland, was 95.0%, while the overall percentage by dry mass was 84.3%. Data for *Passer domesticus domesticus* were obtained from Germany, ○ (Encke 1965a); Poland, ● (Wieloch 1975) and × (Łącki 1962); Romania, □ (Ion 1971); the United States, + (Kalmbach 1940); and New Zealand, ■ (Macmillan and Pollock 1985). For *P. d. indicus*, data were obtained during two seasons in India, ▲ = summer and △ = monsoon (Mathew and Naik 1993).

1985; Wieloch 1975). The coefficient of overlap ($\hat{C}_\lambda$) between the nestling diets of first and second broods in a village in Poland was only 0.132 (Anderson 1984).

The nestling diet also varies with habitat and geographic region. Collinge (1914), for instance, examined the nestling diet in two locations in southern England, one a fruit-growing area and the other a mixed agricultural area. Although the diets were generally similar in the two areas, the particular species utilized by the sparrows differed. Similarly, Anderson (1980) compared the nestling diets at six agricultural locations in the Mississippi River valley in Illinois (USA) and Missouri at which breeding had been experimentally synchronized (see Anderson 1979). Between colonies, which were 2–70 km apart, $\hat{C}_\lambda$ varied from 0.056 to 0.763 (mean = 0.304), and there was no correlation between $\hat{C}_\lambda$ and distance between colonies ($r = -0.35$, $N = 15$, NS). This suggests that very local differences in availability of food influence the composition of the nestling diet.

Encke (1965a) compared the nestling diets of sparrows breeding in a large city and in an agricultural area in Germany. The urban birds fed more homopterans (primarily aphids) and fewer coleopterans, lepidopterans, and grain than did the rural birds. Another study done in an urban habitat in Germany found that a high proportion of the nestling diet consisted of food provided either directly or indirectly by humans (Bower 1999). Arthropods, chiefly aphids (30%), constituted 65% of the diet, and much of the remainder consisted of dog food (13%), bird food (10%), bread (5%), and other human refuse (1%).

Grit

Grit is a prominent component of the diets of both adult and nestling sparrows. Keil (1972) reported that grit comprised 66% of the mass of the stomach contents of the 136 sparrows that he collected in Germany during the winter, and Pinowska (1975) found that only 1 of 1337 sparrows collected in northern Poland, primarily during the breeding season, lacked grit in the stomach. Keil reported that the grit consisted primarily of pieces of quartz (75%) and brick (20%), which usually varied in diameter from 0.2 to 1.5 mm. Best and Gionfriddo (1991a) found grit in the stomachs of 99% of 77 sparrows collected in Iowa (USA), with the number of grit particles in an individual ranging from 0 to 274 (median = 69). The average diameter of grit particles was 0.7 mm. In a subsequent study, Gionfriddo and Best (1995) sampled another 245 sparrows in Iowa and found that the number of grit particles in the stomach ranged from 0 to 3204 (median = 462). Size of grit particles, measured on a subsample of 60 individuals, varied from 0.1 to 2.4 mm (mean = 0.5). There was no difference between the sexes in either the number or the size of grit particles.

The primary function of grit is apparently to aid in the mechanical breakdown of food in the ventriculus (gizzard) (see Chapter 9). Some seasonal variation has been reported in the amount of grit in the gizzard, and the differences have been correlated with changes in diet. In India, grit made up an average of 21% of the mass of the gut contents and was high in every month except February (4%), when the sparrows were feeding heavily on the milk stage of wheat (Saini and Dhindsa 1991). In Iowa, the number of grit particles was higher in months in which there were insects in the diet, compared with the fall and winter months, when there were few or no insects in the diet (Gionfriddo and Best 1995). In an experimental study in which sparrows were maintained on two dietary regimes, one consisting of hard seeds and the other of soft dog chow, sparrows on the hard diet retained significantly more grit particles in their gizzards than did sparrows on the soft diet (Best and Stafford 2002). This suggests that sparrows are able to adjust the period of grit retention in the gizzard, although the mechanism for this response is unknown.

Best and Gionfriddo (1991a) estimated the shape and surface texture (roundness) of grit from sparrows collected in Iowa. The average shape index (length/width) of the grit was 1.7, and the average roundness index, which was based on assignment of each piece of grit to one of five categories (1 = angular, 2 = subangular, 3 = subrounded, 4 = rounded, and 5 = well-rounded), was 2.8. In a subsequent experimental study, Best and Gionfriddo (1994) provided captive sparrows with two shapes of artificial grit after flushing their digestive tracts with saline. After 14 d, the birds had an average of almost 140 pieces of the artificial grit in their stomachs, and 80% of the 30 birds had significantly more angular/oblong grit than expected by chance, while only 10% had significantly more round/spherical grit. The remaining three birds had proportions of grit that did not differ significantly from the proportions available. The observed differences were not due to differences in retention of the two shapes, and Best and Gionfriddo concluded that the observed preference for angular/oblong grit might be based on better grinding ability by grit of this shape.

A second function of grit in the diet may be as a source of nutrients. Some dietary studies have found pieces of chicken eggshells or empty snail shells as grit (e.g., Anderson 1978, 1984; Schifferli 1977), and these, along with some types of stones, may serve as an important dietary source of calcium (Ca) or other nutrients. Pinowska (1975) reported that the amount of grit in the stomachs of female sparrows at various stages of the breeding cycle in Poland tended to parallel the proportion of insects in the stomachs. Egg-laying females had large numbers of grit particles as well as high frequencies of insects. Using the same set of specimens, Pinowska and Krasnicki (1985b) examined the amount of grit in the stomachs of laying females on successive days of the laying cycle. They also examined the Ca and magnesium (Mg) contents of the females' bodies over the same period. The mean weight of grit particles in the stomachs of laying females increased significantly after ovulation of the first egg and remained significantly higher until after ovulation of the fourth egg. Mg content of the females dropped significantly at the time of the first ovulation and remained low throughout the laying period, whereas Ca levels did not change significantly. Pinowska and Krasnicki suggested that the increased intake of grit serves as a source of Mg and Ca for the laying female (see also Chapter 4). Schifferli (1977) obtained similar results on a sample of 235 female sparrows collected during the breeding season in England. He reported that 95% of laying females had snail shells in their stomachs, but only 37.5% of prelaying and 27% of postlaying females had snail shells. He concluded that the snail shells provided an essential source of Ca for the laying female. Grit in the form of pieces of chicken eggshell and snail shells may also serve as a source of essential nutrients for rapidly developing nestlings.

Sparrows apparently have a rapid turnover of grit in the stomach. Gionfriddo and Best (1995) studied the grit retention time of sparrows in the laboratory by changing from quartz to feldspar grit pieces (which were virtually identical in every regard except for microscopic differences) and sampling at intervals after the change. Turnover of grit was very rapid, with almost 40% of the grit turning over in 6 h. After 24 h, less than 25% of the grit of the previous grit type was retained.

Color may play a role in grit selection. Best and Gionfriddo (1996) used small pieces of colored glass, which were readily accepted as girt by captive sparrows, to test for color preferences. Green, white, and yellow pieces of grit were taken most commonly by the sparrows, with each comprising approximately 20% of the grit particles. Brown, red, and clear grit particles were selected less frequently, and blue and black were selected least often. In subsequent experiments in which the sparrows were fed yellow, red, or blue food, food color had no effect on the proportions of colored grit selected.

Neophobia

Animals often avoid novel objects in their environment. For an opportunistic forager like the house sparrow, how does this fear of new things (neophobia) affect the exploitation of novel food items that suddenly appear in the environment? Several studies have been done on neophobia in sparrows, including some that have already been discussed (Best and Gionfriddo 1996; Turner 1965).

The effects of neophobia on the foraging behavior of sparrows may last for several days. Captive sparrows in India showed a strong preference for pearl millet (bajra) over barley (jowar), consuming five times as much of the former (Rana 1989b). When a novel object was placed by the pearl millet feeding tray on the seventh day, however, consumption of barley increased significantly for the next 3 d, and pearl millet consumption decreased by 80%. Consumption of pearl millet did not return to pretreatment levels until the 13th day.

Greig-Smith (1987) added quinine sulphate, to which sparrows have an aversion, to the food of captive sparrows to study neophobia, and to attempt to identify the mechanism or mechanisms responsible for inducing avoidance of a potential food. He dyed turkey starter crumbs red, yellow, or blue and used different combinations of quinine-treated and untreated crumbs to determine whether sparrows use unfamiliar taste (novel taste in a new food), unexpected taste (novel taste in a familiar food), or anomalous taste (previously experienced distasteful substance in a familiar food that normally does not have the taste) in making decisions about potential food items. Prior to the experimental trial, birds were exposed to one of four pretreatments: (1) untreated red versus quinine-treated blue (control), (2) untreated red versus untreated yellow (novel color and taste), (3) untreated red versus untreated blue (unexpected taste, familiar color), or (4) quinine-treated red versus untreated blue (anomalous—familiar taste, wrong color). After the pretreatment, all birds were given a choice between untreated red and quinine-treated blue crumbs. During the trial, control birds consumed almost 65% of the red food, whereas only 49% of the red food was consumed in the novel color and taste group and 32% in the unexpected taste group (both significantly different from the control). The birds in the anomalous group consumed only about 7% of the red food, significantly less than the consumption in the other two experimental groups. The birds also consumed about 6%–12% of the quinine-treated blue food during the trials, an amount that did not differ significantly among the treatment groups. These results indicate that the avoidance response of the birds was as great to a familiar food that was unexpectedly distasteful as to a novel, distasteful food. The results also indicate that the first experience of a novel, distasteful food has a lingering effect.

Chapter 7

SOCIAL BEHAVIOR AND VOCALIZATIONS

> The main and most widely used call is the "chirrup" that gives the bird his name Philip Sparrow—"all sparrows are called Philip, 'phip, phip!' they cry."
>
> —Summers-Smith 1963:26

Virtually every account of the house sparrow emphasizes the gregarious nature of the species. Sparrows forage in flocks (Chapter 6), breed in loose colonies (Chapter 4), bathe in groups (Chapter 5), and form large nocturnal roosting congregations. Such a highly social lifestyle suggests the necessity for a complex communication system to facilitate the frequent interactions among individuals. Sparrows are also socially monogamous breeders (Chapter 4), which requires frequent communication between mates and between adults and offspring. In birds, communication is most often mediated through the visual and auditory sensory modalities. Visual communication can be based on fixed attributes such as plumage coloration (see Chapter 5) or on behavioral displays. The latter are often accompanied by vocalizations. As an oscine passerine, sparrows belong to a group often referred to as the songbirds, so-named because many members of the group have an elaborate song that is learned and that functions in advertisement. Unlike the typical songbird, however, sparrows do not have an elaborate advertisement song. Instead, "Philip Sparrow" seems to use a single, relatively simple vocalization, the "cheep" or "chirrup," in a large number of contexts, including advertisement. Because of this characteristic "chirrup" and the frequency of its use, I can sit in my living room and monitor the occurrence of house sparrows throughout the world while listening to radio or television news reported by journalists broadcasting live from worldwide locations.

The development of the advertisement song in songbirds has been the focus of intense study for the past half-century. One of the major components of that study has been the identification of the roles of genetic determination (instinct) and experience (learning) in song development. An account of song learning by house sparrows in Immanuel Kant's *Über Pedagogik* (1803) shows that the difference between learned and instinctual song has been known for at least two centuries (Wickler 1982). The current understanding of the ontogeny of the species-specific song in songbirds suggests that normal song development is the result of a complex interplay of instinct and experience, but that this normative developmental pathway may be modified by social context, such as naturally occurring contact with conspecific individuals with different songs than previously heard or tutoring by individuals of another species in the laboratory (Baptista and Petrinovich 1984).

Sparrows have a considerable array of displays that are often accompanied by vocalizations (Daanje 1941; Deckert 1969; Summers-Smith 1963), and it is difficult to discuss one without the other. In the following account, I have chosen to first describe the sparrow's vocal repertoire, and then its displays. Some topics that could be included in this chapter are treated in other chapters, including communal sexual display (Chapter 4) and dominance hierarchies (Chapter 6). Some other aspects of sparrow social behavior are also addressed here, after the discussion of vocalizations and displays.

Vocalizations

The scientific study of bird vocalizations was aided greatly by the inventions of the tape recorder and the sonagraph. The former allowed scientists to record vocalizations precisely in the field or laboratory, while simultaneously noting the context of each vocalization. The latter, which produces a printed sonagram plotting sound frequency against time (see Fig. 7.1), enabled them to analyze some components of the vocalizations quantitatively. Until the use of these instruments became common, the study of bird vocalization was based primarily on verbal descriptions of the sounds, a method that was qualitative and highly subjective and permitted few quantitative comparisons. Interpretation of the meaning of a vocal signal has been based primarily on observation of the context in which the signal is given and the behavioral response of another individual (or individuals) to the vocalization. Few experimental studies have been done to test the hypothesized meanings of sparrow vocal signals.

Nomenclature of Vocalizations

Studies based solely on verbal descriptions of sparrow vocalizations have recognized that many of them are structurally similar (at least to the human ear) and are used in multiple contexts (Daanje 1941; Deckert 1969; Summers-Smith 1963). Similar-sounding calls may therefore have different meanings in different contexts. Summers-Smith (1963) identified five basic vocalizations—"chirrup," "quee," "quer," "chree," and "churr"—and recognized several variations in all except the "chree." I have used these five categories in the discussion here, except that I refer to the "churr" as "chatter."

Nivison (1978) performed sonagraphic analyses of recorded vocalizations of marked sparrows in Michigan (USA). In many cases, he knew the pairing and breeding status of the birds and therefore was able to establish a precise context for the recorded calls. Bergmann and Helb (1982) published sonagraphic depictions of four house sparrow vocalizations, all except the "chree," which permitted comparisons between the English and German verbal representations of the calls.

The "chirrup" is undoubtedly the most widely used vocalization by the house sparrow. Bergmann and Helb (1982) referred to the "chirrup" as the "tschilp." Figure 7.1 shows sonagraphic representations of several variations of the "chirrup." The basic pattern of the "chirrup" is that of an inverted "U," with the frequency or pitch initially rising rapidly and then falling at the end. Nivison (1978) described four structural variations based on the number of rising and falling elements in one "chirrup" (i.e., one-, two-, three-, and four-peak "chirrups"). Peak frequencies of the repeated elements tend to decrease with peak number in multipeak "chirrups" (see Fig. 7.1). Duration tends to increase with number of peaks, with the average length of 122 three-peak "chirrups" (0.178 s) being double that of 70 one-peak "chirrups" (0.087 s) (Nivison 1978). Four-peak "chirrups" have the same duration as three-peak "chirrups," and two-peak "chirrups" are intermediate in length between one-peak and three-peak "chirrups." Individual males use all four "chirrup" types, with the two- and three-peak "chirrups" being used most frequently. In addition to the variation in number of peaks, the "chirrup" also varies in the number of harmonics in the individual elements, a form of variation that may permit the birds to use it as a graded signal (Nivison 1978). Change in the interval between successive "chirrups" results in change in "chirrup" rate, which may also contain signal information. Amplitude of the signal may also vary, creating further opportunities for use as a graded signal, but the quantitative study of variation in amplitude is difficult because of the large number of factors that can affect the accurate measurement of amplitude (e.g., distance to recorder, directional orientation of vocalist, wind speed and direction). There is

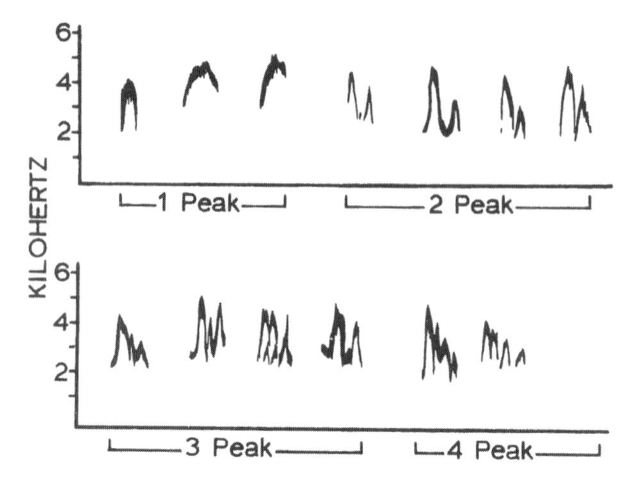

FIGURE 7.1. "Chirrup" vocalization of the house sparrow. Several sonagraphic representations of one-, two-, three-, and four-peak "chirrup" vocalizations recorded in Michigan (USA). From Nivison (1978: Fig. 5), reproduced with permission of the author.

considerable individual variation in the structure of the "chirrup," however, and Nivison concluded that this variation may permit individual recognition by other sparrows.

Males use the "chirrup" much more frequently than do females (Daanje 1941; Nivison 1978; Summers-Smith 1963). The male spends a great deal of time "chirruping" at his nest site throughout the year, but he does so most frequently just before and during the breeding season. Unmated males increase the rate of "chirruping," sometimes to as fast as 2 "chirrups" per second (Summers-Smith 1963). "Chirrup" rate increases to as high as 3/s when a female approaches an unmated male (Deckert 1969; Nivison 1978). The intensity of the "chirrup" also appears to be greater in unmated males, possibly based on more multipeak "chirrups" and a larger number of harmonics in the vocalizations. In this context, it would appear that the "chirrup" functions as a typical passerine advertisement song, serving to attract a mate and, possibly, to repel potential male rivals for the site, but I am unaware of any experimental studies demonstrating this function of the vocalization.

The "chirrup" is used in many other contexts, however. It is used as a general contact call between members of a pair or among individuals during flock foraging or at communal roosts. It is also used by males during communal displays ("sparrow parties") (see Chapter 4). The begging call of older nestlings also sounds like a "chirrup." During foraging, the "chirrup" is used to recruit other sparrows to a food source, particularly if the food is readily divisible or if there is a greater predation risk (Elgar 1986a, 1986b) (see Chapter 6).

The "quee" vocalization—the "dü" of Bergmann and Helb (1982)—is actually a family of apparently related calls that are typically given in interactions between members of a pair (Fig. 7.2). The "quee" call is frequently given during copulation, particularly by the soliciting female, and during the changeover between male and female at the nest during incubation and brooding (Summers-Smith 1963). Nivison (1978) described three variants of the "quee" —"wheea," "quee-1," and "quee-2"—and suggested that each has a different meaning. The "wheea" is a general contact call that serves during breeding to notify the mate on the nest of the arrival of the caller, thereby facilitating changeover at the nest. The "wheea" sounds like a shortened "chirrup" with an average duration of 0.12 s, an average low frequency of 1.6 kHz, and an average high frequency of 3.5 kHz. It is also used outside the breeding season as a contact call among members of a flock. The "quee-1" is used by members of a pair during early courtship and as a contact call between mates when they are in close proximity to each other (in visual contact). It has an average duration of 0.15 s, an average low frequency of 1.9 kHz, and an average high frequency of 3.7 kHz. "Quee-2" is given only during copulation, and although it may be given by both members of the pair, Nivison stated that it is most frequently given by the male. It has an average duration of 0.10 s and average low and high frequencies of 1.9 and 2.8 kHz, respectively.

"Quer" is an alarm call that is often given 1–3 times in potentially dangerous situations (Summers-Smith 1963). Bergmann and Helb (1982) referred to this call as "wäd." Fig. 7.3 shows sonagraphs of the "quer" of five different females when potential predators (cat or human observer) were close to the nest. The large number of harmonics in the "quer" are apparently responsible for the nasal quality of the call (Nivison 1978).

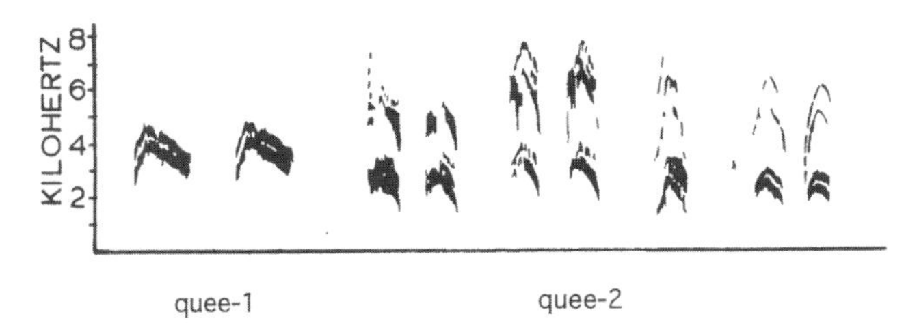

FIGURE 7.2. "Quee" vocalization of the house sparrow. Sonagraphs of "quee-1" and "quee-2" vocalizations recorded in Michigan (USA). From Nivison (1978: Fig. 17), reproduced with permission of the author.

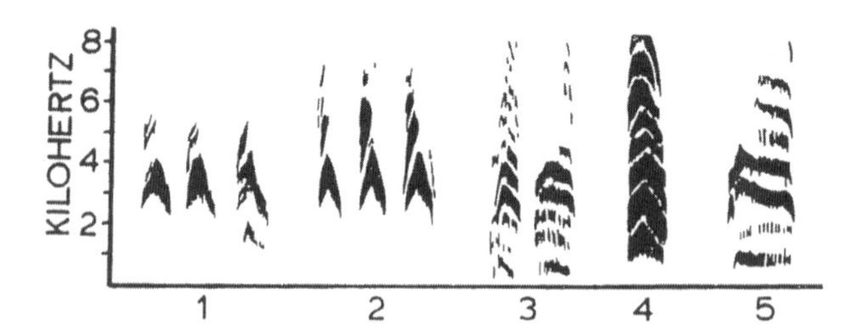

FIGURE 7.3. "Quer" vocalization of the house sparrow. Several sonagraphic representations of the "quer" vocalization recorded in Michigan (USA). From Nivison (1978: Fig. 21), reproduced with permission of the author.

The "chree" is also an alarm call, a shrill cry that Summers-Smith (1963) suggested is indicative of extreme alarm. Fig. 7.4 illustrates two "chree" calls given by the same sparrow. Nivison (1978) stated that the "chree" is longer than the "quer" call, with some of the harmonics eliminated and with consistently high amplitude, resulting in the shrill quality. He also stated that sparrows uttering the "chree" call take cover and assume a crouched, motionless position.

The results of an observational study of foraging sparrows in England (UK) suggest that the two alarm calls may have different meanings (Hull 1998). Two categories of potential predators were recognized in the study: aerially approaching predators (the Eurasian sparrowhawk, *Accipiter nisus*, or the Eurasian kestrel, *Falco tinnuculus*) and ground predators (domestic cats or humans). In all 165 instances of attacks by sparrowhawks or kestrels, the first alarm call heard from the foraging sparrows was a trisyllabic "quer" vocalization; in all 243 approaches by cats or humans, it was a monosyllabic "chree" call. Hull concluded that the "quer-quer-quer" alarm call is specific for aerial predators. It is not clear, however, that Hull's "quer" and "chree" are equivalent to the

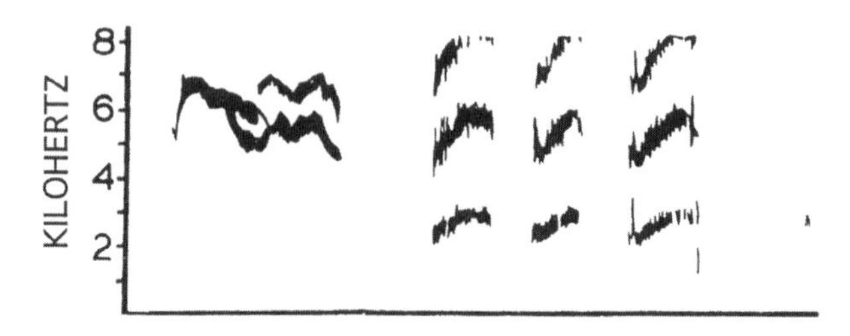

FIGURE 7.4. "Chree" alarm call of the house sparrow. Two sonagraphic representations of the "chree" alarm call recorded in Michigan (USA). From Nivison (1978: Fig. 21), reproduced with permission of the author.

alarm calls of the same names described earlier. She stated that the monosyllabic "chree" was quieter than the trisyllabic "quer," just the opposite of the descriptions of the two calls by Nivison (1978). Daanje (1941) also suggested that sparrows use different alarm calls when they identify a potential predator, a "krüü" call when a sparrowhawk suddenly appears, and a "kewkew" or "kewkewkew" when a cat approaches. Daanje described the former call as soft but piercing, whereas the latter was more nasal. Deckert (1969) suggested that different alarm calls are used for an approaching versus a departing aerial predator. Nivison (1978) reported that breeding females use the trisyllabic "quer" vocalization when cats or humans are close to the nest, suggesting that the same vocalization may have different meanings in different contexts. Without sonagraphic representations of the vocalizations in some of these studies, it is impossible to be certain of the equivalencies of the various calls.

The final vocalization is the "chatter" (referred to by Bergmann and Helb [1982] as "tetetet . . ." (Fig. 7.5). The "chatter" comprises repeated short elements, each with energy over a broad frequency range, indicative of the presence of numerous harmonics. It is apparently a threat signal and is often heard in the vicinity of the nest. The "chatter" is given by both males and females, but Nivison (1978) stated that there is a consistent difference between the sexes. The "chat" elements of the male "chatter" vocalization usually are fewer and of longer duration than those of the female. The average duration of a male "chat" is 0.045 s, whereas that of females is 0.021 s, and 79% of male ""chatters" have three or fewer elements, whereas 61% of female "chatters" contain seven or more elements.

Inspection of the sonagraphic form of the five basic sparrow vocalizations (Figs. 7.1 through 7.5) shows that the elements in all of the vocalizations are of the same general form. This form is the same as that noted earlier for the "chirrup": the inverted U.

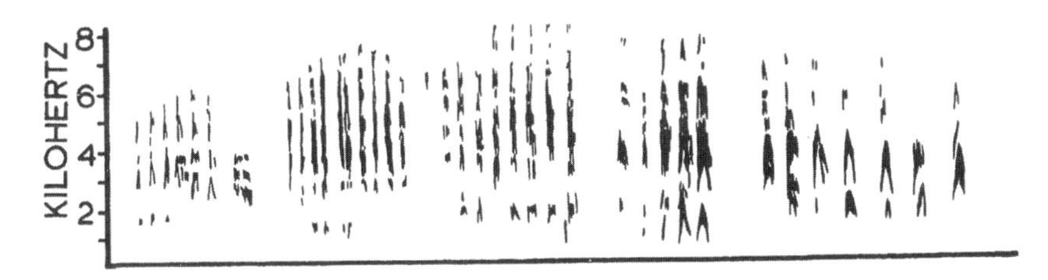

FIGURE 7.5. "Chatter" vocalization of the house sparrow. Five sonagraphic representations of the "chatter" vocalization recorded in Michigan (USA). From Nivison (1978: Fig. 22), reproduced with permission of the author.

Vocal Development

Anecdotal evidence suggests that a considerable amount of learning is involved in the development of sparrow vocalizations. Conradi (1905), for instance, used caged canaries (*Serinus canaria*) as foster parents for two sparrow nestlings and reported that the sparrows sang songs resembling canary song. Wickler (1982) reviewed several other similar reports of sparrows learning to sing songs similar to those of the canary, the linnet (*Caruelis cannabina*), the European goldfinch (*Carduelis carduelis*), or the skylark (*Alauda arvensis*). He also reported accounts of sparrows learning human speech. However, no detailed studies of vocal development in house sparrow nestlings cross-fostered by other species have been performed using modern methods of sound analysis, and therefore the role of learning in the development of song in the species is still unknown.

Nivison (1978) used several approaches to study the development of vocal behavior in Michigan house sparrows from hatching until the age of 30–70 d. He mounted microphones inside nest-boxes and recorded the vocalizations of chicks daily from hatching to fledging in the natural acoustic environment. He also hand-reared chicks in the laboratory, either in acoustic isolation from other individuals or in small groups of nestlings, and again recorded their vocalizations daily. He also transferred house sparrow eggs into canary nests in the laboratory and studied the effect of cross-fostering on vocal development. In addition, he captured free-living fledglings and kept them in the laboratory for up to 3 y to study post-juvenile changes in vocalizations.

Newly hatched chicks give a weak "peep" that is structurally similar to a one-peak "chirrup" except that it has a frequency approximately 2 kHz higher than the "chirrup," a shorter duration, and a frequency range that is much less than that of the "chirrup." "Peep" volume increases rapidly in the first few hours of life, and the "peep" continues to be the dominant vocalization for the first 4 d, although the nestlings occasionally also utter a "chirrup"-like call. On the third or fourth day, the nestlings begin to use a one- or two-peak "chirrup" more frequently, and these calls are structurally very similar to the "chirrups" of adults except that they are about 1.5 kHz higher in pitch. Pitch tends to fall gradually during the nestling period in two-peak "chirrups," reaching adult frequency by fledging. Three-peak "chirrups" also occur in the nestling period; they are infrequent during the first 6 d of nestling life but increase rapidly in number thereafter. They are also structurally similar to adult three-peak "chirrups," except that they are initially higher in pitch; pitch drops slowly during nestling life.

In Nivison's (1978) experiments, hand-reared sparrows used vocalizations that were essentially identical to those of birds reared in the wild. Similar patterns were also observed in acoustically isolated nestlings, except that the proportion of three-peak "chirrups" increased more slowly with age than in nestlings reared in the wild. Nivison concluded that the presence of adults had little impact on vocal development in juvenile sparrows. Two nestlings fostered by canaries died at 25 and 33 d of age, so the long-term impact of fostering on vocal development could not be determined. The one- and two-peak "chirrups" of these nestlings developed normally, however, except that they lacked harmonics present in other nestlings. Birds kept in the laboratory for up to 3 y had individual differences in the structure of their "chirrups" that did not change appreciably over the period of observation. Taken collectively, these results suggest that sparrow vocal development is primarily genetically based, and that learning plays little if any role. The gradual decrease in pitch of the "chirrup" during nestling life is probably a result of developmental changes in the syrinx rather than modification based on learning.

Displays

Efforts to understand the meaning of visual displays in the house sparrow have been based primarily on inferences drawn from identifying the context or contexts in which a display is observed and from observations of the response or responses of the individuals toward whom the displays are directed. Experimental confirmation of the meaning of most of the displays is lacking. This account discusses visual displays of sparrows in the context of their presumed meanings.

Threat

The threat display is often observed during foraging when one bird attempts to supplant another at its feeding station. Unreceptive females also often respond to displaying males with a threat display. The basic threat display in both sexes involves the bird's assuming a crouching position with the head thrust forward (toward the object of the threat) and with the bill sometimes open (gaping) (Summers-Smith 1955). The plumage is sleeked down, the wings are held away from the body, and the tail is held horizontally during this display (Summers-Smith 1963) (see Fig. 5.1b in Chapter 5). The threatening bird also sometimes lunges toward its opponent while maintaining the threat position, an action that occasionally leads to actual fighting between the birds (Summers-Smith 1963).

Fighting involves pecking at the head, neck, wing, or tail of the opponent and sometimes includes aerial combat, in which the birds fly up into the air while continuing to peck at each other (Summers-Smith 1963). Aerial chases lasting several minutes may also occur, particularly between rival males competing for a nest site (Summers-Smith 1955). Occasionally, a bird seriously injures or even kills an opponent. Dawson (1968) described such an attack in New Zealand in which one male continued to peck at the wounds of a seriously injured rival male for an hour (the injured bird had numerous head wounds, and most of the plumage in the head region was missing).

Solicitation

Similar solicitation displays are used by females soliciting copulation and by fledglings begging for food from adults. The bird assumes a crouched position with the neck drawn in, wings drooped and shivering, and tail held horizontally (Summers-Smith 1963). Fledglings utter an insistent "chirrup"-like begging call, whereas soliciting females use frequent "quee" calls (most often the "quee-2" call) (Nivison 1978). Summers-Smith (1963) also described a male solicitation display that was similar in appearance to that of the female but was usually observed near the nest when the young were about ready to fledge. Summers-Smith suggested that the function of this display was to induce the young to leave the nest.

Male Sexual Display

A male displaying to a prospective mate hops stiffly in front of or around the female with his head stretched upward, tail raised and fanned, wings lowered and shivering, and breast feathers puffed out (Møller 1987a; Summers-Smith 1954b, 1955) (see Fig. 5.1a in Chapter 5). Summers-Smith (1963) referred to this display as "standing to attention." The male is usually "chirruping" at a high rate during this display, and he sometimes attempts to mount, but is often rebuffed by an aggressive response from an unreceptive female. Unmated males "chirruping" intensely at their nest sites sometimes perform a modified version of this display, with their tails raised and their wings drooped and shivering (Summers-Smith 1955). Communal displays, or "sparrow parties," involve two to several males displaying to a single female. Communal displays are sometimes accompanied by cloacal pecking, in which one or more of the displaying males pecks at the female's cloaca (see Chapter 4).

Simmons (1951) described a second male sexual display, based on observations in Egypt, that is reminiscent of the solicitation display of the female. In this display, the male assumed a squat position, with legs bent so that the body was parallel to the ground. The neck was contracted, the bill was pointed upward at a 40° angle, the wings were slightly drooped and shivering, and the tail was held horizontally. The bird hopped mechanically around the female for about 4 min in this posture, "chirruping" quietly, and occasionally assuming the more common sexual display posture described earlier.

Agitation

Tail-flicking is the principal agitation display. The bird stands in an upright position with the feathers sleeked and repeatedly depresses and raises the tail (Summers-Smith 1963). The display is often accompanied by the "chatter" vocalization, and it is often used when an intruder or potential competitor approaches the nest.

Vigilance

The principal vigilance behavior of sparrows is the "head-jerk" (Elgar and Catterall 1981), in which the bird raises its head so that the bill is horizontal and rotates the head quickly around the vertical axis. This behavior, which is also referred to as scanning, is seen most frequently in foraging sparrows and is discussed at some length in Chapter 6. The typical "head-jerk" in foraging flocks of sparrows is quite stereotyped and has an average duration of only 0.63 to 0.74 s (Elcavage and Caraco 1983; Elgar and Catterall 1981). The length of an individual scanning event has been observed to be quite variable during preening or perching, or during feeding at a four-sided bird feeder, lasting from just under 1 s to about 10 s (McVean and Haddlesey 1980; Studd et al. 1983). Scanning during foraging may function not only in the detection of potential predators but also in obtaining information from other foraging sparrows (see Chapter 6).

Communal Roosting

Sparrows form large communal roosts, which are used not only for nocturnal roosting but also as sites of social singing during the day, particularly in late afternoon and evening. Communal roosting sites are usually located in trees, shrubs, or vines with dense foliage, and they change locally if deciduous sites lose their leaves. Several tree species were utilized as communal roosting sites in Oklahoma (USA), with tree height (at least 6 m) and the density of the foliage (but not species) apparently being the principal criteria for selection (North 1968). The number of sparrows roosting at a site changes seasonally, with larger numbers present during the nonbreeding season, but with smaller roosts persisting throughout the breeding season (Anderson, personal observation; Summers-Smith 1963). As the breeding season progresses, roosts tend to increase in size, presumably by the addition of young of the year. At a communal roost in Poland, for instance, the number of sparrows increased 3–4 fold between June and September (Gorska 1990). The number of sparrows at communal roosts varies considerably, from a few dozen to several thousand. A roost in Lima, Peru, in August contained approximately 6000 sparrows (Leck 1973); a winter roost (July) in New Zealand numbered more than 14,000 sparrows (Dawson 1967); a late summer roost in London (UK) contained 19,000 sparrows (Summers-Smith 1963); and an autumnal roost in Egypt numbered about 100,000 sparrows (Moreau 1931).

House sparrows sometimes form communal roosts with other species. Heterospecific communal roosts have been observed with Spanish sparrows in Spain (Alonso 1986), with European starlings in Poland (Gorska 1975), with jungle babblers (*Turdoides striatus*) and common mynas (*Acridotheres tristis*) in India (Rana 1989a), and with European starlings and Eurasian tree sparrows in North America (Anderson, personal observation; North 1968). Mahabal and Bastawade (1985) also reported that house sparrows were among several species roosting communally near communal roosts of the common pariah kite (*Milvus migrans govada*) in India.

Sparrows begin arriving at communal roosts up to 2 h before sunset, and from the time the first birds arrive until well after sunset they engage in communal vocalizing (using the "chirrup"). This vocalizing results in an incessant twittering that is interrupted only when the birds are alarmed. The signal that triggers an almost instantaneous halt in the twittering is not known, but if no imminent threat occurs, the twittering soon resumes. Agitation calls are also frequently heard from communal roosts, however, suggesting that there are frequent aggressive encounters among birds at a roost (Summers-Smith 1963). Birds come and go during the evening preroosting period, perhaps leaving for some last-minute foraging before nightfall. Leck (1973) recorded the number of arriving individuals in 5-min intervals between 1615 and 1755 at a 6000-sparrow roost in Peru. The number of arriving sparrows increased rapidly, from 16.8/min at 1615 to 93.4/min at 1645, and then dropped slowly to 20.0/min at 1755. In the morning, sparrows begin vocalizing about 30 min before sunrise, and they usually depart from the roost within 30 min after sunrise (North 1968). Therefore, less time is spent in communal vocalizing in the morning.

At communal roosts with the European starling in Poland, the peak arrival time of sparrows preceded that of the starlings by about 15 min (Gorska 1975). Gorska also reported that a considerable amount of competition took place among individuals of both species for sites within the roost, and that the larger starlings usually obtained sites near the center of the roost (presumably safer roosting sites).

The function or functions of communal roosting in sparrows are poorly understood. The thick foliage at most communal roosting sites suggests that predator avoidance is an important criterion in site selection. The incessant twittering that occurs at communal roosts can be heard at a considerable distance, however, and could provide potential predators with auditory cues to the location of roosts. It seems unlikely, therefore, that predator avoidance is the primary function of communal roosting behavior in sparrows. In fact, the very large roost in Egypt mentioned earlier disintegrated within 1 wk after the appearance of a migrating Eurasian sparrowhawk (Moreau 1931). Instead, communal roosts may serve as important centers for the exchange of information among sparrows. Noting the direction from which individuals with visibly full crops arrive may provide information about the location of good foraging sites. This may account for some of the birds' leaving the communal roosting site during the evening to do last-minute foraging. The fact that communal roosts persist during the breeding season, when most adults are engaged in nesting, suggests that they may also serve for exchange of information about mating status and as recruiting sites for unpaired individuals. The rapidity with which replacement females are recruited to nest sites from which the pair female has been removed (see Anderson 1990) suggests such a possibility.

North (1968) used colored, plastic neck collars to track the movements of sparrows between nesting, foraging, and roosting sites in Oklahoma. Sparrows leaving their roosting sites in the morning often flew to nearby "feeding-staging" areas, where they congregated and fed before flying to more distant sites for daytime foraging. During late summer, the communal roosting sites were located in a town, and the main foraging sites were ripening grain-fields outside the town. The distance between the communal roosting sites and the grain-fields varied from 3 to 6 km, and the birds used indirect routes that included stopovers at the staging areas to move between the two. Both roosting sites and staging areas may therefore provide opportunities for information exchange concerning profitable foraging areas and traveling routes to them.

Chapter 8

POPULATION DYNAMICS AND MOVEMENTS

> Expressed in actual cubic yards of sweepings, they mean that the same number of blocks swept by push cart men in 1911, 1260 cubic yards, in 1914, 1003 cubic yards and in 1919, 474 cubic yards were gathered, all of which indicates that nearly 786 cubic yards of clear horse manure have thereby been gradually subtracted from the daily, food supply of the English Sparrows inhabiting these areas. These data are more or less official and are at least approximately correct and accurate, and illustrate in a convincing manner, the truth of the belief that there has been wrought a great change in urban sparrow's food supply during the past ten years. It would thus appear that there is ample cause alone, in the great diminution of the food supply, to explain the notable decrease of English Sparrows in Denver.
>
> —Bergtold 1921:249

> We cannot afford to dismiss the possibility that the House Sparrow is the modern equivalent of the miner's canary.
>
> —Summers-Smith 1999:386

The breeding bird censuses conducted during the summers of 1968–1972 estimated that there were between 3.5 and 7.0 million breeding pairs of house sparrows in Britain and Ireland (Sharrock 1976). A scant 20 y later, sparrow numbers had declined by 30%–50% (Balmer and Marchant 1993; Glue 1994; Marchant et al. 1990;), a loss in total number of breeding individuals of between 2 and 7 million. The decline in major urban areas may have been even more pronounced, amounting to 85%–98.8% during the last 20–40 y (Summers-Smith 1999, 2000). Not since the Irish Potato

Famine of the mid-19th century have the British Isles witnessed such a major population decline. Sparrows are reported to be declining in numbers in other parts of their extensive range as well (e.g., Bauer and Heine 1992; Bosakowski 1986; Jacobsen 1995; Jones and Wieneke 2000; Vaisanen and Hilden 1993), and in western Europe this decline parallels similar declines in several other farmland bird species (Donald et al. 2001; Krebs et al. 1999). What is responsible for this widespread decline in numbers of house sparrows?

Population ecology is the subdiscipline of biology that deals with the population dynamics of organisms, and one of the most fundamental problems in population ecology is the question of what regulates the population size of an organism (see Newton [1998] for a recent review of the subject for birds). One of the ironies of the history of population ecology is that two landmark works, with fundamentally different ideas about the nature of the factors controlling animal populations, were published in the same year. Half a century after their publication, the issues raised in these two works, *The Natural Regulation of Animal Numbers* (Lack 1954a) and *The Distribution and Abundance of Animals* (Andrewartha and Birch 1954), remain unresolved and continue to generate controversy. Based on the observations or assumptions that (1) all populations have the reproductive potential to experience exponential growth in size, (2) natural selection strongly favors individuals that maximize their reproductive output, and (3) most populations remain relatively constant through time compared with their capacity for increase, Lack argued that populations must be regulated by density-dependent mortality factors. On the other hand, Andrewartha and Birch argued that the concept of density-dependent and density-independent factors represented a meaningless dichotomy and that, in fact, there was no such thing as a density-independent factor. They emphasized the diversity of factors (physical, temporal and biotic) that are known to influence the distribution and abundance of animal populations and suggested that (1) most populations do not reach local densities such that they are limited by competition for resources such as food or space and (2) few natural populations really conform to the "balance of nature" idea, which suggests that populations must be "regulated" somehow to achieve a steady state or dynamic equilibrium.

Population ecologists recognize four fundamental parameters that are responsible for changes in numbers in a population: birth rate, mortality rate, immigration rate, and emigration rate. Each of these parameters must be accurately measured to fully understand the dynamics of a natural population, an assertion that is much easier to state than it is to achieve in the field. Although much data has been collected on birth rates in sparrow populations (see Chapter 4), data on mortality rates, particularly for certain stages in the life cycle, are less well known, and there are very few quantitative data on emigration and immigration rates. Many studies simply assume that immigration and emigration rates offset each other, a dubious assumption at best.

One of the primary tools of population ecology is the life table, an instrument borrowed from human demography that is based on age-specific mortality data and reproductive rates. One important product of a life table is an estimate of the net replacement rate of the population, R_0. If R_0 is 1.0, the average female in the population is just replacing herself during her lifetime, and the population is stable. Values of R_0 significantly greater than or less than 1.0 indicate, respectively, that the population is increasing or decreasing in size.

This chapter begins by describing the evidence for the recent declines in house sparrow numbers in at least parts of its range. It then discusses several types of biotic interactions, including interspecific competition, predation, parasitism/disease, as well as the effects of abiotic factors, on the population dynamics of sparrows. The chapter also discusses the sex-ratio conundrum in sparrows and dispersal and migration in the species.

Decline in House Sparrow Numbers

One of the difficulties in attempting to document wide-scale changes in population numbers of a species is that adequate long-term data are simply not available. One of the fortunate outgrowths of the dawning environmental consciousness of the late 1960s and early 1970s was the awareness of the need for baseline information on the distribution and abundance of common species. Because of the widespread public interest in birds, national and regional ornithological organizations have been able to use thousands of trained amateurs to conduct surveys of various kinds over wide geographic areas. All such surveys have a number of both theoretical and practical problems, including geographic and habitat biases in coverage and differences among observers in both effort and ability. Despite these difficulties, however, their cumulative results provide the most powerful evidence available of the direction, if not the precise magnitude, of population trends on a broad geographic scale for many bird species. This is true for the house sparrow also, despite the fact that some early surveys, such the Common Bird Census in Great Britain before 1976, expressly requested that surveyors ignore sparrows in their counts.

A second difficulty in attempting to determine the extent of population change is the problem of scale. Some studies are based on small, circumscribed areas, and local populations can undergo fluctuations in size that are not necessarily related to changes in population size on a regional or continental scale. In the parlance of population ecology, local populations that have sustained net replacement rates that are less than 1 are known as "sinks" and must be sustained by immigration from other populations in which net replacement rates consistently exceed 1 ("source" populations) (Pulliam 1988). Because of their close association with humans (see Chapter 10); their highly social foraging, breeding and roosting behavior; and their generally short dispersal distances (see later discussion), sparrows tend to have somewhat disjunct populations. Some ecologists deal with this difference in scale by distinguishing between local populations and the metapopulation, the collection of local populations that occupies a larger geographic region within which there may be many source and sink populations. Changes in the size of the metapopulation can be documented only with surveys that cover a wide geographic area.

The first reports of major declines in sparrow numbers during the last 25–30 y came from the eastern United States (Bennett 1990; Bosakowski 1986; Kricher 1983; Wootton 1987). All of these studies documented significant reductions in sparrow numbers in parts of the northeastern United States, based either on Christmas Bird Count data (Bennett 1990; Bosakowski 1986; Kricher 1983) or on both Christmas Bird Count and Breeding Bird Survey data (Wootton 1987). The scales in the four studies varied from regional (northeastern New Jersey) to multistate areas, with Bennett (1990) examining the population trends at three different scales (local, regional, and continental). All of these papers also explored the possible role of interspecific competition with the house finch *(Carpodacus mexicanus)* in contributing to the observed declines in sparrow numbers, a possibility that is addressed below in the section on competition. Two of the studies did note that sparrow numbers were declining outside the range of the house finch as well as in areas where house finch numbers were increasing rapidly, suggesting that other factors might be contributing to the decline (Kricher 1983; Wootton 1987). Other studies at both local and continental scales have also reported major declines in sparrow numbers in North America (Brown 1969; Duncan 1996; Lepage and Francis 2002; Peterjohn et al. 1994; Wells et al. 1998). Breeding Bird Survey results from throughout North America, for instance, recorded a 48.3% ($P < .01$) decline in sparrow numbers from 1966 to 1993 (Peterjohn et al. 1994).

The decline in Great Britain has already been mentioned. Again, the evidence for these declines is based on both local and regional studies and suggests that sparrows have decreased in both abundance and distribution. Marchant et al. (1990) reported a 30% decrease in the breeding population between 1980 and 1988 based on the Common Bird Census (covering the entire British Isles), and Balmer and Marchant (1993) reported a 32% decline in the breeding population between 1976 and 1992, also using the Common Bird Census. Bircham and Jordan (1997) reported a 5.3% decrease in the distribution of sparrows in Great Britain between the Breeding Bird Atlas surveys of 1968–1972 and 1988–1991.

Glue (1994) reported that Garden Bird Feeding Survey data suggest that sparrow numbers during the winter have declined sharply in both suburban and rural gardens in the United Kingdom. In an intensive series of annual surveys of a 1200-km^2 area in Oxfordshire, Easterbrook (1999) reported decreases in both the distribution and the abundance of sparrows. The decline in abundance amounted to 70% and appeared to be more-or-less linear over the 21 y of the study, while distribution decreased by 40%. Summers-Smith (2000) reported that numbers of sparrows in urban parks in London and Glasgow (UK) decreased by 85%–99% in recent years and suggested that the rate of decline may actually have accelerated during the past 5 y.

Numerous locations in western Europe and elsewhere have also recorded significant declines in sparrow numbers. At Lake Constance (Germany), the breeding population decreased by 22% between 1980 and 1990 (Bauer and Heine 1992). In a review of sparrow abundance in three parts of Germany (northwest, northeast, and south) over the period 1850–2000, Engler and Bauer (2002) found major decreases in abundance during the period 1975–2000, particularly in the northwest. Vaisanen and Hilden (1993) reported a steady decline in sparrow numbers during the winter at three rural sites in Finland. And in Denmark, the breeding population decreased by about 40% between 1976 and 1994 (Jacobsen 1995).

As in Great Britain, declines in some urban populations have been particularly pronounced. In two city parks in Hamburg (Germany), numbers declined by 73% between 1983 and 1997 in one (Bower 1999) and by about 50% between 1991 and 1997 in the other (Mitschke et al. 1999). In the most built-up parts of Berlin (Germany), the number of breeding pairs of sparrows decreased 65% between 1979 and 1991 (Witt 2000). Elsewhere, in a tropical city in Australia, sparrow densities decreased by about 40% between 1980 and 1997 in both the wet and dry seasons ($P < .05$ in both cases) (Jones and Wieneke 2000).

The situation in other parts of the house sparrow's range is not known. Recent studies in central and eastern Europe suggest that densities of sparrows remain high. Dyer et al. (1977) summarized data on breeding densities of sparrows in several habitat types throughout Europe. I calculated geometric means for habitat types with 10 or more estimates of breeding density, and the following are means in individuals per square kilometer (number of estimates in parentheses): riparian areas and cemeteries, 85.4 ($n = 10$); old parks in villages and fields with trees, 155.2 ($n = 18$); old parks in larger towns, 179.5 ($n = 28$); villages, 240.3 ($n = 21$); suburban areas with one-family homes, 445.4 ($n = 19$); commercial and shopping areas, 758.5 ($n = 10$); and residential areas with apartment complexes, 999.8 ($n = 21$). The general pattern of increasing density with increasing urbanization is apparent. Data from several recent studies are summarized in Table 8.1. The estimates range from 9.3 to 2300 individuals/

km². The differences in methods among the studies, and the different scales on which the estimates are based (from local study areas of a few hectares to regional or national surveys), make it difficult to draw meaningful conclusions from the data. It does appear that densities may be higher in eastern Europe than in western Europe, a difference that may be related to differences in agricultural intensification in the two regions (Donald et al. 2001).

A major decline in sparrow numbers is not unprecedented. Significant declines in urban populations in the 1910s and 1920s were reported in both North America (Bergtold 1921; Eaton 1924; Rand 1956) and Great Britain (Alexander and Lack 1944). These declines were attributed to decreases in horse populations resulting from the introduction of automobiles, with a concomitant decrease in the food supply of sparrows, which regularly fed on partially digested grain from horse droppings (see epigraph) and on spilled grain from the feeding of horses. This decline was therefore generally attributed to decreases in the food supply for urban populations, which resulted in a downward adjustment in sparrow numbers to a new ceiling imposed by the food supply.

What is responsible for the recent decline? Numerous suggestions have been made to try to explain this decline. These include predation, disease, interspecific competition for food and/or nest sites, pollution, and changes in agricultural practices that have resulted either directly or indirectly in a reduction in food for house sparrows. Abiotic factors such as weather or global climate change have not generally been advanced as candidates for the cause of the declines. Direct evidence for or against any of these proposals is scanty, but most are examined in more detail in the following sections.

A particularly interesting recent study (Siriwardena et al. 1999) examined the effects of survival rates of both adults and juveniles in six species of granivorous birds (including the house sparrow) on the changes in population sizes of the species in Great Britain. Survival rates were calculated for each age group and species based on birds banded in Great Britain between 1962 and 1993 and incorporated allowances for variation in reporting rate, both for age groups and over time. For the house sparrow, birds banded during the 7–mo period March–September were used in the analysis, and the survival rates were calculated based on 984 recoveries of 7448 sparrows banded as adults and 777 recoveries of 8023 sparrows banded as first-year birds. After the annual survival rates for the two age classes were determined, they were compared with indices of population change derived from Common Bird Census data. For the years of relative stability in sparrow numbers (1962–1975), the adult survival rate averaged about 0.59, and during the years of decline after 1976, the average adult survival rate was about 0.52 ($P < .0001$). Siriwardena et al. concluded that there is strong evidence that the general decline in the house sparrow population in Great Britain is linked to changes in survival rates in the species. To the extent

TABLE 8.1. Examples of Estimates of House Sparrow Density Recorded During the Period 1976–2000

Location	Habitat Type	Season	Year	Individuals/km² (SE)	Source
Western Europe/North America					
Rennes, France	Urban/residential	Winter	1995	257 (44)	1
Rennes, France	Urban/residential	Spring	1995	295 (40)	1
Sagunto, Spain	Orange grove	Spring	1986–1988	477 (30)	2
Madrid, Spain	Urban park (edge)	Summer	1998–1999	6.8–8.9	3
Madrid, Spain	Urban park (interior)	Summer	1998–1999	1.7–2.1	3
Bonn, Germany	Urban/suburban	Summer	—	90.6	4
Bodensee, Germany	Suburban/rural	Summer	—	119–127	4
Hamburg, Germany	Urban	Summer	1997–2000	76.0	5
Schleswig-Holstein, Germany	Towns/villages	Summer	1991–1995	715	6
Campania, Italy	Rural	Summer	1985–1990	9.3–13.9	7
Finland	Towns/villages	Winter	—	65–131 (80)	8
Finland	Towns	Winter	2001	65.7 (46.0)	9
Great Britain	Suburban	Summer	1996	380 (44)	10
Great Britain	Urban	Summer	1996	333 (65)	10
Great Britain	Rural	Summer	1996	245 (35)	10
Quebec, Canada	Urban/residential	Winter	1995–1996	326–383 (52)	1
Quebec, Canada	Urban/residential	Spring	1995	251 (42)	1
Ohio, USA	Urban/residential	Summer	1995–1996	300–560	11
California, USA	Urban center	Summer	1992–1993	72	12
California, USA	Golf course	Summer	1992–1993	22	12

Eastern Europe					
Warsaw, Poland	Urban/residential	Spring	1986–1990	289–619	13
Lublin, Poland	Urban parks/cemeteries	Spring	1982–1984	177 (70)	14
Bratislava, Slovakia	Urban parks	Spring	1990–1992	82–778	15
Bratislava, Slovakia	Urban cemeteries	Spring	1992–1995	0–126	16
Czech Republic	Cattle farms	Spring	1985–1987	2300	17
Lvov, Ukraine	Urban	Spring	1993–1995	512	18
Other					
Canberra, Australia	Suburban	Spring		684–818	19

Data are grouped so as to place estimates from North America and western Europe together (where densities have apparently declined significantly in recent years—see text).

Sources: 1, Clergeau et al. (1998); 2, Barba and Gil-Delgado (1990); 3, Fernandez-Juricic (2001); 4, Bezzel (1985a); 5, Mitschke and Baumung (2001); 6, Busche (1999); 7, Goglia and Milone (1995); 8, Jokimäki et al. (1996); 9, Jokimäki and Kaisanlahti-Jokimäki (2003); 10, Gregory (1999); 11, Gering and Blair (1999); 12, Blair (1996); 13, Luniak (1996); 14, Biadun (1994); 15, Mullerova-Franekova and Kocian (1995); 16, Kocian et al. (2003); 17, Honza (1992); 18, Bokotey (1996); 19, Lenz (1990).

that this result is generalizable to other regions in which sparrow numbers have declined in recent years, this means that the causes should be sought in factors that affect the survival rates of juvenile and adult sparrows, not in factors affecting reproductive success.

Predation

Almost everything eats house sparrows (Table 8.2). This includes even humans, who have eaten sparrow soup or sparrow pie in Europe and the near East for centuries, perhaps continuing to this day in some Mediterranean countries (Summers-Smith 1963). This practice dates at least to Biblical times, because Jesus is quoted as saying, "Are not two sparrows sold for a farthing?" But are predators responsible for controlling the size of sparrow populations? The question of whether predators control the populations of their prey has been a hotly debated issue in population ecology for decades. One viewpoint is that predators are generally responsible for removing a "doomed surplus" of individuals that are destined to die anyway due to starvation, lack of suitable habitat, or disease (Errington 1946). Deaths due to predation are therefore compensatory, not additive, and factors other than predation (particularly intraspecific competition for limiting resources such as food or space) are held to be responsible for controlling the population. Others have proposed that predation is responsible for limiting populations below levels at which they would otherwise be controlled by some limiting resource (e.g., Caughley et al. 1980). Newton (1998) considered that predation limits the population size of a prey species only if it results in a reduction in the number of breeding individuals in the population. Unfortunately, the predator exclosure experiments that have been applied to numerous other organisms are impossible to employ for wide-ranging species such as birds, and it is therefore very difficult to actually test whether predation might be limiting species such as the house sparrow.

Two recent trends suggest that increased predation could be a factor in the decreases in sparrow numbers in western Europe and North America. First, population sizes of many raptor species have increased dramatically during the past 20–30 y, many rebounding from very low levels that resulted from the negative effects of DDT and other organochlorine pesticides on their reproductive success. The most intense use of these pesticides occurred in western Europe and North America, where, together with direct human persecution of many raptors, it resulted in the near extirpation of some species. Legal protection and the banning or restricted use of organochlorine pesticides have led to major increases in many raptor populations (see later discussion). Second, many predator species have become increasingly urbanized in the last few

TABLE 8.2. Predators of the House Sparrow

Predator	Location	Source
Reptiles		
Common Garden Snake *(Calotes versicolor)*	India	1
Birds		
Mississippi kite *(Ictinia mississippiensis)*	USA	2
Black kite *(Milvus migrans)*	Egypt, Italy	3,4
Long-winged harrier *(Circus buffoni)*	Argentina	5
Hen harrier *(Circus cyaneus)*	UK	6
Pallid harrier *(Circus macrourus)*	Israel	7
Montagu's harrier *(Circus pygargus)*	Spain	8
Eurasian sparrowhawk *(Accipiter nisus)*	Israel, Sweden, Netherlands, Denmark, Norway, Finland	7,9,10,11,12,13
Sharp-shinned hawk *(Accipiter striatus)*	USA	14,15
Cooper's hawk *(Accipiter cooperi)*	USA	14,16
Red-tailed hawk *(Buteo jameicensis)*	USA	17
Crested caracara *(Caracara plancus)*	Mexico	18
Eurasian kestrel *(Falco tinnuculus)*	UK	19
American kestrel *(Falco sparverius)*	USA, Chile	15,20
Merlin *(Falco columbarius)*	Canada	21
European hobby *(Falco subbuteo)*	Italy, Germany, UK	22,23,24
Lanner falcon *(Falco biarmicus)*	Israel	25
Peregrine falcon *(Falco peregrinus)*	South Africa, Germany, USA	26,27,28
Prairie falcon *(Falco mexicanus)*	USA	29
Water rail *(Rallus aquaticus)*	UK	30
Audouin's gull *(Larus audoouinii)*	Turkey	31
Barn owl *(Tyto alba)*	Ireland, Germany, Hungary, Italy	32,33,34,35
Western screech-owl *(Otus kennicottii)*	USA	36
Eastern screech-owl *(Otus asio)*	USA	36,37
Eagle owl *(Bubo bubo)*	Jordan	38
Snowy owl *(Nyctea scandiaca)*	Canada	39
Little owl *(Athene noctua)*	Greece, UK	40,41
Tawny owl *(Strix aluco)*	Poland, Italy	42,43,44
Long-eared owl *(Asio otus)*	Germany, Italy, Czech Republic	45,46,47
Tengmalm's owl *(Aegolius funereus)*	Finland	48
Pied currawong *(Strepera graculina)*	Australia	49
Loggerhead shrike *(Lanius ludovicianus)*	USA	50
Northern shrike *(Lanius excubitor)*	USA	50
Blue jay *(Cyanocitta cristata)*	USA	51
Black-billed magpie *(Pica pica)*	UK	52
Common grackle *(Quisculus quiscula)*	Canada, USA	53,54

(*continued*)

TABLE 8.2. (*Continued*)

Predator	Location	Source
Mammals		
Brushtail possum (*Trichosurus vulpecula*)	New Zealand	55
Red squirrel (*Tamiaciurus hudsonicus*)	Canada	56
Five-striped squirrel (*Funambulus pennanti*)	India	57
Domestic cat (*Felis domesticus*)	UK, Denmark	58,59

This table contains a partial list of predators that are known to have preyed on free-living house sparrows; it does not include species known to have preyed only on eggs or nestlings.

Sources: 1, Paralkar (1995); 2, Parker (1982); 3, Everett (1992); 4,Sergio and Boto (1999); 5, Bo et al. (1996); 6, Watson (1977); 7, Yosef (1996); 8, Arroyo (1997); 9, Gotmark and Post (1996); 10, Opdam (1979); 11, Frimer (1989); 12, Selas (1993); 13, Solonen (1997); 14, Storer (1966); 15, Errington (1933); 16, Hamerstrom and Hamerstrom (1951); 17, Gates (1972); 18, Rodriguez-Estrella and Rodriguez (1997); 19, Shrubb (1980), Yalden (1980); 20, Jaksic et al. (2001); 21, Oliphant and McTaggart (1977), James and Smith (1987), Sodhi and Oliphant (1993); 22, Sergio and Bogliani (1999); 23, Fiuczynski and Nethersole-Thompson (1980); 24, Parr (1985); 25, Yosef (1991); 26, Jenkins and Avery (1999); 27, Schneider (1995); 28, Ellis et al. (2004); 29, Boyce (1985); 30, Wood (1986); 31, Witt et al. (1981); 32, Smal (1987); 33, Jentzsch (1988); 34, Schmidt (1968); 35, Bose and Guidali. (2001); 36. Bent (1938); 37, Gehlbach (1994); 38, Amr et al. (1997); 39, Gross (1944); 40, Angelici et al. (1997); 41, Hibbert-Ware (1937); 42, Grozczynski et al. (1993); 43, Zalewski (1994); 44, Manganaro et al. (1990); 45, Dathe (1988); 46, Manganaro (1997); 47, Zukal (1992); 48, Korpimaki (1988); 49, Wood (1998); 50, Judd (1898), Bent (1950); 51, Master (1979), Cink (1980); 52, Thomas (1982), Tatner (1983); 53, Davidson (1994); 54, Davis (1944); 55, McLeod and Thompson (2002); 56, Nero (1987); 57, Tiwari (1990), Mathew and Lukose (1995); 58, Churcher and Lawton (1987); 59, Møller and Erritzoe (2000).

decades. This is undoubtedly related to more effective protection of these species, some of which were severely persecuted by humans in the past. The impacts on prey populations resulting from this increasing urbanization of predators may be most evident in human commensals such as the house sparrow.

Three cases of intense predation on sparrows will be discussed here, two involving avian predators—the Eurasian sparrowhawk (*Accipiter nisus*) and the merlin (*Falco columbarius*)—and the third involving the domestic cat (*Felis domesticus*). One of these cases, that of the merlin in the prairie provinces of Canada, represents a natural experiment that could effectively negate the hypothesis that increased predation is responsible for the recent declines in sparrow numbers.

The Eurasian sparrowhawk has shown a dramatic population increase in western Europe and Great Britain in the past 30 y. At Lake Constance (Germany), for instance, sparrowhawk numbers increased by 142.3% between 1980 and 1990 (Bauer and Heine 1992), and the Common Bird Censuses in Great Britain and Ireland indicate that the number of breeding sparrowhawks increased about three-fold between 1972 and 1998 (UK Raptor Working Group 2000). The sparrowhawk is a bird specialist, with birds making up 96.0%–

98.1% of the total dietary items in several studies (Gotmark and Post 1996; Newton 1986; Opdam 1979; Tinbergen 1946).

Although the sparrowhawk is primarily a forest-dwelling species, it has become increasingly common around human settlements, both rural and urban, in recent years. It takes few sparrows when living in its traditional forest habitat. A recent study of sparrowhawks breeding in Oulauka National Park (Finland), for instance, found no sparrows in the diet (Rytkonen et al. 1998), and a similar study in a forested region of southern Norway (only 2% agricultural) found that sparrows made up only 0.24% of the prey items (Selas 1993). In mixed habitats, however, the story is very different. Two studies in the Netherlands reported that the house sparrow is the most frequently preyed-upon species. In his classic study in the early 1940s, Tinbergen (1946) found that sparrows comprised 25.2% of the prey items of sparrowhawks. More recently, Opdam (1979) found that the house sparrow was the most commonly taken species in all three seasons studied, comprising 18.0% of items in winter, 35.0% during the breeding season, and 15.0% in autumn. The proportion of sparrows in the breeding-season diet was strongly negatively correlated with the distance of the sparrowhawk nest from human habitation (Fig. 8.1). Similar results were obtained in Denmark and Finland. In the former, Frimer (1989) found that sparrows were the most common item (22.1%) in the diet of sparrowhawks in urban areas, and the third most common item (9.2%) in rural areas. In Finland, Solonen (1997) found that the house sparrow was the second most common prey item in rural areas (12.2%), but it was an uncommon item (1.3%) in remote areas. Gotmark and Post (1996) studied the diet of sparrowhawks breeding near Goteborg, Sweden, in 1994 and 1995. They also censused the avian community in the vicinity of each sparrowhawk nest and determined the relative predation risk (RPR) of each of 46 prey species. House and tree sparrows were treated together because of the difficulty of distinguishing between them in the samples of prey. The RPR of a species was calculated by subtracting the rank of its frequency in the diet from its rank of abundance among the species in the area; the RPR is greater than 0 for species with high risk of predation and less than 0 for those with low risk of predation. In both years of the study, he house/tree sparrow had the highest RPR value, 22.5 in 1994 and 17.2 in 1995. Even when the censuses were based on open areas near the nests, which were presumed to be favored hunting grounds, the house/tree sparrow had the highest RPR. Gotmark and Post looked for behavioral correlates of high predation risk and concluded that foraging on the ground in exposed areas was a major factor in increasing the risk of predation by sparrowhawks (see Chapter 6). Fledgling house sparrows are particularly vulnerable to predation during the first weeks of their independence. Fledglings constitute a high proportion of the diet of sparrowhawks during the nestling period of the hawks, the peak period of demand during the breeding season (Frimer 1989; Opdam 1979).

The merlin in the prairie provinces of Canada is an example of a raptor that has established urbanized populations in recent years. Merlins began breeding in urban areas such as Saskatoon, Saskatchewan, and Edmonton, Alberta, in the mid-1960s (James and Smith 1987; Sodhi et al. 1992), and their breeding populations in those and other cities have been growing rapidly (see later discussion). Like the sparrowhawk, the merlin is a bird specialist, with 94.0%–

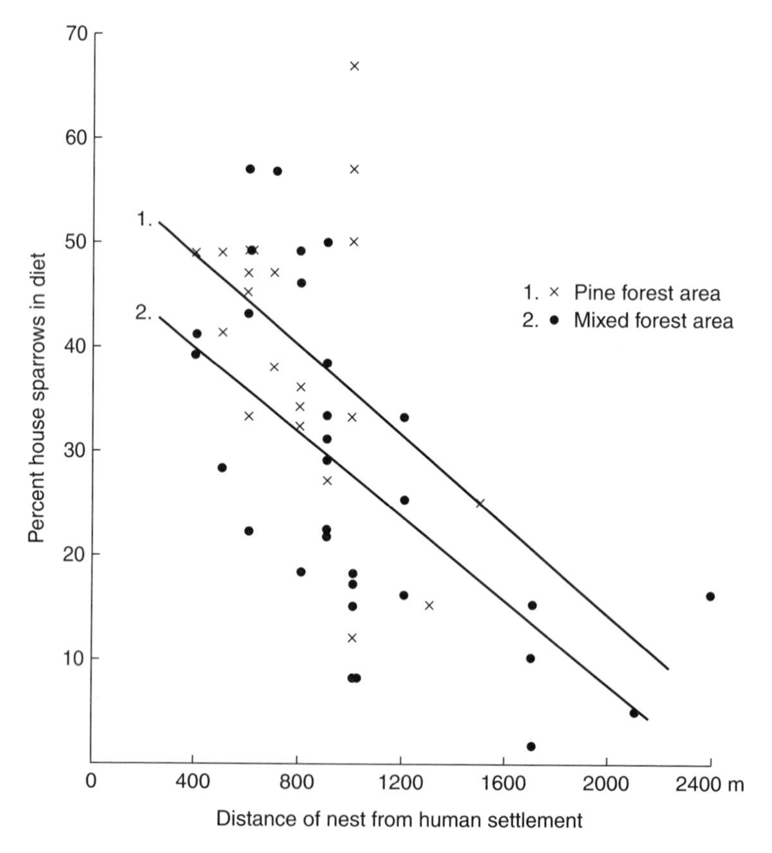

FIGURE 8.1. Relationship between the distance of a Eurasian sparrowhawk nest from human settlement and the percentage of house sparrows in the diet during the breeding season, 1970–1973. Data were collected from a 240 km² area along the German-Dutch border in two habitat types, pine forest (1) and mixed forest (2). The positions of the two regression lines suggest that sparrowhawks nesting in pine forest took proportionately more sparrows than those nesting in mixed forest. From Opdam (1979: Fig. 4).

100% of its prey items being small birds (Becker 1985; Hodson 1978; James and Smith 1987; Oliphant and McTaggart 1977; Warkentin and Oliphant 1990). Studies of the diets of urban merlins have shown that they are virtually house sparrow specialists. Several studies in Saskatoon found that sparrows comprise 65.4%–72.2% of the prey items during the breeding season (Oliphant and McTaggart 1977; Sodhi and Oliphant 1993; Sodhi et al. 1990), and another study found that sparrows represent 72.2% of the merlin's winter diet there (Warkentin and Oliphant 1990). In Edmonton and Fort Saskatchewan, Alberta, sparrows made up 77% of the diet of breeding merlins (James and Smith 1987). Sparrows comprised about 70% of the breeding bird community in residential areas of Edmonton in 1990 (Edgar and Kershaw 1994). Merlins breeding in more natural habitat, such as the shortgrass prairies of southern Alberta and rangeland in southeastern Montana (USA), feed primarily on horned larks (*Eremophila alpestris*) and chestnut-collared longspurs (*Calcarius ornatus*), with sparrows comprising less than 1% of the diet (Becker 1985; Hodson 1978).

Sodhi (1993) used radiotelemetry to obtain the foraging ranges of breeding merlins in Saskatoon and also measured prey density for both house sparrows and all bird species along 1-km transects. Both sparrow density and total bird density showed strong negative correlations with the size of the foraging range of male merlins during both the incubation and the nestling periods (with all but sparrow density during the nestling period being significant at $P < .05$). Male merlins do virtually all of the foraging during the incubation and nestling periods. Female foraging range during the fledgling period showed a significant negative correlation with sparrow density ($P < .05$). Sparrows made up an increasing proportion of the merlin's diet from the incubation period to the fledgling period in both Edmonton and Saskatoon (James and Smith 1987; Oliphant and McTaggart 1977). Sodhi and Oliphant (1993) found that juvenile sparrows were selected as prey in much higher frequency than their occurrence in the population.

Numbers of breeding merlins increased dramatically in Saskatoon between 1971 and 1990, and the breeding density in 1989, 25.4 pairs/km^2, was the highest ever recorded for the species (Sodhi et al. 1992). This increase in numbers and the fact that urban-breeding merlins concentrate so intensively on sparrows mean that local sparrow populations have faced a new source of intense predation. What has happened to house sparrow numbers in Saskatoon since the advent of urban-breeding merlins in the city? If numbers have remained relatively constant or increased, this would represent persuasive evidence that predation is not limiting sparrow numbers. If, on the other hand, numbers have declined, this would provide support for the predator-limitation hypothesis.

The only consistent survey of sparrow numbers in Saskatoon throughout the period of interest is the Christmas Bird Count done annually in December.

Counts have been conducted annually in Saskatoon since at least December 1966 (as far back as I checked), with the center of the 24 km-diameter count circle being in Saskatoon, and with the entire city lying within the census area. Fig. 8.2a presents the estimates of relative abundance of house sparrows in Saskatoon from 1970 to 1998 based on Christmas Bird Count data. The data clearly suggest that sparrow numbers declined during the period of increasing numbers of breeding merlins (1971–1990) and continued to decrease through the 1990s. A similar trend is evident in the Christmas Bird Count data from Edmonton (Fig. 8.2b), where an urban population of merlins has also become established (cf. James and Smith 1987). In Saskatoon, the correlation between the number of sparrows per party-hour and the number of breeding pairs of merlins during the previous summer is marginally insignificant ($r = -0.393$, $N = 20$, $.05 < P < .10$). In a much shorter series of surveys, Sodhi (1992a) censused the breeding birds of the city of Saskatoon in May and June of 1988–1990. Numbers of sparrows declined in both months over the 3-y period, which corresponded to the three peak merlin years (Sodhi et al. 1992). The evidence clearly does not negate the predator-limitation hypothesis, but it provides only weak correlational evidence in favor of it. It appears that urban-breeding merlins may have had a negative effect on sparrow numbers in Saskatoon.

Merlins are not the only falcons that have developed urban breeding populations. The Eurasian kestrel (*Falco tinnunculus*) and the European hobby (*Falco subbuteo*) also breed in urban settings. Studies of the diets of urban-breeding kestrels in Manchester (UK) (Yalden 1980) and of hobbies in Berlin (Fiuczynski and Nethersole-Thompson 1980) found that house sparrows make up a high proportion of the diet in both species. Studies of the diets of other avian predators breeding or wintering in urban areas have also found that they use a large number of house sparrows, even if they are ordinarily mammal specialists. These include the tawny owl (*Strix aluco*) in Rome (Manganaro et al. 1990), in Warsaw, Poland (Grozcynski et al. 1993), and in Toran, Poland (Zalewski 1994), and the long-eared owl (*Asio otus*) in Berlin (Dathe 1988) and Italy (Manganaro 1997). Grozczynski et al. (1993) studied the composition of the diet of tawny owls at 11 sites along a gradient from forest to urban park in Poland and found that the proportion of house sparrows increased significantly as the sites became more urbanized.

Domestic cats are also major predators of sparrows, but it is less clear how they may have contributed to the recent decline in sparrow numbers. Churcher and Lawton (1987) obtained weekly samples of prey items captured by approximately 70 "house" cats in the village of Felnersham, Bedfordshire (UK), for 1 y. House sparrows were also censused during a 10-d period in mid-April, and house sparrow density was estimated at 10.96 sparrows/ha, giving an overall estimate of 340 sparrows in the village. Sparrows made up 16% of the 1090 prey items collected, with mammals comprising more than 50% of the items. Churcher and Lawton estimated that cats took a minimum of 170 sparrows during the year, accounting for at least 50% of the total sparrow mortality in the village.

Møller and Erritzoe (2000) attempted to determine whether the immunocompetence of bird prey was lower than that of nonprey by comparing the spleen mass of birds captured by domestic cats with that of birds killed by accidents (mainly collisions with automobiles or windows). The spleen is a lymphoid organ involved in the production and storage of lymphocytes (see Chapter 9), and its size may therefore represent an indirect measure of an organism's immunocompetence. The study was conducted in Denmark, and 62 house sparrows were among 535 birds that died from either cat predation or accidents. The mean spleen mass of sparrows that were preyed upon by cats was half that of sparrows that were the victims of accidents (0.25 vs. 0.50 g) ($P < .05$). Similar results were obtained for the entire sample of birds, and Møller and Erritzoe concluded that predators prey differentially on parasitized or diseased birds. This result tends to support Errington's hypothesis that predators remove a doomed surplus, and it raises the question of the potential role of parasitism and/or disease in controlling population numbers.

Parasitism and Disease

Disease is a candidate for the cause of population declines such as those observed recently in sparrows. Indeed, there are examples of local disease epidemics leading to significant sparrow mortality (e.g., Cornelius 1969; Stenhouse 1928; Wilson and MacDonald 1967). The number of parasites and disease pathogens that afflict sparrows is truly staggering. The Appendix contains a partial list of these parasites and diseases, and a glance at the list is enough to cause one to wonder how any sparrows survive at all. For some of these organisms, the house sparrow serves as a host for diseases that can infect humans or their domesticated animals (see Chapter 10).

A comprehensive discussion of the pathologies in sparrows caused by the organisms listed in the Appendix is beyond the scope of this book, and indeed would require a book in itself. The fact is, however, that the impact on sparrow populations of most of these organisms are virtually unknown. Indeed, little is known about the impact of parasitism and disease on bird populations in general (see Newton 1998). Therefore, the following account of the effects of parasites and pathogens on sparrow populations is primarily anecdotal.

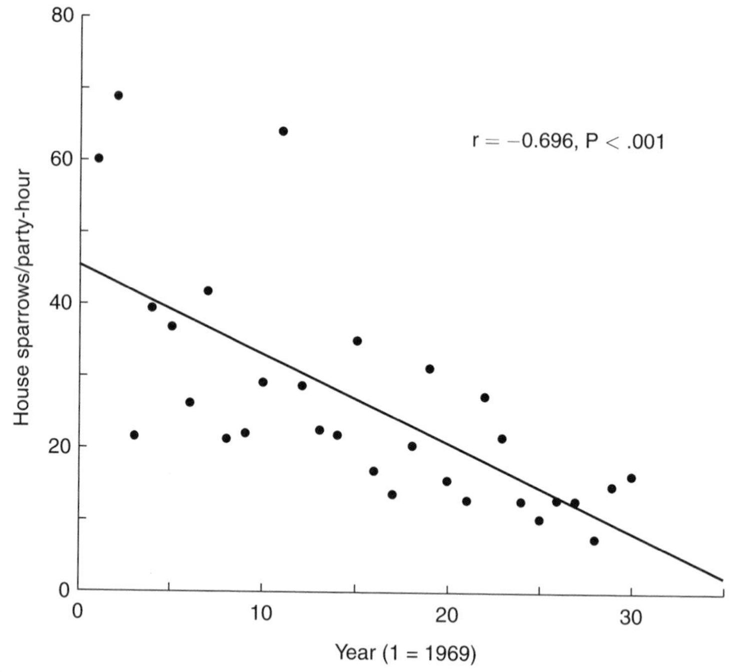

A House sparrow numbers in Saskatoon, 1969–1998

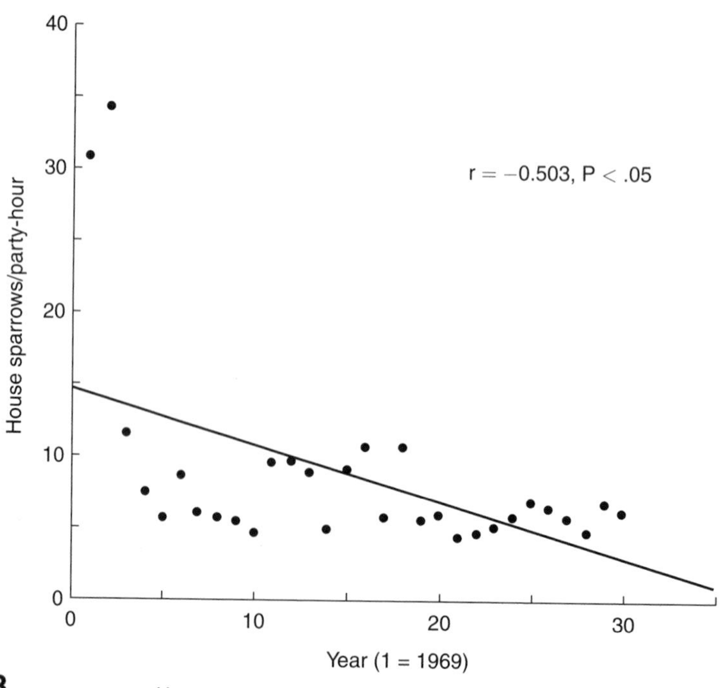

B House sparrow numbers in Edmonton, 1969–1998

Arboviruses

Arboviruses, as anyone knows who has been listening to descriptions of the spread of West Nile virus across North America, are viruses for which the transmitting vector is an arthropod, usually a tick or mosquito. Although the primary hosts of many arboviruses are wild bird species, some of these viruses also cause illness, and sometimes death, in humans or their domestic animals. Because of the medical or economic significance of these latter viruses, it is they about which most is known. Besides the West Nile virus, they include the St. Louis encephalitis (SLE), eastern equine encephalitis (EEE), and western equine encephalitis (WEE) viruses in North America; Venezuelan equine encephalitis virus in South America; and the Japanese B encephalitis and Israel turkey meningoencephalitis viruses in Asia.

Birds appear to be the principal reservoir hosts for many of the arboviruses (Stamm 1963). In sparrows, many of the viruses cause a mild viremia lasting 2–4 d, but there are also accounts of arbovirus epizootics resulting in high rates of mortality. Burton et al. (1966) found a high rate of WEE virus in nestlings in Saskatchewan (Canada) in 1962 and 1963 (during a local epidemic of WEE in horses and humans) and reported that there were many dead nestlings. Many sick and dead sparrows were also reported in the vicinity of an epidemic of EEE in Michigan (USA) (McLean et al. 1985). Williams et al. (1971) reported that most of the sparrows captured in Maryland (USA) in 1968 that tested positive for EEE virus died shortly after being tested. In a laboratory study involving the experimental inoculation of four adult sparrows with the EEE virus, all died within 17–48 h after inoculation (Kissling et al. 1954). Despite these

---

FIGURE 8.2. Change in number of house sparrows per party-hour on Christmas Bird Counts at Saskatoon, Saskatchewan (A), and Edmonton, Alberta (B), 1969–1998. The Saskatoon count area is a 24-km diameter circle with its center on Lome Avenue in Saskatoon, and it comprises 30% city streets and parks, 25% aspen bluffs, 20% fields, 15% river valleys, and 10% game and forest farms. The Edmonton count area is centered on the University of Alberta Farm and consists of 40% woodlands, 30% farmlands, 20% riversides, and 10% residential areas. One significant change in the counting procedure occurred in the late 1970s, when birds counted at birdfeeders were included in the total number of birds counted. Because this would have a major effect on counts of species such as the house sparrow that make extensive use of birdfeeders, the total number of hours spent observing at feeders was added to the total party-hours reported for each count to obtain the denominator (party-hours) for values on the ordinate. Birds counted at birdfeeders were first included in the Edmonton count in 1978 and in the Saskatoon count in 1980. Data from *American Birds*, 1970–1999.

accounts of sparrow mortality due to arbovirus infection, the effects of arboviruses on sparrow populations are completely unknown.

There is evidence that antibodies against specific arboviruses are transmitted transovarially from immune females to the embryo, but there is little evidence that these antibodies confer protection to nestling sparrows, and they may actually increase the intensity and duration of viremia. Sooter et al. (1954) found that 2 of 89 sparrow nestlings in Colorado (USA) possessed antibodies against the WEE virus. Holden, Francy et al. (1973) challenged nestling sparrows being reared in captivity under mosquito-free conditions with WEE virus. Nestlings from both WEE-immune parents and nonimmune parents developed viremias after inoculation with the virus, and all died. There was some evidence that the time to death in nestlings from immune parents was longer than that in nestlings from nonimmune parents, but the susceptibility of both groups to infection and the failure to identify WEE-specific antibodies in serum drawn from 1- and 2-day-old nestlings of immune parents led these authors to conclude that there was little evidence for transovarian transmission of WEE antibody. Ludwig et al. (1986) inoculated adult females with SLE virus to induce development of high SLE antibody titers in the birds, and then bred them in a mosquito-proof aviary. Although all nestlings from SLE-infected females had neutralizing antibodies against SLE virus, these nestlings developed viremias that were more intense and of longer duration than those of control nestlings when they were inoculated with the SLE virus. Ludwig et al. concluded that such viremic enhancement due to the presence of transovarially transmitted antibody might actually play a role both in the amplification of the virus and a more rapid rate of its dissemination.

One problem that has plagued researchers working on arboviruses is the question of how they overwinter and are reintroduced each year into a local bird population. One possibility is that the viruses overwinter in hibernating adults of the vector mosquitoes (Cockburn et al. 1957; Reeves et al. 1958). Some have suggested that the viruses overwinter in swamps (Stamm 1958) or in bird mites (Cockburn et al. 1957; Reeves et al. 1947), and others that they are reintroduced annually by birds returning from the tropics (Cockburn et al. 1957). Recently, evidence has accumulated suggesting that at least some of the viruses overwinter as latent infections in resident bird species such as the house sparrow. Gruwell et al. (2000) found evidence for the recrudescence of SLE virus in house sparrows and house finches in southern California (USA). Support for this hypothesis also comes from the study of the pattern of spread of West Nile virus in the northeastern United States, which appears to be consistent with dispersal by resident bird species, including particularly the house sparrow (Rappole and Hubalek 2003). If the latter is truly the case, it has possible implications both for efforts to control sparrow populations and for the use of the house sparrow as an early-warning sentinel in areas that are especially prone to arbovirus epidemics in humans or domestic animals (see Chapter 10).

Other Viruses

Avian pox is caused by *Poxvirus avium* and is characterized by wart-like growths on the legs or unfeathered skin, often around the eyes. Sparrows have been implicated in the spread of avian pox in Hawaii (USA), where it may have played a significant role in the extinction or threatened extinction of several Hawaiian endemics (van Riper et al. 2002) (see also Chapter 10).

Bacteria and Fungi

Numerous bacterial pathogens, many of which are also pathogens of humans and domestic animals, have been identified in sparrows (see Appendix). One of the most frequently reported bacterial diseases of sparrows is salmonellosis, which is caused by bacteria of the genus *Salmonella* (Bouvier 1968; Macdonald and Cornelius 1969; Pinowska et al. 1976; Schnetter et al. 1968). High mortality rates among infected individuals have been noted (Bowes 1990; Cornelius 1969; Macdonald and Cornelius 1969), but the ultimate effect on the population dynamics of sparrows is unknown. Indeed, even the incidence of *Salmonella* infection in sparrow populations is virtually unknown. Pinowska et al. (1976) tested wintering flocks of sparrows on 36 dairy farms in Poland and found that birds on nine of the farms were positive for *Salmonella* infection, with an average infection rate of 12.9% among the 464 sparrows tested on the nine farms. Infection rates on individual farms ranged from 2% to 50%. Symptoms of salmonellosis in sparrows have been described by several workers. Cornelius (1969) reported that infected individuals stood huddled and hunched, pecked feebly, and could often be picked up by hand. Autopsy of infected individuals reveals that they have enlarged livers and spleens, caseous nodules in the crop, ulceration of the lower digestive tract, and poor general body condition (Bowes 1990; Macdonald and Cornelius 1969; Wilson and Macdonald 1967). Salmonellosis epidemics in sparrows apparently occur primarily during the winter or early spring (Cornelius 1969). However, *Salmonella* and other microbial infections (including *Escherichia coli* and *Candida* sp.) apparently also contribute to embryonic or nestling mortality in the species (Kozlowski et al. 1991; Malyszko et al. 1991; Pinowski et al. 1988, 1994).

The house sparrow may serve as a reservoir for *Salmonella* infections that affect humans or domestic animals (Bowes 1990; Dozsa 1962/63; Schnetter et al. 1968; see also Chapter 10). This may also be true for other bacteria that infect sparrows. One such is *Mycoplasma gallisepticum*, the cause, along with other *Mycoplasma*, of chronic respiratory disease or "avian pneumonia" when it occupies the upper respiratory tract (Kleven and Fletcher 1983). Sparrows have been implicated in the spread of chronic respiratory disease to poultry (Jain et al. 1971). A strain of *M. gallisepticum* is also responsible for a form of avian conjunctivitis that has recently spread through house finch populations in eastern North America and that also infects numerous other bird species, including house sparrows (Hartup et al. 2001).

Other bacteria that have been implicated in embryonic mortality or nestling mortality or morbidity in sparrows include *E. coli*, *Staphylococcus auretus*, *Staphylococcus epidermidis*, *Proteus vulgaris*, *Streptococcus* spp., *Bacillus* spp. and *Yersinia* sp. (Pinowski, Barkowska, and Pinowska 1995; Pinowski et al. 1988, 1994; Stewart and Rambo 2000). Microbial infection of eggs presumably occurs either in the ovary or in the oviduct during egg formation. As described in Chapter 4, Stewart and Rambo (2000) cultured material obtained with sterile cotton swabs from the cloacae of mated pairs of sparrows from Kentucky (USA), and identified microbial groups in the samples. They also found that the ovaries of three additional females were positive for both gram-negative bacteria and *Yersinia* sp., and two were also positive for *Salmonella* sp. and fungi..

Protozoans

Numerous protozoans are known to infect sparrows (see Appendix). In particular, these include parasites that cause avian malaria, toxoplasmosis, and coccidiosis. Chickens, turkeys, and other domesticated birds are often susceptible to these diseases, and sparrows have sometimes been implicated in their spread among domesticated birds (see Chapter 10). Although several protozoans are known to cause diseases in sparrows, the impacts of protozoan parasitism and disease on sparrow populations are poorly understood.

Avian malaria in sparrows is caused by blood parasites of the genus *Plasmodium* (Box 1966; Manwell 1957) and is characterized by anemia and extreme water loss. Rates of *Plasmodium* infection vary seasonally, with higher rates during the summer months at temperate latitudes in North America (Micks 1949) and during the winter months in subtropical India (Singh et al. 1951). *Plasmodium* infection may also be latent, because Hart (1949) found that individuals testing negative at the time of capture subsequently tested positive after being held in captivity in the laboratory.

Toxoplasmosis is caused by intracellular blood parasites, *Toxoplasma* and related protozoans such as *Lankesterella* (= *Atoxoplasma*) (Lainson 1959). Coccidiosis is caused by intestinal parasites of the genus *Isospora* (Boughton 1937a), which occur in virtually all adult sparrows (Boughton 1937b). In Poland, Kruszewicz (1991) found that 11-d-old nestlings with coccidiosis weighed

8.8% less than nestlings that did not have coccidiosis ($P < .02$). Box (1967) reported that caged sparrows exposed to *Isospora* were more susceptible to *Atoxoplasma* parasitism and death. Lainson (1959) reported that in England (UK), juvenile sparrows that were heavily infected with *Atoxoplasma* often died when taken into captivity.

The latter observation may be due to the effect of protozoan infection on the ability of a bird to mount an immune response. Two studies in Spain found that sparrows infected by *Haemoproteus* spp. (intracellular parasites of the red blood cell) had a significantly lowered T-cell response when challenged with phytohemagglutinin (Gonzalez, Sorci, Møller et al. 1999; Navarro et al. 2003) (see also Chapter 9). The intensity of infection had no effect on levels of circulating immunoglobulins, however (Gonzalez, Sorci, Møller et al. 1999).

Other Endoparasites

Sparrows are parasitized by a number of platyhelminths and nematodes, most of which are intestinal tract parasites (see Appendix). Little is known about the effects of these parasites on sparrows, and virtually nothing is known about their impacts on sparrow populations.

Ectoparasites

The most prominent groups of ectoparasites of sparrows are the mites and ticks (Arachnida: Acari) and the feather lice (Insecta: Mallophaga). The Appendix includes organisms found in sparrow nests, as well as those actually isolated from the birds, so some of those listed may have no negative effects on the birds themselves. Indeed, the large number of species identified from sparrow nests (e.g., Bhattacharyya 1990, 1995) suggests that sparrow nests support a community of organisms that includes not only parasites of the birds but also various saprophytic organisms and predators and parasites of these two groups.

Two recent studies have examined the effects of hematophagous nest mites on the growth and condition of nestlings. Weddle (2000) found that the mean parasite load per brood of the mite *Pellonyssus reedi* was negatively correlated with the mean mass of nestlings in the brood on nestling day 11 when both brood size and season were considered as covariates. Individual infestation rate was not related to position in the body mass hierarchy of the brood for broods of either 3 or 4 nestlings. Total parasite load varied significantly with brood size ($P = .04$), with smaller broods tending to have greater parasite loads than larger broods. There was no significant relationship between brood size and total parasite load, however. In a similar study in Hungary, Szabo et al. (2002) examined the effects of two hematophagous mites (*P. reedi* and *Ornithonyssus sylviarum*) and found no relationship between mean parasite load of a brood and size or mass of nestlings on day 12. However, mean parasite load per brood did correlate significantly with number of thrombocytes in the blood ($r = 0.571$, $P = .003$) and with hematocrit ($r = -0.558$, $P = .004$). There was also a tendency for the number of heterophils in the blood to increase with increasing mite infestation.

Behnke et al. (1999) studied patterns of feather mite infestation by *Proctophyllus truncatus* in sparrows captured during April of 1991–1997 in Portugal. Left wing primaries and secondaries were examined, and mite infestation was categorized on a scale of 0–3, with 0 representing no mites observed and 3 representing heavy infestation (>50% of the feathers covered by mites). Microscopic examination of one primary to validate the visual scaling method resulted in a highly significant relationship between the assigned infestation category and the mite count ($r_s = 0.684$, $P < .001$). In some of the years, the right wing primaries and secondaries and the rectrices were also examined, and the results further validated the scaling technique, although there was a small but significant bias toward heavier infestation on the right wing than on the left. Of the 156 sparrows examined during the 7 y, 45.5% were categorized as 0, 28.8% as 1, 17.3% as 2, and 8.3% as 3.

In Massachusetts (USA), the most frequently occurring ectoparasites of sparrows are the mites *P. truncatus* (occurring on 64.7% of individuals, with a median of 11 mites/individual), *Paraneonyssus hirsti* (23.5%), and *Dermanyssus americanus* (14.7%), and the louse *Brueelia cyclothorax* (29.4%, with a median of 20 lice/individual) (Brown and Wilson 1975). Comparisons of the ectoparasitic faunas of introduced populations such as those in North America and New Zealand with those of the source populations have shown an effect, similar to the genetic founder effect, that results in the loss of ectoparasitic species in the introduced populations (Brown and Wilson 1975; Paterson et al. 1999). Paterson et al. referred to this phenomenon as "missing the boat."

Morsy et al. (1999) compared the ectoparasitic mites of sparrows from two governorates in Egypt (Sharkia and Qalyobia). A total of 25 species of mites was identified from the two samples, 23 of which occurred in both samples. There was a highly significant correlation between the incidence rates of the 25 species from the two samples ($r = 0.803$, $P < .001$; calculated from data in Table 2 of Morsy et al. 1999).

There is apparently seasonal variation in the incidence of some ectoparasites. Phillis (1972) found that 100% of the sparrow nests collected during the breeding season in Michigan contained feather mites of the genus *Dermanyssus*, whereas only 66% of the nests in which sparrows were roosting during the winter contained mites.

Brood Parasitism

The house sparrow appears to be little affected by interspecific brood parasites. Although there are records of parasitism by a number of cuckoo and cowbird species (Appendix), the frequency of such parasitism is low. Two reasons for this low rate of parasitism may be the species' close commensal relationship with human settlements (see Chapter 10) and its nest sites and construction (see Chapter 4). Many brood-parasitic species are found primarily in relatively undisturbed habitats, and few, with the exception of the brown-headed cowbird (*Molothrus ater*) in North America, are common in human settlements. The most common nest sites of house sparrows are holes and crevices in buildings or trees, sites that are often inaccessible to the larger brood parasites. Nests that are located in the open branches of trees are domed, which also reduces their accessibility to the parasites (Friedmann 1929).

The house sparrow may also be an unsuitable host species. Although there are records of sparrows successfully fledging parasitic offspring (Friedmann 1929), experimental evidence suggests that this may be a rare occurrence. In an experiment in which seven brown-headed cowbird eggs were transferred into house sparrow nests, six of the eggs hatched, but none of the young fledged successfully (Eastzer et al. 1980). Similarly, two eggs of the shiny cowbird (*Molothrus bonariensis*) transferred into sparrow nests in Argentina hatched, but neither nestling fledged successfully (Mason 1986). This suggests that the house sparrow is an "acceptor species," but that there may be some incompatibility in nestling food or feeding that results in poor fledging success in the parasitic young.

Some of the few records of parasitism may also be spurious. Scott (1988) reported observing both a female house sparrow and a chipping sparrow (*Spizella passerina*) feeding a recently fledged brown-headed cowbird in Ontario (Canada). The chipping sparrow is a common host of the cowbird in Ontario and was the probable host of this fledgling. Scott suggested, therefore, that observations of feeding of fledgling cowbirds by a species do not necessarily imply that the species has raised the parasite.

Interspecific Competition

Interspecific competition is the utilization by individuals of two or more species of resources that are limiting one or more of the species. Limiting resources are those resources which, when utilized by increasing numbers of individuals of a species or its competitors, result in either an increased death rate or a decreased birth rate in the population of the limited species. Two types of competition are often recognized: exploitation competition, in which the resource is "used up" by one individual, making it unavailable to others; and interference competition, in which the behavior of an individual makes a resource inaccessible to other individuals. The first studies attempting to explain the declining populations of the house sparrow in the northeastern United States explored the potential role of competition with the house finch in precipitating the decline (e.g., Bosakowski 1986; Kricher 1983). The many introductions of sparrows throughout the world (see Chapter 1) have brought the house sparrow into contact with many native species, and competition with sparrows has been invoked in explaining population declines in some of these species.

The house finch was apparently deliberately released by a bird dealer on Long Island, New York (USA), sometime around 1940, after the dealer learned that traffic in house finches was no longer legal (Elliott and Arbib 1953). The species became established on Long Island during the 1940s, and then began an explosive range expansion that has taken it throughout the eastern and central regions of the United States and Canada (Hill 1993). Like the house sparrow, the house finch is a sedentary commensal of humans that feeds primarily on seeds and nests in a variety of places, including small ornamental conifers, crevices in buildings, and nest-boxes.

Kricher (1983) was the first to suggest that competition due to increases in numbers of house finches might be contributing to a decline in sparrow numbers in the northeastern United States. Using Christmas Bird Count data from five northeastern states, he found that house finch numbers (= birds per party-hour) increased significantly from 1958 to 1979 in all five states, and house sparrow numbers decreased significantly over the same period. There were also significant inverse correlations between numbers of sparrows and numbers of finches in all five states. Kricher did note that sparrows also showed significant declines in numbers in four southern states that the house finch had not yet colonized, but he concluded that, because the declines were greater in the northeastern states, interspecific competition between the two species was partially responsible for the decrease in numbers of sparrows there. Bosakowski (1986) obtained similar results when he analyzed Christmas Bird Count data for 23 y (1959–1981) from four contiguous, nonoverlapping count areas in northeastern New Jersey.

Wootton (1987) used both Christmas Bird Count data (1968–1983) and Breeding Bird Survey data (1966–1979) from the northeastern United States to test for the potential effects of interspecific competition with the house finch on sparrow numbers. House finch numbers grew exponentially during the period 1966–1979, and their range expanded as well, particularly westward and southward. Sparrow densities decreased significantly with increasing house finch densities in both summer and winter ($P < .005$). Sparrow densities in summer also decreased significantly outside the house finch's range, by about 1% per year, but the negative relationship between sparrow density and finch density was still significant when this decrease was accounted for ($P < .005$). Wootton also examined the possible effects of 76 weather variables on changes in sparrow density and found that only three (January precipitation, annual precipitation, and August average temperature) had significant effects (all negative at $P < .05$). This is about the number of significant results expected by chance at $P = .05$ and suggests no particular weather pattern that might be affecting sparrow numbers. Analyses of gross changes in habitat also showed no relationship to changes in sparrow density. Wootton concluded that further support for the role of the house finch in causing a decline in sparrow populations depended on the fulfillment of three criteria: (1) identification of a shared limiting resource (probably food or nest sites), (2) documentation of a shift in the use of that resource by sparrows inside versus outside the house finch's range, and (3) observation of that shift occurring as the house finch occupies new areas.

Bennett (1990) also used Christmas Bird Count data to test the house finch competition hypothesis. He examined the question at three scales—local, regional, and continental. At the continental scale, he compared changes in sparrow numbers in the central United States (allopatry, where the house finch had not yet colonized) with those in the northeastern United States (recent sympatry) and in the western United States (old sympatry—dating to the early 1900s, when the region was colonized by house sparrows; see Chapter 1). Sparrow numbers had declined in all three regions (Fig. 8.3), but the rate of decline was significantly greater in the regions of sympatry, whereas the rates of decline did not differ between the regions of new and old sympatry. Bennett concluded that interspecific competition with the house finch was not the primary cause of the decrease in sparrow numbers, but that it did contribute to accelerating the rate of decline.

The role of competition with the house finch in the recent declines in sparrow numbers in North America remains unresolved. The declines observed in Saskatoon and Edmonton (Canada), discussed earlier in conjunction with predation by urban-breeding merlins, further illustrate this point. The declines in sparrow numbers in those two cities began long before the arrival of the house finch, which was first noted on the Christmas Bird Count in Saskatoon in 1994 and in Edmonton in 1995. One of the principal problems with invoking house finch competition with the declines is the identification of a limiting resource that is more effectively exploited by house finches than by sparrows. Although there seems to be considerable overlap in both the food and nest sites used by the two species, the house sparrow has been found to be socially dominant to the house finch at winter foraging sites (Giesbrech and Ankney 1998;

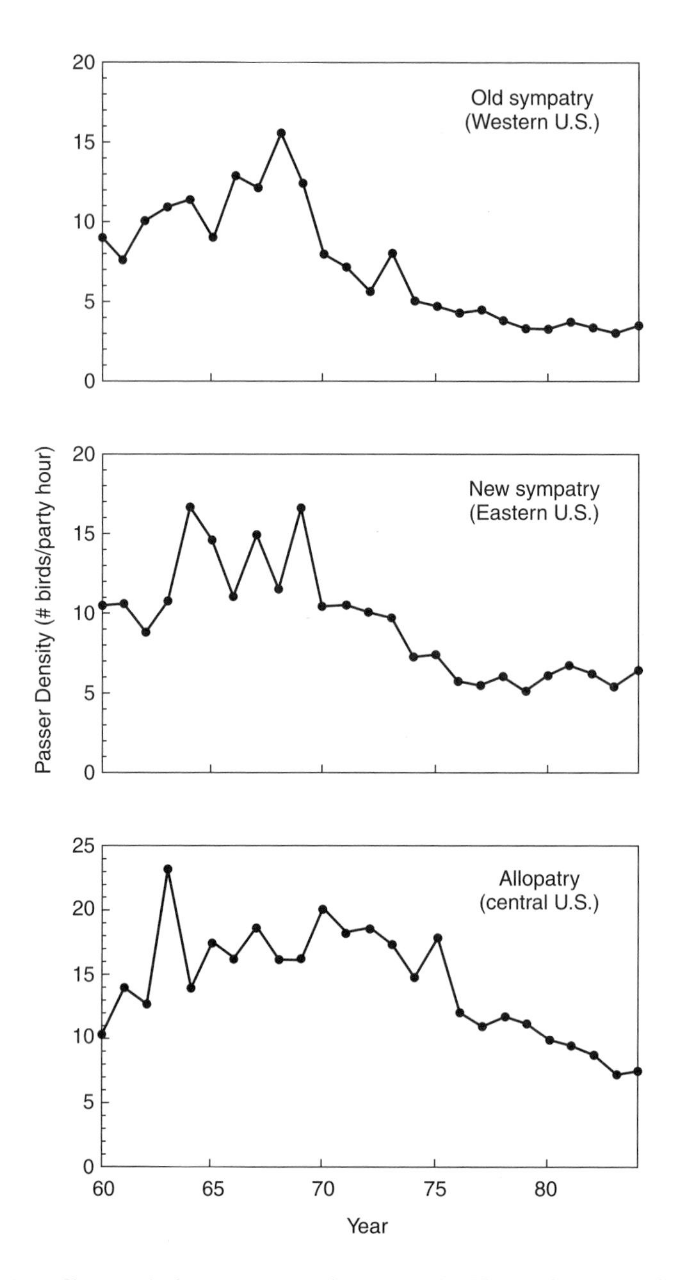

FIGURE 8.3. Changes in house sparrow density in the United States in relation to house finches. Christmas Bird Count data from areas surrounding five cities in each of three regions of the United States were compiled for the 25-y period 1960–1984. The three regions represent old sympatry between house sparrows and house finches (dating to the colonization of the western states by sparrows in late 19th and early 20th centuries), new sympatry in the eastern states (where finches have recently

Kalinoski 1975) and to regularly disrupt house finch breeding attempts and usurp finch nest sites (Bergtold 1913; Evenden 1957). It is therefore difficult to see how competition with the house finch would contribute significantly to the decline in numbers of sparrows.

Competition with feral pigeons (*Columba livia*) has also been suggested as a cause for the sharp decline in urban populations of sparrows in western Europe (cf. Summers-Smith 1999). Although the pigeon remains as a conspicuous presence in many urban areas where the house sparrow has declined, there is virtually no evidence to support the suggestion that competition between the two species has resulted in the decline. Because both species regularly feed on human refuse in urban areas, food would be the principal candidate for a limiting resource that could be mediating a competitive interaction between the two species. Other potential competitors for food in urban areas include several crows (*Corvus* spp.) that have become increasingly urbanized, particularly in North America in the last few decades.

Because sparrows nest primarily in cavities, including holes in trees and nestboxes (see Chapter 4), they have often been thought to compete for nest sites with other secondary cavity nesters. Direct competition for cavity nest sites is often intense and sometimes fatal. Usurpation of cavity nest sites is frequent, particularly at temperate latitudes (Lindell 1996). Competition for nest sites occurs both in the native range of the house sparrow and in much of the area into which it has spread as the result of human introduction. This competition has often been invoked to explain population declines in other secondary cavity nesters, particularly in the latter areas. The evidence for the control of populations of secondary cavity nesters by the availability of cavities has been reviewed by Newton (1998).

Competition for cavity nest sites can be either intraspecific or interspecific. Removal experiments are one technique for detecting whether the availability of nest sites (or other spatial resources) is limiting the numbers of breeding individuals in a population (see Newton 1992). In New Jersey (USA), Clark

---

colonized following their introduction in the early 1940s—see text for details), and allopatry in the central states (where only sparrows occurred during the period). Data are plotted as average number of sparrows observed per party-hour in the five cities from each region. Sparrow numbers declined in all three regions during the period, but the rate of decline in the region of allopatry was significantly different from that in sympatry (two regions combined) ($P < .01$) (Bennett 1990). There was no difference in the rate of decline of sparrows in the two regions of sympatry ($P > .1$). From Bennett (1990: Figs. 3, 5, and 6), reproduced with permission of University of Chicago Press.

(1903) removed five female sparrows from a single nest box between late March and late May, and in North Carolina (USA), Stewart (1973) trapped and killed 14 sparrows (8 females) and 56 European starlings (19 females) from a single nest-box. These anecdotal observations suggest the existence of nonbreeding floaters in the population. Anderson (1990) removed female sparrows after completion of a clutch from 34 nest sites (in nest-boxes and other cavities) in a stable breeding colony in Missouri (USA). By late-May, a total of 56 females had been collected from the 34 sites, with as many as 4 from one site. Although 2-y-old or older females made up half of the first occupants at the nest sites, first-year birds and/or new arrivals to the colony made up the large majority of the replacement individuals. Anderson concluded that there was a shortage of optimal nest sites in this colony, resulting in the existence of a significant number of nonbreeding floaters in the population. Coleman (1974) provided nest-boxes for European starlings in New Zealand and found that the provision of nest sites resulted in a significant increase in the starling population, and possibly also the house sparrow population (sparrows tended to use the sites after the starlings had completed nesting).

One species with which the house sparrow competes for nest sites is the congeneric tree sparrow. The two species are broadly sympatric in their native ranges (including Europe and northern Asia) and show considerable overlap in their ecological niches, being seed-eating commensals of humans. Where the house sparrow is absent in China and other parts of southeastern Asia, the tree sparrow completely fills the "house sparrow niche" and is the common sparrow of urban streets (Summers-Smith 1963). Both species are also cavity or crevice nesters, which creates the possibility of interspecific competition for nest sites. The two species were the most common species using nest-boxes in 21 urban parks in Warsaw (Poland) (Kozlowski 1992), and the correlation coefficient that I calculated for the numbers of pairs breeding in the 20 parks in which either or both species bred was positive ($r = 0.137$, NS), which indicates that there was no strong habitat segregation between the two species.

Anderson (1978) found that the larger house sparrow was more successful at obtaining nest sites that were available to both species in Missouri, where the tree sparrow has also been introduced. He found, however, that nest-boxes with entrance hole diameters of 29 mm were too small for the house sparrow to enter, although large enough for the tree sparrow. Provision of many nest-boxes with the 29-mm entrance resulted in a doubling of the local breeding population of tree sparrows in each of two successive years, 1969–1970, and 1970–1971. The entrance holes of 21 of the nest-boxes were enlarged after the 1971 breeding season, and within 2 y the house sparrow had taken possession of all but 4 of the nest-boxes previously occupied by tree sparrows, and the breeding population of tree sparrows had declined.

Cordero and Rodriguez-Teijeiro (1990) studied the nest-site characteristics of the two species in 59 breeding colonies in both rural and suburban habitats in northeastern Spain (where both species are native). Although there is broad overlap between the species in nest-site type, there is a significant difference in choice of sites ($P < .001$), with house sparrows showing a strong preference for cavities under roof tiles (nearly 50%), in trunks or branches of trees (nearly 20%), in walls (about 15%), and in rafters and joints (about 10%), whereas tree sparrows favored nest-boxes and cavities in walls (about 35% each) and other sites (about 20%). Overall similarity in nest site use was 0.69, and niche breadths of the two species were similar (house sparrow, 3.52; tree sparrow, 3.72). An analysis of the height distribution of nests in cavities in walls led the authors to conclude that the house sparrow had an effect on the nest height selection of the tree sparrow. In a follow-up study examining habitat, food, and nest-site variables at 14 rural breeding colonies (12 of which were occupied by both species), Cordero (1993) found that house sparrow abundance was related primarily to the total number of nest-site cavities available, whereas tree sparrow numbers were related to the total number of nest-boxes at the sites and to the amount of tree cover. Codero concluded that numbers of both species were limited primarily by the number of available nest sites.

Cordero and Senar (1990, 1994) performed two experiments to further explore the competitive relationship between house sparrows and tree sparrows at nest sites in Spain. In the first, Cordero and Senar (1990) placed a model of a house sparrow, a tree sparrow, or a Eurasian siskin *(Carduelis spinus)* on the roof of a nest-box occupied by either a house sparrow or a tree sparrow. The siskin model was ignored by both species, but the models of both house sparrows and tree sparrows elicited aggressive behavior from both species. The house sparrow was more likely to physically attack the model than the tree sparrow was; the tree sparrow tended to defend against the perceived intruder by blocking the nest-site entrance. In the second experiment, Cordero and Senar (1994) studied aggressive encounters between the two species at specially designed nest-boxes (containing two units, each with an entrance). Entrance holes were 13 cm apart. In some cases, both had diameters of 30 mm, and in others, one of the entrances had a diameter of 28 mm (too small for the house sparrow). House sparrows initiated aggression more often than expected by chance and won all 235 interactions that they initiated. Tree sparrows initiated fewer attacks than expected by chance and won only 41% of the 22 they initiated. One tree sparrow was killed by the fierceness of a house sparrow attack. Simultaneous occupancy of both compartments in the double boxes occurred only in cases where the two species each occupied one compartment (no dual occupancies by one species occurred).

There is also broad overlap in the feeding niches of the house sparrow and tree sparrow during the breeding season. In mixed breeding colonies of the two species in Missouri, there was almost complete overlap in both the types and the sizes of foods fed to nestlings (Anderson 1978, 1980). There was some indirect evidence of competition for the nestling food resource, because both species showed an increase in fledging success during that part of the breeding season when there was less overlap in the nestling diets (Anderson 1978). In villages in Poland, there was not as much overlap in the types of foods in the nestling diets of the two species, with house sparrow diets in different villages resembling each other more than they resembled the diets of tree sparrows in the same village (Anderson 1984). In Romania, however, Ion (1971) found that there was broad overlap in the nestling diets. I calculated a coefficient of overlap ($\hat{C}_\lambda$) of 0.894 for the nestling diets of the two species there.

There is also considerable overlap in the diets of the two species during the nonbreeding season in Europe. During a 3-y study in Denmark, Hammer (1948) found that most of the diets of adults of both species were composed of cereal grains and weed and grass seeds. I calculated $\hat{C}_\lambda$ coefficients of 0.971 for cereal grains and 0.863 for weed and grass seeds. The house sparrow took proportionately more cereal grains than did the tree sparrow, however, with about 34% of the dietary items of adult house sparrows being cereal grains, compared with only about 6% of the tree sparrow items. Keil (1972) found a similar pattern in Germany, with house sparrow diets consisting of 48% cereal grains and 38% weed seeds, whereas tree sparrow diets consisted of 9.6% cereals and 79% weed seeds. During the winter in Spain, Sanchez-Aguado (1986) found that the diet of the house sparrow consisted of 82.2% cereal grains (by biomass) and 13.2% weed seeds, whereas that of the tree sparrow was 6.7% cereals and 92.6% weed seeds. These results suggest that there is some segregation of the diets of the two species outside the breeding season. Other studies in Germany (Grun 1975) and Romania (Ion 1992) have found broad overlap in the diets of adults of the two species, however.

There is also considerable overlap in the winter diets of the house sparrow and Spanish sparrow in Spain. In a region of irrigated farmland in west-central Spain, Alonso (1986) found that the winter diets of the two species had a $\hat{C}_\lambda$ of 0.88. The house sparrow diet contained more cereal grains (41% by mass) than did the Spanish sparrow diet (13%). There was also habitat segregation between the two species, with more than 60% of house sparrow foraging flocks feeding within 10 m of the nearest building, while more than 95% of Spanish sparrow flocks fed 100 m or more from the nearest building.

The house sparrow has also been reported to have negative effects on the breeding success of other secondary cavity nesters, both in its native range and in areas where it has been introduced. Its frequent usurpation of nests of other species is described in Chapter 4.

Barba and Gil-Delgado (1990) examined competitive interactions among four species—house sparrows, tree sparrows, great tits (*Parus major*), and black rats (*Rattus rattus*)—utilizing nest-boxes in an orange grove in Spain. There was a significant decrease in great tit occupancy of the nest-boxes over the 3 y of the study ($P < .05$), and a significant increase in both house sparrow and black rat occupancy. House sparrows destroyed two tit nests containing eggs, and a dead tit was found in a house sparrow box containing two eggs. Black rats preyed on nests of the great tit and tree sparrow. Great tit density did not change significantly in a control grove during the 3 y, and Barba and Gil-Delgado concluded that the tits were excluded from access to nest-boxes by house sparrows and black rats. Results of a subsequent study supported this conclusion, because great tits frequently nested in nest-boxes at which house sparrows were discouraged from breeding by repeated removal of their nests (Barba et al. 1995).

In North America, competition with the house sparrow for nest sites has particularly affected bluebirds (*Sialia* spp.) (Gowaty 1984; Miller 1970; Radunzel et al. 1997). During a 6-y study of eastern bluebirds (*Sialia sialis*) breeding in nest-boxes in South Carolina (USA), 28 adult bluebirds were found dead in nest-boxes (Gowaty 1984). The heads and throats of 20 of these birds showed evidence of violent trauma, with 18 having skull fractures and the feathers of the crown removed. Sparrows were present at 18 of the 20 nest-boxes on the visits preceding the deaths, and they were also present at 18 of the 20 after the deaths. On three occasions, sparrows built nests that incorporated a dead bluebird body in the nest. On another occasion, a male sparrow was observed entering a bluebird nest-box repeatedly, and after 20 min the observer interrupted the event and found four of five nestlings with head injuries. Eventually all five nestlings died. Gowaty concluded that there was strong circumstantial evidence that sparrows regularly attack and kill eastern bluebirds at their nest sites. In a 27-y nesting study of eastern bluebirds in Wisconsin (USA), 30% of nest failures were attributed to sparrows (Radunzel et al. 1997). Willner et al. (1983) performed a discriminant function analysis on 24 habitat variables at nest-boxes along two eastern bluebird trails in Maryland and found that the highest habitat overlap among species using the boxes was that for bluebirds and house sparrows.

One species that may outcompete the house sparrow for nest sites is the European starling. In New Zealand, the addition of nest-boxes to an area resulted in increased numbers of both starlings and sparrows, but sparrows tended to use the nest-boxes only after the starlings had completed their breeding (Coleman 1974). During a 10-y study of nest-box occupancy in Turkey involving nest-boxes with either small (30–34 mm) or large (50–80 mm) hole diameters, house sparrows primarily used boxes with small hole diameters, and used only a few of those with large diameters, which were used primarily by starlings (Kiziroglu et al. 1987). In a cottonwood grove in Nevada (USA), sparrows were excluded from breeding during a 5-y period in which starling density was high (Weitzel 1988). These results suggest that starlings are able to outcompete house sparrows for nest sites when the sites are suitable for both species.

In conclusion, there appears to be little evidence that the observed population declines in the house sparrow can be attributed to interspecific competition. In fact, most of the studies of competition involving the house sparrow have focused on its negative effects on other species, particularly its success in excluding other species from nesting cavities. Although changes in agricultural practices in western Europe and North America may have resulted in decreased food availability, particularly during the winter (see later discussion), there is little evidence that competition for this resource has had a particularly negative effect on the house sparrow.

Changes in Agricultural Practices

Several suggestions have been made regarding how changes in agricultural practices in the past 3–4 decades may have had a negative effect on the food supply of sparrows, or more generally on that of the guild of ground-foraging seedeaters to which the house sparrow belongs (Chamberlain et al. 1999; Easterbrook 1999). Increased mechanization in farming has not only virtually eliminated the draft horse from farms in western Europe and North America during the past century, but it has also resulted in larger, more specialized farms, with a loss of small, general farms. Changes in planting practices such as the sharp increases in the United Kingdom in land area planted to winter cereals (wheat and barley), with concomitant declines in the area of spring plantings, means that winter foraging areas may be greatly reduced. Increasingly efficient harvesting machines may also mean much less grain spillage or wastage. Increased use of chemical pesticides and herbicides may also have an indirect effect on sparrows by reducing the amount of insect food available for feeding young (see Chapter 6) and the number of weed seeds available during the winter. The implementation of an European Union hygiene law requiring grain storage to be bird-proof may have had a direct effect on the food supply of sparrows (Easterbrook 1999). Vaisanan and Hilden (1993) cited the decrease in the number of dairy farms in Finland as a possible cause of the sparrow decline there. The scale of these changes in agricultural practices makes it difficult to test the validity of these suggestions as to the causes of the population declines in house sparrows and other ground-foraging seedeaters.

One recent study in the United Kingdom attempted to determine the effects of agricultural practices on the breeding densities of several seed-eating and granivorous species (including the house sparrow) (Robinson et al. 2001). Data from the 1998 Breeding Bird Survey were used to determine the effects at three spatial scales, local (within the 1 km^2 sample plots), regional (in the 121 km^2 area immediately surrounding the sample plots), and landscape (dominant agricultural habitat type in the county containing the sample plot). At the local scale, where the effect of the presence of arable land (annually tilled fields) in predominantly pastoral areas was examined, several of the species (but not the house sparrow) showed a significantly positive effect of the presence of arable land. In the regional analysis, however, sparrow breeding density was significantly affected by the amount of arable land. Sparrow breeding density in sample plots with little arable land was strongly positively affected if 60% of the region consisted of arable land, whereas density in such plots was low if only 20% of the region consisted of arable land. At the landscape scale, agricultural practices had no significant effect on sparrow breeding density. One reason for the lack of clear relationships between the farming landscape and sparrow breeding density may be the virtually obligate relationship between sparrows and human settlement during the breeding season. In another study, using data from the 1996 Breeding Bird Survey, Gregory (1999) found that sparrow breeding density was positively affected only by the presence of human dwellings in suburban, urban, and rural areas. This suggests that if changes in agricultural practices are responsible for the sharp decline in sparrow numbers, the impacts of those changes cut across broad categories of agricultural landscapes.

In an attempt to test the hypothesis that decreases in food supply caused by changes in agricultural practices have contributed to declines in sparrow numbers, Hole et al. (2002) supplied supplemental food to sparrows at two of four farmland colonies in Oxfordshire, England, during the winters of 1998–1999 and 1999–2000. One of the colonies, which had shown a sharp population decline since 1971, had a significant increase in overwinter survival when it was provided with supplemental food.

Survival Rates

Although individual birds may live for several years, with the longest recorded lifespan in the wild being 13 y 4 mo (Dexter 1959; Klimkiewicz and Futcher 1987), the average life expectancy of an adult sparrow is quite short. Estimates of the survival rates of sparrows are based on records of recaptures, resightings (in the case of color-banded individuals), and recoveries of banded birds. There are numerous models for estimating survival rates based on these data, some of which take into account the probability of recapture or resighting to correct for imperfect sampling. Dyer et al. (1977) analyzed several datasets from both local studies and national banding programs using three different models and reported annual survival rates ranging from 0.315 (Missouri) to 0.629 (England). More recent estimates of adult survival rate based on national banding programs included a value of 0.55 for birds banded in Great Britain between 1966 and 1978 (Dobson 1990) and a value of 0.55 for Australia (Yom-Tov et al. 1992). The highest estimate of survival rate (0.678) has been reported for the migratory subspecies *Passer domesticus bactrianus* in Kazahkstan (Gavrilov et al. 1995). These rates are for adult sparrows and do not pertain to the survival prospects of post-fledging juveniles.

Recent studies of sparrow populations living on several small islands in northern Norway (approximately 66°30'N) have provided the most comprehensive knowledge of survivorship patterns in sparrow populations (Ringsby et al. 1998, 1999, 2002; Sæther et al. 1999). The populations on the islands are essentially closed, meaning that there is very little immigration onto or emigration from the islands (but see later discussion), and most of the individuals on the islands were color-banded throughout the period of the study. Nestlings were marked on five islands in the summers of 1993 and 1994, and intense surveys of the populations in November and April permitted estimation of the survival rates of first-year birds from fledging to November (= summer survival) and from November to April (= winter survival). In both years, summer survival rates were significantly lower than winter rates, with the former being about 25%, and the latter about 80%–100% (Ringsby et al. 1998:Fig. 1). The overall first-year survival rates (fledging to April) did not differ significantly in the 2 y, being 0.23 in 1993 and 0.21 in 1994. An extension of the study for two more years resulted in an overall average first-year survival rate of 0.239 for the five islands (Ringsby et al. 1999).

Ringsby et al. (2002) examined the effects of several variables, including island, year, time of hatching, and weather, on the ultimate recruitment into the breeding population (i.e., survival from hatching to 1 April of the next year) on the five islands. Survival from fledging to recruitment was affected by island, year and hatching date, with fledglings from later broods having a higher first-year survival rate. Weather had a significant effect on survival from hatching to fledging (see Chapter 4) but did not affect survivorship from fledging to recruitment. Sæther et al. (1999) presented data on both first-year and adult survival rates on four of the five islands over a 3-y period. First-year survival rates ranged from 0.077 to 0.579 among the islands and years, and adult survival rates ranged from 0.494 to 0.826. I calculated the correlation coefficient between first-year and adult survival rates, and it was positive and highly significant ($r = 0.702$, $P < .001$).

Ringsby et al. (1999) also monitored adult survivorship on eight islands (including the five on which juvenile survivorship was monitored). Adult survivorship on individual islands varied from 0.434 to 0.976 and averaged 0.61 across all eight islands for the 4 y of the study. I calculated the correlation coefficient between juvenile survival and adult survival on the five islands for which there were survivorship estimates for both groups for the 4 y. Although the relationship was not significant ($r = 0.286$, $P > .10$), it was positive. This, together with the significant positive correlation described earlier for a subset of the islands and years, suggests that survivorship in juveniles and adults tended to vary together. This finding is contrary to expectation for closed, stable populations if they are resource-limited, because one would then expect to see a negative relationship between adult and juvenile survivorship. In fact, the sizes of the breeding populations on four of the islands varied by 21%–85% among observation years (Sæther et al. 1999), suggesting considerable lability in population sizes on the islands.

Adult survivorship may vary among habitats. In an intensive 5-y study of survivorship in two color-banded populations in Holland, Heij and Moeliker (1990) found that adult survivorship was higher in the suburban population in all 5 y than in the rural population. The average adult survivorship for the entire period was 0.657 in the suburban population and only 0.433 in the rural population. Survivorship of juveniles was also higher in the suburban population, with the average for the 5 y being 0.414, compared with 0.324 in the rural population. Estimates of juvenile survivorship were based on individuals trapped after becoming independent, rather than on birds banded as nestlings, and therefore are not comparable to the juvenile survival rates from Norway discussed earlier, because they do not account for mortality that occurs after fledging but before independence.

There are also seasonal differences in survival rates. In Great Britain, mortality rates among adults are highest during the breeding season (Summers-Smith 1957), and this is also true for the suburban population in Holland (Heij and Moellker 1990). In the rural population in Holland, however, mortality rates were highest in the fall and winter months (October–January). In an analysis of recoveries of banded birds from throughout North America, Beimborn (1967) reported that the annual survival rate was 0.45 and that the highest mortality rates were during the winter. Senar and Copete (1995) used the Jolly-Seber model to obtain monthly estimates of the annual survival rate in Spain and examined the effects of several weather variables on survival rate. The analysis was based on 2021 sparrows captured in mist nets between 1980 and 1987, and the weather variables examined were (1) number of days below freezing, (2) winter precipitation (October–March), (3) summer precipitation (April–September), (4) sum of mean monthly low temperatures for winter, and

(5) sum of mean summer high temperatures for summer. Survival rate estimates were generally between 0.35 and 0.50+, except for the winter of 1984–1985, which had 12 freezing days in January, with the annual survival rate estimate for the month being about 0.15. Although the studies to date are too few to draw definitive conclusions, it appears that mortality rates in the house sparrow are highest during the breeding season in areas with maritime climates, whereas in regions with more continental climates, mortality rates are highest in winter. Survival rates may also differ between the sexes, a topic that is discussed in the next section.

Table 8.3 presents a life table for the house sparrow in northern Michigan, incorporating data on first-year survival of fledglings from the islands in Norway (Ringsby et al. 1998, 1999). The net replacement rate (R_o) calculated for the population, 1.279, indicates that the population is experiencing rapid growth. The breeding populations on the three dairy farms that served as the study area for the research (see Anderson 1998) did not show such growth, however. This suggests that these sites, which apparently provided abundant food resources throughout the year for sparrows, represent source populations

TABLE 8.3. Life Table for the House Sparrow in Northern Michigan (USA)

x[a]	N_x[b]	D_x[c]	e_x[d]	l_x[e]	b_x[f]	$l_x b_x$[g]
0	1000	867	0.70	1.000	0.00	0.000
1	133	87	1.04	0.133	6.05	0.805
2	46	30	1.04	0.046	6.67	0.307
3	16	10	1.06	0.016	6.67	0.107
4	6	4	1.00	0.006	6.67	0.040
5	2	1	1.00	0.002	6.67	0.013
6	1	1	0.50	0.001	6.67	0.007

Data on age-specific reproductive rates are based on mean clutch size (4.96), mean rates of hatching (0.717) and fledging (0.779) success, and number of broods attempted by marked females (first-year, 2.44; 2-y-old and older, 2.69) in northern Michigan (Anderson 1994, 1998); an average adult survival rate of 0.344 based on 209 breeding females banded between 1987 and 1995 in northern Michigan (Anderson, unpublished data); and the assumption that the primary sex ratio is 1:1 (see Table 8.4). Survival of fledglings to the next breeding season is considered to be 0.239 based on data from islands in Norway (Ringsby et al. 1999).

[a]Age at beginning of interval in years.
[b]Number of females/1000 alive at age x.
[c]Number of deaths/1000 between ages x and x + 1.
[d]Age-specific life expectancy (years).
[e]Proportion of females living to age x.
[f]Age-specific birth rate (number of female offspring produced).
[g]$R_o = \Sigma\, l_x b_x$ = net replacement rate = 1.279.

for other, smaller breeding colonies occupying less favorable sites. An alternative explanation is that the estimate of R_0 may be inflated. Use of the observed first-year survival rate from the Norwegian islands may represent an overestimate of first-year survival in the harsh winter climate of northern Michigan. Annual adult survivorship in Michigan is considerably lower than that on the islands, and first-year survival may also be lower. A second source of error is the assumption that all first-year females breed. In Missouri, Anderson (1990) found that there were numerous nonbreeding floaters in a house sparrow colony, and that first-year females were overrepresented among the floaters. This explanation seems less likely, however, because breeding density in Michigan was considerably lower than in Missouri.

The Sex Ratio Conundrum

Most large-scale, apparently random samples of house sparrow populations indicate that there is a male-biased sex ratio in the species. This male bias was evident in the sample of storm-immobilized sparrows in Bumpus's (1899) classic study on selection (see Chapter 2). Table 8.4 lists 28 such studies, 27 of which showed an excess of males over females, with the difference from an expected 1:1 ratio being significant at the 5% level in 15 of them. Of the remaining studies in which the sex ratio did not differ significantly from unity, 12 of the 13 showed an excess of males, a difference that was itself significant (binomial test, $P = .004$). If an overall mean is calculated for the 28 samples, each taken as an independent estimate of the sex ratio in sparrows, the average percentage of males is 54.76 (SE = 0.71). Is this apparent male bias in the sex ratio real, or is it an artifact of sampling bias? If it is real, is it caused by a difference in the primary sex ratio of the species or by postfertilization differences in survivorship between the sexes?

There are several sources of potential bias in the methods used to sample sparrow populations. One of the more obvious is that counts of individuals feeding at a feeding station (e.g., Nichols 1934) can lead to pseudoreplication due to repeated counting of the same individuals (Beimborn 1976). The two sexes may also show different susceptibilities to baited traps or poisons (Beimborn 1976). Storm deaths of sparrows may also show sex bias if the two sexes have slightly different roosting positions in the communal roosts where most such mortality has been observed. Nichols (1934) suggested that males may forage over a wider area than females, leading to biased estimates of the sex ratio (see also Haukioja and Reponen 1969). Even taking all of these potential sources of sampling bias into account, however, it is difficult to escape the conclusion that the sex ratio is slightly imbalanced in favor of males in most sparrow populations.

TABLE 8.4. Sex Ratio in the House Sparrow

Location	Time of Year	Sampling Method	N	% Male	P	Source
Rhode Island (USA)	February	Storm	136	64.00	.026	1
New York (USA)	Nov–May	Observing feeding	7754	54.72	< .001	2
New Zealand	February	Adult storm death	1150	58.09	< .001	3
Kansas (USA)	July	Storm (juveniles)	129	54.26	.37	4
Germany	Autumn (?)	Poisoning	600	52.17	.31	5
Germany	February	Poisoning	520	46.35	.42 6	6
Azores (Portugal)	Year-round	Unspecified (adults)	305	55.08	.085	7
Azores	Year-round	Mist-netting (adults)	1243	57.52	< .001	8
Kansas (USA)	August	Storm (adults)	553	61.66	< .001	9
Germany	Dec–March	Poisoning	20,931	53.17	< .001	10
Germany	March	Poisoning	1276	50.94	.52	11
Central USA	Oct–Nov	Mist-netting	753	59.36	< .001	12
India	Year-round	Unspecified	501	51.10	.65	13
New Zealand	Year-round	Unspecified	791	56.01	.001	14
Germany	January	Poisoning	4579	53.20	< .001	15
Germany	Dec–Jan	Poisoning	1171	50.38	.82	16
Alberta (Canada)	Spring–Autumn	Mist-netting	187	56.68	.08	17
Germany	Nov–Dec	Poisoning (adults)	473	52.43	.32	18
Missouri (USA)	Oct–Mar	Mist-netting	381	56.17	.02	19
Holland	Year-round	Baited traps	756	53.00	.10	20
Wisconsin (USA)	Year-round	Baited traps	1471	52.82	.032	21
Minnesota (USA)	Year-round	Mist-netting	858	50.58	.76	21
North America	Oct–Nov	Mist-netting	2522	56.34	< .001	22
Holland	Nov–Feb	Observed feeding	4281	55.73	< .001	23
Holland	Feb–Apr	Poisoning	2056	55.01	< .001	23
Oklahoma (USA)	Fall/Winter	Observed/trapped/dead	1260	51.4	.32	24
California (USA)	Year-round	Trapping	442	54.3	.08	25
Kentucky (USA)	Winter	Trapping	162	59.9	< .02	26

Listed are the sex ratios (expressed as percentages of samples that are male) of several large samples of house sparrows observed, collected, or captured in an apparently unbiased way. The two-tailed probability that the sex ratio deviates from the expected 1:1 was calculated using the normal approximation of the binomial distribution. Samples were selected to represent a wide range of seasons, geographic areas, and methods of estimating sex ratio.

Sources: 1, Bumpus (1899); 2, Nichols (1934); 3, Dawson (1967); 4, Ely and Bowman (1969); 5, Nordmeyer et al. (1972); 6, Geiler (1959); 7, Medeiros (1995); 8, Medeiros (1997b); 9, Johnston (1967b); 10, Piechocki (1954); 11, Niethammer (1953); 12, Packard (1967a); 13, Rana and Idris (1986); 14, Baker (1980); 15, Scherner (1974); 16, Grimm (1954); 17, McGillivray and Murphy (1984); 18, Löhrl and Bohringer (1957); 19, Anderson (unpublished data); 20, Heij and Moeliker (1990); 21, Beimborn (1976); 22, Selander and Johnston (1967); 23, Tinbergen (1946); 24, North (1968); 25, McClure (1962); 26, Westneat et al. (2002).

The primary sex ratio is close to unity. Mitchell and Hayes (1973) reported that the male:female ratio was 48:52 among 42 fledglings reared in a captive breeding program in Texas (USA). Cordero et al. (2000) determined the sex of eggs from 34 clutches in Spain by testing for the presence of the *CHDI-W* gene (see Chapter 2) and reported that the sex ratio did not differ from unity (binomial test, $P = 1.0$). Schifferli (1980b) examined the sex ratio in 133 sparrow broods in England. In 36 broods in which all eggs hatched and all young survived until collection, there were 76 males (51.01%) and 73 females ($z = 0.164$, $P = .87$), and for the entire sample there were 202 males (50.88%) and 195 females ($z = 0.30$, $P = .76$). In a similar study, Bosenberg (1958) determined the sex ratio among 397 nestlings from 109 broods in Germany. The nestlings varied in age from less than 5 d to 18 d, and there were 214 males (53.90%) and 183 females ($z = 1.506$, $P = .13$). The sex ratio in the nestlings younger than 5 d old was 1.15:1 in favor of males, which was close to the overall ratio of 1.17:1 for the entire sample. A study in Kentucky in which the sex of 10-d-old nestlings was determined by analysis of the *CHDI* gene found that 53% of the 1162 young sampled over a 5-y period were males, a difference that approached significance ($P = .06$) (Westneat et al. 2002). One of the difficulties in attempting to identify small differences in the sex ratio at a particular stage in the life cycle is the size of the sample required to obtain significance. Schifferli (1980b), for instance, noted that a sample size of 2438 would be necessary to detect (with $P < .05$) a 52:48 sex ratio if it existed.

In the Kentucky study, Westneat et al. (2002) examined several variables that might influence sex-ratio allocation by female sparrows. These included several measures of female quality (e.g., timing of breeding, clutch size, nestling weight, female condition), as well as the badge size (see Chapter 5) of the pair male, that might influence sex-ratio allocation by the females. Although there were significant differences among years in the sex ratio of the offspring, none of the variables associated with female or male quality had a detectable effect on the sex ratio. There was, however, a significant positive relationship between hatching success and the proportion of male offspring.

Assuming, therefore, that the primary sex ratio is close to unity, the observed male bias in most populations of the species must be caused by differential mortality between the sexes. There is no evidence of differential mortality before fledging (but see further discussion later in this chapter), and some studies have indicated that the sex ratio of fledglings and juveniles is closer to unity than the sex ratio of adults. Medeiros (1997b), for instance, found a male:female ratio of 1:1.02 in a sample of 512 juveniles in the Azores, compared with a ratio of 1.35:1 among adults in the same populations (see Table 8.4). This would suggest that the mortality rate of females must be higher than that of males. Studies of post-fledging and adult mortality rates in sparrows have failed to detect such a difference. In fact, most studies have shown that female mortality rates are actually lower than those of males, although not significantly so. Dobson (1987) used British Trust for Ornithology banding records from 1966 to 1978 to study the mortality rate of sparrows in Great Britain. For the entire sample, the annual mortality rate of males was 0.47, and that of females was 0.43, although the difference was not significant. For birds that lived at least 1 y after banding, however, males had a significantly higher mortality rate than females (0.48 vs. 0.39, $P = .04$). Summers-Smith (1957) also reported a higher mortality rate among males than among females in an intensively studied population in England. Sodhi and Oliphant (1993) found that urban merlins took a significantly higher percentage of adult males than adult females in Saskatoon ($P < .001$). The proportion of males in the population was not recorded, however, so the proportion of males taken by merlins may not have deviated significantly from their proportion in the population.

Recent theory has suggested that, during periods of severe resource limitation, individuals are favored that can produce offspring of the sex with the greater variance in individual fitness. For the house sparrow, it is presumably the male that has greater variance in fitness, with some males achieving successful extra-pair fertilizations and others being polygynous (see Chapter 4). Because the female is the heterogametic sex in birds, the ovum is the gamete that is determinative of the sex of the individual, but there is as yet no satisfactory mechanism explaining how females might allocate sex ratio in their offspring. One study reported that male eggs are larger than female eggs (Cordero et al. 2000) (see Chapter 4), suggesting that there may be such a mechanism in sparrows.

Movements

Sparrows are permanent residents throughout most of their range, and they are also remarkably sedentary, returning to the same breeding colony year after year. They forage close to their breeding and roosting sites during most of the year, except that during late summer and early autumn large flocks may feed in ripening grain fields a few kilometers from their breeding or roosting sites. Natal dispersal distances are also quite short, with birds usually breeding within a few kilometers of their natal colony. The rapid spread of sparrows in many localities where they have been introduced (see Chapter 1) suggests, however, that dispersal distances in colonizing individuals may be quite long. Two subspecies, *P. d. bactrianus* and *P. d. parkini*, are migratory (see Chapter 1) and show regular seasonal movements between their breeding grounds and their wintering grounds. The extent of migration of other subspecies is somewhat uncertain.

The principal technique for studying bird movements involves the recapture, recovery, or, in the case of color-banded birds, resighting of individual birds. Unfortunately, this technique has inherent biases that complicate the interpretation of the resulting data, and differences among studies make it difficult to compare results across studies. An example of the former is the much greater probability of recovering a bird at its banding location than at other locations, due to the increased intensity of study that usually occurs at the banding location. An example of the latter is the differences in size of the core study area in different studies, with concomitant differences in the distances over which recoveries are more likely to be made. Despite these difficulties, a considerable amount is known about patterns of movement in sparrows.

Foraging Distance

Foraging distance changes seasonally in sparrows, tending to be short during the breeding season and winter but often longer during late summer and autumn. It also may differ between the sexes, with males having greater average foraging distances than females.

Weaver (1939b) used colored tail plumes to study the daily movements of sparrows in New York and found that they regularly moved 2.4–3.2 km. Cheke (1972) reported that sparrows from both urban and rural populations in England fed primarily within a 0.5–km radius of their breeding or roosting sites, except for visits to regular feeding sites up to 2 km away. North (1968, 1973) studied the movements of sparrows in late summer and autumn from urban/suburban breeding and roosting sites to grain fields outside a city in Oklahoma (USA). Most of the individuals moved distances of 3.2 km or less, although some regularly visited fields up to 5 km from their roosting sites. In three colonies in Holland, Heij and Moeliker (1990) found that the maximum distance moved was 0.6 km, with maximum distances for most individuals being shorter during the breeding season and during the winter. During the winter in Finland, Haukioja and Reponen (1969) found that sparrows regularly moved between two sites 1.5 km apart.

Males may forage over a larger area than females. Haukioja and Reponen (1969) found, for instance, that males were significantly more likely than females to move between the two sites mentioned. In Illinois (USA), Will (1969) reported that 16 of 19 sparrows of known sex that he recaptured more than 30 m from their banding sites were males (binomial test, $P < .005$), with the average distance for males (1.12 km) being greater than that for females (0.75 km). In the Azores, however, Medeiros (1998) found that there was no significant difference between the sexes in the average distance between banding and recapture sites (196 adult males, 0.30 km; 142 adult females, 0.41 km;

39 juvenile males, 0.37 km; 30 juvenile females, 0.46 km). There was also no significant difference based on age.

Natal Dispersal

Natal dispersal is the movement of an individual from its birth site to its breeding site. In sparrows, such dispersal normally takes place when birds first begin to seek potential nest sites in the months immediately after the breeding season during which they were born (Summers-Smith 1958). Once established as breeding birds, sparrows rarely disperse to another colony (Summers-Smith 1963). In one study in Germany, only 7 (5.6%) of 124 adults moved 2 km or farther, whereas 18 (17.3%) of 104 juveniles moved that distance (Rademacher 1951). Only 1 (2%) of 50 adults moved between two colonies located 0.55 km apart in the Czech Republic, whereas 3 (11.5%) of 26 first-year birds moved (Balat 1977). In a larger study involving seven breeding colonies in Kansas (USA), only 1 (6.7%) of 15 adults moved between colonies, while 41 (36.6%) of 112 first-year birds moved (Fleischer et al. 1984).

Dispersal distance has important implications for the genetic structure of the population, as well as for the potential rate of range expansion. Short-range dispersal has been studied using recapture or recovery data from birds banded as nestlings during multiyear population studies with study areas covering a few square kilometers (e.g., Fleischer et al. 1984; Lowther 1979a). A long-term population study on an archipelago in Norway has provided data on intermediate-range dispersal (Altwegg et al. 2000). Estimates of long-range dispersal, however, are generally based on banding returns from national banding programs (e.g., Waddington and Cockrem 1987).

One study of local natal dispersal was done at nine breeding colonies in a rural area of approximately 12 km^2 in Kansas (Fleischer et al. 1984; Lowther 1979a). About 3360 nestling sparrows were banded during the 4 y of the study, and recaptures at least 90 d after banding at a breeding colony other than the natal colony were considered to represent instances of natal dispersal. Possible dispersal distances among colonies ranged from 0.6 to 4.7 km. Females were more likely to disperse than males, with 26 of 48 recaptured females but only 19 of 70 males dispersing ($P < .05$). The average dispersal distance of males (1.98 km) was actually greater than that of females (1.68 km), however, due to the fact that variance in male dispersal distance was significantly greater than that of females ($F_{15,25} = 3.93$, $P < .02$) (calculated from data in Lowther 1979a:Fig. 2). Neither fledging date nor nestling mass at 7 d differed significantly between dispersers and nondispersers (Lowther 1979a). Fleischer et al. (1984) also found that although PCI (an indicator of overall body size) did not differ significantly between dispersers and nondispersers in either males or females, it was significantly negatively correlated with dispersal distance in females ($r = -0.42$, $P < .05$). They also found that heterozygosity at four polymorphic isozyme loci (see Chapter 2) of dispersing sparrows (0.43) was significantly greater than that of nondispersers (0.07) ($P < .025$).

Unpublished data from another long-term study, in Michigan (see Anderson 1994), gave similar results. In this case, however, dispersal distance was based on the measured distance between the natal nest site and a nest site at which an individual was actually breeding. The three colonies in the study were separated by distances ranging from 1.55 to 6.08 km. Females were more likely to leave their natal colony than males, with 15 of 27 females but only 5 of 14 males moving to another colony. Natal dispersal distances of females varied from 6 m to 5.55 km (mean = 1.15 km), and that of males varied from 5 m to 1.85 km (mean = 0.69 km). The geometric mean for females was 0.29 km, and that for males was 0.17 km.

The Norwegian study was done over a 6-y period on an archipelago of 14 islands off the coast of northern Norway (Altwegg et al. 2000). The islands are located 5–40 km from the Norwegian coast and are separated by distances of 2–20 km. Dispersal was defined as movement of an individual from its natal island to another island to breed; individuals that remained on their natal island were considered nondispersers. Females were more likely than males to disperse ($P < .01$), and most dispersers (79 of 81) moved during their first year. Of the 38 dispersers for whom the time of dispersal was known, only 3 moved before October of their birth year; the remaining 35 moved during the winter and early spring after their birth. Among females, the probability of dispersing was not related to fledgling mass or condition, or to hatching date or natal clutch size. Among males, however, the probability of dispersal decreased with hatching date, and the smallest members of a brood were more likely to disperse than the largest ($P < .05$ in both cases). Adult survivorship rates of both sexes were higher for dispersers than for resident birds on all 10 islands for which adult survivorship of resident birds and arriving dispersers could be compared, suggesting that successful dispersers are more fit than residents.

Compilations of recovery data from large-scale banding programs provide information on long-range dispersal. Paradis et al. (1998) used banding records from Great Britain from 1909 to 1994 to estimate natal dispersal distance in sparrows. Natal dispersal distance was determined for 531 sparrows that were banded as nestlings or fledglings and were recovered during a subsequent breeding period; the arithmetic mean was 1.7 km, and the geometric mean was 0.206 km.

Other studies reporting dispersal distances in sparrows, most of which were presumably first-year birds, include those of Preiser (1957) and Przygodda (1960) in Germany and Jenni and Schaffner (1984) in Switzerland. Preiser found that only 23 (14.0%) of 164 birds moved more than 4 km (although one individual moved more than 500 km), and Przygodda found that only 1 (2.6%) of 39 birds moved more than 3 km (that bird moving 60 km). In Switzerland, however, 25 (53.2%) of 47 individuals moved more than 4 km (Jenni and Schaffner 1984). Some of the Swiss sparrows may have been undergoing altitudinal migration (see later discussion).

The conclusions to be drawn from these data are clearcut. Dispersal distance in sparrows is generally very short, with most individuals settling to breed within 2 km of their natal colony. Females are more likely to disperse than males. A small percentage of individuals, however, disperse much greater distances.

Migration

Migration is the seasonal, two-way movement of individuals between their breeding area and a wintering area. Although most sparrow populations are permanent residents, regular seasonal migration occurs in two subspecies, *P. d. bactrianus* and *P. d. parkini* (see Chapter 1). The former, which has received considerable study, migrates more than 1000 km between its breeding grounds in central Asia and its wintering grounds in Pakistan and India (Dolnik 1972), whereas the latter has a more altitudinal migration in the Himalayas. Studies of *bactrianus* indicate that it has physiological adaptations similar to those of other migratory passerines. Limited migration also apparently occurs in some populations of the nominate race, *P. d. domesticus*, in both Europe and North America, but the extent to which this represents true two-way, seasonal movement is uncertain.

Gistsov and Gavrilov (1984) reported the results of a 9-y study of both spring and fall migration of *P. d. bactrianus* through Chokpakskii Pass in Kazakhstan (elevation 1200 m). They captured and banded 76,115 sparrows during the study, of which 88 were recaptured in subsequent years. Individuals tended to use the same route from year to year, and the timing of an individual's migration between years was similar, with 70.5% being recaptured in another year within seven calendar days of their initial capture date. Males migrated an average of 5 d earlier than females, and adults tended to migrate earlier than juveniles in the fall. Migration through the pass was diurnal, and the rate of migration, based on 2-h counts of migrants during the morning, was dependent on wind direction and precipitation. Birds usually flew during calm or against headwinds and ceased flying with tailwinds or during rain. Migratory flights tended to occur in waves (Gistsov and Gavrilov 1984), with flock sizes varying from 1000 to 10,000 birds (Dolnik 1972).

Dolnik and Gavrilov (1975) maintained individuals of both *P. d. bactrianus* and the local resident *P. d. domesticus* in the laboratory under identical conditions (20°C and ambient photoperiod) in Kazakhstan. The annual cycle of body mass in *domesticus* was typical for resident birds, with peaks in mass during the winter and at the beginning of molt (see Chapter 9). In addition to these peaks, however, *bactrianus* showed two more pronounced peaks at the times of fall and spring migration, with masses at those times being about 10% greater than throughout most of the rest of the year. This indicates that hyperphagia and premigratory fattening occur in this migratory subspecies, just as they do in other migratory passerines. In both spring and fall, *bactrianus* individuals had total body fat of about 3.4 g, compared with the normal level of just over 1 g. Dolnik and Blyumental (1967) reported that migrating *bactrianus* individuals in Tadjikistan had an average fat content of 3.80 g, whereas individuals that had just completed the prebasic molt had an average of 1.03 g. Dolnik and Gavrilov (1975) also found that there was increased locomotor activity in laboratory *bactrianus* during the periods of fat deposition in both spring and fall. A Kramer box was used to determine the orientation of hops by the caged birds, and *bactrianus* showed a definite tendency to orient in a northeasterly direction in the spring, whereas *domesticus* showed no orientational preference.

The evidence for seasonal migration by *P. d. domesticus* in Europe and North America is sketchy. Flocks of sparrows have been observed moving along coastlines during the autumn in both the United Kingdom (Lack 1954b) and North America (Broun 1972). During an 8-y study at the Tauvo Observatory on the Finnish coast, peak spring numbers were observed from 21 April to 25 April; in the fall, numbers peaked between 23 September and 17 October (Ojanen and Tynjala 1983). Many more individuals were observed in the fall than in the spring at this observatory. Although most of the evidence for migration comes from northern areas, there is also evidence of migration in more southerly regions. A sparrow arrived on a ship 120 km from the coast in the Mediterranean Sea (de Wavrin 1991), and the colonization of the Dry Tortugas by sparrows was attributed to northward migration from populations in the West Indies (Woolfenden and Robertson 1975). Sparrows have also occasionally been killed during the autumn at a television tower in Florida (USA) (Crawford 1974; Stoddard and Norris 1967), which suggests that they may also migrate nocturnally. There is also some evidence of alpine migration, with Jenni and Schaffer (1984) reporting on seasonal movements of sparrows in Switzerland. The bulk of the records of apparent migration in *domesticus* are from the autumn, and it is unclear whether they represent one leg of true two-way migration or simply southerly-biased dispersal. The repeated captures of a few individuals on the island of Helgoland (Germany) provides some evidence for true two-way migration in Europe (Vaux 1962). Kruger (1944) also recorded both fall and spring migrants in Denmark. Seasonal altitudinal migration has also been reported in the Alps (Bezzel 1985b).

Homing and Orientation

The return of an individual to its capture site from a remote location to which it has been moved is known as homing. Some bird species have remarkable homing abilities, and some of these species (particularly the homing pigeon) have been the subjects of intensive research, especially regarding the external cues and physiological mechanisms that enable them to orient or navigate successfully. House sparrows that are permanent residents do not home from great distances, and, consequently, little work has been done on orientation or navigation in the species. As noted earlier, individuals of the migratory *bactrianus* subspecies tend to orient their movements in a northeasterly direction when held in a Kramer box during the spring (Dolnik and Gavrilov 1975).

In New York, sparrows displaced 16.1, 32.2, or 48.3 km from their winter foraging sites regularly returned, but no sparrows returned from more distant locations (Weaver 1939b). A single female taken from her nest in Poland failed to return from a release location 13 km away (Wojtusiak et al. 1946). In England, all adult sparrows returned from 14 km, some from farther distances, but no first-year birds returned from more than 3.5 km (Cheke 1972). In Texas, 6 of 112 sparrows captured during November and maintained in the laboratory for 47–78 d before release returned to their sites of capture 14.6–35.6 km from the release site (Mitchell and Hughes 1972). In New Zealand, several sparrows accidentally released from a laboratory returned to their capture sites 4.7 and 5.7 km away (Waddington and Cockrem 1987).

One possible orientation cue is the magnetic field of the earth, and house sparrows have been shown to have magnetite particles in their brains that may be used for orientation (Edwards et al. 1992) (see Chapter 9).

Where Have All the Sparrows Gone?

In summary, it is appropriate to return to the question posed at the beginning of this chapter. The short answer to the question about the cause of major declines in sparrow populations in North America and western Europe in recent decades is that there is probably not a single cause, but rather a complex interplay of multiple causes, that accounts for the declines. A multiplicity of causes is suggested by the precipitous decline in urban populations in western Europe alongside the general decline observed in rural populations in both western Europe and North America. In my view, the most likely candidates are changes in agricultural practices and increased predation pressure, both on nests and on free-living sparrows.

There seems to be little evidence that interspecific competition has contributed to the declines. Indeed, some of the species that would be potential competitors for food have themselves experienced major population declines, particularly in Great Britain and western Europe (Bauer and Heine 1992; Donald et al. 2001; Krebs et al. 1999). Even in eastern North America, where early studies proposed that there was a causal link between the rapid range expansion of the house finch and declines in sparrow numbers, the evidence does not appear to support such a link (see earlier discussion).

Disease also does not appear to be the culprit. There are instances of new strains of disease that affect house sparrows, including avian conjunctivitis and West Nile virus in North America (described earlier) and an apparently novel form of encephalitis of unknown etiology in Scotland (Pennycott et al. 2002). However, there is virtually no evidence that a new disease, or even a combination of diseases, could be responsible for the widespread declines in North America and western Europe.

Changes in agricultural practices are the most likely candidate for the wide-scale population declines on the two continents. These changes have probably resulted in lower food supplies, particularly during the winter, which have negative impacts on sparrow survival rates. The result may be that rural populations, which have served as sources for many other, more marginal populations, produce fewer dispersing offspring to sink populations. Increased predation pressure in rural populations may also affect their ability to serve as source populations. Increased predation pressure on urban populations, from both nest predators and predators on free-living sparrows, may have tipped the balance in those areas from being approximately stable to being major sinks. With little immigration from rural populations, such a scenario could result in rapid declines in urban populations. Clearly, however, much work remains to be done to adequately answer the questions related to declines in sparrow populations.

Chapter 9

ANATOMY AND PHYSIOLOGY

So dainty in plumage and hue,
A study in grey and in brown,
How little, how little we knew,
The pest he would prove to the town!
From dawn until daylight grows dim,
Perpetual chatter and scold.
No winter migration for him,
Not even afraid of the cold!
Scarce a song-bird he fails to molest,
Belligerent, meddlesome thing!
Wherever he goes as a guest
He is sure to remain as a King.

—Mary Isabella Forsyth (1840–1914), *The English Sparrow*

The inevitable interplay between form and function means that it is virtually impossible to describe the anatomy of the house sparrow without simultaneously discussing its physiology. Although the most comprehensive studies of avian anatomy and physiology have been done on the domestic chicken (*Gallus domesticus*) and a handful of other domesticated or cage birds, including the mallard (*Anas platyrhychos*), the Japanese quail (*Coturnix japonica*), the pigeon (*Columba livia*), and the budgerigar (*Melopsittacus undulatus*), many studies have also been done on the house sparrow, particularly on its energetics (discussed in this chapter) and on the neuroendocrine control of its circadian rhythm and annual cycle (see Chapter 3). Most physiological studies involve laboratory experiments, although some studies of free-living sparrows have also been done. Laboratory birds show some consistent differences from free-living birds, some of which may have implications when attempting to relate laboratory findings to free-living sparrows. Brockway (1965), for instance, found that uptake of iodine by the thyroid gland is reduced for at least 10 d in wild-caught sparrows maintained in cages, presumably due to stress, and Breuner et al. (2000) found that plasma testosterone (T) dropped below detectable levels during the first 5 d of captivity in breeding-season males. Newly captured sparrows also typically lose body mass during the early period of captivity, with mass stabilizing at about 10%–12% less than that of free-living birds. Davis (1955) found that birds captured in Illinois (USA) lost an average of 11.4% during the first 6–10 d of captivity, with females losing more than males. Some of this loss is presumably the result of flight muscle atrophy due to reduced flight in captivity.

This account of the anatomy and physiology of sparrows is organized on an organ system basis and reviews work done on the species. Because many physiological processes involve multiple organ systems, the placement of the discussion of some physiological processes is necessarily arbitrary.

Integumentary System

The integumentary system consists of the skin and its derivatives. The skin comprises two layers, the outer epidermis and the underlying dermis. The derivatives of the skin, all derived from the epidermis, include the feathers, the rhamphotheca (the horny covering of the bill), the scales and claws on the legs and feet, and the uropygial (preen) and anal glands. The plumage of the house sparrow is treated in Chapter 5. Collectively, the components of the integumentary system perform several major functions, including movement, protection, sensation, reproduction, communication, food acquisition, and thermoregulation. Flight involves not only the flight feathers of the wing and tail but also the propatagium, which is the integumentary membrane stretching between the humerus and radius in the forewing. The protective functions of the integument range from the provision of a physical barrier between the external environment and the vulnerable internal organs to the camouflage provided by the plumage coloration, which presumably reduces the risk of predation by visually hunting predators. General senses such as mechanoreceptors, thermoreceptors, and nociceptors (pain receptors) are widely distributed in the skin, providing for constant monitoring of various elements of the external environment. The brood patch, a bare region of highly vascularized skin on the ventrum of females, provides the heat for incubating the eggs and brooding small nestlings (see Chapter 4). Sexual dichromism of the plumage serves important roles in communicating both within and between the sexes (see Chapter 5). The bill is the primary structure involved in food acquisition. As the interface between the core of the bird and the external environment, the integument plays a major role in thermoregulation.

Flight

Powered flight is a complex activity that involves major contributions from the muscular and skeletal systems as well as the integumentary system. The primary and secondary feathers of the wings (see Fig. 5.3 in Chapter 5) provide thrust and lift, both of which are necessary for powered flight. Although sparrows are generally quite sedentary (see Chapter 8), flight is an integral part of their daily lives, particularly flights between their roosting or breeding areas and foraging sites. In a review of flight speeds in birds, Meinertzhagen (1955) measured the flight speed of migrating *Passer domesticus bactrianus* in Kashmir as being between 14.8 and 15.6 m/s. He also reported two other records of flight speed in sparrows, both from the United Kingdom, as being 10.7 m/s and 14.3–14.8 m/s. Schnell (1965) used Doppler radar to measure the flight speed of 84 sparrows near their nesting sites in Michigan (USA) and recorded speeds ranging from 2.2 to 17.4 m/s, with an average of 7.9 m/s. Schnell and Hellack (1978) also used Doppler radar to measure the flight speed of seven sparrows in Kansas (USA) and recorded an average speed of 12.6 m/s. Evans and Drickamer (1994) measured the flight speed of sparrows at two times of day in Illinois, and the average flight speeds at midday (10.2 m/s) and evening (12.3 m/s) were not significantly different.

Experiments with sparrows have helped to elucidate the roles of the flight feathers of the wing (primaries and secondaries) and the propatagium in powered flight (Brown and Cogley 1996; Brown et al. 1995). The propatagium, as noted earlier, is a triangular fold of skin that stretches between the humerus and radius and forms the leading edge of the proximal part of the wing. Brown and Cogley (1996) shortened the three proximal primaries (P1–P3) and all nine secondaries (by clipping them at their bases) of 40 wild-caught sparrows to test the effects of the propatagium and the distal six primaries (P4–P9) on flight. They established eight experimental groups of five sparrows each. The remaining six primary feathers were either left intact or shortened by 0.8, 1.6, or 2.4 cm (in each case reducing the remaining wing surface by about 15%); in half of the groups, the propatagium was left intact, and in the rest the propatagium was severed 0.75 cm distal to the elbow. The birds were then forced to fly in a 40-m long corridor, and the mean of six trials for each bird was taken as the bird's flight distance. Among the birds in which the six primaries were intact or were shortened by only 0.8 cm, there were no significant differences in mean flight distance between birds with severed propatagia and those with intact propatagia ($P > .05$). Among birds with primaries shortened 1.6 and 2.4 cm, however, there was a significant decrease in flight distance of birds with severed propatagia compared to those with intact propatagia ($P < .001$). For instance, in sparrows with primaries reduced by 1.6 cm, the flight distance was about 27 m with intact propatagia but only about 6 m with severed propatagia. Birds with primaries shortened by 1.6 cm were also incapable of unassisted takeoff. Brown and Cogley concluded that the propatagium, which is cambered and represents only 20% of the length of the secondaries in the airfoil of the proximal wing surface, nevertheless contributes the highest percentage of the total lift to the proximal wing. They also concluded from this experiment that the primaries of sparrows are capable of providing lift as well as thrust for the bird's flight. Brown et al. (1995) examined the role of a ligament that runs through the propatagium from the head of the humerus to the distal end of the radius (the ligament propatagiale) in sparrows, turkey vultures (*Cathartes aura*), and great horned owls (*Bubo virginianus*). They concluded that the ligament provides passive support to the distal wing against drag forces created at normal flight speeds.

Blem (1975b) examined wing-loading (body mass divided by wing area) in sparrows collected during the winter (November–February) at 11 sites in North America. Wing area was estimated from an empirically determined linear regression of wing area (A in cm^2) on wing chord (C in mm): $A = 1.66C - 25.86$. Average wing-loading varied at the 11 sites from 0.262 to 0.332 g/cm^2 in males and from 0.269 to 0.339 g/cm^2 in females. Wing-loading increased significantly with latitude in both sexes, and average wing-loading of females was greater than that of males at 9 of the 11 sites. Blem also tested the flying ability of experimentally weighted birds and found that difficulties in flying were first evident at wing-loadings of 0.33–0.35 g/cm^2, with severe difficulties occurring when wing-loading exceeded 0.36 g/cm^2. Average wing-loading from the most northerly sites therefore approached the values at which sparrows begin to show negative effects on their flying abilities.

Thermoregulation

Sparrows are homeothermic and endothermic, meaning that they maintain a constant internal body temperature by the metabolic production of heat. The maintenance of this constant body temperature involves not only the integumentary system but also the respiratory, muscular, circulatory, and nervous systems. The daytime body temperature of sparrows is approximately 41°C (Hudson and Kimzey 1964; Mills and Heath 1970), although average daytime temperatures as high as 43°C have been reported (Binkley et al. 1971; Riley 1936a). Several methods have been employed to measure body temperature, which may account for some of the differences among studies. Thermometers are sometimes inserted into the lower esophagus (Riley 1936a), into the proventriculus (O'Connor 1975d), or into the cloaca (Palokangas et al. 1975; Seel 1969). For continuous monitoring of body temperature in caged sparrows, Hudson and Kimzey (1966) inserted an intramuscular thermocouple into the pectoralis muscle, and Binkley et al. (1971) implanted a temperature-recording telemeter encased in paraffin into the abdominal cavity. There is a circadian rhythm in body temperature (see Chapter 3), with nighttime temperatures averaging about 3°C lower than daytime temperatures (Binkley et al. 1971; Hudson and Kimzey 1966; Riley 1937).

The plumage is the principal integumentary component involved in thermoregulation. It acts as an insulative covering that greatly reduces heat exchange between the bird and the environment. Development of thermoregulatory ability in nestling sparrows coincides with the emergence of feathers between 7 and 10 d after hatching (Seel 1969). Nestlings isolated from their nest and nest-mates during the first 6 d after hatching lost body temperature rapidly, but by 10 d they were able to maintain a constant high temperature (also see Chapter 4). Similar results were obtained experimentally with adult sparrows in Finland. Palokangas et al. (1975) determined the insulative quality of the plumage by measuring the cooling rate of the body with and without the plumage. The cooling constant of intact sparrows collected during the summer (1.18°C/h) was significantly greater than that in autumn (0.97°C/h) or winter (0.96°C/h). Plucked birds had a cooling constant of 2.06°C/h, 74.6% greater than that of intact summer birds and 114.6% greater than intact winter birds. The significantly lower cooling constant in fall and winter birds is due to the presence of more contour and down feathers during those seasons, which is reflected in the differences in average plumage mass between summer (1.23 g), autumn (2.07 g), and winter (2.02 g) (calculated from data in Palokangas et al. 1975) (see also Chapter 5). Insulative quality of the plumage can also be affected behaviorally by the extent to which the feathers are fluffed.

One potential avenue of significant heat loss in sparrows could be the legs below the ankle (the joint between the tibiotarsus and tarsometatarsus), which are not covered by feathers but by scales. The lower legs lack muscles and instead are moved by long tendons that are attached to muscles in thecalf. A steep temperature gradient exists in thisregion in sparrows, with the legs below the ankle tending to approach the environmental temperature (Heisler 1978). This steep gradient is caused by countercurrent heat exchange between blood in the major artery serving the lower leg and blood in several veins that lie in close apposition to the artery (Heisler 1978), which results in significant heat conservation.

Metabolic heat production in cold environments is discussed later in the section on energetics. The principal avenue for heat loss in warm environments is evaporative cooling from the respiratory system. Panting has been observed in both nestling and adult sparrows maintained at high ambient temperatures. O'Connor (1975c) placed individual nestlings into chambers maintained at either 35°C or 40°C and observed open-mouth breathing or panting in nestlings of all ages at 40°C. Kendeigh (1944) observed a rapid increase in water loss from adult sparrows maintained at temperatures greater than 35°C (Fig. 9.1). He suggested that, although most was respiratory water loss, some water might also be lost across the skin, despite the lack of sweat glands in sparrows.

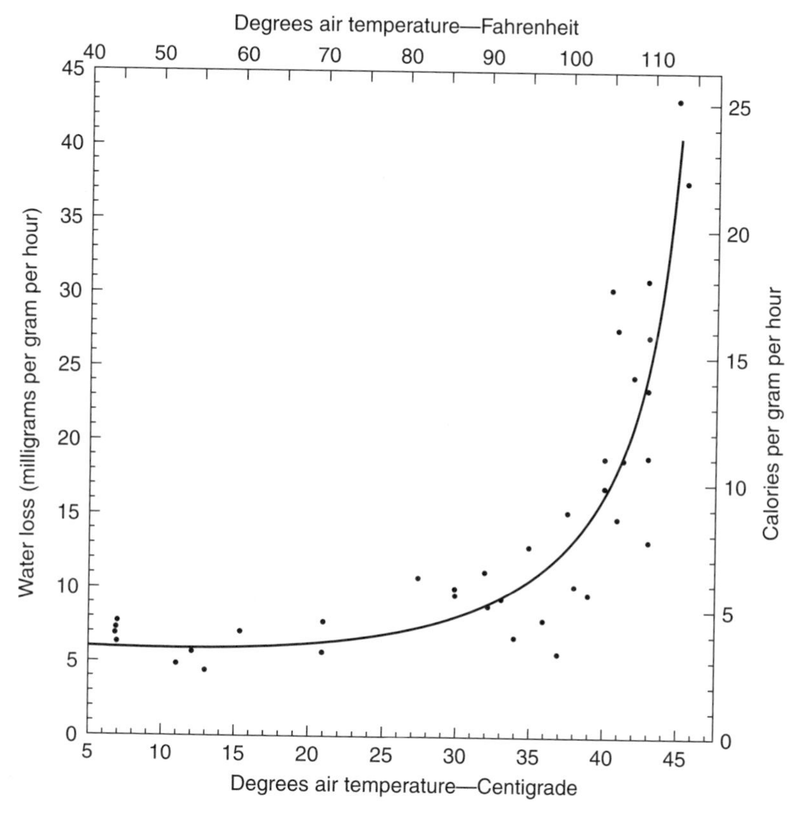

FIGURE 9.1. Water loss from house sparrows as a function of ambient temperature. Data are for sparrows from Illinois (USA) maintained in the laboratory under constant conditions with various ambient temperatures. Water loss is recorded as milligrams per gram of total body mass (TBM), and approximate energetic expenditure (in cal/g per hour) is also indicated. From Kendeigh (1944: Fig. 2), reproduced with permission of *Journal of Experimental Zoology*.

Uropygial Gland

The uropygial gland, or preen gland, is a bilobed, oil-producing gland located in the skin of the middorsum at the base of the tail. Preening is generally regarded as having a function in feather maintenance, maintaining feather flexibility and integrity and helping to control the plumage microflora (Jacob and Ziswiler 1982). When preening, the bird wipes its bill across the uropygial gland orifice, which is located on a papilla (or nipple), and then runs its feathers through the bill, particularly the flight feathers of the wing and tail (Summers-Smith 1963). In the congeneric tree sparrow, the gland is heart-shaped with a wart-like papilla that extends from the gland at approximately right angles to the plane of the lobes, and comprises 0.24% of total body mass (TBM) (Jacob and Ziswiler 1982).

Jacob and Zeman (1970) determined the chemical structure of the uropygial secretion of the house sparrow. The acidic components consisted entirely of 3-methylated ("methyl-substituted") fatty acids with chain lengths of C_{11}–C_{17}, and the alcohol components were *n*-alkanols with chain lengths of C_{12}–C_{19}. They found no differences in the composition between males and females. As noted in Chapter 1, Poltz and Jacob (1974) found that the chemical structure of the uropygial secretion of the house sparrow is similar to those of the fringilline and emberizine finches but differs from that of the ploceine finches (weaverbirds).

Anal Glands

Anal glands are small epidermal glands located near the cloacal orifice. The glands secrete mucus, which may facilitate the mechanics of internal fertilization, although the function of the secretion is not fully understood (Quay 1967). In sparrows, the glands are located either just inside the cloacal orifice, just outside it, or both (as in Fig. 9.2). Quay (1967) examined the anal glands in eight sparrows and found them to be intermediate in size compared to those of other species, and to not differ between the sexes.

Bill

The bill is not only the primary foraging apparatus of sparrows, but it is also used in many other activities, including preening, display, and fighting. The upper bill is the horny sheath covering the premaxilla; it includes the dorsal ridge (the culmen) and the lateral cutting edges (the tomia). The lower bill is the horny sheath covering the mandible (dentary); it also possesses lateral cutting edges (tomia) and a ventral ridge (gonys). The bill has the conical shape typical of seed-eating birds. Bill dimensions are often measured either in the field or on museum specimens, and the typical measurements recorded are bill

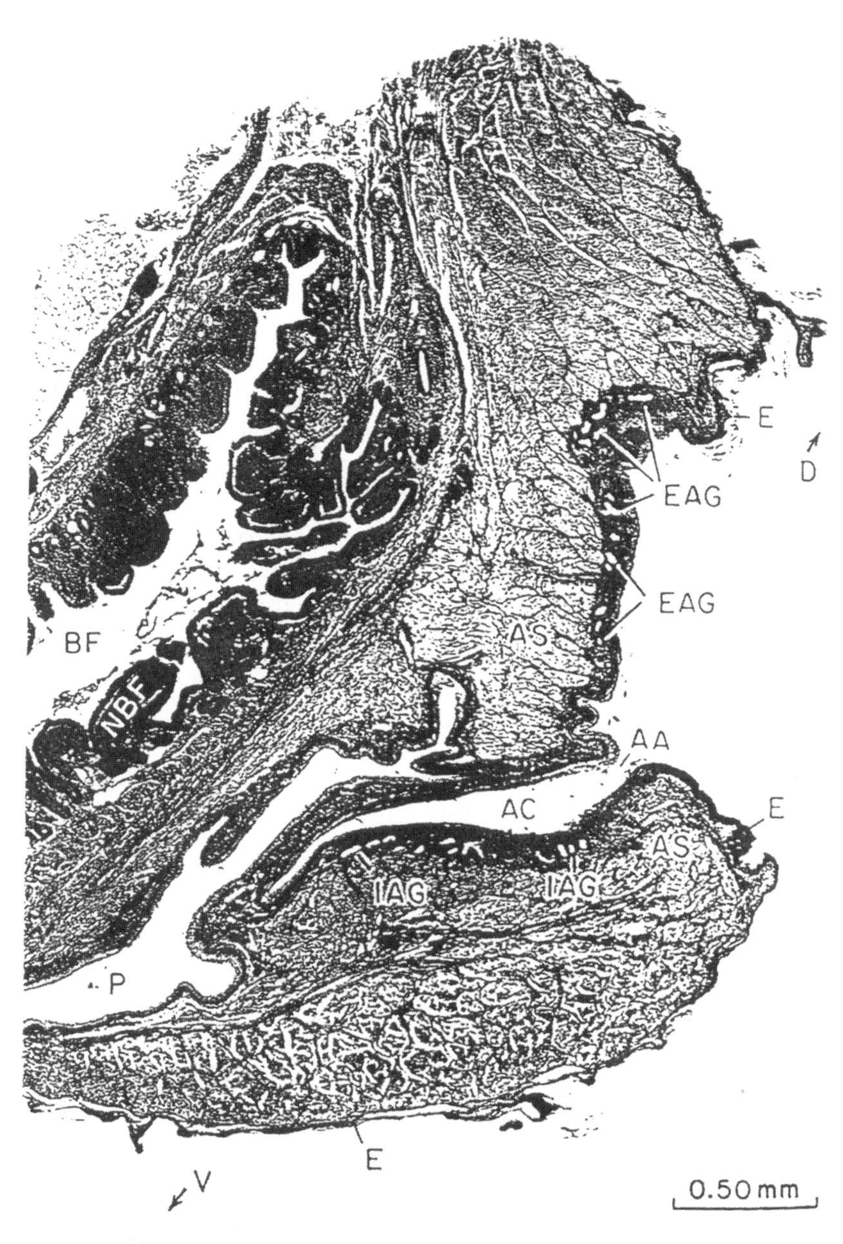

FIGURE 9.2. Anal glands of the house sparrow. Photomicrograph of a parasagittal section through the cloacal region, showing the anal canal (AC) and anal aperture (AA). Anal glands are located both internally (IAG) and externally (EAG). Other structures include the bursa of Fabricius (BF) and lymphoid nodules of the bursa (NBF), the anal sphincters (AS), epidermis (E), and inner proctodeum (P). The dorsal (D)-ventral (V) axis is indicated by the accompanying arrows. From Quay (1967: Fig. 3), reproduced with permission of the American Ornithologists' Union.

length (usually culmen length from the anterior margin of the nostril to the tip), bill width (width at the base of the tomia on the upper bill), and bill depth (depth of closed bill at its base). Bjordal (1983, 1984) studied the effects of freezing, freeze-drying, skeletonization, and traditional skin preparation on bill length and width in house sparrow specimens. Bill length increased significantly after 11 mo of being frozen but decreased significantly after being freeze-dried or after traditional skin preparation. Both bill length and bill width decreased after skeletonization, probably due to destruction of the horny covering by the chemical agents used. Bill length changes seasonally in some populations, probably as a result of seasonal changes in the hardness of the food and its effect in abrading the horny sheath (see Chapter 2).

Bill coloration also changes seasonally, particularly in males (see Chapter 3). The darkening of the bill in males during the breeding season is caused by the effect of T acting directly on the basal cells of the epidermis, which results in melanin deposition in the developing horny sheath (Kirschbaum and Pfeiffer 1941). The complete darkening of the bill requires 20–25 d (Keck 1933; Witschi 1936), suggesting that the horny covering of the bill may be completely replaced every 3 wk.

Ziswiler (1965) examined the bills of several seed-eating finches, including house sparrows, and also described their methods of seed-husking. The palatal surface of the upper bill of sparrows has a midventral ridge running two-thirds of its length from just proximal to the tip (Fig. 9.3 left). Sparrows husk seeds by crushing them between the tomia of the upper and lower bills (Fig. 9.3 right). As noted in Chapter 1, sparrows, in these characteristics, are more similar to fringilline and emberizine finches than to ploceine finches.

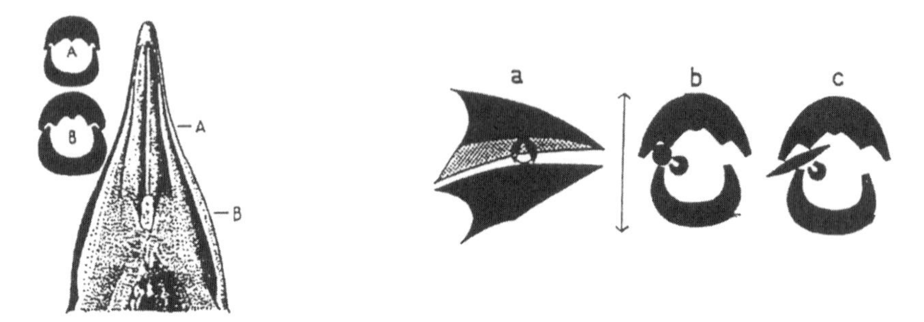

FIGURE 9.3. Bill of the house sparrow. Left: Ventral view of the upper bill, with cross-sectional views through two areas of the upper and lower bill at A and B. Right: Lateral view of the bill a, with cross-sectional views, b and c, showing sequential steps in the seed-husking mechanism. From Ziswiler (1965: Figs. 12 and 13), with permission of Deutsche Ornithologen-Gesellschaft.

Skeletal System

The skeletal system comprises the bones of the skeleton and their articulations, and its functions include support, protection, movement, storage, respiration, and hematopoiesis. As the structural framework of the body, its role in providing support is obvious, as is its role in protection, with the bones that make up the ribs, the vertebrae, and the braincase providing protection to vulnerable underlying or encased organs such as the heart, lungs, spinal cord, and brain. The bones of the skeleton also serve as sites of attachment for the skeletal muscles, and the joints control the direction of movement, both major contributions to flying and other movements in sparrows. Bone tissue serves as a storage site for calcium (Ca) and phosphate. Some hollow bones possess air sacs connected to the complex system of air sacs that are part of the respiratory system. The synthesis of erythrocytes also occurs in bone marrow cavities. The skeleton comprises about 4% of the TBM of sparrows; it was found to weigh an average of 1.01 g in sparrows from Israel with an average TBM of 25.53 g (Yalden and Yalden 1985).

Calcium Metabolism

The bones of the skeleton serve as a major repository for Ca. Normal Ca metabolism is under the control of hormones secreted by the parathyroid and ultimobranchial glands (see later discussion). Shell deposition in laying females greatly increases the demand for Ca, however. Is this Ca obtained from exogenous food sources during egg-laying, or does the female store Ca before the initiation of egg-laying? The answer, it appears, is both.

Pinowska and Krasnicki (1985a) collected 247 female sparrows during the breeding season in Poland and assigned each to a breeding-stage category based on an examination of the ovary and oviduct: 0 = non-breeding, 1 = prebreeding, 2 = egg-laying, 3 = early incubation, 4 = late incubation, and 5 = feeding young. The carcasses were dried and homogenized, and the levels of Ca, magnesium (Mg), copper (Cu), nitrogen (N), and phosphorus (P) were obtained. The relative levels of Ca did not change among the six stages, but because total dry mass increased significantly during stages 1 and 2, so did the total body Ca (TBC). Both Ca and Mg levels increased rapidly in the days preceding the initiation of laying and dropped during laying. In a similar study in Ontario (Canada) using similar breeding-stage categories, Krementz and Ankney (1995) found that average TBC increased significantly between pre-reproductive (0.283 g) and pre-laying (0.342 g) stages in females. TBC did not change significantly during the pre-laying and egg-laying stages but decreased after the end of egg-laying. Kirschbaum et al. (1939) found that female sparrows that were laying eggs had spicules in the medullary cavities of their long bones, a phenomenon they called hyperossification, but that females feeding young ready to fledge had no such spicules. Tiemeier (1950) also reported the presence of endosteal bone in the medullary cavities of the tiny os opticus bone in the eyes of female sparrows with enlarged ovarian follicles, whereas females with small follicles showed no such bone. Hyperossification was experimentally induced in both male and female sparrows by the intramuscular administration of estrogen (Pfeiffer et al. 1940). Therefore, some Ca is apparently stored by female sparrows in preparation for egg-laying.

Other evidence, however, suggests that most, if not all, of the Ca required for shell production is provided by the diet. Pinowska and Krasnicki (1985b) determined the dry mass of the grit found in the digestive tracts of breeding females in Poland and found increased grit contents in laying females. They concluded that increases in grit consumption during laying resulted in increased amounts of Ca and Mg, either by serving as a direct source of the nutrients or by increasing digestive efficiency or nutrient extraction from other dietary items. Krementz and Ankney (1995) tested for the presence of calciferous material in the contents of the digestive tracts of breeding sparrows in Ontario and found that calciferous food was present in a low percentage of diets in all stages except the laying period, when 88% of females had calciferous material in their guts. Indirect evidence also supports the proposition that dietary sources of Ca are sufficient for eggshell production during laying. Sparrows are removal indeterminate layers (see Chapter 4), and Anderson (1995) found that the average dry shell mass of supernumerary eggs (eggs 7+ in the laying sequence) was not significantly different from that of the first six eggs of indeterminate laying females in Michigan. This suggests that dietary Ca is sufficient to produce normal shells in the supernumerary eggs. Krementz and Ankney (1995) concluded that exogenous sources of Ca are sufficient for egg production in sparrows but proposed that dietary Ca is actually used to replenish Ca stored in medullary bone, which serves as the direct source of Ca used in the shell gland for shell production.

Muscular System

The muscular system consists of the skeletal muscles and the tendons and ligaments that attach them to bones. Although the primary function of the muscular system is movement, the system also plays roles in postural maintenance, thermogenesis, and storage. The numerous interactions between the skeletal and muscular systems in providing support to the body and in movement are so extensive that the two systems have sometimes been considered together, as the skeletomuscular system (e.g., Bock 1974). Muscle and nervous tissues are two of the most metabolically active tissues in the body, and the heat produced by their catabolic activities contributes substantially to the amount of internally generated heat required to maintain a constant body temperature. In addition, shivering thermogenesis plays a major role in generating heat in cold environments. Muscle tissue also serves as a reservoir for fat and protein to meet the need for additional energy or metabolic building blocks under conditions of high demand, such as low temperatures or egg formation in females.

The skeletal muscles can be divided into several functional groups, which include the jaw musculature (muscles associated with the beak), the wing musculature (muscles of the pectoral girdle associated with flight), the leg musculature (pelvic girdle muscles associated with hopping and other leg movements), and the trunk musculature. The principal flight muscles are the large *musculus pectoralis* and *m. supracoracoideus*, both of which originate on the keel of the sternum and insert on the humerus. Reference to the pectoral muscles in some studies may refer to both muscles together, but other studies clearly differentiate between the two. In any case, the pectoral muscles constitute a significant part of the body, with one study finding that the *m. pectoralis* made up 14.9% of the TBM of free-living sparrows (Dubach 1981). There is often considerable individual variation in muscle structure. Berman et al. (1990) described variation in the hindlimb musculature of 40 sparrows collected in Wisconsin (USA). In two muscles, the *m. puboischiofemoralis* and the *m. flexor hallucis longus*, about half of the individuals showed a variant of the normal pattern. In each case, the variant occurred bilaterally within an individual.

Muscle Tissue

Individual skeletal muscles are surrounded by a layer of connective tissue (the epimysium) and are composed of bundles (fasciculi) of individual muscle cells (muscle fibers or myofibers). The fasciculi are separated by thin sheets of connective tissue (the perimysium), and the individual muscle fibers are separated by even thinner connective tissue sheets (the endomysium). Muscle fibers in birds (including sparrows) are of at least two types, twitch fibers and tonic fibers; these may, however, represent extremes in a continuum of muscle fiber variation (Bock 1974). Twitch fibers contain many long, cylindrical myofibrils, which are surrounded by cytoplasm (sarcoplasm) and contain the myofilaments. The myofilaments are made up of the contractile proteins responsible for muscle contraction. Tonic fibers lack myofibrils and instead have clumps of myofilaments (pseudofibrils) embedded in the sarcoplasm. Twitch fibers have a very rapid contraction response and a very short duration of contraction, whereas tonic fibers have a slower contraction response and a longer duration of contraction. Electron microscopic examination of the myofibrils of muscle fibers from *m. pectoralis* of sparrows indicated that they are typical for vertebrate twitch muscles (Jones 1982). Tonic fibers tend to be involved in postural maintenance, including joint stabilization, whereas twitch fibers are involved in movements such as flight. Twitch fibers can also be either white or red, depending on the amount of myoglobin they contain. Red fibers contain large amounts of myoglobin, which is a heme-containing pigment that stores oxygen for oxidative metabolism. White fibers lack significant quantities of myoglobin; therefore, they often face oxygen depletion during activity and must rely on anaerobic metabolism.

One technique for identifying muscle fiber types is based on metabolic differences that can be characterized histochemically by assaying their myofibrillar adenosine triphoshatase (mATPase) activity after acidic and alkaline preincubation of the tissue (Rosser and George 1986). The *m. pectoralis* of sparrows is composed entirely of red twitch fibers, and the average diameter of the myofibers is 35.0 μm (Chandra-Bose and George 1965; Rosser and George 1986). The red fibers of *m. pectoralis* have large quantities of dihydrolipoic dehydrogenase, which indicates the use of lipids as an energy source during activity (Cherian and George 1966). However, biochemical characterization of *m. pectoralis* myofibers indicates that they are actually of two types, with the peripheral fibers being able to metabolize either carbohydrates or fats and the deeper fibers metabolizing primarily fats (George et al. 1964). The peripheral fibers are therefore able to respond either under aerobic or anaerobic conditions, whereas the deeper fibers depend on aerobic conditions. A similar arrangement has been observed in the *m. triceps brachii*, which is also involved in flight. The origin region of *m. triceps brachii* has both twitch fibers (83%) and tonic fibers, with the latter being found only in the peripheral fasciculi of the muscle (Geyikoglu and Özkaral 2000). The presence of the tonic fibers in *m. triceps brachii* suggests that this muscle is also involved in supporting the shoulder when the bird is not flying. Two fiber types have also been observed in the skeletal musculature in the proximal esophagus (Geyikoğlu et al. 2002).

Marquez and Braun (2002) used a functional classification system for the distal wing musculature of sparrows based on the number of glucose transporter (GLUT) molecules in the muscle fibers. They recognized three fiber types: fast glycolytic (FG), slow oxidative (SO), and fast oxidative glycolytic (FOG). Two muscles, *m. extensor digitorum communis* and *m. extensor metacarpi ulnaris*, were composed of homogeneous FOG fibers, whereas the *m. extensor metacarpi radialis* was composed primarily of FOG fibers, but with a significant component of SO fibers. The *m. superior* contained 71% SO fibers and 11% FOG fibers. More recently, efforts have been made to identify the location of GLUT vesicles in sparrow muscle fibers (Sweazea and Braun 2003).

Lipid and Protein Storage

Lipids are stored in two locations in muscle tissue of sparrows, in the perimysium (intermuscular) and within the muscle fibers (intracellular) (Jones 1980a). Lipids in *m. pectoralis*, which comprise 2.63% of the wet mass of the muscle in sparrows, apparently serve as an energy source during flight (George and Naik 1960). Protein storage appears to be intracellular, with fluxes in protein content occurring in both the sarcoplasmic and myofibrillar fractions of the muscle fibers (Jones 1990).

Jones (1980a) examined changes in content and distribution of lipids in *m. pectoralis* of sparrows captured in late winter in 1976 and 1977 in England. Birds were captured in the early evening and maintained without food under outdoor conditions until they were sacrificed at 1900 on the day of capture, or at 0100, 0700, or 1300 on the following day. After sacrifice, sections of the left *m. pectoralis* were collected, frozen, and subsequently stained for lipid content and for succinic dehydrogenase activity. The right *m. pectoralis* was dissected from each specimen, dried, and weighed, after which lipids were extracted, and the residue was dried and weighed to obtain lean dry mass. Lipid content of the *m. pectoralis* decreased overnight in the fasting sparrows in both years, but the decrease was significant only in 1976. The sparrows apparently stored less lipid in 1977 than in 1976, with the amount of lipid in the *m. pectoralis* at 1900 in 1977 (0.012 g) being less than half that in 1976 (0.028 g). Intermuscular lipids decreased overnight in both years (with the amount being lower in 1977 than in 1976), whereas the intracellular lipids showed a variable pattern, actually increasing substantially between 0100 and 0700 in one year (1976). Intracellular lipid distribution also showed a mosaic pattern, with some muscle fibers having high lipid levels and others having little or no lipids. Jones proposed that this mosaic distribution of intracellular lipid was associated with certain motor units' being responsible for shivering thermogenesis during the night.

Jones (1980b) also examined the overnight loss of protein from the flight muscles of fasting sparrows. She captured birds in three different months (October 1975, March 1976, and March 1977), and used a protocol similar to that just described. The flight muscles (*m. pectoralis* and *m. supracoracoideus*) were dissected from the carcasses after sacrifice, dried, and weighed, after which the lipids were extracted to obtain lean dry mass of the muscles. A flight muscle index (lean dry mass of the flight muscles divided by the diagonal, the distance from the posterior end of the sternum to the anterior surface of the coracoid bone), was calculated for each specimen to correct for differences in structural size. Flight muscle index consistently declined during the night in all three samples, and the average loss of lean dry mass during the night (1900–0700) was about 0.15 g. Although this loss could be due to the loss of either carbohydrate or protein, estimates of carbohydrate loss from other studies, according to Jones, indicate that it is not sufficient to account for the observed loss in lean dry mass; she therefore concluded that some protein is being lost overnight from the flight muscles of sparrows.

Chaffee and Mayhew (1964) compared the mean wet mass of the pectoral muscles of wild-caught sparrows captured in late winter in California (USA) with that of sparrows maintained for 6 wk in the laboratory on a short photoperiod (9L:15D; see Chapter 3) and at one of three temperatures: 35°C (heat-treated), 23°C (control), or 1°C (cold-treated). There was no significant difference in the average wet mass of the cold-treated, control, and wild-caught groups, but mean pectoral mass was significantly lower in the heat-treated birds. Chaffee and Mayhew also compared the activity of succinoxidase (an oxidative enzyme) in the muscles of the four groups and found that average succinoxidase activity was significantly greater in the wild-caught birds than in all of the experimental groups, which did not differ significantly from each other. In a parallel study involving only the control and cold-treated birds, however, Chaffee et al. (1965) found that there was a highly significant increase in the myoglobin content of the cold-acclimated birds. The average myoglobin content was 62.5% higher than that in the control group. They concluded that cold-acclimated sparrows thermoregulate by shivering, and that increased myoglobin content in the pectoral muscles facilitates oxygen transport and storage in the muscles. On the other hand, reduced size of the large pectoral muscles in the heat-acclimated birds presumably reduces heat production.

Acclimation to heat or cold also has an effect on the types of fatty acids that make up the muscle lipids. Zar (1977) measured the relative proportions of fatty acids in the *m. pectoralis* of sparrows maintained in the laboratory on a constant photoperiod (12L:12D) and at one of four temperatures: −20°C, 5°C, 20°C, and 35°C. The principal fatty acids found in the *m. pectoralis* at 20°C were palmitic (16:0) acid (30.0%), oleic (18:1) acid (28.2%), stearic (18:0) acid (15.1%), and linoleic (18:2) acid (12.5%). Birds maintained at −20°C had proportionally less palmitic, oleic, and palmitoleic (16:1) acids and proportionally more stearic and docosahexaenoic (22:6) acids. Starved sparrows showed a similar pattern of change in fatty acid composition of the muscle lipids. Birds maintained at 35°C had proportionally greater amounts of palmitic and oleic acids and less stearic acid. Zar also assayed the fatty acids in the brains of the sparrows and found that palmitic (27.2%), stearic (21.8%), oleic (16.8%), docosahexaenoic (14.2%), and arachidonic (20:4) (7.8%) acids predominated at 20°C. Birds held at −20°C and 35°C showed no changes in brain fatty acid composition, but the proportion of arachidonic acid was significantly greater in the brains of starved birds. Because these findings are based on changes in the proportions of fatty acids in the muscle and brain lipids and not on the absolute amounts of the various fatty acids, they are somewhat difficult to interpret. The lack of change in the fatty acids in the brain suggests that brain lipids are largely structural, with the longer-chain fatty acids contributing to the phospholipids that make up the cellular membranes. The observed changes in fatty acid composition of the muscle lipids, particularly in the cold-treated birds, suggest that lipids stored as a reserve energy source in the muscles may be partially depleted at low ambient temperatures. In a study of the fatty acids in the bodies of sparrows collected in winter from several sites in North America, Blem (1973) found that oleic acid tended to increase and palmitic acid tended to decrease with increasing latitude, a finding that is only partially consistent with the change in fatty acid proportions in laboratory birds held at −20°C.

The uncoupling protein that is involved in nonshivering thermogenesis from "brown" fat in mammals is not found in the *m. pectoralis* of sparrows, nor in the subcutaneous lipid stores or the liver (Saarela et al. 1991). This protein is also not present in several other bird species.

Changes in lipid and glycogen storage in the muscles are apparently caused by ambient temperature and not by photoperiod. Farner et al. (1961) examined the lipid and glycogen contents of the pectoral and thigh muscles of sparrows maintained at 22°C on either a short (8L:16D) or a long (20L:4D) photoperiod. They found no difference in either lipid or glycogen content of the muscles between the two groups.

Nervous System

The nervous system is undoubtedly the most complex system, both structurally and functionally, in the avian body. The principal functions of the nervous system are the monitoring of both internal and external conditions, processing and integration of this flow of information, and coordination and control of the organism's responses to changing environmental conditions. In performing these functions, the nervous system interacts extensively with the other control system in the body, the endocrine system, by stimulating or inhibiting the production and release of hormones from some of the endocrine glands, or by responding to some of the hormones directly, thus forming a closed circuit that is sometimes referred to as the neuroendocrine system.

Structurally, the nervous system is divided into the central nervous system, comprising the brain and spinal cord, and the peripheral nervous system, comprising the cranial and spinal nerves and the sense organs. The sense organs include both the special senses, such as the eye and ear, and the widely distributed general senses. The large majority of studies of the nervous system in sparrows deal with either the brain or the eye, with only a few studies addressing other elements of the nervous system.

Brain

The house sparrow brain weighs approximately 1 g (average fresh masses between 0.922 and 1.014 g) (Graber and Graber 1965; Quiring and Bade 1943b; Rehkamper et al. 1991), about 3.5% of TBM. It is composed mostly of water (77.3%), with a lipid content of 9.7% (Graber and Graber 1965). Adult brain mass shows no diurnal or seasonal change (Danilov et al. 1969; Graber and Graber 1965) and does not differ between the sexes (Quiring and Bade 1943b). Whitfield-Rucker and Cassone (2000) also found no difference in total brain size in adult males after experimental manipulations of photoperiod in the laboratory. Graber and Graber (1965) found that brain mass increased approximately linearly from about 85 mg at hatching to about 1.05 g at fledging (15 d). After fledging, brain mass continued to show a slow increase to about 1.1 g at 35–40 d, after which there was a slow linear decrease to the average adult mass (0.922 g).

The nascent avian brain consists of five regions, which develop from three enlargements at the anterior end of the embryonic neural tube that gives rise to the central nervous system. The three enlargements are the prosencephalon (forebrain), which subsequently divides to form the telencephalon and diencephalon; the mesencephalon (midbrain); and the rhombencephalon (hindbrain), which divides to form the metencephalon and myelencephalon. The telencephalon, in turn, gives rise to the cerebral hemispheres of the mature brain; the diencephalon, to the thalamus and hypothalamus; the mesencephalon, to the optic lobes (optic tectum) and pineal gland; the metencephalon, to the tegmentum ("pons") and cerebellum; and the myelencephalon, to the medulla oblongata (Fig. 9.4). The ventricles of the mature brain arise from the lumens of the telencephalon (lateral ventricles), diencephalon (third ventricle), and metencephalon (fourth ventricle). Rehkamper et al. (1991) reported the following volume percentages for the parts of the sparrow brain: cerebral hemispheres, 68.1%; cerebellum, 10.0%; tegmentum and medulla oblongata, 9.0%; optic tectum, 6.7%; hypothalamus and thalamus, 4.8%; and optic tracts, 1.4%. Most studies of the brain have examined the cerebral hemispheres, the hypothalamus, or the pineal gland.

Major functions of the cerebral hemispheres include processing of visual and auditory information, learning and motor control of vocalization, and organization of spatial information and memory (Gahr et al. 1993; Saldanha et al. 1998). Two areas in the cerebral hemispheres that have received particular attention are the hippocampus and the cerebral components of the song control system. The hippocampus is located in the dorsomedial region of the hyperstriatum. The song control system comprises several regions in the brain, including the higher vocal center (HVC, the ventral hyperstriatal nucleus of

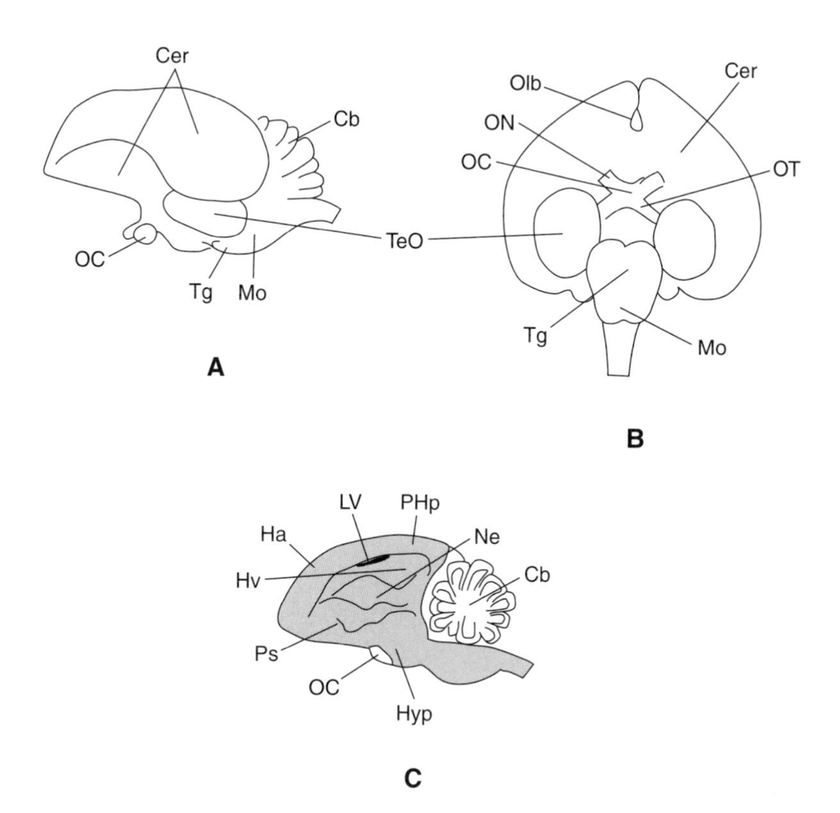

FIGURE 9.4. The house sparrow brain. (A) Lateral view of the left side of the brain. (B) Ventral view of the brain. (C) Medial view of a parasagittal section through the right side of the brain. Cb, cerebellum; Cer, cerebrum; Ha, accessory hyperstriatum; Hv, ventral hyperstriatum; Hyp, hypothalamus; LV, lateral ventricle; Mo, medulla oblongata; Ne, neostriatum; OC, optic chiasma; Olb, olfactory bulb; ON, optic nerve; OT, optic tract; PHp, parahippocampus; Ps, paleostriatum; TeO, optic tectum; and Tg, tegmentum. Redrawn from figures in Huber and Crosby (1929), Cobb (1960), Mills and Heath (1970), Cassone and Moore (1987), Metzdorf et al. (1999), and Whitfield-Rucker and Cassone (2000).

the neostriatum), the robust nucleus of the archistriatum (RA), the magnocellular nucleus of the anterior neostriatum (MAN), and area X of the parolfactory lobe, all located in the cerebral hemispheres (Fig. 9.5). Nuclei in the brain are clusters of nerve cell bodies that are morphologically similar and share staining properties, indicating that the cells are functionally related.

Studies of the avian hippocampus have focused on its role in spatial learning and memory, particularly in species that cache food. Krebs et al. (1995) compared the relative size of the hippocampus of house sparrows (which do

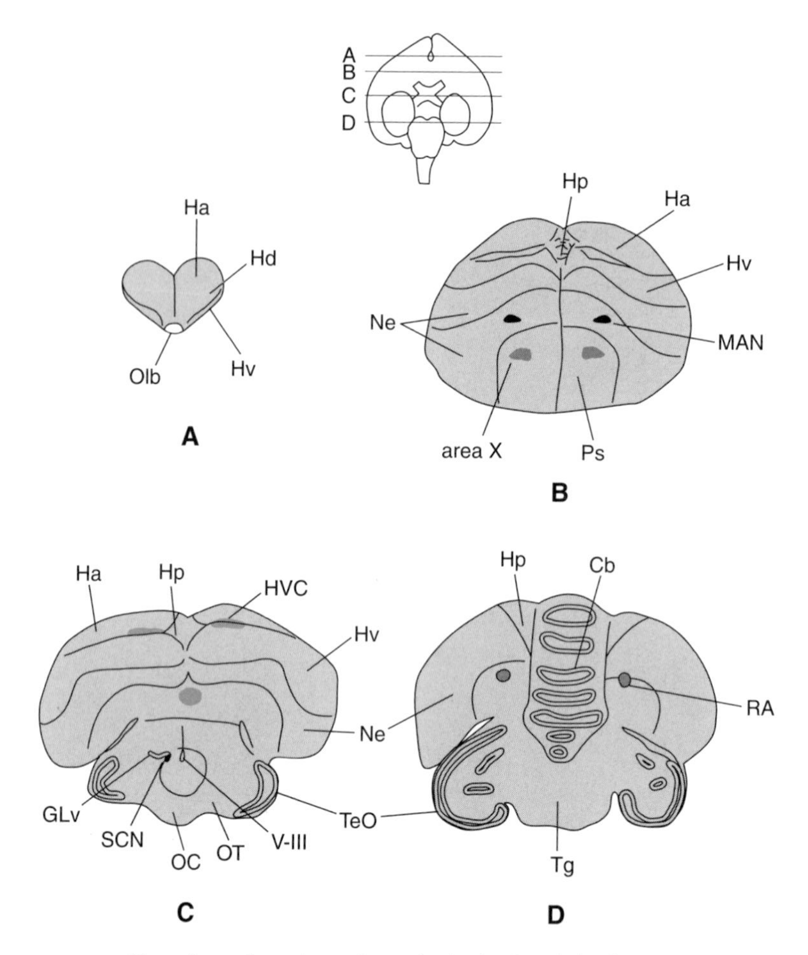

FIGURE 9.5. Four frontal sections through the brain of the house sparrow from anterior-most (A) to posterior-most (D) (see top of figure for approximate positions of the frontal sections). (A) Section through the olfactory bulb (Olb). (B) Section through the cerebral hemispheres at the level of the magnocellular nucleus of the anterior neostriatum (MAN) and area X. (C) Section through the optic chiasma (OC). (D) Section through the robust nucleus of the archistriatum (RA). Cb, cerebellum; GLv, ventral lateral geniculate nucleus; Ha, accessory hyperstriatum; Hd, dorsal hyperstriatum; Hp, hippocampus; Hv, ventral hyperstriatum; HVC, higher vocal center; Ne, neostriatum; OT, optic tract; Ps, paleostriatum; SCN, suprachiasmatic nucleus; TeO, optic tectum; Tg, tegmentum; V-III, third ventricle. Redrawn from figures in Huber and Crosby (1929), Rehkamper et al. (1991), Cassone et al. (1992), Cassone (2000), and Whitfield-Rucker and Cassone (2000).

not cache food) with that of the food-caching black-capped chickadee (*Parus atricapillus*) under two photoperiodic regimes in the laboratory. Hippocampal volume (including the parahippocampus) was determined relative to total telencephalic volume, and the log of hippocampal volume was positively correlated with the log of telencephalic volume ($P < .002$); the relationship did not change with photoperiodic treatment. Average hippocampal volume of 10 sparrows, estimated from Fig. 2 in Krebs et al. (1995), was 15.1 mm^3, and the average telencephalic volume was 459.1 mm^3. The relative size of the sparrow hippocampus was approximately half that of the chickadee.

Saldanha et al. (1998) used a radiolabeled androgen (androstenedione) to test for the presence of aromatase (ARO) in the hippocampus of the house sparrow and four other species (two other songbirds and two nonpasserines). ARO is an enzyme catalyst responsible for the conversion of androgens such as T into locally produced estrogens. After incubation with the androgen substrate, radiolabeled estrogens were observed in the hippocampi of the house sparrow and the other songbirds but were not detected in the hippocampi of the nonpasserines. The average rate of estrogen production in sparrows was 12.9 fM/mg of substrate per minute and did not differ between the sexes (Saldanha et al. 1998). In contrast to the hippocampus, all five species showed ARO activity in the preoptic area of the hypothalamus, and Saldanha et al. suggested that locally produced estrogens may play a role in the development and functioning of the hippocampus in songbirds. A second study compared ARO, 5α-reductase, and 5β-reductase activities in the hippocampi of four songbird species, two food-caching species and two noncaching species (including the house sparrow). 5α-Reductase and 5β-reductase catalyze the reactions that produce 5α-dihydrotestosterone (DHT) and 5β-DHT (an inactive androgen), respectively, from T. High ARO activity was observed in all four species, whereas the activity of 5α-reductase was comparable and that of 5β-reductase was low, compared with other parts of the cerebrum (Saldanha et al. 1999). 5α-Reductase activity was significantly higher in the two food-caching species than in the two noncaching species, however, suggesting a possible role for both locally produced estrogens and DHT in spatial learning. The presence of estrogen receptors was demonstrated in cells in the hippocampi of sparrows and other birds, both passerines and nonpasserines, with the use of a monoclonal antibody (H222Spy from humans) to immunostain for estrogen receptors (Gahr et al. 1993), and a second study using complementary RNA (cRNA) probes to identify cells containing messenger RNA for ARO (ARO mRNA) found that hippocampal cells of both passerines (including sparrows) and nonpasserines contained ARO mRNA (Metzdorf et al. 1999). This suggests that the hippocampus of both passerines and nonpasserines may be capable of responding to circulating androgens by converting them to estrogens using locally produced ARO.

The development and functioning of the song control system of songbirds also appears to be affected by circulating androgens. Studies that have included sparrows among the subjects suggest that the organization and functioning of the song control system of the house sparrow is similar to that of other songbirds (Gahr et al. 1993; Metzdorf et al. 1999; Saldanha et al. 1999; Smith et al. 1996), despite the fact that the males do not sing a typical songbird song (see Chapter 7). Gahr et al. (1993), using the monoclonal antibody H222Spy to identify estrogen receptors in cell nuclei in the brain, found intense labeling in the caudal neostriatum including the HVC, in areas along the dorsal surface of the RA, and in an area of the rostral forebrain anterior to the anterior MAN in songbirds (including sparrows), but not in nonsongbirds. Several other regions of the brains of both songbirds and nonsongbirds had estrogen receptors, however, including the hippocampus and amygdala (taenia nucleus) of the forebrain, the preoptic area of the hypothalamus, and areas in the mesencephalon (intercollicular nucleus) and rhombencephalon. Smith et al. (1996) used a polyclonal rabbit antibody (PG-21) to detect the distribution of androgen receptors in the brains of three songbird species (including sparrows). Androgen receptors were found in the HVC, the RA, and the MAN of the forebrain, as well as in the tuberal, preoptic, and magnocellular paraventricular nuclei of the hypothalamus; the intercollicular nucleus of the mesencephalon; and the tracheosyringeal portion of the hypoglossal nerve (cranial nerve XII).

A third study used hybridization with cRNA probes to examine the distribution of mRNAs for the androgen receptor (AR mRNA), the estrogen receptor (ER mRNA), and the enzyme aromatase (ARO mRNA) in the brains of both nonsongbirds and songbirds (including sparrows) (Metzdorf et al. 1999). The presence of mRNA indicates that a cell is actively synthesizing the corresponding protein. The distribution of ARO mRNA-expressing cells in the brains of songbirds and nonsongbirds was similar except for certain regions of the forebrain. Regions of the forebrain with ARO mRNA-expressing cells only in songbirds included the caudomedial neostriatum (adjacent to, but not including, the HVC), the medial anterior MAN, and regions of the hyperstriatum. In songbirds, the forebrain distribution of AR mRNA-expressing cells overlapped that of ARO mRNA-expressing cells in the caudal neostriatum and medial MAN but also included the lateral MAN, the RA, and, most notably, the HVC. ER mRNA-expressing cells showed little overlap with ARO mRNA cells and were most prominent in the HVC. The presence of ARO mRNA-expressing cells in the caudal neostriatum adjacent to the HVC, with its ER mRNA-expressing cells, suggests the possibility that the caudal neostriatum may act as a local estrogen-producing gland, with the locally produced estrogen diffusing to the adjacent HVC (Metzdorf et al. 1999).

Components of the song control system of sparrows also have melatonin-binding sites. Whitfield-Rucker and Cassone (1996) used radiolabeled melatonin (2-[^{125}I]iodomelatonin) to identify melatonin-binding sites in the brains of reproductively active sparrows of both sexes. Melatonin binding was observed in regions of the hypothalamus associated with the control of the circadian rhythm (see Chapter 3) in both sexes. Only male sparrows showed melatonin binding in the HVC, the RA, and area X of the forebrain, whereas both sexes showed melatonin binding in the MAN. The rate of melatonin binding was also examined in males that were made reproductively photorefractory, photosensitive, or photostimulated using manipulation of photoperiod in the laboratory (see Chapter 3). The HVC and RA had significantly lower rates of melatonin binding in photorefractory sparrows than in photosensitive or photostimulated birds, whereas the MAN showed significantly lower melatonin binding in photostimulated birds than in photosensitive birds, and area X had significantly lower binding in photostimulated than in photosensitive or photorefractory birds. The rate of melatonin binding did not show any consistent differences in the HVC or RA of castrated or sham-operated males, however. Whitfield-Rucker and Cassone suggested that seasonal changes in the vocal control system of sparrows may be influenced by the circadian mechanism as well as by seasonal changes in the levels of circulating gonadal hormones.

In a second set of experiments, Whitfield-Rucker and Cassone (2000) examined changes in size of components of the song control system of male sparrows under various photoperiodic regimes. Both the HVC and the RA of photostimulated birds were significantly larger than those of photorefractory or photosensitive birds, with the HVC of photostimulated birds being almost twice as large as that of photorefractory birds and approximately 50% larger than that of photosensitive birds. The MAN and area X showed no significant differences in volume with photoperiodic treatment, however. Similar results were obtained in a second experiment involving castrated male sparrows and sham-operated controls: there were no significant differences between them in the volume of the HVC or the RA. This very interesting result suggests that seasonal changes in the song control system and singing behavior may be mediated in part by the circadian clock mechanism, as well as by seasonal changes in circulating levels of gonadal hormones (Whitfield-Rucker and Cassone 2000). This control could be mediated by effects of melatonin on cellular activity in the nuclei of the song control system. One caveat to unqualified acceptance of this interpretation is based on the fact that both castrated and control birds in the second experiment were housed together, with the consequence that social interactions among individuals may have had an effect on the results.

The hypothalamus is derived from the diencephalon and forms the floor and ventral-most lateral walls of the third ventricle (see Fig. 9.5c). It has multiple functions, including control of the autonomic nervous system, the circadian rhythm, thermoregulation, and the pituitary gland. Although most of the studies on the hypothalamus in sparrows have concerned its central role in controlling the circadian rhythm, several other aspects of hypothalamic function have also been studied in the species. The hypothalamus can be subdivided into three regions: the preoptic area, the tuberal area and the mamillary area (Kuenzel and van Tienhoven 1982). The preoptic area lies anterior to the optic chiasma, and the tuberal area lies above the optic chiasma and posterior to the preoptic area. The tuberal area contains several nuclei that receive visual stimuli and are involved in the control of the circadian rhythm (see Chapter 3). The mamillary area includes the median eminence, which is located medial to the optic tracts (which transmit visual stimuli into the optic tecta).

One of the functions of the preoptic area in sparrows is thermoregulation, presumably via its control of the autonomic nervous system. Mills and Heath (1970) tested the thermoresponsiveness of the preoptic area of sparrows by surgically inserting an ultrasmall thermode into the region and then monitoring body temperature and resting metabolic rate of the birds when the temperature of the thermode was changed by passing water of different temperatures through it (either 18°C–21°C or 44°C). The birds increased their body temperature by an average of 0.9°C and increased their metabolic rate by 15% when the thermode temperature was 18°C–21°C, and decreased their body temperature by 0.5°C and decreased their metabolic rate by 12% when the thermode temperature was 45°C. Mills and Heath concluded that the preoptic area is thermosensitive and helps to control body temperature by regulating metabolic rate. The preoptic area also contains neurons that stain for gonadotropin-releasing hormone I (GnRH-I) (Bentley et al. 2003) (see later discussion).

The median eminence is involved in the control of the pituitary gland, to which it is attached by the infundibulum, which contains the hypothalamo-hypophysial portal system. Bern and Nishioka (1965) performed histological studies of the median eminence of sparrows and reported that about 25% of the axons of median eminence neurons terminate in the basement membrane adjacent to portal system capillaries. The axons also contain many neurosecretory vesicles, with an average diameter of about 400 Å, and granules similar to those found in the neurohypophysis. Oehmke et al. (1969) reported the presence of two types of vesicles (one about 500 Å, and the other about 1000 Å in diameter) in several regions of the hypothalamus and hypothalamo-hypophysial tract of sparrows. These neurosecretory vesicles may be involved in transporting the releasing factors such as GnRH that regulate pituitary gland secretions. On the other hand, Priedkalns and Oksche (1969) also observed vesicles of two sizes

(500 Å and 800–1000 Å in diameter) at axodendritic and axosomatic junctions in the infundibular region in sparrows, and found that the smaller vesicles were located in the axon terminals, whereas the larger ones were associated with the postsynaptic membranes.

A recent study used immunocytochemistry to identify three GnRHs in the brains of house sparrows based on reactivity of neurons to antibodies to chicken GnRH-I, chicken GnRH-II, and lamprey GnRH-III (Bentley et al. 2004). In addition to being located in the median eminence, neurons reactive to anti-lamprey GnRH-III were found in the preoptic area of the hypothalamus and in several telencephalic areas of the song control system, including the HVC, the RA, and area X. Another study identified cells in the preoptic area that stained for GnRH-I (Bentley et al. 2003). This study also found that cells in the paraventricular area of the sparrow hypothalamus stained for gonadotropin inhibitory hormone (GnIH), a newly identified neural hormone that inhibits pituitary secretion of gonadotropins.

The pineal gland (or pineal body), a dorsal outgrowth from the roof of the mesencephalon, lies in the transverse fissure between the cerebral hemispheres and the cerebellum. Quay and Renzoni (1963) found that the gland originated from an outgrowth of the right side of the third ventricle. Its role in the control of the circadian rhythm of sparrows involves its secretion of melatonin and has been discussed at length in Chapter 3. Oksche and Vaupel-von Harnack (1966) examined the pineal of sparrows with the electron microscope and found no evidence of photoreceptor cells, but did report that the gland received autonomic nervous system innervation. Korf et al. (1982) examined the gland after perfusion with horseradish peroxidase and reported that neuronal processes from the pineal projected to periventricular regions of the hypothalamus. They also reported the presence of synaptic ribbons in pineal cells similar to those found in photosensitive cells in the eye. In a pair of experiments, Quay and Renzoni (1963, 1966) examined the cell structure of the pineal gland of sparrows after exposure to either very short (1L:23D) or long (14L:10D, increasing to 18L:6D) photoperiods. Neurosecretory cells at the base of the pineal stalk were significantly more numerous in the short-photoperiod birds (Quay and Renzoni 1966). There was also evidence of a daily rhythm in the number of neurosecretory cells, with more being present in the evening. This is consistent with the nocturnal production of melatonin described in Chapter 3.

The Eye and Vision

There are excellent early descriptions of both the anatomy and the development of the house sparrow eye (Slonaker 1918, 1921). The two eyes weigh 0.445–0.499 g, approximately 2% of TBM and 45% of brain mass (Quiring and Bade

1943b; Slonaker 1918). The axis of vision of each eye has an angle of 65° with the median plane of the head, and there is no binocular vision involving the foveae of the two eyes (Slonaker 1918). The following description of the eye is based primarily on Slonaker (1918) unless otherwise noted.

The sparrow eyeball, like that of other vertebrates, is composed of three layers surrounding an internal, fluid-filled cavity: the external fibrous tunic, the intermediate vascular tunic, and the internal nervous tunic or retina. The internal cavity actually comprises two cavities separated by the lens, the small anterior cavity being filled with aqueous humor and the large posterior cavity with vitreous humor. The aqueous humor, lens, and vitreous humor all act as refracting elements responsible for focusing incoming light onto the surface of the retina. The diameter of the sparrow eyeball is about 7.5 mm, and its depth (from cornea to fovea) is about 7.0 mm, with the depth of the anterior cavity being about 0.6 mm, lens diameter about 2.8 mm, and posterior cavity depth about 3.3 mm (Lord 1956). Average dry mass of the lens of 45 adult sparrows was 6.17 mg and did not differ with sex (Payne 1961).

The fibrous tunic includes the transparent cornea, which covers virtually all the visible portion of the intact eye, and the sclera, which forms the outer layer of the remainder of the eyeball. The cornea, which is about 0.07 mm thick at its center, is covered by the conjunctiva, which also lines the inside of the eyelids. The eyelids close only when the bird is asleep, and the conjunctiva is moistened and cleaned by a translucent nictitating membrane that regularly sweeps across its surface. In dim light in the laboratory, the nictitating membrane closes at a rate of 40 times/min, whereas in bright light it closes at about twice that rate. The fluid that moistens the conjunctival surface is produced primarily by the harderian gland, which is located on the posterior surface of the eyeball. The sclera is comprised primarily of cartilage, with 14 small, thin bones (scleral plates) forming a ring around the eyeball at the anterior edge of the cartilaginous region. Another small bone, the os opticus, surrounds the base of the optic nerve as it exits the posterior surface of the eyeball (Tiemeier 1950).

The vascular tunic comprises the choroid layer posteriorly and the ciliary body anteriorly. The latter forms a ring around the anterior cavity and supports the lens. The iris, which is brown in sparrows (Lord 1956), extends from the ciliary body into the anterior cavity and controls the size of the pupil. As the name of this tunic implies, it contains the blood supply to the interior of the eyeball, which is received from the ophthalmotemporal artery, a branch of the external ophthalmic artery, which, in turn, is a branch of the internal carotid artery. The ciliary body also contains muscles that help to regulate the shape of the lens. The choroid layer gives rise to the pecten ocelli, a highly vascularized structure that extends from the optic disc into the vitreous humor of the posterior cavity. The pecten of the house sparrow consists of a thin membrane that is folded accordion-like 20 times and measures 4.0 mm along its base (on the optic disc), 2.75 mm along its free edge (in the vitreous humor, and 2.5 mm high (Lord 1956). Comparable measurements given by Slonaker (1918) are basal length, 2.7 mm; length of free edge, 1.5 mm; and height, 1.2 mm. Detailed descriptions of the microstructure of the pecten capillaries of sparrows have been provided by Jasinski (1973). The function of the pecten is to supply oxygen to the cells of the sensory layer of the retina.

The nervous layer of the eyeball, the retina, actually consists of two layers, the outer pigmented layer and the inner sensory layer, and covers the interior of the eyeball forward to the posterior margins of the scleral plates. The pigmented layer of the retina also extends anteriorly to cover the inner surface of the ciliary body and the posterior surface of the iris. The sensory layer of the retina, which has a total area of about 85 mm^2 in the sparrow eye, is composed of a layer of photosensitive cells adjacent to the pigmented layer, an intermediate layer of bipolar neurons, and a layer of ganglion neurons adjacent to the vitreous humor of the posterior cavity. The processes of the ganglion neurons pass over the inner surface of the retina and converge at the optic disc to form the optic nerve. The photosensitive cells are of two types, rods and cones, the former sensitive to dim light and the latter to bright light. The number of cones exceeds that of rods in the diurnally active house sparrow. On the posterior surface of the retina opposite the lens is the area centralis, containing an oval depression, the fovea centralis. This region of the retina is tightly packed with cones and is the part of the retina providing the most acute vision. The fovea centralis has outside dimensions of 0.432 mm (horizontal diameter) by 0.300 mm (vertical diameter) and narrows to a truncated medial end with a horizontal diameter of 0.016 mm and vertical diameter of 0.008 mm. Cone density in the fovea averages 55.6 cones per 500 μm^2 and drops to an average of 8.2 cones per 500 μm^2 in the retina outside the area centralis (Lockie 1952).

The eyeball is moved by the six extrinsic muscles of the eye: the superior, inferior, external and internal rectus muscles and the superior and inferior oblique muscles. Two other muscles, the pyramidalis and quadratus muscles, are responsible for moving the nictitating membrane. Slonaker (1918) conducted experiments on sparrows in which one eye had been surgically removed and concluded that the birds were able to rotate the eyeball a maximum of 40°. This does not permit binocular vision in the two foveae, but it does permit binocular vision in the temporal regions of the general retina. Although some birds have a second fovea in this region, sparrows do not; however, they do have a greater blood supply to this region than to other parts of the general retina, suggesting that the temporal region is used in binocular vision, probably during flight (Slonaker 1918).

The wavelengths of light to which the sparrow eye are sensitive are not known, but it is known that the eye is sensitive to near-ultraviolet (UV) radiation. Parrish et al. (1984) tested five sparrows using both positive (food reward) and negative (electric shock) conditioning to determine whether they were capable of detecting near-UV light (approximately 360 nm wavelength). The birds had a 91% success rate in both experimental protocols, with only a 3% error rate when a UV filter was used as a control. Possible functions for maintaining the ability to detect UV light include detection of insect prey that possess UV coloration and orientation using polarized light from clear parts of the sky (Parrish et al. 1984).

Slonaker (1921) provided a detailed description of the development of the house sparrow eye. The retina of the eye, both pigmented and sensory layers, and the optic nerve develop from neural ectoderm, whereas the lens and conjunctiva are derived from somatic ectoderm. All other parts of the eye are derived from mesoderm. Payne (1961) studied growth in the size of the lens in nestling and post-fledging sparrows. Dry mass of the lens grew rapidly from about 1 mg at 6 d to almost 6 mg at 2 mo, and Payne concluded that lens size could not be used to age young sparrows older than 2 mo.

Other

The role of olfaction in birds is poorly understood. Sparrows have a single olfactory bulb located anteromedially on the ventral surface of the brain (see Fig. 9.4). Tucker (1965) exposed the olfactory nerves of several bird species (including sparrows) and monitored electrical activity in the nerves in response to pulses of various odorants. All species showed electrical activity in the olfactory nerves that was coincident with inspiration, and they showed graded responses to varying concentrations of amyl acetate. The role of olfaction in sparrows is not known, but in homing pigeons it apparently plays a role in orientation by means of the creation of an "olfactory map" (Baldaccini et al. 1975).

The ability to detect electromagnetic fields is another sensory modality that may play a role in orientation behavior in birds. Although little is known about the physiological basis for such an ability, Edwards et al. (1992) determined both the natural and induced remanence in the head and neck of several bird species (including sparrows). Remanence is the fixed magnetic reading in iron-containing materials that is independent of external magnetic fields and could serve as the basis for detecting such fields. The average natural remanence of 10 sparrows was 8.854×10^{-12} tesla, with an average orientation 90° to the right of the longitudinal axis. When placed in a magnetic field of 1500×10^{-4} tesla for 10 s, the average induced remanence was 586.6×10^{-12} tesla. Edwards et al. concluded that the remanence was consistent with the presence of the magnetic iron oxide magnetite in the head and neck of the birds, which could provide a basis for detecting magnetic fields.

Endocrine System

The endocrine system consists of a structurally diverse set of glands that are related primarily by their mode of action. They secrete hormones that are usually transported through the circulatory system to remote sites in the body, where they affect the functioning of other glands or organs. Hormones tend to be involved in one or more of three broad functions: growth and development, homeostasis, and reproduction. Two categories of hormones are usually recognized, water-soluble hormones and fat-soluble hormones. Water-soluble hormones, which are usually proteins or protein derivatives, generally exert their effect by binding with specific receptor sites on the cell membranes of target organs, thereby affecting the rate of specific metabolic activities in the cells. This effect is often mediated by cyclic nucleotides (such as cAMP or cGMP) acting as second messengers (see later discussion). Fat-soluble hormones, which are usually lipid or cholesterol derivatives but which also include the thyroid hormones, generally exert their effect by entering the target cells and forming complexes with specific receptors, often in the nucleus, that affect rates of transcription in the cells, thereby resulting in accelerated protein synthesis.

Pituitary Gland

The pituitary gland actually consists of two endocrine glands with independent embryonic origins, the anterior pituitary (adenohypophysis) and the posterior pituitary (neurohypophysis). The anterior pituitary is located ventral to the median eminence of the hypothalamus, just posterior to the optic chiasma, and is connected to the hypothalamus by a portal system. The posterior pituitary is located on the posterior surface of the median eminence. Hormones secreted by the anterior pituitary include three gonadotropins involved in the control of the annual reproductive cycle, follicle-stimulating hormone (FSH), luteinizing hormone (LH), and prolactin, and four other hormones, thyrotropin (TSH), somatotropin (growth hormone), adrenocorticotropin (ACTH), and melanotropic hormone (MSH) (Tixier-Vidal and Follett 1973). Secretion of anterior pituitary hormones is regulated by hormones (releasing and inhibiting factors) that are produced in the hypothalamus and transported to the anterior pituitary through the portal system. The posterior pituitary secretes two hormones, antidiuretic hormone and oxytocin (Kobayashi and Wada 1973).

The gonadotropins FSH and LH play a major role in controlling the annual cycle in sparrows (see Chapter 3). The secretion of both FSH and LH is regulated by GnRH release from the hypothalamus (Farner and Wingfield 1980), although the recent identification of a GnIH produced in the paraventricular area of the hypothalamus (Bentley et al. 2003) suggests that gonadotropin secretions are regulated by an interplay of releasing and inhibiting hypothalamic hormones. The primary target of FSH in the male is the testis, where it stimulates the initiation of spermatogenesis; in the female, FSH stimulates the development of ovarian follicles. The testes and the ovary are also the primary targets of LH, which stimulates production of androgens and androgen derivatives in both gonads. The annual cycles of circulating LH in free-living sparrows of both sexes are shown in Fig. 3.5. Average plasma levels of LH are about 2 ng/ml during the winter in both sexes, and they more than double at the beginning of the breeding season.

Prolactin has multiple physiological and behavioral effects in birds, some of which involve synergistic interactions with other hormones (Tixier-Vidal and Follett 1973). These effects include induction of incubation behavior and parental care, stimulation of the development of the brood patch (see Chapter 5), and possibly playing a role in the initiation of molt (see Chapter 5) and in both gonadal recrudescence and photorefractoriness (see Chapter 3). In migratory species, and presumably in the migratory *P. d. bactrianus*, prolactin influences premigratory fattening and migratory restlessness. Goodridge (1964) found that prolactin significantly increased *in vitro* glucose uptake by mesenteric fat tissue from sedentary sparrows taken in Michigan. Prolactin did not, however, increase lipid synthesis in the tissue, and it in fact inhibited lipid synthesis at high acetate concentrations.

TSH has the thyroid gland as its primary target, where it stimulates the secretion of thyroid hormones. Likewise, ACTH primarily targets the adrenal cortex, where it stimulates the production of corticosteroid hormones. MSH regulates the activity of melanocytes in the integumentary system and presumably plays a role in feather pigmentation.

Both antidiuretic hormone and oxytocin are involved in maintaining water balance in birds. Antidiuretic hormone, as its name implies, is involved in water retention. Avian antidiuretic hormone is arginine vasotocin (AVT) (Kobayashi and Wada 1973). Goldstein and Braun (1988) found that captive sparrows maintained on *ad libitum* water had an average plasma level of AVT of less than 25.8 pg/ml, whereas birds deprived of water for 30 h had circulating levels of AVT nearly 8 times greater (207.4 pg/ml). The main target of AVT is the kidney, where it reduces water loss primarily by decreasing the glomerular filtration rate (see later discussion). Oxytocin acts as a diuretic in birds.

Thyroid Gland

The thyroid glands lie just lateral to the trachea near its distal end. Structurally, the thyroid glands are composed of numerous follicles, each consisting of a secretory membrane surrounding a colloid-containing lumen (Assenmacher 1973). The thyroid secretes two hormones, thyroxine (T_4) and triiodothyronine (T_3). T_4 has broad effects on body metabolism, increasing heat production, stimulating breakdown of glycogen in the liver leading to hyperglycemia, and promoting growth (Assenmacher 1973). T_4 is apparently converted to T_3 peripherally, and circulating levels of T_3 have been found to be positively correlated with resting metabolic rate in free-living sparrows (Chastel et al. 2003). T_3 is also involved in the development of the skin epidermis and feathers, and it may be involved in controlling molt (but see Chapter 5). Three studies in the United States reported the average mass of the thyroid gland as being, respectively, 3.9 mg, 0.016% of TBM (Quiring and Bade 1943b); 2.4 mg or 0.010% (Hartman 1946); and 2.8 mg or 0.011% (Kendeigh and Wallin 1966). In Hungary, Péczely (1976) reported that the average mass of the thyroid gland was 3.6 mg and did not vary significantly throughout the year, although it was lower in January and February (3.3 mg) than in March through October (3.6–3.9 mg).

Kendeigh and Wallin (1966) studied seasonal and sexual differences in the thyroid gland of sparrows (and other bird species) in Ohio (USA). Thyroid glands were removed shortly after the birds were killed, fixed, and stained for histological examination. The lumen diameters of 115 thyroid follicles from eight sparrows were measured, as were the cell heights of five of the secretory (epithelial) cells from each lumen. Secretory cell height was inversely correlated with lumen diameter, and follicles were smaller in diameter in winter (mean volume, 4.76 mm^3) than in summer (5.46 mm^3). Thyroid mass showed a similar difference (mean mass of the thyroids plus the parathyroids was 4.6 mg in winter and 6.7 mg in summer), and there was no difference between the sexes. Kendeigh and Wallin concluded that thyroid activity is directly related to cell height and inversely related to follicle size and thyroid mass, and therefore that thyroid activity in the sparrow is greater in winter than in the summer. They attributed this increase to the need for greater metabolic activity due to increased thermoregulatory costs in winter.

A more recent study using radioimmunoassay to study annual variation in the levels of circulating thyroid hormones failed to confirm this conclusion. Smith (1982) maintained sparrows in outdoor aviaries in Washington (USA) and collected blood samples at biweekly intervals (Fig. 9.6). Levels of T_3 did not change throughout the year, averaging about 4 ng/ml of plasma. Levels of T_4 were 3–5 times higher than those of T_3 and were highest during molt

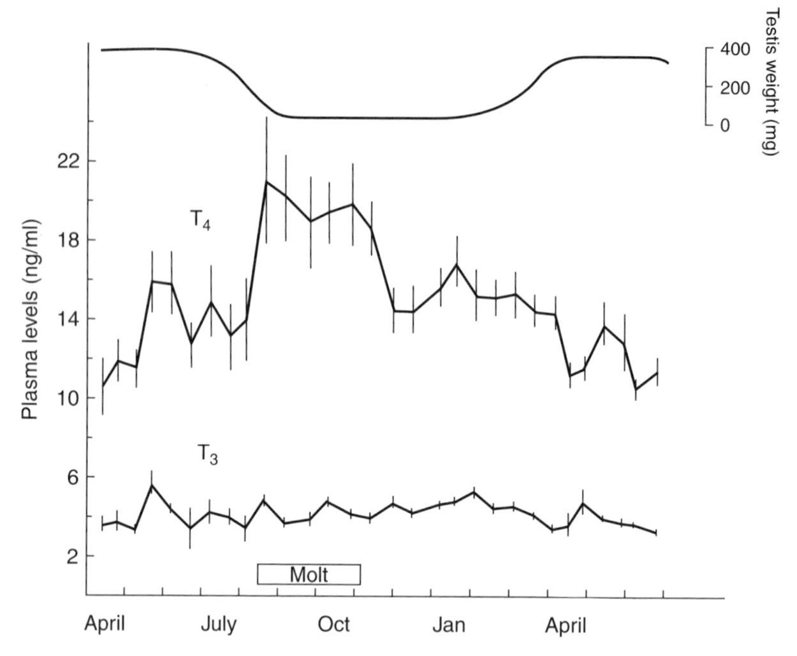

FIGURE 9.6. Annual variation in the levels of circulating thyroid hormones in house sparrows maintained in outdoor aviaries in Seattle, Washington (USA) (48°N). Blood samples were collected from each bird (10 males and 10 females) at approximately biweekly intervals, and levels of T_3 and T_4 were obtained by radioimmunoassay. There was no difference in levels of thyroid hormones between males and females. The timing of the molt was determined by inspection during the biweekly collection of blood samples. Testis mass of the males was estimated by laparotomy and by comparison with a reference collection. From Smith (1982: Fig. 1), with permission. Copyright © 1982 Cooper Ornithological Society.

(see Chapter 5). Although T_4 levels were somewhat higher during the winter months, the differences between winter levels and those of spring and summer were not significant. Thyroid hormone levels of wild-caught sparrows taken during the same period did not differ from those of the captive birds. There was a tendency for levels of T_4 to vary inversely with ambient temperature in the wild-caught sparrows ($r = -0.34$), but the relationship was not significant.

Increased thyroid activity during the winter in Ohio (Kendeigh and Wallin 1966) and increased levels of circulating T_4 during the molt in Washington (Smith 1982) both point to the role of thyroid hormones in stimulating increased metabolic activity. One possible explanation for the failure to observe a significant increase in circulating T_4 during the winter in Washington is that winter temperatures in Ohio are considerably colder than those in Washington (Seattle).

Other general effects of the thyroid hormones are evident in the results of thyroidectomy experiments. In India, surgically thyroidectomized male sparrows with fully developed testes showed a rapid decrease in testis size (Lal and Thapliyal 1982). The thyroidectomized birds also showed a rapid loss of black pigmentation in the bill despite receiving exogenous T. This suggests that the thyroid hormones are essential in the process of T-induced pigmentation of the bill (see Chapter 3). Dawson (1998b) used radioactive iodine implanted near the thyroids to thyroidectomize 40 male sparrows that were captured in England and being maintained on short days (8L:16D). Intact males transferred to 16L:8D showed rapid increases in hypothalamic GnRH and in testis mass, but thyroidectomized birds had significantly lower hypothalamic GnRH and showed no significant increase in testis mass. After 25 weeks, thyroidectomized birds on long days had GnRH levels that did not differ significantly from those of either intact or thyroidectomized males kept on 8L:16D. Testis size of the short-day, intact males had increased significantly, however, whereas that of both short-day and long-day thyroidectomzed birds remained unchanged. These results indicate that thyroid hormones are required for both the development and maintenance of the testes under hypothalamopituitary stimulation.

Parathyroid and Ultimobranchial Glands

A pair of parathyroid glands lie just posterior to each thyroid gland. The parathyroid glands secrete parathormone (PTH), which has as its primary target the osteocytes in the bones, stimulating the reabsorption of Ca ion from the bone matrix, resulting in increased circulating levels of Ca (hypercalcemia) (Assenmacher 1973). A second target of PTH is the kidney, where it acts to reabsorb Ca and secrete phosphate in the nephron tubules (Wideman 1987). A recent study in which bovine PTH was administered to segments of nephron tubules and collecting ducts from sparrows found that 10^{-6} M PTH resulted in increased adenylyl cyclase activity in both the thin descending and thick ascending limbs of the loop of Henle (see later discussion) (Goldstein et al. 1999). The difference between the combined mass of the thyroid and parathyroid glands and the average thyroid mass observed by Kendeigh and Wallin (1966) suggests that the mass of the sparrow parathyroid glands is about 2.8 mg.

The ultimobranchial glands are paired glands that lie just posterior to and in line with the thyroid and parathyroid glands (Assenmacher 1973). They secrete the hormone calcitonin, which tends to reduce blood Ca levels (causing hypocalcemia); they thereby serve, with PTH, to regulate blood Ca levels. I found no studies on the structure or function of the ultimobranchial glands in sparrows.

Pancreas

Three hormones are secreted by the islet cells of the avian pancreas. Glucagon, secreted by the α-cells, has the hepatocytes of the liver as its primary target; in these cells, it stimulates the conversion of glycogen into glucose, thereby tending to increase blood sugar levels. It also stimulates lipolysis in the liver. Insulin, secreted by the β-cells, targets cells throughout the body, facilitating glucose absorption by the cells, as well as the hepatocytes, where it stimulates the conversion of glucose into glycogen. These effects tend to lower blood sugar levels. Somatostatin, secreted by the δ-cells, has its primary effect in regulating the secretions of the α- and β-cells. Collectively, therefore, the pancreatic hormones regulate sugar metabolism. Glucagon may also have some effect in the activity of the thick ascending limbs of the loops of Henle of kidney nephrons (Goldstein et al. 2001) (see later discussion).

Goodridge (1964) studied the effects of insulin and glucagon on *in vitro* glucose uptake and fatty acid synthesis by mesenteric and furcular fat tissue taken from sparrows captured in Michigan. Insulin had no detectable effect on the rate of incorporation of ^{14}C from labeled glucose into fatty acids, or on the rate of triglyceride synthesis in either mesenteric or furcular fat. Glucagon also had no effect on glucose uptake by the mesenteric fat tissue, but it did significantly inhibit lipid synthesis and increase the release of free fatty acids from the tissue. Insulin did significantly increase the rate of uptake of glucose by smooth muscle tissue taken from the ventriculus of the sparrows, and it also significantly reduced *in vivo* plasma glucose concentrations.

Skeletal muscle tissues of sparrows have numerous vesicles containing the glucose transporter protein GLUT (Sweazea and Braun 2003). These vesicles apparently release GLUT in the presence of insulin, facilitating rapid glucose uptake by the cells.

Adrenal Gland

The adrenal glands, located just anterior and medial to the cephalic lobes of the kidneys, are, as in other vertebrates, actually composed of two parts that are of independent embryonic origin and secrete different hormones. Moens and Coessens (1970) described the adrenal glands of sparrows as consisting of alternating strands of interrenal tissue (= adrenal cortex) interspersed with islands of chromaffin cells (= adrenal medulla). Quiring and Bade (1943b) reported a mean adrenal gland mass of 5.9 mg (0.024% of TBM) for 101 sparrows. Hartman (1946), however, found that the average mass of the adrenals was 2.43 mg (0.010% of TBM) and reported that there was no difference in adrenal mass between the sexes, although individual adrenal masses varied by a factor greater than 2. In India, Bhattacharyya and Ghosh (1965) reported that the monthly average adrenal mass varied from 1.80 to 2.90 mg (0.009%–0.015% of TBM) in male sparrows and tended to be larger during the breeding season. They also reported that the proportion of interrenal tissue ("cortex") varied from 0.60 to 0.74, and tended to be greater in the breeding season.

Two of the major hormones secreted by the adrenal cortex (interrenal tissue) are corticosterone and aldosterone. Although corticosterone apparently targets primarily the liver, where it stimulates glucogenesis, lipogenesis, and protein degradation (Assenmacher 1973), it also has many other sites of activity, including parts of the brain (Breuner and Orchinik 2001). It may also affect the strength of the immune response (Evans et al. 2000). The principal effect of aldosterone is the stimulation of sodium (Na) retention by the excretory system (Assenmacher 1973). Both hormones could therefore be considered homeostatic, with corticosterone helping to regulate the metabolic rate (see Buchanan et al. 2001) and aldosterone contributing to maintenance of Na balance.

The adrenal medulla (chromaffin cells) secretes epinephrine and norepinephrine, which stimulate lipolysis and moderate hyperglycemia (elevated blood sugar levels) and the higher blood pressure caused by increased rate and strength of contraction of the heart.

Corticosterone release apparently shows a rapid response to stress, and Hegner and Wingfield (1990) found that there was a highly significant correlation between level of circulating corticosterone (B, measured in nanograms per milliliter of plasma) and time after capture at which a blood sample was drawn (t in minutes) for a sample of 444 sparrows in New York (USA): $B = 5.11 + 0.96t$ ($r = 0.55$, $P < .01$). Similar stress responses have been observed in both laboratory (Rich and Romero 2001), and wild-captured sparrows (Romero and Romero 2002). In the former study, corticosterone levels of sparrows maintained at two different photoperiods (11L:13D and 19L:5D) increased rapidly from basal levels of about 7 ng/ml plasma to about 24 ng/ml at 15 min after capture, with no significant difference between the sexes. Similar results were obtained in male sparrows maintained in the laboratory in Scotland (UK), with basal levels 1 min after capture averaging about 10 ng/ml, and levels 10 and 30 min after capture averaging about 30 and 45 ng/ml, respectively (Buchanan et al. 2003). Basal corticosterone levels of sparrows captured during the winter in New Mexico (USA), from which blood samples were drawn within 3 min of capture, were about 5 ng/ml plasma, whereas sparrows held for 15–45 min after capture had average levels exceeding 30 ng/ml (Romero and Romero 2002). In the laboratory study, basal corticosterone levels also showed a consistent daily rhythm, with the highest levels being found at night and the lowest during the day in both photoperiods (Rich and Romero 2001). A plasma protein with high corticosterone affinity has been found in the house sparrow

(Wingfield et al. 1984); it was subsequently identified as corticosteroid-binding globulin (CBG) (see Breuner and Orchinik 2001).

Hegner and Wingfield (1990) monitored levels of circulating corticosterone in free-living male and female sparrows in a population in New York (see Fig. 3.5 in Chapter 3). Corticosterone levels were lowest (approximately 1 ng/ml plasma) in both sexes in September and gradually increased to about seven times that level by midwinter (peaking in January in males and in December in females). During the breeding season, corticosterone levels in both sexes reached peaks of 8–12 ng/ml at the beginning of each breeding attempt, and then dropped to about 1–3 ng/ml during the incubation period. In males, these changes tended to parallel changes in levels of circulating LH and T (see Fig. 3.5a), and in females they tended to parallel changes in LH, estradiol (E_2), and DHT (see Fig. 3.5b). Because high levels of corticosterone coincided with high or rising levels of T, DHT, and/or E_2 in both males and females, Hegner and Wingfield (1986a, 1986c) concluded that neither sex was experiencing undue stress but that high corticosterone levels simply coincided with periods of high metabolic demand. A possible link between corticosterone and prolactin in initiating gonadal recrudescence in sparrows has been suggested, however (see Chapter 3).

Breuner and Orchinik (2001) also observed a seasonal change in both basal and stress-induced levels of circulating corticosterone in free-living sparrows in Arizona (USA). Sparrows captured during the breeding season and in winter had significantly higher basal levels of circulating corticosterone than did those captured during the molt, and breeding birds had a significantly higher level 30 min after capture (stressed) than did those captured during the molt or in winter. These seasonal differences in plasma corticosterone levels may be modulated by differences in amount of circulating CBG, which is also significantly higher during the breeding season than during molt and/or winter. Breuner and Orchinik also identified two corticosterone-binding sites in cytosol isolated from the brains of sparrows, as well as one binding site located in cell membranes from the brain. There were significant seasonal changes in the numbers of both low-affinity and high-affinity cytosolic receptors, with the numbers being significantly lower in the winter than during breeding or molt. On the other hand, the number of membrane binding sites was significantly lower during the breeding season than during molt or winter. Taken together, these results suggest that there is a complex pattern of seasonal change in the central nervous system response to corticosterone secretion.

The seasonal changes in corticosterone levels may be related to differential activity of various parts of the adrenal gland. Moens and Coessens (1970) examined histological preparations of adrenal glands of sparrows collected from November to May in Belgium. They divided the glands along anterior-posterior and lateral-medial axes and estimated the activity of the cortical (interrenal) cells by measuring nuclear size (with the assumption that nuclear size is positively related to cell activity). There was no difference in nuclear size in the anterior-posterior axis, but there were significant differences along the lateral-medial axis of the gland. From November to January, there were significantly larger lateral nuclei (= higher cortical activity) than central and medial nuclei in both males and females. In February, there were no significant differences in either sex, but after February, all three regions tended to show increases in nuclear size, with the central region particularly increasing in both sexes.

The relationship between corticosterone level and metabolic demands is illustrated by the effects on plasma corticosterone levels of the experimental manipulation of brood size (see Chapter 4). Hegner and Wingfield (1987a) either added or removed two nestlings from 17 sparrow broods in New York 4–6 d after hatching. Corticosterone levels increased with brood size in females ($r = 0.321$, $P = .06$), but not in males.

The relationship between corticosterone levels and T levels in males is unclear. Two studies have examined the effects of experimentally manipulated T levels on levels of circulating corticosterone. Both studies used subcutaneously implanted Silastic tubes containing T (or a T antagonist, flutamide, in one study) to manipulate plasma T levels. Hegner and Wingfield (1987b) implanted either T or flutamide into free-living males with nestlings 4–6 d old in New York. The T levels of males captured during the next breeding attempt varied significantly among the treatment groups, with males implanted with T having significantly greater levels than those of control or flutamide-implanted males. Corticosterone levels, which averaged 10.5 ng/ml, did not differ among the three groups, however. In the second experiment, Evans et al. (2000) performed implants on caged males captured in Scotland. Three groups of males were castrated and implanted with Silastic tubes containing either high T, low T, or no T, and a fourth group was sham-operated. Blood samples were drawn at intervals, and T and corticosterone levels were both measured by radioimmunoassay. Plasma levels of both T and corticosterone varied significantly among the treatment groups, and they tended to vary in concert.

The latter study also found that corticosterone levels apparently have an effect on humoral immunity. Evans et al. (2000) estimated the immune competence of the individuals by challenging the blood with sheep erythrocytes. Antibody titer was determined with serial dilution, by observing the lowest concentration at which agglutination of the erythrocytes occurred. High levels of corticosterone tended to lower the immunocompetence of the birds.

The role of aldosterone in maintaining blood Na levels has been tested by manipulating dietary Na levels in captive sparrows. Goldstein (1993) captured sparrows in Ohio, sampled some immediately for plasma aldosterone concentrations, and placed others on diets with either low Na (LS: 0.02 mEq Na^+ per gram of food) or high Na (HS: 0.2 mEq/g) content. After a week, the birds on the LS diet were switched to the HS diet, and those on the HS diet were either switched to the LS diet or kept on the HS diet with the addition of 1% NaCl drinking water (HSS). Some HS birds were also deprived of water for 22 h (dehydrated). Blood samples were taken at intervals after transfer to the new diet (1, 3, and 7 d for LS; 4 h, 24 h, and 4+ d for HS; and 4+ d for HSS), and plasma aldosterone levels were determined. The average plasma aldosterone level of the wild-caught birds was 289.8 pg/ml. Plasma levels in LS birds climbed to 296.2 pg/ml on day 3, then dropped to 182.7 after 7 d, both of which were significantly higher than levels in HS sparrows after 4 d (136.3). Aldosterone levels in HS birds dropped from 224.5 pg/ml after 4 h to 136.3 after 4 d. HSS birds had aldosterone levels of 45.9 pg/ml after 7 d, and dehydrated HS birds had plasma levels of 86.1. Plasma Na level in the wild-caught birds was 151.3 mEq/L, and there were no significant differences from this level among the treatment groups, including the dehydrated sparrows.

Ovary

Ovarian hormones are secreted primarily by cells associated with the ovarian follicles, with the granulosa cells secreting estrogens and the thecal cells secreting progesterone (Farner and Wingfield 1980). The most common estrogen, E_2, is a hormone with numerous morphological, physiological, and behavioral effects (see later discussion), and its plasma concentration shows a regular pattern of seasonal change in free-living female sparrows in New York (see Fig. 3.5b in Chapter 3). Little is known about the effects of progesterone in sparrows, however. It does apparently enhance the development of cutaneous edema in the brood patch of female sparrows, although E_2 is primarily responsible for the development of the brood patch (see Chapter 5) (Selander and Yang 1966). Sparrows do have a plasma protein with high affinity for progesterone, but there is no such plasma protein for E_2 (or for androgens) (Wingfield et al. 1984).

Early studies involving the injection of E_2 into nonbreeding females had no observable effect on the ovary but did result in the development of the oviduct (Kirschbaum et al. 1939; Ringoen 1940). Tewary and Ravikumar (1989a) examined the effects of three different doses of exogenous E_2 (5, 25, or 100 µg, administered at 2-d intervals) on the ovary and oviduct of photostimulated female sparrows in India. They observed significant positive effects on the mass of the oviduct, but there also were significant and dose-dependent decreases in the size of the ovary, which they attributed to the negative effects of E_2 on secretion of gonadotropins by the anterior pituitary gland. Exogenously administered E_2 has negative effects on testis growth in male sparrows as well, which was also attributed to its negative effect on pituitary gland secretion (Pfeiffer 1947).

E_2 also has effects on female reproductive behaviors, including nest building and solicitation of copulation (Farner and Wingfield 1980; Møller 1988). Møller (1990) also reported that male sparrows are apparently able to discriminate between females that may be receptive (based on having E_2 implants) and those that are unreceptive (controls), but the basis for this apparent discrimination is unknown. As noted earlier, E_2 also has an effect on Ca metabolism during egg-laying.

Testis

The principal hormones secreted by the sparrow testis are the androgens T and DHT, but the testis also produces some progesterone (Fevold and Eik-Nes 1962). The Leydig cells of the interstitium are the primary sites of androgen production, although T is also apparently produced by cells of the germinal epithelium (Farner and Wingfield 1980; Pfeiffer and Kirschbaum 1943). Fevold and Eik-Nes (1963) studied the synthesis of T in *in vitro* tissue homogenates of the sparrow testis and found that progesterone was the precursor and 17α-hydroxyprogesterone and androstenedione were intermediate metabolites in the biosynthetic pathway of T. As described in Chapter 3, LH secreted by the pituitary gland is primarily responsible for interstitial development and androgen production in the sparrow testis. After experimental photostimulation, T levels increase rapidly in the testes, but they increased in the plasma more slowly, suggesting that T is initially sequestered in the testes (Donham et al. 1982).

T has multiple effects on the morphology, physiology, and behavior of males, particularly in the development of secondary sexual characteristics and reproductive behaviors. Some of these effects are discussed in detail in other chapters, specifically the effects on bill coloration (Chapter 3) and possible effects on badge size (Chapter 5). The results of a study involving experimental elevation of yolk T levels suggested that T levels during development also influence the dominance behavior of females (Strasser and Schwabel 2004). Fig. 3.5a shows the levels of plasma T in free-living males throughout the year in New York.

T can either stimulate or inhibit the development of the seminiferous tubules and spermatogenic activity in the testes, with the difference in effect apparently being a function of testis size. Pfeiffer (1947) found that daily intramuscular injections of T stimulated rapid development of the testes if the initial testis length was greater than 2–3 mm, but they had no effect on smaller testes. He suggested that the 2–3 mm length represents a threshold size after which the testes respond directly to T, whereas at prethreshold sizes T acts by inhibiting secretion of pituitary gonadotropins, resulting in no increase, and sometimes a decrease, in testes size. The effect of T on the testes is also dose dependent. Turek, Desjardins, and Menaker (1976) implanted T-containing Silastic tubes of different lengths into males (tubes delivered T doses of approximately 1.5 μg/d per millimeter). High doses of T caused full testicular development in males that were captured in December in Texas (USA) with fully regressed testes and maintained on a nonstimulatory photoperiod (8L:16D). High doses administered to breeding-season males resulted in maintenance of active testes, whereas low doses resulted in a marked decrease in testis size and termination of spermatogenesis. Turek, Desjardins, and Menaker concluded that low doses of T inhibit spermatogenesis by negative feedback through the hypothalamus, resulting in reduced gonadotropin release from the pituitary gland. Similar results were obtained by Wolfson and Stahlecker (1950) and by Haase (1975), with the former finding that rapid testis enlargement occurred with daily injections of 2.5 mg of T propionate and the latter observing no increase in size with daily injections of 0.4 mg.

T also stimulates the development of the vas deferens and the glomus vesicles, as well as enlargement of the cloacal protuberance (Haase 1975: Hegner and Wingfield 1987b). Enlargement of the latter may be due in part to an increase in glomus length. Haase (1975) found that T-treated males had a mean glomus length of 143.1 μm, whereas untreated controls had a mean of only 48.7 μm ($P < .001$). In New York, the length of the cloacal protuberance was low (mean, approximately 1 mm) until February, after which it enlarged rapidly, to a mean of greater than 3 mm in April, and remained high throughout the breeding season (Hegner and Wingfield 1990) (see Fig. 3.5A).

As indicated earlier, T exercises some of its effects by influencing the functioning of the nervous system. Smith et al. (1996) used polyclonal rabbit antibody (PG-21) against amino acids 1–21 of the rat androgen receptor (AR) to identify ARs in the brains of male sparrows, some of which were breeding and others of which were non-breeding (with the latter group being either injected with 10 mg/kg of T or maintained as controls). Males in all three groups consistently showed staining for ARs in several hypothalamic regions, including the tuberal, preoptic, and magnocellular paraventricular nuclei. In addition, breeding and T-injected nonbreeding males showed heavy staining for ARs in several structures associated with singing behavior in songbirds, including the HVC, the RA, the MAN, and intercollicular nuclei of the forebrain, and the hypoglossal nerve (cranial nerve XII). The syringeal musculature also stained for the presence of ARs in these groups. The staining was confined to the nuclei in the stained cells. The presence of these ARs indicates that T influences the activity of these regions of the nervous system and the musculature of the syrinx. The fact that staining in song control areas occurs only in breeding males and T-injected nonbreeding males also indicates either that T stimulates the upregulation of the expression of AR genes, or that it influences the synthesis or degradation of AR protein (Smith et al. 1996). Circulating T may also affect the hippocampus of the sparrow brain by being used for the local synthesis of estrogens (Gahr et al. 1993; Metzdorf et al. 1999; Saldanha et al. 1999; see earlier discussion).

T may also have immunosuppressive effects (Folstad and Karter 1992), although evidence for such effects in sparrows is equivocal. In Spain, juvenile males receiving T implants showed a significant decrease in leukocyte count compared with controls, but T-implanted adults actually showed an increase in leukocyte count, although the increase was not significant (Puerta et al. 1995). Similar results were obtained in a subsequent study in which the different types of white blood cells were identified, and adult males with T implants had significantly more lymphocytes than control males did (Paz Nava et al. 2001). In Scotland, the immune competence of castrated males receiving either high, low, or no T supplements (as well as that of sham-operated controls) was tested by challenge with sheep erythrocytes, with the antibody titers being determined by serial dilution (Evans et al. 2000). Antibody response varied significantly among the groups, with the highest titers in the sham control and low-T groups, and lower titers in the high-T and no-T castrates. After the effect of corticosterone levels was removed, with increasing corticosterone levels causing a lower antibody response in all groups (see earlier discussion), the effects of T treatment remained significant, but the highest titers were now in the sham control and high-T groups, the lowest response in the no-T castrates, and an intermediate response in the low-T group. In a subsequent study using a similar design, Buchanan et al. (2003) reported that high T during the breeding season had a negative effect on the humoral immune response, but had no effect on the cell-mediated immune response (measured by wing-web swelling in response to injection of phytohemagglutinin, as discussed later). Greenman et al. (2005) also found that male sparrows implanted with T-containing Silastic tubes had a cell-mediated response that did not differ from that of controls with empty tubes ($P > .3$).

The metabolic rate of sparrows is also affected by T. Buchanan et al. (2001) measured the resting, postabsorptive metabolic rate (BMR; see later discussion) of male sparrows in four treatment groups: (1) castrated, with high T (subcutaneous implants); (2) castrated, with low T; (3) castrated with no T; and (4) intact controls. Circulating levels of T and corticosterone were measured by radioimmunoassay after determination of the BMR. BMR varied significantly among the three castrated groups, with the highest average BMR being in the high-T group (about 57 kJ per bird per day) and the lowest being in the no-T group (about 38 kJ/d); the low-T group was intermediate (about 48 kJ/d). BMR was also significantly positively correlated with circulating T levels within each of the three castrated groups, and it was positively correlated with circulating T in the intact group, although this relationship was marginally nonsignificant ($P = .078$). The significant positive relationship between BMR and circulating T in the castrated groups remained after the positive effect of circulating corticosterone was removed.

Regular changes in the level of circulating T in male sparrows during the course of a single breeding attempt also correlated with marked differences in behavior. Hegner and Wingfield (1986a) monitored both circulating levels of T and behavioral activities of males in a breeding colony in New York. During each breeding attempt, plasma T was maximal during the prebreeding and egg-laying periods (approximately 5–6 ng/ml), dropped to a low during incubation and the early nestling period (about 1 ng/ml), then rose rapidly to peak levels again during the late nestling period. The rate of intrusions by other males was maximal during the prebreeding and egg-laying periods, dropped during incubation, but increased again during nestling feeding, whereas intrusion rates by potential nest predators did not vary during the breeding cycle. Mate-guarding behavior by the male was determined by two behavioral measures, percent of time that both members of a pair were at the nest and percent of time the male spent following the female, and was highest during the prebreeding and egg-laying periods. Males contributed approximately half of the feeds to the young during the first 10 d of the nestling period, but this proportion dropped to about 20%–25% during the last 5 d of the nestling period, and the difference was significant ($P < .005$). This decline in feeding by the male occurred at the same time that T levels were rising again to peak levels. Hegner and Wingfield concluded that high levels of T are associated with behaviors related to high levels of competition for nesting sites and mates.

To experimentally test the effect of T levels on the parental behavior and aggressive behavior of breeding male sparrows, Hegner and Wingfield (1987b) captured males during their first breeding attempt of the season in a breeding colony in New York and gave them subcutaneous implants of Silastic tubes containing one of the following: (1) T, (2) flutamide (a T antagonist that effectively lowers plasma T levels), or (3) nothing (control). The behavior of the males was then monitored during the remainder of the first breeding attempt and during the second breeding attempt. Feeding rates of the males were considered a measure of parental investment, and singing, nest-site defense, and mate guarding were considered aggressive behaviors. During the first breeding attempt, T-treated males fed 8–10 d old nestlings at about half the rate (1.4 feeds/h) of control or flutamide-treated males (2.9 and 2.5 feeds/h, respectively), but the difference was not significant. However, the proportion of total feedings by T-treated males was significantly lower than the proportion of feedings by control males for all ages (8–13+ d old), and the proportion of feedings by flutamide-treated males was significantly higher than that of control males for nestling days 11–13+. Rates of nest-defense behavior were higher in T-treated males at nestling days 8–10 ($.05 < P < .10$), and higher in T-treated males and lower in flutamide-treated males at nestling days 11–14 ($P < .01$). The males were recaptured during the second brood, and mean plasma T level of T-treated males (10.6 ng/ml) was significantly greater than those of control males (3.7 ng/ml) or flutamide-treated males (1.3 ng/ml), and males in all three groups showed the normal pattern of increase in T levels with age of nestlings. Mazuc, Chastel, and Sorci (2003) also found that males implanted with Silastic tubes containing T had significantly lower provisioning rates for both 5- and 10-d-old nestlings in France.

Circulatory System

The circulatory system has two primary functions, the transportation of materials throughout the body and protection from disease. These two functions are jointly performed by two subsystems that are structurally and functionally related, the cardiovascular and lymphatic systems. The former comprises the heart and a closed network of arteries and veins that transport blood away from and toward the heart, respectively. The lymphatic system consists of the lymphatic vessels, blind vessels originating in the tissues that return fluid from the tissues to the systemic venous circulation; lymphatic organs such as the thymus, spleen, and bursa of Fabricius; and various lymphatic tissues.

Cardiovascular System

The avian heart, like that of mammals, is a four-chambered, double-barreled structure that maintains a complete separation between the systemic circulation and the pulmonary circulation. Blood from the systemic circulation returns to the right atrium of the heart through the anterior and posterior vena cavae. Blood then passes from the right atrium into the right ventricle, from which it is pumped into the pulmonary circulation through the pulmonary artery. After the blood is oxygenated in the lungs (discussed later), it returns to the left atrium of the heart though the pulmonary veins, then passes into the left ventricle before being pumped into the systemic circulation through the aorta. Quiring and Bade (1943b) reported that the average mass of the heart for 101 sparrows from Ohio was 0.38 g (1.59% of TBM), and that heart mass correlated strongly with TBM

($r = 0.805$). Tucker (1968) found that the average mass for 5 sparrows from North Carolina (USA) was 0.30 g (1.34% of TBM), and Dubach (1981) found an average of 0.32 g (1.33% of TBM) for 16 sparrows in Switzerland.

The rate of heart contraction is controlled primarily by the autonomic nervous system and shows considerable variation depending particularly on the oxygen demand. Odum (1935) found that the basal heart rate of seven sparrows (postabsorptive, inactive, and being maintained in the laboratory at approximately 31°C) was 350 beats/min. Tucker (1968) found that the average heart rate for five active sparrows being maintained at 5°C and 760 mm Hg was 649 beats/min, and that average heart rate increased to 722 beats/min at 344 mm Hg (see later discussion). Goecke and Goldstein (1997) reported average heart rates of 626.0 and 607.5 beats/min for two control groups of sparrows held at 25°C in the laboratory, and they also found that heart rate tended to decrease with increasing doses of AVT (avian antidiuretic hormone, discussed earlier), although the decreases were not significant.

Cardiac output, the rate at which blood is pumped from the heart, can be estimated as the product of heart rate and stroke volume (the volume of blood pumped during one contraction). Stroke volume can, in turn, be estimated by using the following relationship: O_2 consumption = heart rate × stroke volume × AV difference (where AV difference is the difference between the O_2 content of arterial and venous blood) (Tucker 1968). Tucker estimated that the stroke volume of sparrows held under conditions simulating those at 6100 m altitude would be 2.6 ml/kg TBM and the cardiac output would be 1.88 L/kg/min. For a 28-g sparrow, this would mean that stroke volume under conditions of O_2 stress could be as high as 0.073 ml with a cardiac output of 52.64 ml/min. With a total blood volume of approximately 2.9 ml (Stangel 1986), the heart is therefore pumping the equivalent of 18 times the total blood volume each minute under these conditions!

Blood

Blood is the transport medium that carries materials throughout the body, and it moves through the systemic and pulmonary arteries as the result of a gradient of hydrostatic pressure produced by the heart. It also plays a major role in protection against disease. It is composed of blood cells, the so-called formed elements of the blood, and blood plasma, a complex fluid matrix. Blood samples are usually obtained by venipuncture of the brachial vein lying along the ventral surface of the humerus, and Stangel (1986) found that the taking of samples as large as 0.2 ml has no apparent effect on the health of the individual.

The blood cells are of three types, erythrocytes, leukocytes, and thrombocytes. Hematocrit is the volume proportion of the formed elements in the blood and is obtained by separating the formed elements from the plasma by centrifugation. The average hematocrit of free-living sparrows has been recorded to vary from 0.394 to 0.524 (Baumann and Baumann 1977; Gavett and Wakeley 1986a; Maina 1984; Palomeque et al. 1980; Puerta et al. 1995), whereas in laboratory birds it varies from 0.37 to 0.488 (Goecke and Goldstein 1997; Goldstein and Zahedi 1990; Puerta et al. 1995; Tucker 1968). Puerta et al. (1995) found no age difference in hematocrit between first-year sparrows and adult sparrows captured in June in Spain (0.404 and 0.394, respectively). The hematocrit of both juveniles and adults increased significantly after 3 mo in captivity, however. Gavett and Wakeley (1986a) found no significant difference between the hematocrit of urban (0.474) and rural (0.483) sparrows collected in Pennsylvania (USA). They also found no difference in hematocrit with age or sex, but they did observe a significant effect of the interaction of sex and age. Kruszewicz (1994) found that hematocrit was positively correlated with stage of molt ($r_s = 0.446$, $P < .01$) but negatively correlated with the rate of erythrocytic *Plasmodium* infection ($r_s = -0.452$, $P < .005$) in first-year sparrows in Poland.

The avian erythrocyte is an oval-shaped, nucleated cell containing large quantities of hemoglobin (Hb) and therefore has as its primary function the transport of respiratory gases. The volume of an erythrocyte is about 100 μm^3, and Puerta et al. (1995) found that average cell volume of adults, 100.5 μm^3, was significantly larger than that of juveniles, 94.6 μm^3. Considerable variation has been reported in the number of erythrocytes in sparrow blood. Nice et al. (1935) reported a mean of 5.2 million cells/mm^3 (range, 4.2–5.8) for six sparrows from Ohio, whereas Palomeque et al. (1980) reported a count of 7.3 million/mm^3 for one individual from Spain. Puerta et al. (1995) found that the average for 23 adults in Spain (4.1 million/mm^3) was significantly lower than that for 25 juveniles (4.6 million/mm^3), but there was no difference between June and September in either group. The sparrow erythrocyte possesses membrane agglutinogens that result in the agglutination of the cells in the presence of human ABO and other blood group agglutinins (Bocchi et al. 1960; Norris 1963).

The Hb content of sparrow blood is about 11–13 g/dl (Baumann and Baumann 1977; Puerta et al. 1995), although Palomeque et al. (1980) recorded a value of 17.3 g/dl for one sparrow from Spain. The binding of O_2 to Hb shows the typical sigmoidal pattern as the partial pressure of oxygen (pO_2) increases (Tucker 1968) (Fig. 9.7). Lutz et al. (1974) also determined the O_2 dissociation curve under similar conditions (41°C, pH = 7.5, pCO_2 = 36 mm Hg) for a pooled sample of suspended erythrocytes from six sparrows in North Carolina. They found that the curve was displaced leftward from that in Fig. 9.7, indicating a higher affinity for O_2, and attributed this difference to the use of a methodology that eliminated artifacts caused by O_2 utilized by the erythrocyte nuclei. Their estimate of half-saturation pressure (P_{50}) was at pO_2 = 41.3 mm Hg, compared

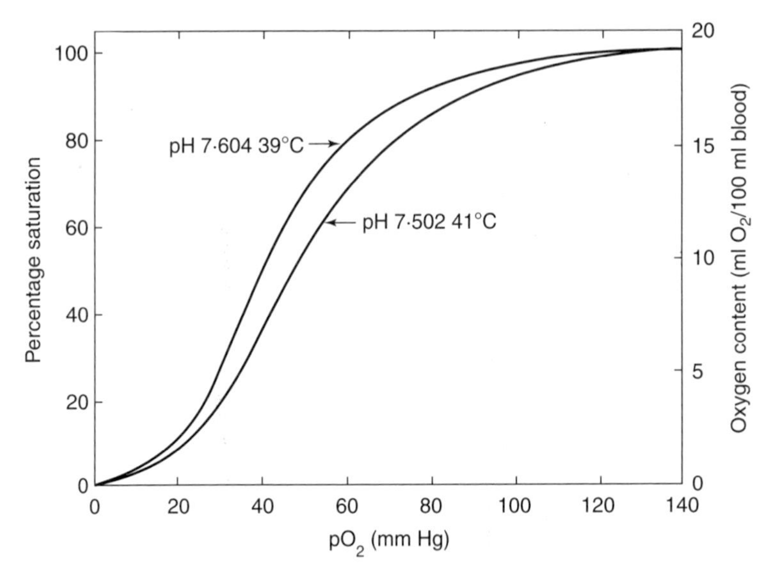

FIGURE 9.7. Oxygen dissociation curve of house sparrow blood, with % saturation of O_2 on the left ordinate and O_2 content of blood (ml O_2/100 ml blood) on the right ordinate plotted against pO_2 (mm Hg). The figure also illustrates the Bohr effect (ml O_2/100 ml blood). From Tucker (1968: Fig. 1), with permission of Company of Biologists, Ltd.

to an estimate of 47.4 mm Hg by Tucker (1968). Fig. 9.7 also illustrates the Bohr effect for sparrow Hb, in that the oxygen dissociation curve shifts to the right as pH decreases. Tucker (1968) found that for sparrow Hb this effect could be described by the equation, $\log P_{50} = 4.571 - 0.386$ pH (SE $= 0.085$, $n = 10$), where P_{50} is the pO_2 (mm Hg) at which 50% of the Hb is saturated as pH ranges from 7.346 to 7.900. A second study of the Bohr effect in sparrow blood estimated the slope of the regression to be -0.51 (Baumann and Baumann 1977). Tucker found that the pH of the blood of sparrows being held at 760 mm Hg was 7.502. Both studies also estimated the oxygen carrying capacity of sparrow blood at pH 7.5, with Tucker finding it to be 19.1 ml O_2 per deciliter of blood, and Baumann and Baumann finding it to be 16.2 ml/dl.

Avian leukocytes are of several types, including granulocytes, monocytes, and lymphocytes. Lymphocytes, which are involved in disease resistance, include both T-lymphocytes and B-lymphocytes (see later discussion). In sparrows captured in June in Spain, Puerta et al. (1995) found that the leukocyte count for first-year birds (29,500 cells/mm^3) was significantly greater than that for adults (21,800 cells/mm^3). This difference was primarily due to a significantly greater number of lymphocytes in first-year birds compared with adults (11,800 vs. 6,500 in adults) (Paz Nava et al. 2001).

Thrombocytes are nucleated cells that may represent an arrested stage in erythrocyte poiesis and are involved in blood clotting. I found no reference for the number of thrombocytes in sparrow blood.

Blood plasma is a complex water-based solution of proteins and other organic molecules and various inorganic ions. It is the medium within which the transport of virtually all materials except most of the O_2 and some of the CO_2 occurs. These materials include the basic organic building blocks required by all cells, such as sugars, amino acids, and triglycerides, as well as hormones, vitamins, and organic waste products such as urea, uric acid, and other nonprotein nitrogen. They also include significant quantities of CO_2, some of which is in solution in the plasma and some of which is transported in solution as bicarbonate ion. Among the proteins are the albumins and immunoglobulins, the latter involved in disease resistance, as well as numerous protein enzymes.

Parrish and Mote (1984) analyzed many of the serum components of adult female sparrows that were captured during the winter in Kansas and maintained in captivity for 2–4 wk. Total serum protein averaged 3.90 g/dl, with albumins averaging 1.12 g/dl and globulins, 2.77 g/dl. The average serum cholesterol level was 211.0 mg/dl, and the glucose concentration was 528.60 mg/dl. Average urea and uric acid concentrations were 2.79 and 19.87 mg/dl, respectively, and the Ca content was 8.44 mg/dl.

Gavett and Wakeley (1986a) found that the mean total blood cholesterol of urban sparrows (249.0 mg/dl) was significantly greater than that of rural sparrows (221.9 mg/dl) in Pennsylvania. Albumin was 1.66 g/dl for urban and 1.59 g/dl for rural birds and differed significantly with age, with adults having higher albumin levels than juveniles. Uric acid content was 17.3 mg/dl for urban and 16.6 mg/dl for rural birds and did not differ significantly with habitat, sex, or age. Average blood urea nitrogen differed significantly with habitat, being 3.87 mg/dl in urban and 2.46 mg/dl in rural sparrows. Puerta et al. (1995) reported that the protein content in 40 sparrows from Spain was 3.6 g/dl of plasma, and that it did not differ between adult and first-year birds. Mean plasma concentration of triglycerides was 224 mg/dl and also did not differ significantly between adults and juveniles.

Lymphatic System and Immunity

The two major functions of the lymphatic component of the circulatory system are protection from disease and the return of interstitial fluid from the tissues to the cardiovascular system. Several lymphatic organs, but particularly the thymus and the bursa of Fabricius, produce several types of lymphocytes that are involved in protection from disease.

The thymus of the house sparrow consists of a variable number of lymphatic lobes located adjacent to the jugular vein in the neck. Typically, the two largest lobes are found at the angle of the jaw (Hohn 1956). The principal function of the thymus is the production of T-lymphocytes. The thymus is usually larger in juveniles than in adults, in which it tends to become involuted. The lobes of the juvenile thymus have a distinct cortical region consisting primarily of lymphocytes, and a medullary region having islands of epithelial cells and fewer lymphocytes (Hohn 1956). The average mass of the thymus in juvenile sparrows in Alberta (Canada) was about 30 mg, with no difference between the sexes (Hohn 1956). Adults have a much smaller thymus throughout most of the year, averaging about 7.5 mg, except during the summer months. During June–August in males and July–September in females, thymus mass is significantly higher in both sexes, averaging 11.5 mg in males and 36.7 mg in females (Hohn 1956). This seasonal enlargement of the thymus may be caused by increased production of lymphocytes to accommodate an increased blood volume during the molt.

The bursa of Fabricius is a dorsal diverticulum of the posterior cloaca that is present in juveniles but disappears after sexual maturity (see Fig. 9.2). Its primary function is the production of a group of lymphocytes called B-lymphocytes, which are involved in the production of antibodies in the humoral immune response. In Alberta, no bursae were found in adult sparrows of either sex, whereas the mean mass of the bursa in juvenile males was 31 mg, and in juvenile females it was 19 mg (Hohn 1956). Møller et al. (1996) measured the length and width of the bursae of 145 sparrows collected between late September and late March in Denmark. The mean length for all specimens was 5.95 mm, and the mean width was 3.88 mm. Bursa volume was estimated from the linear measures and did not differ between the sexes. There was a significant negative correlation of bursa volume with $\log_{10}$ date (days beginning with 1 May) ($R^2 = 0.52$). Møller et al. also found that the number of Mallophaga feather mites (determined by counting the number of holes made in the primaries and secondaries of both wings) was positively related to bursa size.

The spleen is located along the dorsal surface of the right lobe of the liver, dorsal to the proventriculus. It comprises both white pulp and red pulp, the former containing numerous lymphocytes and the latter storing excess blood. Quiring and Bade (1943b) found that the average mass of the spleen in sparrows from Ohio was 42 mg (0.2% of TBM). Møller and Erritzoe (2000) found that the mean spleen mass of sparrows captured by domestic cats in Denmark (approximately 25 mg) was significantly less than that of sparrows killed in accidents (approximately 55 mg). They used spleen mass as an indication of disease state and concluded that predators prey differentially on diseased individuals (see Chapter 8).

The two principal disease-fighting mechanisms, both associated with the lymphatic system, are humoral immunity and cell-mediated immunity. The former involves the production of antibodies by derivatives of B-lymphocytes, and the latter involves the production of cells, primarily T-lymphocytes, that attack disease organisms. Recent studies of the immune responses of sparrows have used wing-web (propatagium) swelling in response to the injection of phytohemagglutinin (PHA), a mitogen capable of stimulating a local inflammatory response without causing infection, to identify and quantify a cell-mediated response. In most cases, the PHA is injected intradermally into the propatagium of one wing, and the other wing is injected with physiological saline, with each individual serving as its own control. Propatagial thickness is then monitored in both wings, and the difference between the two wings (wing-web swelling) is interpreted as representing the strength of the cell-mediated response (Bonneaud et al. 2003; Buchanan et al. 2003; Gonzalez, Sorci, and de Lope 1999; Gonzalez, Sorci, Møller et al. 1999; Navarro et al. 2003). Other studies used the difference between thickness of the propatagium before injection and that 24 h after injection to indicate the strength of the response (Greenman et al. 2005; Martin et al. 2004). Martin et al. (2003) used two sets of individuals to test for the energetic cost of the cell-mediated response (see later discussion).

Two techniques have been used to measure the strength of the humoral response in sparrows. Both involve stimulation of antibody production by the injection of sheep erythrocytes (SRBCs), followed by a challenge with a second injection of SRBCs. The first technique then attempts to measure the antibody response by isolating plasma proteins, separating them electrophoretically, and comparing the immunoglobulin-containing fraction densimetrically with total plasma protein (Gonzalez, Sorci, Møller et al. 1999). The second technique uses a radiolabeled antibody to the SRBC antibody to directly measure the strength of the response (Buchanan et al. 2003).

Martin et al. (2004) examined latitudinal differences in immune response in sparrows. Sparrows from Panama (9°1'N) showed significantly greater wing-web swelling in response to PHA injection than did sparrows from New Jersey (USA) (40°21'N) during the early part of the breeding season, but New Jersey birds had significantly greater responses during the late breeding and nonbreeding seasons. This difference persisted after 2 y in a common garden experiment in which New Jersey and Panama birds were held in the laboratory in New Jersey. The adaptive significance of such a latitudinal difference in the nonspecific inflammatory response is unclear.

Respiratory System

The avian respiratory system is the most efficient among animals, permitting some birds to fly at altitudes in excess of 7500 m (Tucker 1968). Under conditions simulating those at 6100 m, sparrows remain active, maintain their body temperature within the normal range, and can even fly short distances (Tucker 1968). During 1 h under the same conditions, a comparably sized laboratory mouse (*Mus musculus*) becomes totally inactive and loses more than 10°C in body temperature. The difference in response of the two animals is primarily due to the ability of the sparrow to sustain oxidative metabolism by extracting sufficient O_2 from air that contains less than half as much O_2 as air at sea level. This ability rests primarily on the exceedingly efficient movement of O_2 across the respiratory exchange surfaces in the lung, which emphasizes the close functional relationship between the respiratory system and the circulatory system.

The avian respiratory system consists of the nasal cavities, the pharynx, the trachea, the bronchi, the lungs, and two groups of air sacs, anterior and posterior. In addition, the syrinx, which is responsible for producing vocalizations, is located at the junction of the trachea and the primary bronchi. The principal function of the respiratory system is the exchange of respiratory gases with the environment, obtaining O_2 that is required for oxidative energy production and eliminating excess CO_2. The respiratory system, as the producer of vocal signals, also plays a major role in communication. In addition, it plays a role in thermoregulation, with evaporative cooling through the respiratory system serving as the major means of dissipating excess heat. Even nestlings that are not yet capable of generating sufficient heat to thermoregulate at low ambient temperatures are capable of dissipating excess heat at high ambient temperatures by open-mouth breathing and panting (O'Connor 1975c).

External Respiration

Air enters the system through the external nares and passes through the nasal cavities and internal nares into the pharynx (posterior buccal cavity). It then passes through the slit-like glottis into the trachea, with the glottis being protected by the larynx, a valve-like structure surrounding the glottis that opens and closes with each breathing cycle.

Unlike the saccular, bellows-like mammalian lung, the avian lung has a fixed size, and air flows through it into the air sacs rather than into and out of it. The primary bronchus on each side passes through the entire lung, giving rise to numerous secondary bronchi that are in turn connected to each other by a dense, parallel network of parabronchi. The parabronchi and some of the secondary bronchi are surrounded by a mantle of exchange tissue consisting of closely apposed air capillaries arising from the parabronchi and blood capillaries from the pulmonary circulation (Maina 1984). Respiratory gas exchange takes place across the air–blood barrier separating air in the air capillary from the erythrocytes in the blood capillary. Quiring and Bade (1943b) reported that the average mass of the lungs of 101 sparrows from Ohio was 0.40 g (1.65% of TBM), whereas Tucker (1968) found that the average for 5 sparrows from North Carolina was 0.17 g (0.74% of TBM), and Dubach (1981) found that the average for 16 sparrows from Switzerland was 0.27 g (1.14% of TBM). The more than two-fold difference among the studies is probably caused by differences in tissue preparation and degree of blotting.

The seven air sacs of the house sparrow are depicted in Fig. 9.8 and include three anterior sacs (cervical, interclavicular, and anterior intermedial) and four posterior sacs (paired posterior intermedial and paired abdominal). The clavicular and anterior intermedial sacs are connected medially and are sometimes treated as a single sac. Dubach (1981) prepared silicone casts of the respiratory system (beginning at the anterior end of the trachea) of 16 sparrows from Switzerland. The air sacs comprised an average of 85.6% of the total volume of the respiratory system, with the lung making up 12.0%, the trachea 1.2%, and pneumatic spaces 1.2%. The volume of the trachea, 0.08 ml, was very similar to the 0.07 ml calculated by Hinds and Calder (1971) for five sparrows from Arizona.

Dubach (1981) and Maina (1984) used both light and electron microscopy to describe the internal anatomy of the sparrow lung and to estimate the total respiratory exchange surface. Dubach found that 11.4% of the lung volume consisted of primary and secondary bronchi, 83.9% of parabronchi, and 4.7% of large blood vessels. The parabronchi, in turn, consisted of 30.6% parabronchial vessels, including the atria into the air capillaries, and 69.4% blood and air capillary net. Maina reported that 1.0% of the lung was primary bronchus, 6.4% large blood vessels, 36.9% secondary bronchi and parabronchi, and 55.7% exchange tissue (blood and air capillaries). The latter figure corresponds well with the 58.2% obtained from multiplying the percentage of exchange tissue in the parabronchi reported by Dubach (69.4%) by the percentage of the lung occupied by parabronchi (83.9%). Within the exchange tissue, Maina reported that 45.7% of the area consisted of air capillaries, 38.0% blood capillaries, 12.9% tissue of the air–blood barrier, and 3.4% tissue not involved in gas exchange. He also found that the total effective surface area of the air capillaries was 0.170 m^2 (96.0% of the total surface area of the air capillaries), which meant that there was 63 cm^2 of effective surface area per gram of TBM. Dubach found that the total effective surface area was 0.150 m^2 (84.3% of the total air capillary surface area), or 59 cm^2 per gram of TBM. Both studies also measured

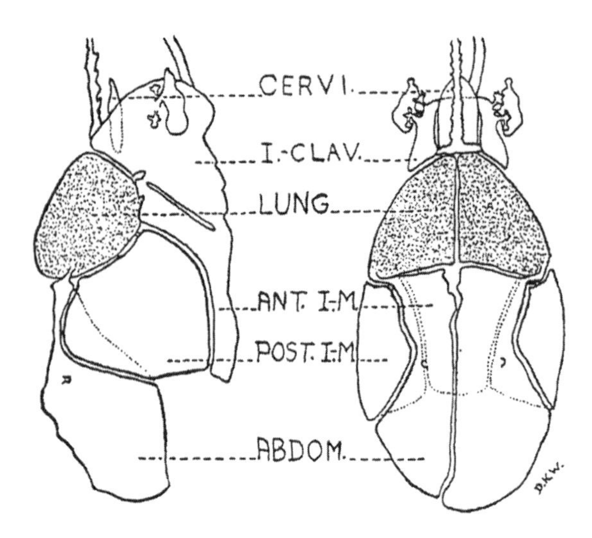

FIGURE 9.8. Air sacs of the house sparrow. Lateral (left) and dorsal (right) views of the respiratory system showing the seven major air sacs—cervical (CERVI..), interclavicular (I.-CLAV.), anterior intermedial (ANT. I.-M.), paired posterior intermedial (POST. I.-M.), and paired abdominal (ABDOM.)—in relation to the lungs (LUNG). From Weatherbee (1951), with permission of the American Ornithologists' Union.

the thickness of the air–blood tissue barrier, with Dubach reporting a harmonic mean of 0.118 μm and Maina reporting a harmonic mean of 0.096 μm.

The ventilation of the respiratory exchange surfaces of the lung is an exceedingly complex process. During inhalation, inspired air passes through the lungs in the primary and some of the secondary bronchi to the posterior air sacs, while air already present in those sacs is being forced anteriorly through the respiratory exchange apparatus of the lung to the anterior air sacs. During exhalation, air from the anterior air sacs is expired and some of the air in the posterior sacs passes anteriorly through the exchange apparatus to the anterior sacs. The result is an almost continuous flow of air through the air capillaries, which permits gaseous exchange between the air capillaries and the blood capillaries. Countercurrent flow of air and blood permits even more efficient exchange.

Arens and Cooper (2005) measured daytime and nocturnal respiratory rates and tidal volumes of sparrows on the day of capture during both summer and winter in Wisconsin. The mean summer daytime respiratory rate (61.4 breaths/min) was significantly higher than the nocturnal rate (52.5 breaths/min) ($P = .01$). Tidal volumes averaged about 1.0 ml and did not differ significantly with either time of day or season.

As noted earlier, Tucker (1968) examined some respiratory and circulatory system responses of sparrows being held for 1 h under conditions simulating those at approximately 6100 m altitude (atmospheric pressure 344 mm Hg, temperature 5°C). In addition to the general behavioral observations described previously, respiration rate, heart rate, body temperature, and O_2 consumption were monitored during the trial and compared to the same measures for the birds held at 760 mm Hg and 5°C. Both respiration and heart rate increased with the decrease in atmospheric pressure, the former increasing from 81 to 112 breaths/min (a 38% increase that was probably significant), and the latter from 649 to 722 beats/min (an 11% increase that probably was not significant). O_2 consumption increased from 4.52 to 5.54 ml/g/h (a 23% increase). Tucker calculated the effective ventilation (milliliters of gas actually reaching the respiratory surfaces of the lungs per gram of TBM per hour) under the two conditions and concluded that it increased by 77% in the high-altitude conditions. Despite this increase in effective ventilation, however, the total amount of O_2 that reaches the respiratory surfaces of the lung at 6100 m is only 74% of that at sea level. The sparrow lung is therefore utilizing 7% of the O_2 in the effectively ventilated air at sea level and 12% at 6100 m. Tucker also concluded that sparrows at 6100 m could tolerate severe hypoxia, with the arterial blood containing only 24% of the O_2 held at sea level. The combination of these responses enables sparrows to remain active, and even fly, under conditions that are inimical to mammals.

Syrinx

The syrinx of the house sparrow closely resembles that of other oscine passerines (Ames 1971). It is composed of cartilaginous elements in both the trachea and bronchi, internal membranes attached to those elements, and external muscles, also attached to the elements. The muscles include two pairs of extrinsic muscles (*musculus tracheolateralis* and *m. sternotrachealis*) and four pairs of intrinsic muscles (*m. bronchotrachealis posticus, m. bronchotrachealis anticus, m. bronchilis posticus*, and *m. bronchilis anticus*) (Ames 1971). Vocalizations are produced by vibrations in the internal membranes of the syrinx, with the musculature modulating the sound by altering the tension in the membranes.

Digestive System

The digestive system consists of the alimentary tract, which begins at the mouth and ends at the cloacal orifice, and various glands derived from and associated with the tract. These glands include the salivary glands, liver, gall bladder, and pancreas. The principal functions of the alimentary tract are the ingestion, transport, mechanical breakdown, chemical breakdown (digestion), and absorption of food and the elimination of indigestible material. Food is the source of energy and nutrients for growth, development, and maintenance metabolism. Because growth is determinate in sparrows, once growth and development are complete, the intake of nutrients and energy must balance their expenditure over the long haul in adults. The house sparrow has been one of the principal species in which studies of avian energetics have been done.

Alimentary Tract

The alimentary tract includes the oral cavity, pharynx, esophagus, proventriculus, ventriculus, small intestine, large intestine (rectum), and cloaca. The oral cavity and pharynx of a house sparrow are depicted in Fig. 9.9, which also includes a sagittal section through the tongue. The bill (derived from the skin) is the primary structure involved in the acquisition of food (as noted earlier), but the tongue also plays a role in food acquisition, as well as in the manipulation of food before swallowing. The tongue in the house sparrow has a well-developed seed cup, which is supported by a partially ossified endochondral bone, the preglossale (Fig. 9.9). During swallowing, food passes from the pharynx into the esophagus, the first part of the digestive tract. The average mass of the digestive tract in 101 sparrows from Ohio was 2.60 g (10.8% of TBM) (Quiring and Bade 1943b). Gier and Grounds (1944) and Klem et al. (1982, 1983) provided detailed descriptions of the gross anatomy and histology of the digestive tract of sparrows, and the following summary account of alimentary canal morphology is based on their descriptions.

The esophagus consists of three parts: an anterior esophagus with a diameter of about 3.2 mm and a length of about 7 mm; the crop (ingluvies), which is about 17.5 mm long and bag-like; and the posterior esophagus, which is about 16.5 mm long with a diameter of about 2.2 mm. The crop has distensible walls that permit considerable variation in diameter. The function of the anterior and posterior parts of the esophagus is the transport of ingested food from the pharynx to the proventriculus, and that of the crop is temporary storage of food.

The proventriculus (or glandular stomach) is about 5.5 mm long and cone-shaped, with the diameter adjacent to the entrance from the esophagus being about 2.2 mm and that at the entrance to the ventriculus about 4.4 mm. Externally it resembles the esophagus, but internally it has a narrow lumen surrounded by spongy-appearing, multifolded walls. It is lined with simple columnar epithelium, with uniformly distributed papillae that mark the openings of proventricular glands. The external muscle layers are thin, which readily distinguishes the proventriculus from the ventriculus.

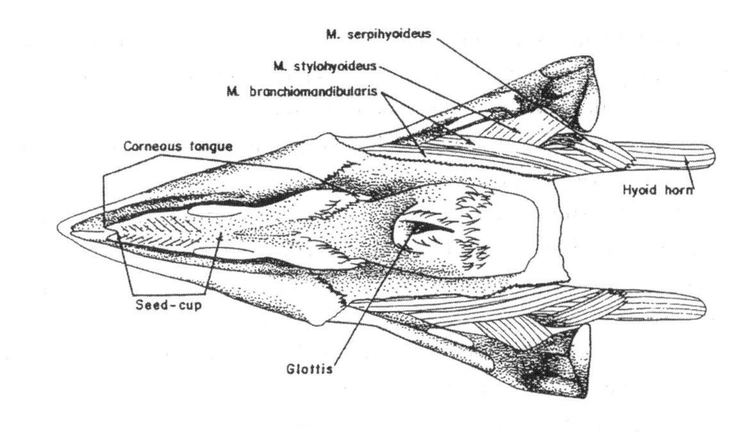

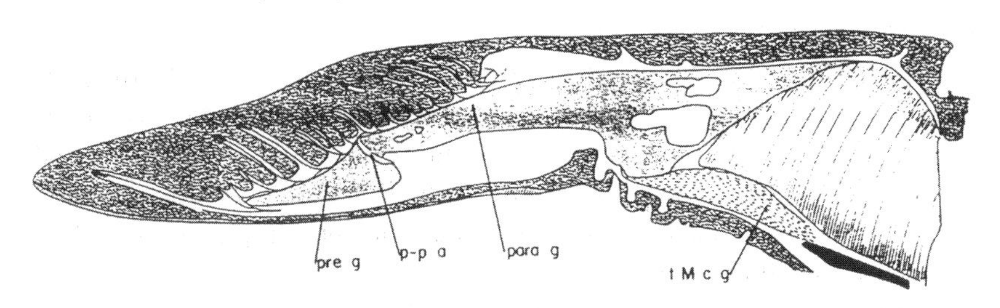

FIGURE 9.9. Mouth and tongue of the house sparrow. Top: Dorsal view of the lower mandible showing the seed cup in the tongue and the position of the glottis. Bottom: Sagittal section through the right side of the tongue showing the preglossale (pre g), the paraglossale (para g), and the articulation between the two bones (P-p a) and the tendon of the *m. ceratoglossus* (t M c g). From Bock and Morony (1978b: Figs. 1 and 4A), with permission of Wiley-Liss, a subsidiary of John Wiley & Sons, Inc.

The ventriculus (or gizzard) is oval-shaped; the proventriculus enters at the anterior end, and the duodenum of the small intestine leaves the ventriculus laterally to the left. Its maximum length and width are about 17.5 mm and 15.9 mm, respectively, and its dorsoventral depth is about 10.0 mm. It is lined internally with a hard cuticle that ranges in thickness from 50 to 425 μm. The inner epithelium consists of a layer of cuboidal or columnar cells overlying a row of densely packed tubular glands that secrete the cuticle. The external muscle layer can be separated into four semiautonomous muscles radiating from a central tendon, and together they make up the bulk of the ventriculus.

The primary function of the ventriculus, which often contains large amounts of grit (see Chapter 6), is the mechanical breakdown of food by rhythmic contractions of the thick muscle layers. Some chemical breakdown of food is also performed in the stomach, with the digestive enzymes being secreted primarily by the proventricular glands.

The small intestine varies in length from 115 to 138 mm and comprises three regions, the duodenum, jejunum, and ileum. Externally there is no marked difference among the three regions, but the small intestine gradually decreases in diameter from about 3.5 mm anteriorly to about 2.5 mm posteriorly. Internally the small intestine is lined with villi, which are arranged in a zigzag pattern and decrease in length, from 325–500 μm anteriorly to 40–205 μm posteriorly. The epithelium of the small intestine is the simple columnar type with a brush border, whereas the submucosa is composed of layers of glands of Lieberkuhn. The duodenum receives two bile ducts and three pancreatic ducts. Most of the chemical breakdown of food and most of the absorption of those breakdown products and mineral nutrients take place in the small intestine.

The juncture between the small and large intestines is marked by the presence of a pair of elliptical ceca, which are composed primarily of lymphatic tissue. The large intestine is about 10.5 mm long, 3.0 mm in diameter, and structurally similar to the small intestine. The villi of the large intestine are 25–130 μm long and have about twice as many goblet cells per unit length as those of the small intestine.

The cloaca forms a chamber with a larger diameter than the lumen of the large intestine. Anteriorly it is similar in structure to the large intestine, with villi 50–85 μm in height, and lined with simple columnar epithelium with many goblet cells. A sphincter at midcloaca separates this region (the coprodeum) from the posterior cloaca, which is lined with stratified squamous epithelium. The posterior cloaca receives the ureters and the oviduct or the vasa deferentia. The primary function of the large intestine is apparently the absorption of water from the intestinal contents, but reflux of urine into the coprodeum and large intestine may also result in reabsorption of Na as well as water from the urine (see later discussion).

Panicker and Acharya (1994) described the development of the ventriculus, proventriculus, and intestine of sparrows in India. They collected nestlings 1, 5, 10, and 15 d old, as well as adults, and prepared sections of the three structures for histological examination. The size of the compound glands of the proventriculus grew throughout the period, with the greatest increase occurring between 10 and 15 d of age. The glands grew 8%–9% between 15 d and adulthood. The muscular mucosa of the proventriculus also increased throughout development, although the only significant increase (52%) occurred between 5 and 10 d. The ventriculus, which comprised 3.6% of adult body mass, was 7.5% of the mass of 1-d-old chicks and 4.5% of the mass of 15-d-olds. The muscular layer of the ventriculus increased significantly in thickness during each interval, with the largest increases occurring in the early nestling stages. The villi in the intestine increased in length throughout the growth period, with the greatest increase being between the 15-d-old nestlings and the adults. The relative size of the alimentary tract at hatching is large in sparrows, as in other altricial species, but the developmental changes in internal morphology suggest that the digestive and absorptive efficiencies improve considerably during the nestling period, and even after fledging (however, see later discussion).

Digestion and Absorption

As noted previously, most of the digestion and absorption of food takes place in the small intestine. Most digestive enzymes are produced either in the pancreas or in the lining of the small intestine, and enzyme release is controlled by the autonomic nervous system and by a series of hormones produced within the digestive tract. Absorption of the products of digestion through the walls of the small intestine is either passive or active, with the latter being mediated by Na^+-dependent transport sites.

The Adaptive Modulation Hypothesis, an hypothesis related to digestive and absorptive efficiency, is based on the assumption that the maintenance of enzyme production sites and of specific cell-mediated absorption sites is energetically costly to the individual. The hypothesis was first formulated to explain regulation of absorption mechanisms caused by seasonal changes in diet or physiological state (Karasov and Diamond 1983). The hypothesis predicts that, in an omnivorous species such as the house sparrow (see Chapter 6), the sites responsible for enzymatic production and mediated absorption should be upregulated by the presence of the specific substrate upon which they act. Upregulation of these sites means that the presence or concentration of a particular substrate determines the activity of the substrate-specific sites of enzyme production or mediated absorption.

Recent studies on sparrows attempting to test the predictions of the Adaptive Modulation Hypothesis have obtained equivocal results (Caviedes-Vidal and Karasov 1995, 1996; Caviedes-Vidal et al. 2000). The studies used modifications of a semisynthetic diet for the house sparrow first described by Murphy and King (1982), who found that the diet provided sufficient energy and nutrients to sustain body mass and normal molt schedule in molting birds. The diet was modified to produce a high-carbohydrate diet (HC = 60.5% corn starch content), a high-protein diet (HP = 60.3% casein and amino acid mixture), or a high-lipid diet (HL = 40.0% corn oil).

Caviedes-Vidal and Karasov (1996) used the everted-sleeve technique to measure *in vitro* absorption rates of tritium-labeled D-glucose and two amino acids, L-leucine and L-proline. Small intestines of sparrows maintained for 3 wk on one of the three diets (HC, HP, or HL) were removed, and 1-cm segments of either the proximal 25% or the distal 75% of the intestine were everted over stainless steel rods and incubated with varying concentrations of one of the labeled substrates. The rate of D-glucose absorption increased with increasing glucose concentration, indicating the presence of mediated uptake mechanisms for the absorption of glucose. Contrary to the prediction of the adaptive modulation hypothesis, however, the rate of mediated glucose absorption in birds on the HC diet was significantly lower than that in birds on the HP diet. Concurrent studies of the absorption rate of R-glucose (the stereoisomer of D-glucose), which is not metabolized by sparrows and for which there are no transport sites, indicated that about 80% of the absorption of glucose in the small intestine is passive. High rates of passive absorption of carbohydrates in the sparrow intestine were confirmed in *in vivo* studies of the absorption of two other stereoisomers, (D-mannitol and L-arabinose) (Chediack et al. 2001). The L-leucine results conformed to the expectations of the hypothesis, with the rate of leucine uptake being significantly higher in the HP group than in the HC and HL groups (Caviedes-Vidal and Karasov 1996). The uptake of L-proline was tested only in the distal segments and did not differ among the three dietary groups. The absorption rates of both D-glucose and L-leucine were higher in the distal segments than in the proximal segments, which indicates that more absorption of these nutrients occurs in the jejunum and ileum than in the duodenum.

Caviedes-Vidal et al. (2000) used a similar protocol to test for the regulation of enzyme activity in the small intestine. Sparrows were sacrificed after 10 d on an HC, HP, or HL diet, and their small intestines were removed and divided into three equal-length segments (proximal, middle, and distal). The activities of three intestinal carbohydrate enzymes (maltase, sucrase, and isomaltase) and one protein enzyme (aminopeptidase-N) were assayed in whole tissue homogenates of the intestinal segments. The results of the experiment are depicted in Fig. 9.10. Two of the carbohydrate enzymes (maltase and sucrase) showed significantly decreased activity from proximal to distal segments in the intestine, but isomaltase activity did not vary significantly with position. Aminopeptidase-N, on the other hand, showed significantly increased activity distally. Aminopeptidase-N activity was also significantly higher in the birds on the HP diet. The activities of the carbohydrate enzymes did not differ significantly among the dietary groups, although there was a tendency for birds on the HL diets to have lower levels of both maltase and sucrase than

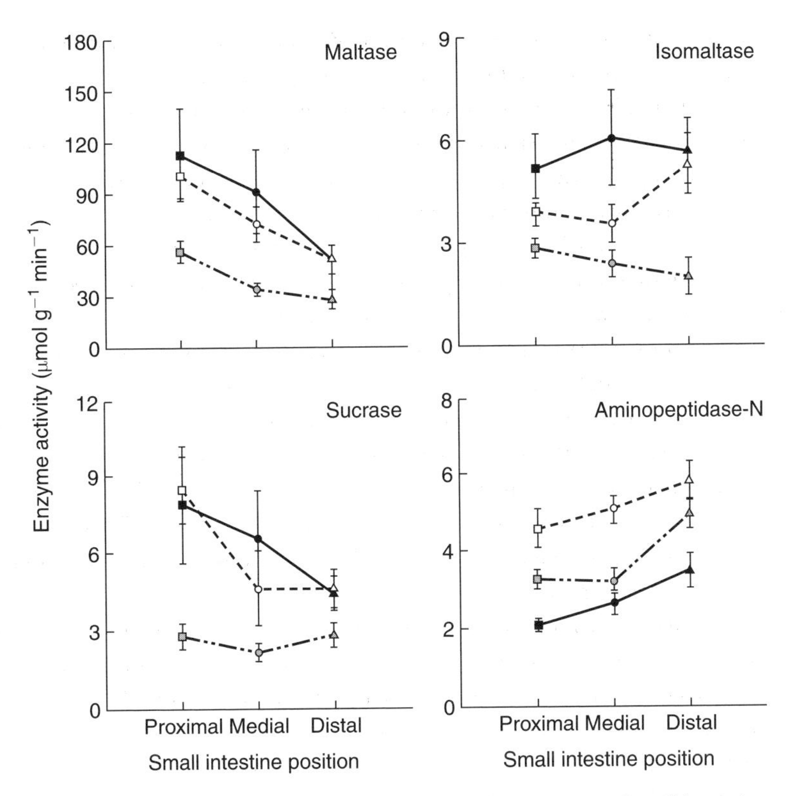

FIGURE 9.10. Enzyme activities in the house sparrow intestine. Small intestines of sparrows were removed after 10 d of maintenance on one of three diets: high carbohydrate (HC, solid line), high protein (HP, dashed line), or high lipid (HL, dashed and dotted line) (see text). Enzyme activities of three intestinal carbohydrate enzymes (maltase, sucrase, and isomaltase) and one intestinal protein enzyme (aminopeptidase-N) were measured in tissue homogenates of the proximal, middle, and distal sections of the intestine. From Caviedes-Vidal et al. (2000: Fig. 2), with permission from Elsevier.

birds on the other diets ($P < .10$). Similar results were obtained by Caviedes-Vidal and Karasov (1995) for pancreatic enzyme assays of sparrows held for either 10 or 45 d on one of the three diets. The ratio of amylase to trypsinogen or chymotrypsinogen was higher in HC birds than in HP birds, but the difference was not significant. The ratios were significantly higher in both of those groups than in the HL birds, however.

Collectively, these results indicate that regulation of protein enzyme production by the small intestine and pancreas and of mediated-absorption sites for amino acids in sparrows may conform to the Adaptive Modulation Hypothesis, but that this is not true for carbohydrate enzymes or absorption sites. One possible explanation for this failure to conform to the predictions of the hypothesis is that the house sparrow, despite its great dietary opportunism, tends to have a high proportion of grain in its diet at all times of the year (see Chapter 6). Because most grains have a relatively high carbohydrate content, sparrows virtually always have a high-carbohydrate diet, making modulation of carbohydrate digestion and absorption superfluous. One interesting corollary observation was that the average length and average mass of the small intestine in HC birds were both significantly lower than those of HP birds ($P <$ .001) (Caviedes-Vidal and Karasov 1996). Caviedes-Vidal et al. (2000), however, found no difference in intestinal mass and only a trend ($P = .07$) toward a shorter intestine in HC birds. The difference between the two studies is possibly related to the different periods of time on which the birds had been maintained on the different diets—21 d and 10 d, respectively. These results suggest that there may be other ways in which the intestine can modulate its structure or physiology to adapt to different diets.

The passive absorption of carbohydrates, for instance, may be modulated by the rate of active transport of glucose or other nutrients. This absorption apparently occurs along a paracellular pathway for water-soluble molecules that are unable to cross the intestinal cell membranes (Chediack et al. 2003). Two studies found that the rate of passive absorption of nonmetabolizable carbohydrates increased in sparrows in which active absorption was occurring. In one study, the rate of L-glucose absorption increased in the presence of mediated absorption of D-glucose *in vivo* (Chang et al. 2004), and in the other study, rates of passive absorption of L-arabinose, L-rhamnose, perseitol, and lactulose increased significantly in fed birds and in birds in which D-glucose had been introduced by gavaging (Chediack et al. 2003). In the latter study, the rate of absorption was also found to decrease with increasing molecular weight of the carbohydrates, indicating that the paracellular pathway may limit the size of molecule that can be passively absorbed.

Caviedes-Vidal and Karasov (2001) examined developmental changes in the digestive physiology of nestling sparrows in Wisconsin by monitoring the activities of both intestinal and pancreatic enzymes from hatching to fledging (12-d-old nestlings). For all seven enzymes assayed (intestinal aminopeptidase-N, sucrase, maltase, and isomaltase and pancreatic trypsin, chymotrypsin, and amylase), activities increased significantly with age from hatching to fledging, with a 10-fold increase in maltase and sucrase activities and a 100-fold increase in amylase activity. Caviedes-Vidal and Karasov suggested that these developmental changes in digestive efficiency may limit growth rate in developing sparrows (see Chapter 4).

Liver

The liver is functionally one of the most diverse organs in the body. It performs multiple metabolic functions, both anabolic and catabolic; stores glycogen, lipids, and some nutrients; secretes bile salts that aid in emulsification of fats in the intestine; and participates in nitrogen excretion by synthesizing uric acid. It consists of two lobes, each drained by a bile duct. The bile duct from the left lobe drains directly into the duodenum, and that from the right lobe branches, with one branch leading to the gall bladder located on the dorsal surface of the lobe and the other leading to the duodenum. The gall bladder stores bile and apparently also concentrates it.

The average wet mass of the liver for 101 sparrows from Ohio was 1.10 g (4.6% of TBM), and liver mass was significantly correlated with TBM ($r = 0.45$, $P < .001$) (Quiring and Bade 1943b). In California, however, the average liver mass of 13 free-living sparrows was only 0.55 g (2.2% of TBM) (Chaffee and Mayhew 1964), and in India it was only 0.52 g (2.3% of TBM) (Asnani 1984). Danilov et al. (1969) found that the liver weighed 1.64 g (4.8% of TBM) in January and 1.11 g (3.6% of TBM) in June in samples of male sparrows in Russia. Female sparrows did not show a similar seasonal difference in the relative mass of the liver, with their summer value being similar to that of winter males. During the breeding season in Australia, the dry mass of the liver was significantly greater in female sparrows than in males (Chappell et al. 1999). Danilov et al. (1969) also observed a similar seasonal change in the size of the alimentary tract, whereas kidney, heart, adrenal gland, and brain showed no such change.

The seasonal changes in both the absolute and relative sizes of the liver in Russian sparrows suggest that the size of the liver may change to meet changes in metabolic demands. Chaffee and Mayhew (1964) examined the liver masses of sparrows maintained in the laboratory for 7 wk at constant temperatures of 1°C, 23°C, or 35°C. The average liver mass of the birds maintained at 1°C was significantly greater than that of birds held at 23°C or 35°C. An earlier study, however, found no significant difference in liver mass between sparrows maintained for 7 wk at 1°C versus 23°C (Chaffee et al. 1963). The effect of temperature on energetic demands (see later discussion) must play a direct role in controlling the size and composition of the liver. Farner et al. (1961) examined the glycogen and fat content of the liver and *m. pectoralis* in sparrows maintained in the laboratory at a constant temperature (22°C) on either short-day (8L:16D) or long-day (20L:4D) photoperiods. Birds held on the short-day photoperiod had significantly higher fat content in the liver than long-day birds, as well as higher liver glycogen content, although the latter difference was not significant. There were no significant differences in lipid or glycogen content of *m. pectoralis*. One interpretation of these results is that the liver adjusts its energy storage to the nocturnal demand for energy and therefore stores more energy for a 16-h night than for a 4-h night. A second explanation is that short-day photoperiods play a role in stimulating increased energy storage due to the colder temperatures normally experienced during the winter. Experiments using skeleton photoperiods (see Chapter 3) might permit discrimination between these two alternatives.

Asnani (1984) studied changes in the relative size and composition of the liver during nestling development in Indian house sparrows. On day 1, the liver comprised about 3% of the TBM of the hatchlings and had a high content of water (75%) and fat (16%). Relative liver mass tended to increase during the first 10 d, to about 4.5% of TBM, and it then decreased to about 3.6% before fledging. Water content tended to decrease monotonically during development, whereas fat content initially decreased rapidly to about 6% on day 5 (presumably with the depletion of yolk lipids) and then increased to about 10.5% by fledging. Asnani suggested that the high lipid content of the liver at fledging may represent an adaptation to enhance post-fledging survivorship. He also reported that the liver of adult sparrows was approximately 66% water, 25% protein, and 9% fat. Caviedes-Vidal and Karasov (2001) found in Wisconsin that liver mass increased until nestlings were 9 d old.

Haarakangas et al. (1974) monitored the liver contents of four mineral nutrients (Ca, Cu, Mg, and Zn) in sparrows collected throughout the year in Finland. Ca content was significantly higher in females than in males during the breeding-season months of May and June, and it was significantly lower in both sexes during the peak of molt (September) than at other times of the year. Mg content was lower in both sexes from late autumn through early spring (November–April) and was significantly higher in females during the breeding season. Liver content of both Cu and Zn was highest in males and females in August, at the beginning of molt, and tended to be lowest during the winter. Zn content was also significantly higher in females than in males at the beginning of the breeding season. Haarakangas et al. attributed the high liver contents of Ca and Zn in females in May to egg formation, and particularly to the synthesis of Ca-binding protein complexes in the liver and the increased transcription and translation in the liver, which may be Zn-limited. They attributed the increases in Cu and Zn during molt to the high content of these metals in feathers (11–24 and about 150 μg/g, respectively) and to pigmentation requirements during molt, which may be Cu-mediated.

Pancreas

The pancreas is ribbon-like and lies in the curvature of the duodenum (Gier and Grounds 1944). The pancreas has both endocrine (see earlier discussion) and exocrine functions, with the products of the latter, digestive enzymes, emptying into the duodenum through three pancreatic ducts. The islets of Langerhans, which produce the hormones, are separated from other pancreatic tissue by a layer of connective tissue (Gier and Grounds 1944). Quiring and Bade (1943b) reported that the average mass of the pancreas of sparrows in Ohio was 158 mg (0.65% of TBM).

Energetics

The house sparrow has probably been the primary wild species used in energetic studies of birds (see Kendeigh et al. 1977). Energetic studies attempt to quantify the rate of metabolic energy production required for maintenance, thermoregulation, and activities such as feeding, digestion and absorption, locomotion, reproduction, and molt. Most energetic studies have been performed in the laboratory under tightly controlled conditions, but a few have attempted to determine the energetic requirements of free-living sparrows. Use of different techniques as well as different measures of metabolic rate complicate comparisons of results across the many energetic studies on sparrows, and in the analyses that follow I have attempted to ensure that there is meaningful comparability among the studies. For comparative purposes, have also converted all measures of metabolic activity to a single unit, Joules per gram per hour (J/g/h).

Physiologists have used a plethora of terms to describe various components of the energy metabolism of birds (e.g., Kendeigh 1976), only some of which will be defined and used in this account. *Resting metabolism* refers to the metabolic rate of inactive, postabsorptive individuals and can be separated into basal metabolic rate (BMR) and standard metabolic rate (SMR). BMR is defined as the minimal energy required to maintain the individual at rest (no locomotor costs), in a postabsorptive state (no digestive costs), and at thermal neutrality (no thermoregulatory costs). SMR includes the energetic costs of thermoregulation outside the range of thermoneutrality. *Existence metabolism* includes the energetic costs of feeding (including limited locomotion) and digestion as well as thermoregulation. *Productive energy* is the difference between existence metabolism and the maximum sustainable metabolism, usually considered to be the rate of existence metabolism at a temperature just above the lower lethal temperature (the low temperature at which a bird is unable to produce sufficient energy to sustain thermoregulation). *Maximal energy production* ("summit metabolism") is the rate of energy production during strenuous activities such as flight; it is a rate that can normally be maintained only for short periods of time.

Several techniques have been used to attempt to measure metabolic rate, all involving indirect methods of estimating actual metabolic energy production.

Laboratory studies normally involve maintaining individual birds under constant temperature and photoperiodic regimes, although one study used multiple individuals maintained in cages in an outdoor aviary with fluctuating temperatures and natural photoperiod (Weiner 1972).

One frequently employed method of estimating existence metabolism is the energy balance technique, which involves monitoring the energy intake (by determining the energy content of the food consumed) and the energy lost in the feces (which includes undigested and excreted material) (Blackmore 1969; Blem 1973, 1976; Davis 1955; Dolnik and Gavrilov 1975; Kendeigh 1949; Kendeigh et al. 1969; Seibert 1949; Weiner 1972). Blem (1968) found that the energy content of sparrow excreta was about 14.9 J/g of excreta and did not differ for birds held at two different temperatures, −4°C and 20°C. The difference between ingested and excreted energy represents the energy metabolized during the time period involved; provided the bird maintains a constant mass, it represents the energy required to sustain all activities performed during the period.

Another set of techniques involves measurement of the rate of O_2 consumption (Arens and Cooper 2005; Chappell et al. 1999; Dutenhoffer and Swanson 1996; Hudson and Kimzey 1966; Koteja 1986; Miller 1939; Mills and Heath 1970; O'Connor 1975d; Palokangas et al. 1975), the rate of CO_2 production (Kendeigh 1944, Teal 1969), or the rates of both O_2 consumption and CO_2 production (Buchanan et al. 2001; Chastel et al. 2003; Martin et al. 2003; Quiring and Bade 1943a; Walsberg and Wolf 1995). One of the difficulties of estimating metabolic energy production from the rate of either O_2 consumption or CO_2 production is the variation in these parameters based on what organic substrate or substrates are being metabolized. The ratio of CO_2 production to O_2 consumption is called the respiratory quotient (RQ), and RQ varies considerably depending on the organic substrate being catabolized. The theoretical limits of RQ vary from 1.00 for carbohydrates to 0.71 for lipids, with intermediate values representing mixed substrates of carbohydrates, lipids, and proteins (see Walsberg and Wolf 1995).

Another technique for estimating the metabolic rate under varying conditions, including potentially the metabolic rate of free-living birds, involves the use of doubly-labeled water ($^3H_2{}^{18}O$). This technique is based on the assumption that the turnover rate of tritium is equivalent to the turnover rate in body water, and the turnover rate of ^{18}O is equivalent to the turnover rate of body water plus CO_2 production. The difference between the two is therefore equivalent to CO_2 production. Williams (1985) used one sparrow among individuals of several species to test the adequacy of the technique for measuring the metabolic rate in small birds. Existence metabolism of each bird was simultaneously measured using the energy balance and CO_2 production techniques, as well as the doubly-labeled water technique, and the three techniques provided similar estimates of existence metabolism. The doubly-labeled water technique requires that injected tritium reach equilibrium with the body water, and Williams found that this occurred within 1 h in three sparrows tested.

As noted earlier, one of the principal problems with estimating metabolic rate from O_2 consumption or CO_2 production is variation in these parameters based on differences in the organic substrate being catabolized. This problem is particularly acute for estimates based on CO_2 production, because energy production varies from 20.9 J/ml CO_2 produced for carbohydrate substrates to 27.8 J/ml for lipid substrates, with protein substrates being intermediate (similar values for O_2 consumption vary only from 18.7 to 20.9 J/ml O_2) (Walsberg and Wolf 1995). Measurements of RQ can improve the estimates of energy production based on CO_2 production by indicating what the primary substrates are during the period of measurement. Walsberg and Wolf (1995) determined the RQ of inactive sparrows at 30-min intervals after termination of feeding. At night, RQ dropped steadily, from about 0.82 at 30 min after termination of feeding, to a steady value of about 0.71 by 6 h after feeding stopped, indicating that after 6 h the birds were metabolizing primarily lipids. During the daytime, however, the initial RQ (30 min after termination of feeding) differed significantly based on diet, with sparrows on a seed diet having an average RQ of 0.93 and those fed mealworms an average of 0.75. Average RQ dropped quickly after termination of feeding and averaged 0.66 in both groups beginning about 3 h after feeding stopped. This value of RQ was significantly lower than the theoretical minimum value of 0.71 for a purely lipid substrate and suggests that birds may somehow be sequestering CO_2 during these periods of daytime fasting (Walsberg and Wolf 1995), something that would further complicate the determination of metabolic rate based on CO_2 production.

Fig. 9.11A summarizes the relationship between resting metabolism and temperature from 11 studies on house sparrows. Although there is considerable variation among the results from different studies, the overall pattern is similar to that found in intensive studies such as that of Kendeigh (1944). Despite the variation, it appears that there is a zone of thermoneutrality, from about 22.5°C to about 35°C, although resting metabolic rate does decrease slightly but nonsignificantly between those two temperatures. The slope of the line for diurnal resting metabolism (Y) between 22.5°C and 35°C (X), $Y = 118.8 - 1.30X$, is greater than that including nocturnal values of BMR (Fig. 9.11A), suggesting that Kendeigh's (1969) conclusion that there is no daytime thermoneutral zone in the house sparrow may be correct. Resting metabolic rate increases linearly both below and above those critical temperatures, and both thermoregulatory increases are significant.

Nocturnal BMR is lower than daytime BMR (Miller 1939), a fact that is related to the daily fluctuation in body temperature (see earlier discussion).

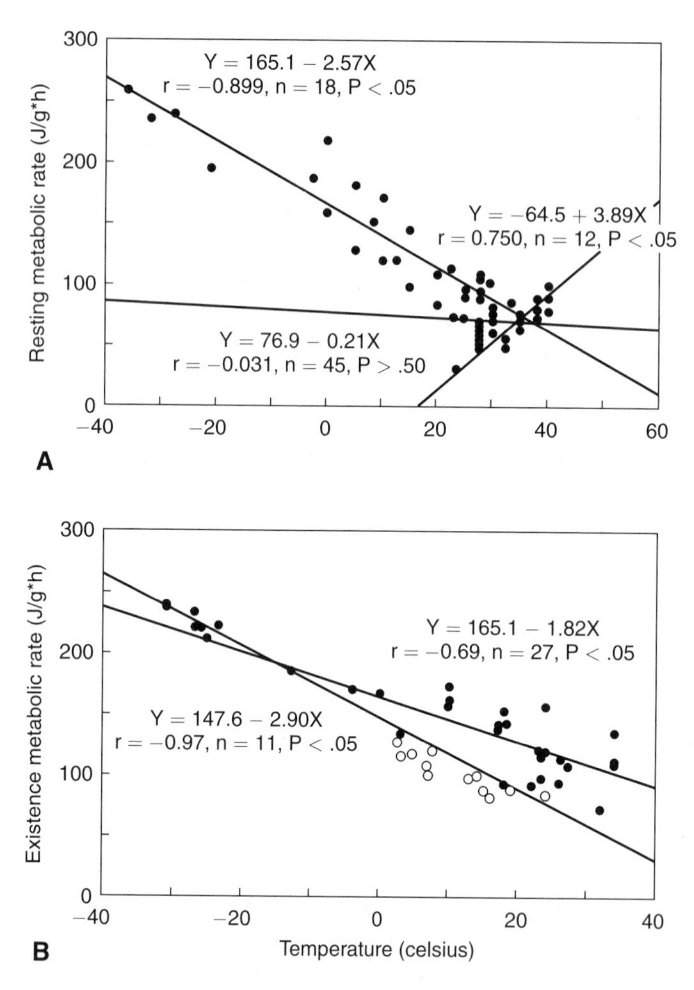

FIGURE 9.11. The relationship between ambient temperature and metabolic rate in the house sparrow. (A) Resting metabolism based on mean values from 11 studies (Buchanan et al. 2001; Chappell et al. 1999; Chastel et al. 2003; Dolnik and Gavrilov 1975; Dutenhoffer and Swanson 1996; Hudson and Kimzey 1964, 1966; Kendeigh 1944; Miller 1939; Palokangas et al. 1975; Quiring and Bade 1943b). In cases in which ambient temperature was reported only as being in the zone of thermoneutrality (i.e., Hudson and Kimzey 1966, Dolnik and Gavrilov 1975), a temperature of 27.5°C was used (the thermoneutral zone of the house sparrow is often considered to be 20°C–35°C). Regression lines in the figure are based on ambient temperatures up to and including 22.5°C ($Y = 165.1 - 2.57X$), temperatures from 22.5°C to 35°C inclusive ($Y = 78.9 - 0.21X$), and temperatures 35°C and higher ($Y = -64.5 + 3.89X$). The slope of the regression line for temperatures of 22.5°C–35°C does not differ significantly from zero, whereas the slopes of the other two regressions do differ significantly from zero. (B) Existence metabolism

Indeed, the circadian rhythms in BMR and body temperature are undoubtedly causally related (see Chapter 3). Hudson and Kimzey (1964, 1966) found that daytime BMR exceeded nocturnal BMR in the same individuals by approximately 33% (range 28%–38%). BMR also varies seasonally, with lower rates during spring and summer than during the late autumn and winter. In Iowa (USA), Miller (1939) found that BMR averaged 50% higher from October to February than from March to September. Arens and Cooper (2005) also found that summer nocturnal BMRs were generally lower than winter BMRs for sparrows tested at different ambient temperatures on the day of capture in Wisconsin. The slope of the regression line relating winter BMR to ambient temperature was significantly shallower than that for summer BMR ($P < .01$), indicating seasonal acclimatization in metabolic rate.

Miller (1939) also reported that the BMR of males tended to be higher than that of females, particularly in the period from February to April. This may be due in part to the effect of T on BMR (Buchanan et al. 2001). Chastel et al. (2003) found that nocturnal BMR in March in France was positively correlated with circulating levels of the thyroid hormone T_3 in both males and females but found no correlation between T_3 levels and either T levels in males or E_2 levels in females. Resting metabolism of juveniles in summer is higher than that of adults (Palokangas et al. 1975), presumably due in part to the lower insulative quality of the juvenal plumage.

The resting metabolic rate also varies geographically (see Chapter 2). Hudson and Kimzey (1966) found that the BMR of sparrows from Texas was significantly lower than that of sparrows from three other North American locations at higher latitudes. The Texas birds had an average nocturnal BMR that was 19% lower than the average nocturnal BMR at the other three sites, and diurnal resting metabolic rates at three different temperatures were also consistently lower in the Texas birds. In central Asia, the nocturnal BMR of individuals of the sedentary subspecies *P. d. domesticus,* averaged 23% higher than that of the migratory subspecies *P. d. bactrianus* (Dolnik and Gavrilov 1975). Hudson and Kimzey (1966) proposed that the lower BMR of Texas sparrows is an

---

based on mean values from seven studies (Blackmore 1969; Davis 1955; Kendeigh 1949; Kendeigh and Blem 1974; Seibert 1949; Weiner 1972; Williams 1985). Regression lines in the figure are based on temperatures between −31°C and −5°C ($Y = 147.6 - 2.90X$) and between −5°C and 34°C ($Y = 165.1 - 1.82X$). Existence metabolic rates based on groups of birds (○, from Weiner 1972) are ignored in the latter equation due to lack of comparability: the individuals may have huddled or interacted in other ways to reduce heat loss. The slopes of both lines are significantly different from zero.

adaptation that reduces vulnerability to heat stress in the high ambient temperatures that these birds periodically experience.

Disease apparently affects resting metabolic rate. Diseased individuals require additional expenditure of energy, both to mount an immune response and to repair damaged tissues. Martin et al. (2003) injected sparrows with the mitogen PHA, which stimulates a cell-mediated immune response without causing disease, enabling them to determine the energetic cost of the immune response alone. Nocturnal BMRs of PHA-injected sparrows 60 and 84 h after injection were significantly greater than nocturnal BMRs of saline-injected controls. The peak difference, 60 h after injection, was approximately 84% greater in PHA-injected birds, indicating that there is a significant energetic cost in mounting a cell-mediated immune response. This result is supported by the observation that male sparrows injected with lipopolysaccharide (a mitogen from the cell wall of *Escherichia coli*) lost 75% more mass overnight than did saline-injected controls (Bonneaud et al. 2003).

Fig. 9.11B summarizes data on existence metabolism of sparrows from seven studies. The slope of the regression equation for temperatures from −5°C to 34°C (−1.82) is shallower than that for resting metabolism (−2.57), but the regression lines converge at 0°C, whereas the slope of the regression equation for existence metabolism at less than −5°C (−2.90) is somewhat higher than that for resting metabolism. This suggests that the energy production associated with ingestion, digestion, and absorption of food and limited locomotion is at least partially compensatory, providing heat for thermoregulation below the lower critical temperature (Kendeigh 1969).

Blem (1973) measured existence metabolism during both winter and summer in sparrows from 11 North American sites spanning latitudes from 28°N to 59°N. A stepwise multiple regression analysis of the data (Blem 1977) revealed that, in addition to the expected relationship between existence metabolism and ambient temperature, it also varies significantly with mass in both seasons, with sex, with mean January and mean July temperatures during the summer, and with isophane during the winter. The relationship with mass (M) is the usual negative correlation with mass to the 0.75th power $(M^{0.75})$, an allometric relationship also observed in interspecific comparisons of metabolic rates of both passerines and nonpasserines (see Kendeigh et al. 1977) that is widely interpreted to reflect the effects of the surface-to-volume relationship (i.e., the rate of heat loss is a function of surface area, whereas the rate of heat generation is a function of volume). Females had significantly higher rates of existence metabolism in summer than males, which Blem suggested might be due to the energetic costs of egg production. The effects of mean January and July temperatures and isophane possibly represent adaptations to local climate conditions similar to those identified by Hudson and Kimzey (1966) for BMR (see earlier discussion).

Two techniques have been used to measure maximal energy production of sparrows. One involves measurements taken during enforced strenuous activity such as flight (Teal 1969, Chappell et al. 1999), and the other employs a helium-oxygen atmosphere to stimulate maximal energy production over short time intervals (Koteja 1986, Dutenhoffer and Swanson 1996). The latter technique depends on the much greater heat conductivity of helium (He) compared to atmospheric nitrogen (N_2), which results in a more rapid rate of heat loss from the bird, stimulating rapid increases in thermoregulatory metabolism. The two studies based on enforced activity suggested that the maximal metabolic rate of sparrows is between 7 and 11 times the nocturnal BMR (= factorial scope) (Teal 1969, Chappell et al. 1999), and the one study based on He-induced thermoregulation suggested that the factorial scope is about 8 (Chappell et al. 1999). Koteja (1986) found that the maximal metabolic rate of females was significantly higher than that of males during the summer in Poland. This may be due, in part at least, to the fact that females have significantly larger liver mass than males in the summer (see earlier discussion).

Chappell et al. (1999) attempted to determine whether maximal energy production is constrained by central or peripheral organ size. Using 36 adult and 30 juvenile sparrows captured during the summer in Australia, they determined both BMR and summit metabolism for each bird and obtained the dry organ masses of several internal organs (gut, liver, gizzard, heart, lung, and kidney) and two peripheral organs (flight and leg muscles). BMR was significantly positively correlated with liver, heart, lung, and flight muscle size in adults and with heart and flight muscle size in juveniles, whereas summit metabolism was significantly positively correlated with heart and flight muscle size in the combined sample of adults and juveniles. Chappell et al. concluded that the results tended to support the synmorphosis model, which states that the organs are optimally scaled so that no one organ or organ system constrains metabolic performance.

The increased BMR of sparrows during winter (described earlier) is one indication of the physiological acclimation that occurs at different times of the year. Maximum sustainable metabolism also differs with season, being 23%–38% higher in winter-acclimated than in summer-acclimated sparrows in Illinois (Kendeigh 1969; Kendeigh et al. 1977). In addition to this increase in potential energy production, the insulative quality of the plumage is greater in winter than in summer (see Chapter 5). These differences mean that sparrows can survive at much lower ambient temperatures during the winter than during the summer. In Illinois, for instance, the lower lethal temperature in summer is about 0°C (Davis 1955), but in winter-acclimated sparrows it is about −37°C (Kendeigh 1944). Blem (1973) found similar differences in sparrows from eight North American sites and also reported that the lower lethal temperatures of sparrows from more northern sites were significantly lower than those from southern sites.

Behavioral adaptations can affect the ability of sparrows to survive at extremely low temperatures (Kendeigh 1976). Roosting in well-insulated cavities during the winter represents one such adaptation. Kendeigh (1961) simultaneously monitored temperatures inside and outside an occupied nest-box during winter nights in Illinois. The differential between nest-box temperature and ambient temperature (Y in °C) increased significantly with decreasing ambient temperature (X): $Y = 4.7 - 0.19X$. Kendeigh estimated that, for each 1°C increase in nest-box temperature above ambient temperature, the bird conserved 1.26 kJ of energy over a 15–h roosting period. This energetic savings presumably also accounts for the construction and utilization of roost nests in some areas (see Chapter 4).

Blem (1975a) determined both resting metabolic rate and existence metabolism of nestling sparrows in Virginia (USA). Resting metabolism was determined for single nestlings held in the dark in the laboratory at 36°C, but not necessarily in a post-absorptive state. Resting metabolism increased from an average of 34.2 J/g/h for nestlings 1–2 d old to a peak of 68.7 J/g/h for nestlings 9–10 d old, and decreased slightly thereafter. No difference was noted in the resting metabolism of nestlings measured at night or during the day. Existence metabolism was twice resting metabolism for nestlings 1–2 d old, dropped rapidly to about 1.5 times resting metabolism for 4-d-old nestlings, then increased to 1.7–1.8 times resting metabolism for nestlings 6–8 d old, before decreasing again to 1.5 times resting metabolism for nestlings 11–17 d old. The initial decrease may be due to an increase in digestive efficiency, and the subsequent increase may be related to the rapid development of thermoregulatory ability by the nestlings (see Chapter 4).

Excretory System

The principal excretory organ is the kidney, which, however, has numerous functions in addition to the excretion of metabolic waste products. Broadly defined, the kidney is the primary organ responsible for water and electrolyte balance in the body. The kidneys of sparrows are located retroperitoneally in depressions in the synsacrum, and each consists of three divisions: a large posterior lobe, a small middle lobe, and a somewhat larger anterior lobe (Casotti and Braun 2000). Each lobe consists of an outer cortex surrounding an inner medulla, with the former comprising 78.6% of the total kidney volume, and the latter 7.4%; the remaining 14.0% consists of the major blood vessels (Casotti and Braun 2000). Neither total kidney size nor sizes of the three major components of the kidney change seasonally (Casotti 2001a). Urine produced in the kidneys passes through the ureters to the posterior cloaca, where it mixes with the feces from the digestive tract before being vacated through the cloacal orifice. Quiring and Bade (1943b) reported that the average mass of the kidneys was 338 mg (1.4% of TBM) for 101 sparrows from Ohio, whereas Johnson (1968) found an average of 293 mg (1.0% of TBM) for 5 sparrows from the United States, and Danilov et al. (1969) found an average mass of 283 mg (0.9% of TBM) for 39 male sparrows from Russia. Goldstein and Braun (1986b) found that the average mass was 204 mg (0.8% of TBM) for 15 sparrows from Arizona being maintained in the laboratory on *ad libitum* water.

The functional unit of the kidney is the nephron, of which there are two types in birds. Cortical nephrons (reptilian-type nephrons) are confined to the cortex and consist of a renal corpuscle, a proximal tubule, and a distal tubule, the latter tubule emptying into collecting ducts that converge in each kidney to form the ureter. Juxtamedullary nephrons (mammalian-type nephrons), in addition to the components present in the cortical nephrons, have a loop of Henle between the proximal and distal tubules that extends down into the medulla. In a sample of sparrows from Arizona, Goldstein and Braun (1986b) found that about 18% of the approximately 71,000 nephrons in the two kidneys were juxtamedullary. The renal corpuscle is the site of filtration, the movement of blood plasma minus proteins from the glomerulus (a thin-walled region of a renal arteriole) into the lumen of the nephron. The glomeruli of cortical nephrons are smaller than those of the juxtamedullary nephrons, and they are generally located more peripherally in the cortex (Goldstein and Braun 1986b). The nephron, and to a lesser extent the collecting duct and lower intestinal tract (see later discussion), then selectively reabsorb most of the filtrate to form the final excretory product.

The medulla consists of numerous cones through which the collecting ducts pass before converging to form the ureter. A steep osmotic gradient in each medullary cone, with a higher osmotic potential in the medullary tissue at the apex of the cone, establishes a countercurrent multiplier system that enables the kidney to reabsorb most of the water filtered in the renal corpuscle and, under certain circumstances, to produce a urine that is hyperosmotic to the blood plasma. The medullary cones of sparrows are organized as in other bird species, with the loops of Henle consisting of thin descending limbs and thick ascending limbs. The descending limbs, which are at the core of each cone, are surrounded by a ring of collecting ducts; these are, in turn, surrounded by the ascending limbs (Casotti and Braun 2000).

Glomerular filtration rate (GFR), the rate at which fluid passes into the nephrons, has been estimated using tritiated polyethylene glycol (^{3}H-PEG) (Goldstein and Braun 1988, Goldstein et al. 2001). PEG is neither reabsorbed nor secreted by the nephrons, so a comparison of the PEG concentrations in plasma and urine, and a measure of the rate of urine production, permits an estimate of GFR. The average GFR of sparrows from Arizona maintained in the laboratory on *ad libitum* water was 7.66 ml/h, whereas the average urine flow rate (UFR) was 0.207 ml/h (Goldstein and Braun 1988). Sparrows from Ohio maintained under similar laboratory conditions had an average GFR of 5.7 ml/h and an average UFR of 0.26 ml/h (Goldstein et al. 2001). This means that 95.4%–97.3% of the water that is filtered in the renal corpuscles is reabsorbed in the kidney, with a higher percentage of water reabsorption in the sparrows from the more xeric environment (Arizona).

Sparrows potentially face several environmental challenges whose effects are mitigated by the kidneys. These include water deprivation (in xeric environments or during seasonally dry conditions) and salinity (in environments with brackish water). Thomas (1997) also noted that, because of their primarily granivorous diet (see Chapter 6), house sparrows might have problems with retention of sufficient sodium (Na^+) and chlorine (Cl^-), and with elimination of potassium (K^+). Experiments exposing sparrows to some of these conditions have provided valuable insights into the functioning of the sparrow kidney. Two ways in which the kidney can respond to such challenges are by altering GFR or by changing patterns of reabsorption in the nephron tubules or collecting ducts or both. Changes in GFR are due to changes in blood flow into the glomeruli (vascular responses), whereas changes in rates of reabsorption are due to changes in permeability or active transport in parts of the nephron or in the collecting ducts (tubular responses). Some studies have indicated that the lower intestinal tract also plays a role in selective reabsorption during such environmentally-induced stresses (Casotti 2001b; Goldstein 1993; Thomas 1997).

Sparrows require more water at higher temperatures, presumably not only because of the much higher saturation vapor pressure of warmer air, but also because of the greater water loss incurred by the necessity for evaporative cooling through the respiratory system. Seibert (1949) measured the water consumption of sparrows held in the laboratory at four different ambient temperatures. For sparrows maintaining constant mass, the average daily water consumption (measured in milligrams per gram of body mass per day) at the four temperatures were 163.2 mg/g at 0°C, 172.8 mg/g at 23°C, 266.4 mg/g at 34°C, and 328.8 mg/g at 37°C. Sparrows deprived of water lose mass rapidly (Goldstein 1993; Goldstein and Braun 1988; Goldstein and Zahedi 1990), as do sparrows provided only salt water (Minock 1969). Minock found that sparrows maintained at an average temperature of 22°C could not maintain their mass at salt concentrations greater than about 0.30 M NaCl (1.8%). Other factors that can affect the water requirements of sparrows include the water content of the food and the water vapor pressure of the air, or relative humidity. The former varies depending on the type of food but is generally low for seeds and grains, whereas the latter affects the rate of respiratory water loss.

The osmotic potential of the blood plasma, plasma osmolality (P_{osm}), is maintained homeostatically within fairly narrow limits. Goldstein and Zahedi (1990) measured P_{osm} of free-living sparrows throughout the year in Ohio. The average P_{osm} was 341.0 mOsm/kg. P_{osm} was negatively correlated with ambient temperature ($r = -0.22$, $P < .005$), as was hematocrit ($r = -0.35$, $P < .001$). Hematocrit, the volume proportion of the blood cells in whole blood (see earlier discussion), varies inversely with total blood volume over short time intervals and is therefore an indicator of total blood volume. Goldstein and Zahedi collected urine samples from the ureteral orifices of some of the sparrows within 30 s of capture and measured urine osmolality (U_{osm}). U_{osm} varied from 132 to 658 mOsm/kg and was positively correlated with ambient temperature ($r = 0.21$, $P < .05$) but was not significantly correlated with P_{osm}. The relationships of P_{osm} and U_{osm} with ambient temperature are difficult to explain but may reflect a tendency for sparrows to remain better hydrated during warm weather. P_{osm}, U_{osm} and hematocrit showed no relationship to sex or to time of day.

Goldstein and Zahedi (1990) also monitored the P_{osm} and hematocrit of sparrows being maintained on *ad libitum* water in the laboratory. The average P_{osm} was 352 mOsm/kg and did not differ significantly from that observed in free-living sparrows. Both within and among individuals, coefficients of variation were low (1.9 and 1.7, respectively), indicating the narrowness of the range within which P_{osm} is normally maintained. The average hematocrit for the laboratory sparrows was 0.48, not significantly different from that of the free-living sparrows, but with within and among individuals the coefficients of variation (7.4 and 6.9, respectively) were greater than those for P_{osm}. Sparrows deprived of water for 24 h showed significant increases in both P_{osm} (average increase, 41.8 mOsm/kg) and hematocrit (average increase, 0.028) and lost mass rapidly. Goldstein and Zahedi found no differences in P_{osm} or hematocrit for birds alternately placed on a diet with low Na (seeds with 0.02 mEq Na^+ per gram of dry mass) or high Na (crickets with 0.2 mEq Na^+ per gram).

Similar results were obtained by Goldstein and Braun (1988), who compared several physiological measures of hydrated (*ad libitum* water) and dehydrated (deprived of water for 30 h) sparrows being maintained in the laboratory at 26.5°C and 28% relative humidity. Dehydrated birds had a higher P_{osm} than hydrated birds (368.4 vs. 326.8 mOsm/kg, $P < .01$), higher hematocrit (0.539 vs. 0.439, $P < .01$), and moderately higher plasma Na^+ concentration (180.2

vs. 165.8 mEq/L, $P < .05$). Plasma K^+ levels were unaffected by dehydration (hydrated mean, 6.05 mEq/L). The changes in blood chemistry occurred despite compensatory changes in the kidney. GFR decreased in dehydrated sparrows (from 7.66 to 3.54 ml/h, $P < .01$), as did UFR (from 0.207 to 0.033 ml/h, $P < .01$), whereas U_{osm} increased dramatically (from 325.3 to 825.5 mOsm/kg, $P < .01$). Thus, U_{osm} went from being essentially isosmotic to P_{osm} in the hydrated birds to being highly hyperosmotic during dehydration. The estimate of daily urine production for the hydrated sparrows at 26.5°C (4.97 ml/d) was intermediate between the water consumption rates observed by Seibert (1949) at 23°C and 34°C (3.94 and 6.07 ml/d, respectively), assuming that the hydrated birds weighed an average of 22.8 g (cf. Goldstein and Braun 1988). As noted previously, Goldstein and Braun (1988) also measured the level of circulating antidiuretic hormone (AVT), and this increased significantly in the dehydrated sparrows. They concluded that the primary target of AVT was vascular, resulting in the much lower GFR and UFR observed in dehydrated sparrows.

In a subsequent study attempting to further clarify the role of AVT in conserving water in the kidney, Goecke and Goldstein (1997) infused anesthetized sparrows intravenously with saline solutions containing AVT or AVT plus one of two AVT analogs. These two chemicals were developed as analogs to mammalian antidiuretic hormone, and each blocks one of two receptor sites normally responsive to antidiuretic hormone: dPTyr(Me)AVT is an antagonist at V_1 receptors, and $d(CH_2)_5$[D-Ile^2, Ile^4, Ala-Mh_2]AVP is an antagonist at V_2 receptors. In mammals the effect of antidiuretic hormone on V_1 receptors is vascular, reducing UFR by lowering GFR, and the effect on V_2 receptors is tubular, reducing UFR by increasing tubular reabsorption of water. The average UFR of control sparrows (infused with 25 mM/L of NaCl [0.0015%] at a rate of 0.6 ml/h) was 0.74 ml/h (not significantly different from the infusion rate), whereas the average GFR was 8.2 ml/h. Therefore, 91.0% of the water filtered in the renal corpuscles was being reabsorbed in the nephrons or collecting ducts. Compared with the 95.4%–97.3% reabsorption observed in sparrows maintained on *ad libitum* water (see earlier discussion), this suggests that the infusion rate into the anesthetized sparrows was introducing a slight excess of water. Addition of AVT to the saline infusion significantly reduced both GFR and UFR, whereas addition of AVT plus the V_1 receptor antagonist resulted in no change in either GFR or UFR from control levels. The addition of AVT plus the V_2 receptor antagonist reduced GFR from the level observed in controls, although the difference was not significant, and significantly reduced UFR compared with both controls and AVT alone. These results suggest that AVT reduces water loss in sparrows primarily by lowering GFR (vascular response) and that it has little effect on a tubular response.

This conclusion was called into question by the results of an experiment by Goldstein et al. (1999). They isolated segments of medullary collecting ducts and of thin descending and thick ascending limbs of the loop of Henle of juxtamedullary nephrons of sparrows and assayed them for activation of adenylyl cyclase based on accumulation of cyclic nucleotides. They tested several hormones (including AVT, PTH, glucagon, norepinephrine, and atrial natriuretic peptide [ANP]) for their effects in stimulating adenylyl cyclase activity in the various segments. AVT (at a concentration of 10^{-6} M/L) increased adenylyl cyclase activity by more than 4 times in the collecting duct segments and almost 2 times in the thick ascending limb segments but had no effect in the thin descending limb. PTH at the same concentration increased adenylyl cyclase activity by about 2.5 times in the thick ascending limb and by 2 times in the thin descending limb but had no effect in the collecting duct. Glucagon, norepinephrine, and ANP had no effect on adenylyl cyclase activity in any of the segments. These results suggest that AVT may have tubular effects on water reabsorption, possibly by enhancing tubular ion reabsorption in the thick ascending limbs or by altering permeability in the collecting ducts (Goldstein et al. 1999). As noted earlier, PTH apparently acts to increase reabsorption of Ca and excretion of phosphate ion in the loop of Henle.

One of the principal excretory products voided in the urine of birds is uric acid, the main amine-containing byproduct of protein catabolism. Goldstein et al. (2001) studied the effect of a high-protein diet on kidney morphology and function in sparrows from Ohio. Birds were maintained in the laboratory (23°C, 11L:13D) for at least 4 wk on either a high-protein (30% casein) or a low-protein (8% casein) diet. Hematocrit (0.484 and 0.469, respectively) and P_{osm} (352 and 345 mOsm/kg, respectively) did not differ significantly between the two diets, but plasma concentration of uric acid was more than twice as high in birds on the high-protein diet (1.09 vs. 0.39 mM/L, respectively; $P < .01$). GFR did not differ significantly between the two diets (5.9 and 5.7 ml/h, respectively), indicating that there was no vascular response to increased plasma uric acid. UFR almost doubled, however, in birds on the high-protein diet (0.49 vs. 0.26 ml/h, $P < .05$), and the concentration of uric acid in the urine was more than five-fold higher in high-protein birds (525.0 mM/L) than in low-protein birds (103.0 mM/L) ($P < .01$). Average U_{osm} did not differ between the two groups, however, being 449.5 mOsm/kg in high-protein birds and 515.9 mOsm/kg in low-protein birds, a fact no doubt reflective of the low solubility of uric acid in water. One morphological difference in the kidneys of the two groups was the significantly greater relative mass of the medulla in high-protein birds (9.4% of total kidney mass vs. 8.0% in low-protein birds, $P < .05$). This suggests that the tubular response to increased uric acid load may be mediated in the loops of Henle of the juxtamedullary nephrons or in the collecting ducts. However, the authors found no difference in the adenylyl cyclase activity of isolated segments of either the collecting ducts or the thick ascending limbs of the loops of Henle from birds on the two diets.

Some reabsorption of water and possibly Na^+ may also take place after the urine has entered the cloaca. Goldstein and Braun (1986a) used *in vivo* flow-through perfusion of the lower intestine to study absorption of water and Na^+ in the large intestine and coprodeum of sparrows. The ureters of anesthetized sparrows were severed to prevent urine flow, and plastic tubes were inserted at the beginning of the large intestine and in the cloacal orifice. Water with various concentrations of salts was then perfused through the large intestine to determine absorption rates of water and Na^+. The perfused intestine absorbed Na^+, from all perfusion concentrations, and excreted K^+. These experimental results were further confirmed by differences in the concentrations of Na^+ and K^+ between the ureteral urine and the voided urine of hydrated sparrows. Ureteral concentrations were 86.5 and 60.5 mEq/L, respectively, both significantly different from the concentrations in voided urine, 59.8 and 89.9 mEq/L. Goldstein (1993) measured the electrical potential difference and short-circuit current across the walls of the large intestine (*in vitro*) of sparrows after the dietary Na manipulations described earlier (see discussion of the adrenal gland). The potential difference across the intestinal walls was lumen-negative for virtually all of the LS birds but tended to become lumen-positive for HS and HSS birds. Recent histological studies of the ceca, large intestine (rectum), and coprodeum of sparrows have revealed that the epithelial lining of the ceca and large intestine are densely packed with villi and microvilli, suggesting that they are involved in absorption (Casotti 2001b). Collectively, these results support the conclusion that the lower intestinal tract of sparrows is important for reabsorption of water and Na^+. Thomas (1997) suggested that urinary reflux into the lower intestine occurs facultatively, except when sparrows are dehydrated or Na^+-loaded.

Reproductive System

The reproductive system includes the gonads and the associated ducts and glands. There are marked seasonal changes in gonad size in both male and female sparrows, with the gonads normally being small and inactive during autumn and winter and enlarged and active during the spring and summer (see Chapter 3). Similar seasonal changes take place in the reproductive ducts.

Female

The female possesses a single ovary, the left ovary, which is located in the abdominal cavity anterior and ventral to the left kidney. When the ovary is active, it not only produces the female gamete, the ovum, but also serves as a major endocrine gland, producing primarily E_2 and progesterone, but also some androgens. After ovulation, a highly differentiated oviduct transports the ovum from the ovary to the cloaca and, during this process, adds the albumen, the shell membranes, and the shell to complete the development of the egg. Riley and Witschi (1938) identified four stages of ovarian and oviductal development in sparrows (Fig. 9.12).

Egg development begins in the ovarian follicles, each of which contains an ovum. As the follicular cells proliferate and differentiate under the influence of FSH from the anterior pituitary (see earlier discussion), some of the cells begin to produce hormones and others contribute to the maturation of the ovum. Vitellogenesis, or the addition of yolk proteins and lipids to the egg, is a major part of this maturation process (as noted previously, the synthesis of these

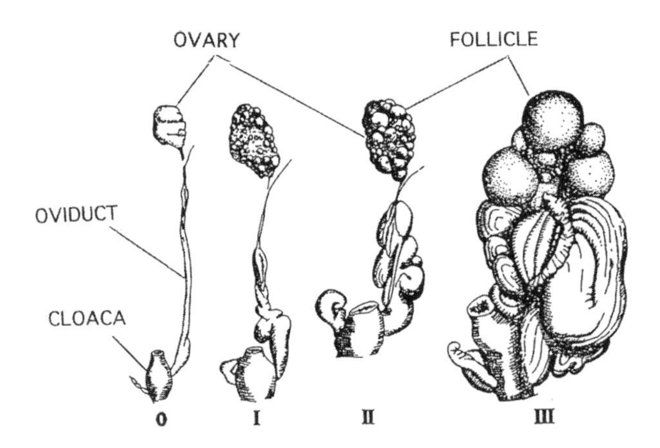

FIGURE 9.12. Developmental stages of the left ovary and oviduct of the house sparrow in Iowa (USA). The four stages recognized by Riley and Witschi (1938) are as follows: 0, quiescent stage with average ovarian mass about 9 mg; I, developing ovary and oviduct with visible follicles in the ovary and ovarian mass of about 27 mg; II, oviduct considerably enlarged with increase in follicular size and ovarian mass of about 40 mg; and III, fully functional ovary with greatly enlarged follicles, an egg in the oviduct, and ovarian mass of about 290 mg. From Riley and Witschi (1938: Fig. 1), with permission. Copyright © 1938 The Endocrine Society.

proteins and lipids takes place in the liver). In sparrows, vitellogenesis and the accompanying rapid follicular enlargement require about 3–5 d (Pinowska 1979; Schifferli 1980a) (see Fig. 4.6). After vitellogenesis is completed, the ovum is ovulated into the oviduct. The proximal part of the oviduct, the magnum, adds the albumens to the egg, and the shell membranes are added in the next part of the oviduct, the isthmus. Finally, the shell and its pigments are added in the uterus, or shell gland. A short vagina connects the uterus to the cloaca. Passage through the oviduct requires about 24 h (Schifferli 1980a).

Numerous sperm storage glands are located in the lining of the oviduct at the junction between the uterus and the vagina. Shugart (1988) measured 21 such glands from a single laying female and found that the average length of the glands was 240 μm and their average diameter was 58 μm. All of the glands, which were tubular with an enlarged distal end, contained sperm.

Male

Males possess two testes located in the abdominal cavity anterior and ventral to the kidneys, each drained to the cloaca by a vas deferens. Both the testes and the vasa deferens undergo marked changes in size during the annual cycle (see Chapter 3). Active testes consist of many seminiferous tubules, in which spermatogenesis occurs, and interstitial tissue that produces androgens (see earlier discussion).

Spermatogenesis occurs only at night in temperate resident sparrows, although it apparently occurs throughout the day in subtropical India (Mehrotra 1951). Foley (1928) was the first to note that breeding males captured throughout the day in Wisconsin and Louisiana (USA) showed no active spermatogenesis, and that only males captured before 0750 in March showed all phases of meiotic divisions I and II. In males captured in Iowa in March, meiotic figures were found only at night (from 2230 to 0515, with the most active spermatogenic activity occurring between 0145 and 0400) (Riley 1937). Spermatogenic activity began about 4 h after sunset, and no meiotic figures were found during the daylight hours.

The daily cycle of spermatogenic activity in sparrows is apparently controlled by body temperature. Riley (1936a) monitored body temperature over the 24-h cycle and found that sparrows had an average daytime body temperature of 43.2°C, which dropped to 39.5°C at night. When he experimentally reversed the light and dark periods, the birds had elevated temperatures during the apparent day and decreased temperatures during the darkness; spermatogenesis also occurred during the subjective night (real late afternoon) (Riley 1937). Castrated males showed the same daily fluctuation in body temperature. Experimental manipulation of body temperature (lowering it during the day by feather clipping and cold temperatures, and increasing it at night by using a revolving cage to force the birds to be active at night) indicated that spermatogenic activity was stimulated by lowered body temperature and inhibited by increased body temperature.

In a detailed histological study of spermatogenesis in the testes of sparrows collected in India, Kumar (1995) described 12 stages (I–XII). The heads of sperm cells in the maturation phase (stages XI and XII) were embedded in the syncytium of Sertoli cells, and the average dimensions of the mature sperm were as follows: head, 0.137 μm; middle piece, 0.583 μm; and tail, 0.162 μm. Goés and Dolder (2002) described six stages in the maturation of spermatids in sparrows from Brazil. In the latter stages (S4–S6), both the acrosome and the sperm nucleus elongate and take on a helical structure, and the mitochondria spiral around the central flagellar fibers in the long middle piece. Goés and Dolder reported that the average length of S6 cells was 62.0 μm, with the acrosome length being 9.0 μm. I have no explanation for the approximately 2 orders of magnitude difference in the size of the sperm reported in these two studies.

Sperm produced in the testes are stored in the seminal glomera (the coiled distal ends of the vasa deferentia) until copulation, and the number of sperm in the glomera is positively correlated with testes mass (Birkhead et al. 1994). Some leakage of sperm occurs from the glomera and vasa deferentia, because Quay (1987) found sperm both in the feces and in lavage samples from the cloaca of males maintained in the laboratory. Peak numbers of sperm in the lavage samples occurred predawn (0400), and minimal numbers were found in the late daytime, findings consistent with the observation that sperm production occurs only at night in the species. Average daily sperm production during the egg-laying period in Spain has been estimated as 31×10^6 sperm per day (Birkhead et al. 1994).

Body Mass and Composition

Body mass of sparrows changes both daily and seasonally. Most of these changes are caused by changes in the lipid content of the body, with daily fluctuations due to individuals' storing sufficient fat reserves to provide energy for surviving the period of nocturnal fasting, and with seasonal differences due primarily to increased fat levels during the winter (although plumage mass is also greater during the winter) (see Chapter 5). Increased subcutaneous fat reserves in the winter serve the dual functions of providing additional energy during the long, cold nights and increasing the insulative qualities of the integument, thereby reducing heat loss (Blem 1973).

Two studies at north temperate localities showed similar patterns of monthly changes in mass throughout the year (Folk and Novotny 1970, Anderson 1978). The highest average mass of males was in the winter months, whereas females had their largest average mass during the breeding season. The average mass of males tended to be greater than that of females except during the breeding season, when female mass was consistently greater than that of males.

Body size of sparrows changes geographically in a manner that is generally consistent with Bergmann's ecogeographic rule (see Chapter 2), and the composition of the body changes geographically as well. Blem (1973) performed carcass analyses of winter-captured sparrows from eight sites in North America. Average TBM of the birds varied from 26.9 g (Arizona) to 33.5 g (Saskatchewan, Canada), and lean dry mass (LDM) varied from 5.9 g (California) to 7.2 g (Manitoba, Canada). Both TBM and LDM were negatively correlated with lowest mean monthly temperature at the sites and positively correlated with isophane (all P values were $<.05$). Water as a percentage of TBM tended to decrease with increasing latitude (64.5% in Arizona to 61.0% in Saskatchewan and Minnesota [USA]), but the trend was not significant. Average lipid content of the birds varied from 1.74 g (Arizona) to 4.05 g (Saskatchewan). Lipid mass was positively correlated with both TBM and LDM (P values $< .05$) when one northern sample was omitted (the Churchill, Manitoba, sample was omitted because sparrows apparently survive the long, extremely cold nights only by roosting in grain elevators). Lipid indices (lipid mass/LDM) were significantly higher at most of the northern sites (Churchill excepted). Blem also analyzed the body composition of summer-captured sparrows from three sites (two of which had also been used in the winter analysis). Water as a percentage of TBM tended to be higher in the summer (66.2%–7.2%), and lipid masses tended to be lower (1.47–1.90 g). Blem (1975a) has also examined changes in body composition of nestling sparrows during development (see Chapter 4).

Because lipid serves as the primary short-term energy reserve, not only for overnight survival but also for egg formation in females (see Chapter 4) and for migration (see earlier discussion), it would be useful to have a way of estimating lipid content of living birds. A noninvasive technique that has been used to obtain estimates of lipid content of living organisms in the field is the measurement of total body electrical conductivity (TOBEC). The TOBEC technique can be calibrated to estimate lean mass of the organism, which can then be subtracted from TBM to obtain an estimate of lipid content. Two studies on sparrows using two different TOBEC instruments have shown that, although there is a high correlation between measured TOBEC values and lean mass, the resulting estimates of body lipid content are too imprecise to be used for meaningful comparisons among individuals (Asch and Roby 1995, Barkowska et al. 2000).

Chapter 10

HUMAN COMMENSALISM AND PEST MANAGEMENT

> Concerning this unmitigated pest we have little to say, further than to bewail the misfortune of its introduction, and to plead for its extermination. It is in every respect a first-class nuisance, to be classed along with the house-rat and other noxious vermin.
>
> —Ridgway 1889:248

One of the most frequently remarked attributes of the house sparrow is its close, some say obligate, commensal relationship with humans. Commensalism is defined as an interaction between two species in which one species benefits from the relationship while the other is unaffected by it. As the statement by the pioneer American ornithologist, Robert Ridgway, suggests, however, the relationship between humans and sparrows is more complex than that implied by the term commensalism, and there are significant negative effects for humans caused by their avian commensals. The destruction of grain crops by foraging sparrows is probably the most frequently cited negative effect, but numerous other "transgressions" have also been ascribed to them. Despite their many transgressions, however, many people have a deep fondness for the species, a fact that is attested to by the concerns expressed about its recent population collapses in some regions (see Chapter 8).

Numerous studies have attempted to determine the economic impacts of sparrows on humans (e.g., Barrows 1889; Dearborn 1912; Kalmbach 1940; Southern 1945), with most focusing on their impacts in agroecosystems. During the 5-y (1968–1972) global study of ecosystems under the umbrella of the International Biological Program (IBP), the Working Group on Granivorous Birds was formed with the express purpose of quantifying the effects of granivorous bird species on agroecosystems throughout the world (Pinowski 1967). The principal species selected for intensive study by that group were the house sparrow and its congener, the tree sparrow, although several other granivorous species, including pigeons and doves (Columbiformes), weaverbirds (Ploceidae), and New World blackbirds (Icteridae), were also studied. Several monographs have been published under the auspices of the Working Group (Kendeigh and Pinowski 1972; Pinowski, Kavanagh, and Pinowska 1995; Pinowski and Kendeigh 1977; Pinowski and Summers-Smith 1990; Pinowski et al. 1991). A journal entitled *International Studies on Sparrows* was initiated in 1967 and has compiled an extensive bibliography of sparrow literature.

The purpose of this chapter is to explore the nature of the complex relationship between sparrows and humans. Included in this discussion is the identification of negative impacts of sparrows on humans, as well as possible positive effects that the species has on its human hosts. The chapter also discusses the numerous attempts to limit or minimize the negative effects of sparrow populations.

Obligate Commensal?

Although Johnston and Klitz (1977) concluded that the house sparrow evolved its species status as a commensal of sedentary humans, thereby suggesting that this relationship is obligatory, other evidence suggests that the species has a much longer history than do sedentary human communities (see Chapter 1). This would mean that its status as an obligate commensal is also open to question. Summers-Smith (1963) discussed this question at considerable length in his monograph on the species. He concluded that some of the best evidence supporting its status as an obligate commensal comes from historical records of its occurrence on some of the smaller islands of the British Isles, and he cited several instances in which the species disappeared from such islands after the departure of humans.

Craggs (1967, 1976) recorded the 22-y history of the species on one such island. Hilbre Island is a 5-ha island located 2.4 km from the coast and from the nearest town (West Kirby, Cheshire, UK); it was occupied during the period of the study only by a caretaker's family. A small population of sparrows (3–7 pairs) bred on the island from 1953 to 1957, while the caretaker kept a horse and several chickens. Sparrows disappeared from the island in 1958, after the horse was replaced by an automobile and the number of chickens dwindled to none (Craggs 1967). After a new caretaker arrived in 1968 with two horses and about a dozen chickens, sparrows recolonized the island, with the first breeding attempt in 1971 (Craggs 1976). By 1974, there were seven breeding pairs on the island.

Although this and other similar accounts from other islands (see Summers-Smith 1963) suggest that the relationship between sparrows and humans is obligate, at least one subspecies does not show such an obligate reliance on human settlements. The migratory subspecies *Passer domesticus bactrianus*, which breeds in central Asia, where its range overlaps that of recently established populations of the sedentary subspecies *P. d. domesticus* (see Chapter 1), sometimes breeds in colonies that are associated with *domesticus* (and hence closely associated with human settlements) and sometimes breeds alongside Spanish sparrows, which are less closely tied to humans (Gavrilov 1965). This suggests that, in this subspecies at least, commensalism with humans is facultative rather than obligate.

Pest Status

The identification of the house sparrow as a pest is usually based on one or both of the following: (1) its consumption of agricultural products and (2) its involvement in transmission of pathogens of humans or their domesticated animals. Other attributes that are often cited as contributing to its status as a pest include the fouling of its environment and its displacement of native bird species in areas where it has been introduced. In at least two cases, introduced sparrows have also been identified as one of the principal dispersal agents in the spread of other undesirable species, the bridal creeper (*Asparagus asparagoides*) in South Australia (Attiwill 1970), and an ectoparasitic bug that is a vector for Chiagas' disease (a frequently fatal form of human trypanasomiasis) in Brazil (Smith 1973).

Crop Destruction

Although sparrows have remarkably catholic tastes, grain forms a significant component of their diet at all times of the year (see Chapter 6). In India, sparrows are considered serious pests of standing wheat, pearl millet, sorghum, and rice crops (Chahal et al. 1973; Dhindsa et al. 1984; Rana and Idris 1991a), and they also consume substantial numbers of rice and pearl millet seedlings (Chahal et al. 1973; Sarwar and Murty 1982). In Europe, wheat, barley, rye, and oats are frequently damaged (Havlín 1974), and in North America, sparrows have been reported to damage wheat, oats, rye, barley, corn (maize), sorghum, millet, and sunflower crops (Barrows 1889; Bullard 1988; Royall 1969). Wheat and barley crops are damaged in New Zealand (Dawson 1970), and Smith (1973) reported that sparrows regularly feed on rice laid out to dry in Brazil. In addition to the consumption of standing grain, at least one study has suggested that estimates of actual grain loss should be twice the amount actually consumed by the sparrows, due to stem breakage and other damage caused by foraging sparrows (Havlín 1974).

Although cereal grains are usually the principal crops damaged, sparrows have also been reported to damage numerous other cultivated plants. Flower buds of cultivated fruit trees are frequently eaten (Barrows 1889; Dawson and Bull 1970; Fryer 1939; Mann 1986; Smith 1973), and some fruits are also consumed in large quantities (Chahal et al. 1973; Dawson and Bull 1970). Some vegetable crops are also damaged by sparrows, including peas, sweet corn, and lettuce (Barrows 1889; Chahal et al. 1973).

Despite the frequent assertion that sparrows cause significant economic losses to grain, fruit, and vegetable growers, few studies have attempted to quantify the actual extent of such losses. Dawson (1970) sampled 4 barley fields and 11 wheat fields at Hawke's Bay (New Zealand), where the house sparrow was the only species feeding regularly. The wheat fields lost 1%–25% of their grain, with the geometric mean loss being 4.6% (95% confidence interval, 2.6%–8.4%). Losses in the barley fields varied from 6% to 19%, with a geometric mean of 9.3% (95% confidence interval, <1%–150%). Sparrows preferred the "milk" stage but continued to feed on the grain until it was harvested. Some compensatory increase in grain mass was evident, especially if the losses occurred during the "milk" stage. Dawson also found that the sparrows showed a strong preference for feeding near the edge of the field and near cover.

Havlín (1974) sampled grain losses due to sparrows (house sparrows and tree sparrows) during two growing seasons in a 1040–ha area in the Czech Republic. In large fields the losses varied from 0.005% to 0.01% of the yield, whereas in small fields losses varied from 0% to 12.5% of the yield. He estimated that the total grain loss in the area was 2% to 4% of the yield, with most of the losses occurring in wheat and barley fields.

Dhindsa et al. (1984) estimated grain losses in randomly selected fields of pearl millet and grain sorghum in India. Losses due to bird depredations averaged 45.6% in the eight millet fields and 40.6% in the sorghum fields. House sparrows were the most common bird species feeding in both types of grain fields, comprising 48.8% of the individuals observed feeding in the millet fields and 45.9% of those in the sorghum fields. Assuming that the losses due to sparrows were proportional to the number of sparrows feeding in the fields, the losses due to sparrows would be 22.3% and 18.6% in millet and sorghum fields, respectively.

Another approach to estimating the deleterious effects of sparrows on cereal grain production was developed by Wiens and Dyer (1977) as a part of the program of the Working Group on Granivorous Birds (IBP). They developed a model of the energetic demands of sparrow populations based on empirical data on their population dynamics, empirical data on the energetic demands of individual sparrows (see Chapter 9), and knowledge of the food consumption patterns of sparrows. They used the model to estimate that the loss of cereal grains to a sparrow population with a density of 500 individuals/km^2 was 7.2% of the cereal grain production in a region in central Poland (i.e., they consumed an estimated 1615 kg/km^2 per year in an area with an annual production of rye and oats of 22,492 kg/km^2). One difficulty with comparing the predictions of the model with studies of consumption of standing grain crops is that the model predicts total annual grain consumption without partitioning its source into standing grain, stored grain, grain fed to livestock, or waste grain. Nevertheless, it corroborates the findings of other studies suggesting that grain consumption by sparrows can have significant deleterious effects on cereal crop production.

The fact that losses averaged over large areas may be only about 5% does not fully account for the seriousness of sparrow depredation on agricultural crops, however. Because of their flock foraging (see Chapter 6) and communal roosting (see Chapter 7) behaviors, local losses or losses to individual farmers may be overwhelming. For instance, sparrows were reported to have caused 100% loss in a newly planted filed of pearl millet in India (Sarwar and Murty 1982).

Disease Transmission

Sparrows have been implicated in the spread of numerous diseases of humans and their domesticated animals. In some cases, sparrows are involved in the direct transmission of the pathogen; in others, they serve as a reservoir for the pathogen, which is transmitted to humans or their animals by another vector such as mosquitoes or mites. However, in a review of the epizootic role of sparrows, Kruszewicz (1995a) concluded that sparrows have not been shown to play a major role in the transmission of human pathogens and instead raised the question of the impact of human-spread pathogens on sparrow populations. The Appendix lists many of the parasites and pathogens that are known to occur in sparrows or their nests. Most of the pathogens that have been well studied are those that are associated with diseases of humans or their domesticated animals.

The arboviruses are one group of viruses that cause intermittent epidemic outbreaks among humans and some of their domestic animals for which sparrows serve as an important reservoir. The recent appearance of West Nile virus in North America has focused renewed attention on the role of birds in the transmission cycle of arboviruses. Although arthropods, primarily mosquitoes, are the direct vectors causing human and animal infection with these viruses, birds, particularly nestlings of altricial species, appear to be the primary reservoirs and amplification centers for the viruses. Viruses for which the house sparrow is considered to participate importantly in the transmission cycle include St. Louis encephalitis (SLE) (Chamberlain et al. 1957; Gruwell et al. 2000; Holden, Hayes et al. 1973; Lord et al. 1973; McLean et al. 1983, 1993), western equine encephalitis (WEE) (Burton et al. 1966; Hayes et al. 1967; Holden, Francey et al. 1973; Stamm 1958), and eastern equine encephalitis (EEE) (McLean et al. 1985; Stamm 1958).

In some areas in which the house sparrow is the most common bird species, it is apparently the primary reservoir for one or more of these viruses and consequently plays a major role in the amplification of viral numbers that can lead to epidemics. Human epidemics of SLE occurred in three Texas cities in the years 1964, 1965, and 1966 (Lord, Calisher and Doughty 1974). Transect censuses during the epidemics found that sparrows were the most common bird species in all three cities (comprising 57.0%–76.1% of all birds counted). Antibody tests of captured sparrows revealed a high rate of SLE infection in the species in all three cities, and the authors concluded that the house sparrow was the most important bird species in the local transmission cycle of the disease. Similarly, after human epidemics of WEE and SLE in Hale County (Texas) in both 1965 and 1966 (Holden, Hayes et al. 1973), monthly transect surveys over a 3-y period at four locations in the county found that sparrows made up 81.6%–89.9% of the birds counted (Mitchell et al. 1984). In an intensive 10-y study after a 1984 epidemic of SLE in southern California (USA), Gruwell et al. (2000) concluded that the recrudescence of latent SLE infections in overwintering sparrows was the principal route for reintroducing the virus annually into the region. This was due to the correspondence between the timing of recrudescence of the SLE virus in sparrows and the local appearance of the mosquitoes (*Culex* spp.) that serve as its principal vectors. Lord, Calisher and Doughty (1974) suggested that population control measures against sparrows in at-risk urban areas might be an effective means of preventing human epidemics.

Although crows (Corvidae) are apparently the primary reservoir for West Nile virus, sparrows are known to be infected by the virus in Europe and Africa (Juricova et al. 1998; Taylor et al. 1956), from whence the virus was introduced into North America. A recent study of the spread of West Nile virus in North America since its first appearance in 1999 suggested that the house sparrow may be a major agent of dispersal of the disease in North America (Rappole and Hubalek 2003).

Sparrows have also been found to be infected with the tickborne encephalitis virus in Europe (Juricova et al. 1998, 2000), but the status of their involvement in the transmission of this rare human illness is unknown. In Sweden, sparrows have been found to be infected by the Ockelbo virus (Francy et al. 1989; Lundstrom et al. 1992), an arbovirus that causes Ockelbo disease in humans (Lundstrom and Niklasson 1996). Sparrows are also apparently involved in the transmission of the turkey meningoencephalitis virus to poultry in Israel (Nir et al. 1967, 1969).

Sparrows have also been implicated in the transmission or amplification of several other viruses afflicting humans or their animals. In Europe, these include Paramyxovirus 2 (the cause of Newcastle's disease in poultry) (Maldonado et al. 1994) and Aphthovirus sp. (the cause of foot-and-mouth disease in domesticated mammals and humans) (Kaleta 2002). Sparrows have also been identified as agents of transmission of H5N1 avian influenza virus (responsible for Hong Kong avian influenza in poultry) (Perkins and Swayne 2003).

Several bacterial agents responsible for human or animal disease occur commonly in sparrows, and sparrows have been implicated in their transmission. Seymour et al. (1994) identified drug-resistant *Helicobacter* B52, a bacterium that can cause disease in humans and swine, in sparrow feces from Massachusetts (USA). In India, sparrows have been identified as carriers of *Micoplasma* spp. (Jain et al. 1971) and *Chlamydia psinaci* (Chahota and Katoch 2000), bacteria responsible for chronic respiratory disease and chlamydiosis, respectively, in poultry. Laboratory studies have demonstrated that *Micoplasma* spp. can be transmitted from sparrows to chickens (Kleven and Fletcher 1983). Sparrows are also commonly infected with *Escherichia coli*, (Pawiak et al. 1991), *Salmonella* spp. (Bowes 1990; Dozsa 1962/63; Schnetter et al. 1968), and other bacteria that cause disease in humans and livestock (see Chapter 8), but their role in the transmission of these diseases is largely unknown.

The same may be said for several protozoan pathogens. Boughton (1937b) suggests that sparrows may be involved in the transmission of coccidiosis, which is caused by the intestinal parasite *Isospora*, in Massachusetts. In New Zealand, sparrows were found to be infected with *Giardia* spp. on a farm where both human and livestock infection had occurred (Chilvers et al. 1998). Sparrows were also found to be infected with *Cryptosporidium* spp., another human pathogen. They have been implicated in the transmission of toxoplasmosis, but recent studies in which sparrows were experimentally infected with *Toxoplasma gondii* found that sparrows are resistant to infection and hence may not play an important role in its transmission (Literak et al. 1999).

Sparrows may also play a role in the spread of ectoparasites to poultry. Rosen et al. (1984) found the fowl mites, *Ornithonyssus sylvarum* and *O. bursa*, in sparrow nests on a farm in Israel where the chickens were infected by the mites. They also found *Dermanyssus gallinae* in a sparrow nest at a second farm with infected chickens. They suggested that sparrows may be a major vector in the spread of these ectoparasites among poultry farms.

Negative Effects on Native Bird Species

Severe declines in the population sizes of native species have been attributed to the introduction and subsequent spread of sparrows. Many of the affected species are secondary cavity nesters, and competition for nest sites with the aggressive house sparrow has been one of the major problems for these species (see Chapter 8). There are also some suggestions that sparrows have been involved in the transmission of introduced avian diseases that have led to the extinction or endangerment of native species.

In North America, the most severe negative impact of sparrows on native bird species is that on bluebirds (*Sialia* spp.), particularly the eastern bluebird (*S. sialis*) (Gowaty 1984; Radunzel et al. 1997). Among more than 70 species that Barrows (1889) identified as being "molested" by sparrows in North America, the eastern bluebird was by far the most affected species, with 377 reports of attacks by sparrows. Other species with more than 100 attacks included martins (*Progne* spp.), American robins (*Turdus migratorius*), and wrens (*Troglodytes* spp.).

Introduced sparrows may also have played a role in the transmission of diseases to native bird species. In Hawaii, for instance, the introduction and spread of avian pox and avian malaria have apparently contribued to the extinction and endangerment of numerous endemic bird species (van Riper et al. 1986, 2002). Although the disease organisms themselves were presumably introduced by the arrival of domestic chickens, sparrows may have played a crucial role in their spread from poultry to native bird species because of their close commensal relationship with humans.

Pest Management

The three main methods employed to reduce the negative impacts of pest species on humans have been reducing population size, developing pest avoidance of seeds or crops, and frightening pests away from sensitive areas (Chahal et al. 1973). Most of the pest management work on sparrows has focused on the first two methods. At least two studies, however, have examined the effects of ingestion of 4-aminopyridine on house sparrows (Frank et al. 1981; Shafi et al. 1984). At the dosages administered, 4-aminopyridine causes birds to become disoriented, and distress calls emitted by the disoriented individuals are supposed to frighten off other birds. Ingestion of 4-aminopyridine can also be fatal to sparrows (Frank et al. 1981). Dermally encountered fenthion ("Rid-a-Bird") is also supposed to discourage birds from visiting a treated site, but one study found that fenthion-treated sparrows were more vulnerable than untreated controls to predation by American kestrels (*Falco sparverius*), suggesting that the use of this deterrent may actually have lethal consequences (Hunt et al. 1992).

Population Reduction

Undoubtedly the most widely employed method of reducing sparrow numbers has been the use of strychnine-treated grain (Neff 1959). Large strychnine poisoning programs were conducted in Germany during the 1950s and 1960s (Geiler 1959; Grimm 1954; Niethammer 1953; Nordmeyer et al. 1972), and tens of thousands, or perhaps hundreds of thousands, of individuals were destroyed. Piechocki (1954), for instance, collected biometrical measurements on more than 20,000 sparrows killed in one region of Germany between 1952 and 1954. Strychnine poisoning has also been employed in the United States (Neff 1959) and in Pakistan (Hussain et al. 1993). In Great Britain, stupefying poisons have been used to capture and kill large numbers of sparrows (Thearle 1968). Two problems with the use of strychnine to control sparrow populations are the killing of individuals belonging to nontarget species and the need for continual use of the poison to prevent rapid recovery of the population. Preiser (1957) found that one area in Germany in which the sparrow population had been reduced by 60% was rapidly recolonized by individuals dispersing from nearby areas.

Numerous other poisons have also been tested against house sparrows. Various pesticides, mostly organophosphates, have been tested for their effectiveness (e.g., Chahal et al. 1973; Morris 1969; Schafer et al. 1973). Some have been found to be stupefying or narcotic for sparrows, resulting in markedly increased mortality (Morris 1969; Rana 1992). Fryday et al. (1994) examined the effects of lowered acetylcholinesterase (AChE) levels in the brains of sparrows exposed to organophosphates in the diet. They found that brain AChE activity was inversely correlated with organophosphate dose ($r = -0.84$, $P < .005$) and that brain AChE levels lower than 70% of normal resulted in an identifiable depression in seed-handling efficiency in the birds. Pesticides that have been reported to show promise in reducing sparrow populations include $\propto$-chloralose (Morris 1969, Rana 1992); azodrin, carbofuran, dasanit, optunal, and parathion (Schafer et al. 1973); and endrin and carbofuran (Chahal et al. 1973).

One study examined the effects of the anticoagulant brodifacoum on sparrows (Sandhu et al. 1985). Sparrows maintained in the laboratory were fed for 1 d with 10 g of pearl millet treated with 0.005% brodifacoum. The birds consumed an average of 3.5 g of the poisoned grain, and all birds tested died within 11 d after the treatment (mean = 7.3 d).

Large-scale trapping operations have also been employed against sparrows (Bösenberg 1957; Dearborn 1912; Royall 1969). Trapping has one advantage over poisoning, in that nontarget species are not affected, because individuals of nontarget species can be released unharmed. Trapping is labor intensive, however, and, like poisoning, it must be employed continuously to prevent recolonization of the protected area.

Chemosterilization is another technique that has been tested as a method of reducing sparrow numbers (Sanders and Elder 1976). In a laboratory study, sparrows fed for 10 d on treated grain (0.1% diazacholestenol dihydrochloride, "Orntrol") ceased nest-building and other reproductive behaviors shortly after the treatment. Although Sanders and Elder concluded that the technique could prove effective against sparrows, they also expressed concerns about its potential effects on nontarget species.

Deterrents

Two major methods have been employed to deter sparrows from feeding on grain seeds or crops. The first is the development of resistant strains of seed crops that are less accessible or less palatable to sparrows. The second is the application of repellent substances to developing seeds that discourage sparrows from feeding on the seeds.

The development of resistance in crops can involve either morphological changes that reduce the accessibility of the seeds to foraging birds or chemical changes that make the seeds less palatable (Bullard 1988). Tipton et al. (1970) examined the characteristics of grain sorghum (*Sorghum bicolor*) hybrids that affected the rate of sparrow exploitation in Louisiana (USA). Both morphological and chemical differences among the hybrids contributed to reduced utilization. Open-headed variants experienced lower exploitation rates than closed-headed variants, and sparrow-resistant hybrids had eight times more tannic acid content and four times more total astringents than sparrow-susceptible hybrids. In North Dakota (USA), Parfitt (1984) examined the morphological variants of sunflower (*Helianthus annuus*) that reduced exploitation by sparrows and red-winged blackbirds (*Agelaius phoeniceus*). Variants with concave heads and down-turned heads were exploited less by sparrows.

Various fungicides, insecticides, and other chemicals have been tested for their efficacy in reducing or eliminating crop depredation by sparrows. Babu (1988) performed laboratory tests on 10 fungicides and other chemicals and found that sparrows showed complete avoidance of seeds treated with 500 ppm of either Daxon or Fytolon. A laboratory study testing five chemicals as potential deterrents to sparrow foraging found that all five significantly reduced consumption of the treated food (turkey starter crumbs) (Greig-Smith and

Rowney 1987). The study also examined the effect of color on the development and persistence of the aversion. For two of the deterrent chemicals, the sparrows developed an aversion more rapidly when the treated food was blue, and aversions that did develop to all of the chemicals persisted for 1 wk after removal of the chemical from blue-colored food. The results of another laboratory experiment involving a choice between natural and blue-coated seeds suggested that sparrows may have an innate aversion to blue-colored food (Pawlina and Proulx 1996).

Positive Impacts of Sparrows

The alarm expressed over the recent catastrophic collapses of some populations of house sparrows, particularly those in European cities (see Chapter 8), is evidence that humans have a latent fondness for this pesky avian commensal despite its negative impacts. Sparrows have made numerous positive contributions to their human hosts and have the potential to continue to do so. One such contribution is their role as avian "laboratory rats" in studies of human and livestock disease. Sparrows have also been used as a sentinel species, signaling the potential for epidemic disease outbreaks. Their close association with humans and their domesticated animals suggests that they may also play a role as a bioindicator species in the detection and quantification of environmental hazards, thereby serving as so-called canaries in the mine. For instance, one innovative study suggested that the significant decrease in both body mass and tarsus length in sparrows in Israel between 1960 and 1998 might be indicative of global warming (Yom-Tov 2001). There are few reports of sparrow foraging directed against pest species (e.g., Moore and Hanks 2000), although such was the reason for their initial introduction into the United States (see Chapter 1).

Sparrows played an important role in early studies elucidating the transmission cycles of various arboviruses (Chamberlain et al. 1956, 1957; Hammon et al. 1951a, 1951b). As noted earlier, birds are typically the primary hosts of these viruses. Mosquitoes, particularly those in the genus *Culex*, serve as the major vectors for disease transmission among the avian hosts and are also responsible for transmission to humans or domesticated animals. Laboratory studies involving experimentally infected sparrows have demonstrated that mosquitoes can obtain a sufficient number of viral particles from viremic sparrows to transmit the virus to other hosts. Similar studies have recently been conducted with sparrows to understand the amplification and transmission of West Nile virus (Komar et al. 2003) and H5N1 avian influenza virus (Perkins and Swayne 2003).

Sparrows have served as sentinel species for a number of arboviruses that infect humans and domesticated animals. In North America, these include the

SLE virus (Lord, Calisher et al. 1974; McLean et al. 1983, 1988) and the EEE and WEE viruses (Williams et al. 1971). The typical protocol has been for local public health departments to monitor the frequency of virus-specific antibodies in samples of sparrows trapped in middle to late summer in areas with a history of encephalitis epidemics among humans or livestock. High numbers of antibody-positive sparrows indicate an increased risk of a local disease epidemic and lead to mosquito-control or other measures to reduce the risk. The development of increasingly sensitive tests for the presence of the viruses themselves has resulted in a transition by many public health departments from using sparrows as sentinels to sampling mosquitoes directly.

The fact that sparrows live in such close association with humans means that they share many potential environmental hazards with their hosts, as well as with other species that are common in human-altered habitats. Several studies have attempted to use sparrows as bioindicators of potential hazards to the health of other wildlife species caused by environmental contaminants. In some cases, these contaminants also pose potential health risks to humans. Pinowski, Aukowski et al. (1995) reported that levels of pesticide residues and polychlorinated biphenyls (PCBs) in sparrows in Poland were similar to the levels observed in humans in the area of study.

Agricultural use of various pesticides, herbicides, and other biologically active compounds has increased exponentially over a wide geographic scale in the past 60 y. The impacts of these compounds and their residues on wildlife are poorly understood. Numerous studies on sparrows have attempted to identify and quantify the potential impacts on farmland bird species.

Hill (1971) determined the median lethal dose (LD_{50}) for house sparrows of DDT and three other insecticides used against mosquito larvae. LD_{50} values ranged from 47 ppm for Abate to 1000 ppm for Gardona, with DDT having an intermediate value of 415 ppm. Subsequent studies indicated that sparrows were capable of detecting the presence of lethal doses of DDT, Gardona, and Bromophos in treated food, and most individuals survived exposure if provided with alternative sources of untreated food (Hill 1972).

Schafer et al. (1973) determined the LD_{50} values for adult sparrows of both oral and dermal applications of 17 pesticides (primarily organophosphates). Oral LD_{50} values varied from 0.32 to 18 mg/kg of total body mass (TBM), whereas dermal values were invariably higher and varied from 1.0 to more than 100 mg/kg TBM.

Gupta and Saxena (1999) studied the effects of both acute exposure (single dose of 15 mg/kg TBM) and chronic exposure (1 mg/kg/d) to the carbamate insecticide Mexacarbate on the activities of liver and serum phosphatases in sparrows. Both hepatic and serum acid phosphatases showed significant increases in activity after acute exposure and after 15 d of chronic exposure.

Hepatic alkaline phosphatase had a significant decrease in activity after acute exposure and after 5 and 10 d of chronic exposure. Gupta and Saxena concluded that Mexacarbate had significant toxicity for sparrows.

Sublethal effects of actual field applications of pesticides can be studied by examining levels of enzyme activity in exposed birds. Thompson et al. (1992) examined the temporal pattern of change in the activities of serum glutamate oxaloacetate transaminase (GOT), serum carboxyesterase (CbE), serum cholinesterase (ChE), and brain acetylcholinesterase (AChE) in sparrows exposed to demeton-*S*-methyl. In one year of the study, serum ChE and brain AChE activities dropped after pesticide application, whereas serum CbE activity increased for the first 2 d after application before dropping. In the second year, brain AChE activity remained relatively constant, but serum GOT, serum ChE and serum CbE activities all decreased. Liver abnormalities were also observed in the exposed birds. In another study, McInnes et al. (1996) examined the effects of the agricultural application of an organophosphorus insecticide (Chlorpyrilos) on nestling sparrows in Minnesota (USA). Surface washing and flushing of the gastrointestinal tracts of eight nestlings demonstrated that one bird had detectable levels of the insecticide on the day after application. Exposed nestlings showed a significant increase in brain AChE activity.

Nestling sparrows have also been used to identify the risks associated with the application of various chlorohydrocarbon insecticides and the industrial release of PCBs. A series of studies in various habitats near Warsaw (Poland) examined the presence of insecticide residues and PCBs in the eggs, nestling food, and nestling tissues (particularly liver and brain). Use of DDT was banned in Poland in 1974, and a ban on lindane was instituted in 1988–1990 (Niewiadowski et al. 1998). In 1988, the brains of nestling sparrows in a village near Warsaw contained 1.006 ppm DDT residues, 0.095 ppm lindane residues, and 6.118 ppm PCBs, with some individuals having levels that could impair development or even cause mortality (Karolewski et al. 1991). Levels of total insecticide residues and PCBs in the livers of nestlings at three rural sites and one urban site paralleled those found in the nestling food. Nestlings from suburban sites, where DDT use had been widespread, had higher levels of DDT residues than did nestlings from urban parks in Warsaw (Pinowski, Łukowski et al. 1995). Follow-up studies on the brains of fledglings from some of the same locations in 1995 showed greatly reduced levels of insecticide residues and PCBs (Niewiadowski et al. 1998). Assays of insecticide residues and PCBs in the lipids of adult sparrows from the same area in 1994–1996 showed significant quantities of both, however, and there was no significant difference in either DDT residues or PCBs between sparrows from a suburban site and those from a rural site (Pinowski et al. 1999).

Agricultural insecticides are sometimes applied to fields in granular form, and sparrows have been used to identify ways of reducing risks to birds by minimizing the ingestion of these granular insecticides as grit particles (see Chapter 6). Best (1992) found that the size and shape of five common granular pesticides for corn rootworms (*Diabrotica* spp.) used in Iowa (USA) were similar to the size and shape of grit particles used by sparrows (and many other avian species). Studies of the delivery media of the granular pesticide found that several of them were rapidly broken down in the sparrow intestinal tract, with only silica particles maintaining their integrity for prolonged periods (Best and Gionfriddo 1991b, 1995). This rapid breakdown led Best and Gionfriddo to speculate that the birds' "appetite" for grit would not be satisfied by the particles, possibly resulting in higher rates of consumption of the particles and consequently increasing exposure to the pesticide. Best and Gionfriddo (1996) also examined the role of color in grit selection by sparrows (see Chapter 6) and found that blue and black particles were taken least often, suggesting that use of one of these colors for pesticide granules might reduce their ingestion by birds.

Sparrows have also been used as bioindicators for environmental hazards such as heavy metals. A series of studies were conducted on the concentrations of heavy metals in eggs and nestlings of both house and tree sparrows in Poland (Pinowski, Kavanagh, and Pinowska 1995; Pinowski et al. 1991). Liver concentrations of the toxic metals lead (Pb) and cadmium (Cd) tended to be higher in nestlings from urban parks in Warsaw than in those from suburban areas near the city (Sawicka-Kapusta et al. 1995). There was little difference in the liver concentrations of iron (Fe) and zinc (Zn) at the two areas, however. Fe and Zn are required nutrients, and their absorption and excretion are presumably physiologically regulated. Pollution from heavy automobile traffic adjacent to the parks apparently contributed to the higher levels of toxic heavy metals. Pinowski et al. (1993) examined the Pb and Cd contents of the livers of dead, sickly, and healthy sparrow nestlings from six sites in the vicinity of Warsaw. Dead nestlings had significantly higher Pb levels than did sickly or healthy nestlings. There was a significant negative correlation between liver Pb content of nestlings and deviation from mean mass ($r = -0.285$, $P < .00001$). Liver Cd content did not differ among the three groups of nestlings. Levels of Fe and Zn in the liver also differed significantly between dead, sickly, and healthy nestlings, with Fe being significantly lower and Zn levels significantly higher in dead nestlings (Romanowski et al. 1991). Lipid and protein reserves in nestlings with high Pb contents in the liver (>4 μg/g of dry mass) were significantly lower than those in nestlings with low Pb levels, and the lipid reserves were frequently marginal for overnight survival of the nestlings (Gorzelski et al. 1995). No consistent differences in liver concentrations of the metals were observed between urban and suburban sites (Pinowski, Sawicka-Kapusta et al. 1995). Sparrow eggs may also be used to identify potentially harmful heavy metal contamination. In Poland, Jopek et al. (1995) found unusually high concentrations of Pb in sparrow eggs from urban parks during one year.

Even when sparrows are not the target of control measures, they may still be used as "guinea pigs." Stahl et al. (2002) used house sparrows and piegions to test for the efficacy of a new method for quantifying the presence of the avicide 3-chloro-*p*-toluidine hydrochloride (CPTH) in the gastrointestinal tract following administration of different numbers of CPTH-treated rice grains.

Sparrow eggs are often available in quantity, and, because sparrows are unprotected legally in many areas, these eggs can be used for research. In North America, sparrow eggs have been used in artificial nests to study rates of nest predation in island versus mainland birds (George (1987), forest edge versus forest interior nests (Degraaf et al. 1999), and open versus cavity nests (Purcell and Verner 1999).

Sociology of Sparrows

As noted earlier, the relationship between sparrows and their human hosts is complex. No doubt because of their familiarity to their human hosts, representations of sparrows have frequently been used by artists, poets, playwrights, composers, theologians, and sociologists in their efforts to portray the human condition. In a particularly interesting study, Doughty (1978) explored the complex and paradoxical relationship between 19th-centry Americans and the imported house ("English") sparrow. She defined tropophilia as the bond between people and place and suggested that animals figure prominently in myth, superstition, and folklore of landscapes. Although the introduction of the house sparrow was originally welcomed and strongly encouraged in the American press, attitudes changed over time, and Doughty suggests that sparrows began to represent the large number of Irish, Italian, and eastern European immigrants who were thronging to North America. She writes, "In fact, the bird's multiplication and spread caused consternation to many observers whose Victorian ideas of behaviour and decency were upset by the conduct of the sparrow . . . [The] lessons from the 'Book of Nature' taught by English sparrows alarmed and disgusted many" (Doughty 1978:5).

As I watch live television news from Baghdad, Gaza, Jerusalem, or Kosovo and hear sparrows chirping in the background, I sometimes wonder what opinion, if any, the house sparrow has about the havoc wreaked by its human hosts.

Appendix

PARASITIC AND DISEASE ORGANISMS OF THE HOUSE SPARROW

This table contains a partial list of the parasitic and disease organisms that have been identified from house sparrows or their nests. Available information on the target (e.g., blood, intestine) and/or reported percentages of infection or infestation is summarized. (In the case of detection using the presence of antibodies to a particular disease organism, the values represent the percentage of individuals that test positively for the presence of antibody.) Although I have attempted to identify as many sources as possible, the list is far from exhaustive. In some cases where a review was obtained (e.g., Brown and Wilson 1975, Peirce 1981), I made no effort to consult the original sources.

Parasite or Disease Organism	Target and % Infected/Infested	Location	Source
Endoparasites and Disease Organisms			
Viruses			
Arboviruses			
Eastern equine encephalitis	9.3–18.9	USA	1,2,3
Western equine encephalitis	2.3–72.4	Canada, USA	2,3,4,5.6,7,8,9,10
St. Louis encephalitis	0.3–41.9	USA	6,7,10,11,12,13,14,15,16
Highlands J virus	0.5	USA	1
Alphavirus sp. (Buggsy Creek virus)	n.a.	USA	17
Flanders virus	0–2.2	USA	18
Flanders Hart Park virus	0.6–0.9	USA	6,7,14
Turlock virus	1.9	USA	6,7
West Nile flavivirus	2.8–42	Czech Republic, Poland, Egypt, Israel	19,20,21,22,23
Sindbis alfavirus	1.1–2.2	Czech Republic, Poland, Israel	19,20,22,23
Tick-borne encephalitis flavivirus	1.1–1.8	Czech Republic, Poland	19,20,22
Tahyha bunyavirus	2.8–14.7	Czech Republic, Poland	19,20,22
Calovo bunyavirus	1.1–2.2	Czech Republic, Poland	19,20,22
Ockelbo virus	5.5–50	Sweden	24
Israel turkey menigoencephalitis virus	n.a.	Israel	23
Poxviruses			
Avian pox (Poxvirus avium)	7.4	Hawaii (USA)	25
Picornaviruses			
Foot-and-mouth disease virus (Aphthovirus sp.)	n.a.	Europe	26
Paramyxoviruses			
Paramyxovirus 2	68.6	Spain	27
Paramyxovirus 3	8.6	Spain	27

Bacteria			
Bacillus subtilis	egg, n.a.	Poland	28
Bacillus spp.	nestling intestine or liver, 29.3	Poland	29,30,31
Chlamydia psinaci	16.7	India	32
Citrobacter freundi	egg, n.a.	Poland	28
Citrobacter diversus	egg, n.a.	Poland	28
Citrobacter spp.	nestling intestine, 0.8–2.8	Poland	30,31,33
Corynebacterium spp.	nestling intestine, 1.1	Poland	30,31
Edwardstella ictaluri	egg, n.a.	Poland	28
Edwardstella sp.	nestling intestine, 0.9	Poland	33
Enterobacter intermedius	egg, n.a.	Poland	28
Enterobacter aerogenes	egg, n.a.	Poland	28
Enterobacter spp.	nestling intestine, 4.4–8.9	Poland	30,31,33
Escherichia coli	pathogenic strains: 65.2–72.2	Poland	28,29,30,31,33
Escherichia vulneris	egg, n.a.	Poland	28
Escherichia fergusonii	egg, n.a.	Poland	28
Escherichia spp.	nestling intestine, 68.7	Poland	33
Hafnia alvei	egg, n.a.	Poland	28
Hafnia sp.	nestling intestine, 5.6	Poland	33
Helicobacter sp.	7.7	USA	34
Klebsiella spp.	nestling intestine or liver, 1.6–3.3)	Poland	28,30,31
Kluyvera ascorbata	egg, n.a.	Poland	28
Lactobacillus sp.	88	USA	35
Micrococcus spp.	nestling intestine, 1.6	Poland	30,31
Morganella spp.	nestling intestine, 1.9	Poland	33
Mycoplasma gallisepticum	trachea, 6.3	India	36
Proteus mirabilis	nestling intestine/liver, 0.8	Poland	30,31
Proteus morgani	nestling intestine/liver, 5.5	Poland	30

(*continued*)

Parasite or Disease Organism	Target and % Infected/Infested	Location	Source
Proteus rettgeri	egg, n.a.	Poland	31
Proteus vulgaris	nestling intestine/liver, 4.6	Poland	28,29,30,31
Proteus sp.	egg, 1.5, nestling intestine, 3.0–3.7	Poland	30,31,33
Providentia retiigeri	egg, n.a.	Poland	28
Providencia stuarti	egg, n.a.	Poland	28
Pseudomonus aeruginosa	nestling intestine, 0.3	Poland	30,31
Salmonella typhimurium	19.5	UK, Switzerland, Germany, Poland, Canada	37,38,39,40,41,42,43,44,45
Salmonella dublin	n.a.	Poland	38
Salmonella paratyphi B	n.a.	Poland	38
Salmonella anatum	0.4	UK	43
Salmonella sp.	2.5–60	USA, Poland	30,35
Sarcinia spp.	nestling intestine/liver, 0.8	Poland	30
Serratia plymuthica	egg, n.a.	Poland	28
Serratia sp.	nestling intestine, 3.3	Poland	33
Staphylococcus auretus	egg, n.a.	Poland	28,29,31
Staphylococcus epidermidis	n.a.	Poland	28,29,30,31
Streptococcus faecalis	egg, n.a.; nestling intestine/liver, 20.5	Poland	28,31
Streptococcus lactis	egg, n.a.	Poland	31
Streptococcus thermophilus	egg, n.a.	Poland	28
Streptococcus viridans	egg, n.a.	Poland	31
Streptococcus spp.	55.9–81	Poland, USA	29,30,31,35
Tatumella sp.	nestling intestine, 1.9	Poland	33
Yersenia sp.	31	USA	35
Anaerobic bacteria	100	USA	35

Gram-negative enterics dark lactose fermentors	50	USA	35
Gram-negative enterics red lactose fermentors	50	USA	35
Gram-positive enterics	nestling intestine/liver, 73.2	Poland	30
Fungi			
Aspergillus flavus	nestling lungs, 0.7	Poland	46
Aspergillus fumigatus	egg, 1.5; nestling lungs, 21.3	Poland	31,46
Aspergillus glaucu	egg, 1.5; nestling lungs, 3.0	Poland	31,46
Aspergillus niger	egg, 1.5; nestling lungs, 5.2	Poland	31,46
Candida albicans	egg, n.a.; nestling intestine, 4.6	Poland	28
Candida tropicalis	nestling intestine, 2.5	Poland	33
Candida pseudotropicalis	nestling intestine, 0.8	Poland	33
Candida krusei	nestling intestine, 5.1	Poland	33
Candida lipolytoca	nestling intestine, 1.7	Poland	33
Candida quillermondi	nestling intestine, 2.5	Poland	33
Candida utilis	nestling intestine, 1.7	Poland	33
Candida parapsilosis	nestling intestine, 0.8	Poland	33
Candida clausseni	nestling intestine, 1.3	Poland	33
Candida pelliculosa	nestling intestine, 0.8	Poland	33
Candida spp.	9.3–34.2	Poland	29,31,46
Cryptococcus mesentroides	nestling intestine, 0.4	Poland	33
Penicillium spp.	egg, n.a.; nestling lungs, 38.4	Poland	31,46
Rhodotorula rubra	nestling intestine, 0.4	Poland	33
Torulopsis inconspicula	nestling intestine, 0.8	Poland	33
Protista: Protozoa			
Cryptosporidium spp.	n.a.	New Zealand	47
Entamoeba sp.	nestling intestine, 9.7	Poland	33

(continued)

Parasite or Disease Organism	Target and % Infected/Infested	Location	Source
Giardia spp.	n.a.	New Zealand	47
Haemogregarina sp.	blood, 1.0	USA	48
Haemoproteus passeris	blood, 41	Western Europe, Israel	49,50
Haemoproteus spp.	blood, 6.8–43.2	Western Europe, India, USA, Spain	49,51,52,53,54
Isospora lacazei	intestine, 10.5–100	UK, USA, Poland	55,56,57
Isospora chloridis	n.a.	UK	55
Isospora sp.	intestinal coocidiosis	USA	58
Lankesterella garnhami	blood, 3–34	USA, UK	59,60
Lankesterella (Atoxoplasma) spp.	blood, 5.7–100	India, UK, USA, Western Europe	48,49,51,53,56,61,62
Leucocytozoan spp.	blood, 0.4–18.2	India, USA	51,53,63
Plasmodium relictum	blood, 7–52	USA, India, Western Europe, Hawaii (USA)	16,48,49,52,53,55,63,64,65
Plasmodium cathemerium	blood, 1.1	USA, Western Europe	48,49,64
Plasmodium circumflexum	blood, n.a.	USA	48
Plasmodium elongatum	blood, 0.5–0.9	USA	48,53,63
Plasmodium spp.	blood, 3.9	Poland, Western Europe, USA	49,64,66
Trypanosoma sp.	blood, n.a.	Western Europe	49
Toxoplasma gondii	1.5–25.3	Poland, Czech Republic	67
Toxoplasma sp.	blood, n.a	Western Europe	49
Platyhelminthes			
Trematodes			
Echinostoma revolutum	0.7	Brazil	68
Eumegacetes medioximus	cloaca, 1.4	Brazil	68
Leucochloridium parcum	cloaca, 4.9	Brazil	68
Tamaisia inopina	13.3	Brazil	68

Cestodes			
Anonchotaenia globata	n.a.	Turkmenistan (?)	69
Choanotaenia paserina	intestine-21.1	Brazil, Spain	69,70
Mogheia domesticus	intestine, n.a.	India	71
Nematodes			
Acuaria subula	n.a.	Turkmenistan (?)	69
Capillaria tridens	n.a.	Spain	72
Dispharynx nasuta	esophagus-4.2	Brazil	68
Diplotriaena ozouxi	n.a.	Turkmenistan (?)	69
Tetrameres minima	proventriculus, 20.4	Brazil	68
Acantocephala			
Mediorhynchus papillus	intestine, 1.4	Brazil	68
ECTOPARASITES			
Bacteria			
Bacillus licheniformis	23.3	USA	73
Mycoplasma gallisepticum	n.a.	USA, Canada	74
Arachnida: Acari (Mites and Ticks)			
Acaropsis docta	n.a.	India	75
Acaropsis sollers	n.a.	India	75
Acarus immobilis	n.a.	India	75
Acheles meyerae	n.a.	India	75
Amblyomma maculatum	n.a.	North America	76
Amblyseius alstoniae	n.a.	India	75
Amblyseius largoensis	n.a.	India	75
Analgidae	n.a.	UK	77
Analges chelopis	n.a.	Europe	76
Analges mucronatus	n.a.	Europe	76

(continued)

Parasite or Disease Organism	Target and % Infected/Infested	Location	Source
Analges passerinus	n.a.	Europe	76
Analges sp.	2.9	USA, UK	76
Argas persicus	n.a.	Europe	76
Asca pseudospicata	n.a.	India	75
Austroglycyphagus geniculatur	n.a.	India	75
Bakericheyla chanayi	0.2–3.3	Czech Republic, Egypt, USA	78,79,80
Bdella bakeri	n.a.	India	75
Blattiscoius dentricus	n.a.	India	75
Blattiscoius keegani	n.a.	India	75
Blomia sp.	n.a.	India	75
Boyalaia nigra	2.2	Egypt	79
Brevipalpus californicus	n.a.	India	75
Chaunoproctus sp.	n.a.	India	75
Cheilostigmaeus midnapurensis	n.a.	India	75
Chelacaropsis moorei	n.a.	India	75
Cheletonella sp.	n.a.	India	75
Cheletopsis anax	n.a.	India	75
Cheyletus malaccensis	3.6–6.7	Egypt	79
Cheyletus eruditus	4.4–6.1	Egypt	79
Cheyletidae sp.	n.a.	UK	76
Ctenoglyphus plumiger	n.a.	India	75
Cunaxa capreolus	n.a.	India	75
Dermanyssus americanus	11.3–14.7	USA	76,80,81,82
Dermanyssus hirundinis	5.0	USA	80,81
Dermanyssus gallinae	6.7–8.2	UK, Egypt, Europe, North America	76,77,78,79,

Dermanyssus passerinus	14.6	Czech Republic, Europe, North America	76,78
Dermanyssus longipes	n.a.	Europe, North America	76
Demataphagoides passericola	n.a.	Europe	76
Dermoglyphus elongatus	n.a.	UK, North America	76
Dermoglyphus columbae	1.2–1.5	Egypt	79
Dermoglyphus sp.	0–56.0	USA	80
Epidermaptes bilobatus	21.2–22.4	Egypt	79
Eremulus flagellifer	n.a.	India	75
Erythraeidae sp.	n.a.	India	75
Eugamasus butleri	n.a.	India	75
Eupodes sp.	n.a.	India	75
Falculifer rostratus	4.8–6.7	Egypt	79
Gabucinia delibata	n.a.	India	75
Glycyphagus domesticus	0.9–6.7	Egypt	79
Haemaphysalis concinna	n.a.	Europe	76
Haemaphysalis leporis-palustris	n.a.	Europe	76
Haemolaelaps casalis	3.0–18.9	Egypt, North America	6,79,80
Harpirhinchus holopus	n.a.	Europe, North America	76
Harpirhinchus ovalis	n.a.	Europe	76
Harpirhinchus pilirostris	0–56.0	Europe, North America	76,80
Harpirhinchus sp.	1.8	Egypt	79
Hypoaspis vacua	n.a.	India	75
Indostigmaeus rangatensis	n.a.	India	75
Ixodes ricinus	0.9	Czech Republic, North America, Europe	76,78
Ixodes arboricola	0.2	Czech Republic	78
Ixodes canisuaga	n.a.	Europe	76
Ixodes auritutus	n.a.	North America	76
Ixodes bruneus	n.a.	North America	76

(continued)

Parasite or Disease Organism	Target and % Infected/Infested	Location	Source
Ixodes hexagonus	n.a.	Europe	76
Kleemania bengalensis	n.a.	India	75
Kleemania plumosus	n.a.	India	75
Kleemania sp.	n.a.	India	75
Knemidocoples passeris	n.a.	Europe	76
Kuzinia sp.	n.a.	India	75
Lamellobates palustris	n.a.	India	75
Lardoglyphus zacheri	0.9–7.8	Egypt	79
Lasioseius sp.	n.a.	India	75
Liponyssus silvarium	n.a.	USA	82
Macrocheles merdarius	n.a.	India	75
Macroynssus bursa	17.9–23.3	Egypt	79
Macroynssus psittacula	n.a.	India	75
Mealia sp.	n.a.	India	75
Megninia columbae	20.0–31.8	Egypt	79
Megniniella sp.	n.a.	India	75
Michaelia microcorbonis	n.a.	India	75
Microlichus avus	2.9	USA, Europe	76
Neocunaxidae sp.	n.a.	India	75
Neocypholaelaps stridulans	n.a.	India	75
Oligonychus indicus	n.a.	India	75
Oligonychus mangiferus	n.a.	India	75
Ololaelaps venata	n.a.	India	75
Ornithocheyletia lawrenceae	n.a.	India	75
Ornithonyssus sylviarum	4.0–5.8	USA, UK, Israel, Hungary	76,80,83,84

Ornithonyssus bacoti	n.a.	India	75
Ornithonyssus bursa	n.a.	USA, Israel, Europe 76,83	
Ornithonyssus iheringi	n.a.	North America	76
Paraneonyssus (Neonyssus) hirsti	23.5–31.0	USA	76,80
Parasitus consanguineus	n.a.	India	75
Parlagas sp.	1.5	Egypt	79
Passeroptes dermicola	n.a.	USA	80
Pellonyssus passeri	18.3–25.8	India, Egypt, North America	76,79,80,85,86
Pellonyssus nidicolus	5.6–7.0	Egypt	79
Pellonyssus reedi	n.a.	USA, Hungary	84,87
Pellonyssus sp.	n.a.	India	75
Pergalumna sp.	n.a.	India	75
Proctophylodidae	n.a.	UK	77
Proctophyllodes amelides	n.a.	Europe	76
Proctophyllodes bruniceps	n.a.	India	75
Proctophyllodes orientalis	n.a.	India	75
Proctophyllodes passeris	n.a.	Europe	76
Proctophyllodes pinnatus	n.a.	Europe, North America	76
Proctophyllodes profusus	n.a.	Europe	76
Proctophyllodes truncatus	50–100	USA, UK, Europe, Portugal	76,80,88
Pronematus fleschneri	n.a.	India	75
Protokalumma sp.	n.a.	India	75
Protolichus sp.	n.a.	India	75
Pterodectes sp.	n.a.	India	75
Pterolichus obtusus	14.2–14.4	Egypt	79
Ptilonyssus hirsti	11.1–34.5	Egypt	79
Ptilonyssus nudus	1.3	UK, Europe, North America	76,80
Ptilonyssis passeri	8.9–11.1	Egypt	79

(continued)

Parasite or Disease Organism	Target and % Infected/Infested	Location	Source
Rhinonyssidae sp.	n.a.	India	75
Rhipicephalus sanguineus	n.a.	Europe	76
Rhizoglyphus sp.	n.a.	India	75
Rivoltasia dermicola	n.a.	Europe	76
Speleognathus sturni	n.a.	North America	76
Spinibdella cronini	n.a.	India	75
Sternostoma cryptorhynchum	n.a.	Europe, North America	76
Sternostoma tracheocolum	1.0	Europe, North America	76,80
Sturnidoecus ruficeps mite/tick?	2.5	Czech Republic	78
Suidasia nesbitti	n.a.	India	75
Syringophiloidus minor	2.9–7.8	USA, UK, Canada, Egypt, Europe	76,79,80,89
Syringophilus bipectinatus	2.4–5.6	Egypt	79
Syringophilus passerinae	4.8–6.7	Egypt	79
Tenuipalpus fici	n.a.	India	75
Tetranychus ludeni	n.a.	India	75
Trichouropoda sp.	n.a.	India	75
Trombicula autumnalis	n.a.	Europe	76
Trombidium sp.	n.a.	India	75
Trouessartia lonchura	n.a.	India	75
Trouessartia sp.	n.a.	North America	76
Typhloderomus sp.	n.a.	India	75
Typhlodromus sp.	n.a.	India	75
Tyroglyphus farinae	7.9–10.0	Egypt, India	75,79
Tyrophagus putrescentia	3.6–5.6	Egypt, India	75,79
Zygoribatula sp.	n.a.	India	75

Arachnida: Pseudoscorpionida (Pseudoscorpions)			
Apocheiridium indicum	n.a.	India	90
Cheiridium museorum	n.a.	India	90
Chelifer sp.	n.a.	India	90
Chemes sp.	n.a.	India	90
Minniza loyolae	n.a.	India	90
Tyrannochthonius madrasensis	n.a.	India	90
Insecta: Mallophaga (Feather Lice)			
Brueelia cyclothorax	3.0–29.4	Czech Republic, USA, UK, New Zealand	76,78,91
Brueelia subtilis	24–92	Europe, North America	76,80
Brueelia vulgata	n.a.	Europe, North America	76
Brueelia sp.	n.a,	UK	77
Colpocephalum scopularium	n.a.	Europe	76
Lipeurus heterographus	n.a.	Europe	76
Menacanthus annulatus	1.7	Europe, North America	76,80
Menacanthus eurystemus	n.a.	New Zealand	91
Menacanthus sp.	1.3	UK	76
Myrsidea quadrifasciata	n.a.	Europe	76
Philopterus aeneas	n.a.	Europe	76
Philopterus fringillidae	1.9–19.5	UK, Czech Republic	76,77,78
Philopterus subflavescens	n.a.	Europe	76
Rostrinirimus refractariolus	n.a.	Europe	76
Insecta: Hemiptera (Bugs)			
Oeciacus hirudinis	n.a.	Europe	76
Oeciacus vicarius	n.a.	USA	17
Triatoma sordida	n.a.	Brazil	92
Insecta: Diptera (Flies)			
Apaulinaenea herperia	n.a.	North America	76

(continued)

Parasite or Disease Organism	Target and % Infected/Infested	Location	Source
Apaulinaenea hirudo	n.a.	North America	76
Apaulinaenea sp.	n.a.	North America	76
Boreallus caerulae	n.a.	Europe	76
Carnus hermapteras	n.a.	Israel	93
Lipoptena cervi	n.a.	Europe	76
Neottiiphilum praeustum	n.a.	Europe	76
Ornithomyia avicularia	n.a.	Europe	76
Ornithomyia fringillina	n.a.	Europe, North America	76
Ornithomyia lagopdis	n.a.	Europe	76
Ornithoica vicina	n.a.	North America	76
Ornitophia metallica	n.a.	Israel	93
Protocalliphora azurea	20	Poland	94
Insecta: Siphonoptera (Fleas)			
Ceratophyllus borealis	n.a.	Europe	76
Ceratophyllus celsus celsus	n.a.	USA	17
Ceratophyllus fringillae	12.7–38.2	Czech Republic, Poland, Europe	76,95,96,97,98
Ceratophyllus gallinae	50.9–53.9	Czech Republic, UK, Poland, Europe, North America	76,77,95,96,97,99
Ceratophyllus garei	n.a.	Czech Republic	95
Ceratophyllus hirundinis	n.a.	Europe	76
Ceratophyllus niger	n.a.	North America	76
Ceratophyllus rusticus	n.a.	Europe	76
Ceratophyllus tribulis	n.a.	Germany	99
Dasypsyllus gallinulae	n.a.	Czech Republic, Europe	76,97
Echidnophaga gallinacea	n.a.	North America	76
Hectopsylla psittaci	n.a.	Europe	76
Monopsyllus sciurorum	n.a.	Czech Republic	97
Nosopsyllus fasciatus	n.a.	Czech Republic	95,97
Tarsopsylla octodecimdentata	n.a.	Czech Republic	97

Brood Parasites		
Cuculidae (Cuckoos)		
Cuculus canorus (common cuckoo)	Europe	100
Cuculus pallidus (pallid cuckoo)	Australia	101
Cuculus pyrrhophanus (fan-tailed brush cuckoo)	Australia	101
Chrysococcyx basalis (Horsfield's bronze cuckoo)	Australia	101
Chrysococcyx lucidus (golden bronze cuckoo)	Australia	101
Eudynamys taitensis (long-tailed cuckoo)	New Zealand	102
Icteridae (cowbirds)		
Molothrus ater (brown-headed cowbird)	North America	103
Molothrus bonariensis (shiny cowbird)	South America	104

n.a., not available.

Sources: 1, McLean et al. (1985); 2, Stamm (1958); 3, Williams et al. (1971); 4, Burton et al. (1966); 5, Holden (1955); 6, Holden, Hayes et al. (1973); 7, Hayes et al. (1967); 8, Sooter et al. (1954); 9, Cockburn et al. (1957); 10, Reeves et al. (1964); 11, Kokernot et al. (1969a), McLean et al. (1993); 12, Lord, Calisher and Doughty (1974), Lord, Calisher et al. (1974); 13, Kissling et al. (1954); 14, Lord et al. (1973); 15, McLean et al. (1988); 16, Reeves et al. (1952); 17, Hopla et al. (1993); 18, Kokernot et al. (1969b); 19, Juricova et al. (1999); 20, Juricova et al. (1998); 21, Taylor et al. (1956); 22, Juricova et al. (2000); 23, Nir et al. (1967, 1969); 24, Francy et al. (1989), Lundstrom et al. (1992); 25, van Riper et al. (2002); 26, Kaleta (2002); 27, Maldonado et al. (1994); 28, Koz<lx>owski et al. (1991); 29, Pinowski et al. (1988); 30, Kruszewicz, Kruszewicz et al. (1995); 31, Pinowski, Mazurkiewicz et al. (1995); 32, Chahota and Katoch (2000); 33, Malyszko et al. (1991); 34, Seymour et al. 1994); 35, Stewart and Rambo (2000); 36, Jain et al. (1971); 37, Macdonald and Cornelius (1969); 38, Pinowska et al. (1976); 39, Schnetter et al. (1968); 40, Wilson and MacDonald (1967); 41, Bouvier (1968); 42, Dozsa (1962/63); 43, Cornelius (1969); 44, Bowes (1990); 45, Macdonald (1977); 46, Kruszewicz, Pinowski et al. (1995); 47, Chilvers et al. (1998); 48, Manwell (1957); 49, Peirce (1981); 50, Paperna and Gill (2003); 51, Kruszewicz (1989); 52, Singh et al. (1951); 53, Hart (1949); 54, Navarro et al. (2003); 55, Anwar (1966); 56, Box (1967); 57, Kruszewicz (1995a); 58, Boughton (1937a, 1937b); 59, Box (1966); 60, Lainson (1959); 61, Peirce (1980); 62, Wetmore (1941); 63, Hunninen and Young (1950); 64, Micks (1949); 65, van Riper et al. (1986); 66, Kruszewicz (1994); 67, Literak et al. (1997); 68, Brasil and Amato (1992), Brasil et al. (1991); 69, Meredev and Golovkova (1978); 70, Illescas-Gomez and Lopez-Roman (1980); 71, Jadhav et al. (1990); 72, Rocio et al. (1981); 73, Burtt and Ichida (1999); 74, Hartup et al. (2001); 75, Bhattacharyya (1995); 76, Brown and Wilson (1975); 77, Polani et al. (2000); 78, Machacek (1977); 79, Morsy et al. (1999); 80, McGroarty and Dobson (1974); 81, Phillis (1972); 82, Reeves et al. (1947); 83, Rosen et al. (1984); 84, Szabo et al. (2002); 85, Clark and Yunker (1956); 86, Kaul et al. (1978); 87, Weddle (2000); 88, Behnke et al. (1999); 89, Bochkov and Galloway (2001); 90, Bhattacharyya (1990); 91, Pilgrim and Palma (1982); 92, Smith (1973); 93, Singer and Yom-Tov (1988); 94, Draber-Molko (1997); 95, Nemec et al. (1995); 96, Cyprich et al. (2002); 97, Jurik (1974); 98, Kaczmarek (1991); 99, Kutzscher (1993); 100, Wyllie (1981); 101, Brooker and Brooker (1989); 102, McLean (1988); 103, Friedmann (1929); 104, Friedmann (1963).

LITERATURE CITED

Abraham, U., U. Albrecht, and R. Brandstätter. 2003. Hypothalamic circadian organization in birds. II. Clock gene expression. Chronobiology International 20:657–669.

Abraham, U., U. Albrecht, E. Gwinner, and R. Brandstätter. 2002. Spatial and temporal variation of passer Per2 gene expression in two distinct cell groups of the suprachiasmatic hypothalamus in the house sparrow (*Passer domesticus*). European Journal of Neuroscience 16:429–436.

Abraham, U., E. Gwinner, and T. J. Van't Hof. 2000. Exogenous melatonin reduces the resynchronization time after phase shifts of a non-photic zeitgeber in the house sparrow (*Passer domesticus*). Journal of Biological Rhythms 15:48–56.

Alatalo, R. 1975. [On the breeding of the house sparrow at Oahu.] Lintumies 10:1–7. (In Finnish, English summary).

Albrecht, J. S. M. 1983. Courtship behaviour between tree sparrow and house sparrow in the wild: a possible case of hybridisation. Sandgrouse 5:97–99.

Al-Dabbagh, K. Y., and J. H. Jiad. 1988. The breeding biology of house sparrow in central Iraq. International Studies on Sparrows 15:22–43.

Alexander, W. D., and D. Lack. 1944. Changes in status among British breeding birds. British Birds 38:42–45.

Allende, L. M., I. Rubio, V. Ruiz-del-Valle, J. Guillen, J. Martinez-Laso, E. Lowry, P. Varela, J. Zamora, and A. Arnaiz-Villena. 2001. The Old World sparrows (genus *Passer*) phylogeography and their relative abundance of nuclear mtDNA pseudogenes. Journal of Molecular Evolution 53:144–154.

Allender, C. 1936a. Microscopical observations on the gonadal cycle of the English sparrow. Transactions of the American Microscopical Society 55:243–249.

Allender, C. 1936b. Seasonal gonadal cycle of the English sparrow, *Passer domesticus* (L.). Ecology 17:258–262.

Alonso, J. C. 1984a. Kreuzung spanisher Haus- (*Passer domesticus*) und Weidensperlinge (*Passer hispaniolensis*) in Gefangenschaft. Journal für Ornithologie 125:339–340.

Alonso, J. C. 1984b. Zur Mauser spanisher Weiden-und Haussperlinge (*Passer hispaneolensis* und *deomesticus*). Journal für Ornithologie 125:209–223.

Alonso, J. C. 1985. Descriptions of intermediate phenotypes between *Passer hispaniolensis* and *Passer domesticus*. Ardeola 32:31–38.

Alonso, J. C. 1986. Ecological segregation between sympatric Spanish sparrows (*Passer hispaniolensis* Temm.) and house sparrows (*Passer domesticus* [L.]) during winter. Ekologia Polska 34:63–73.

Al-Safadi, M. M., and M. Kasparek. 1995. Breeding observations on the birds of the Tihamah, Yemen. Zoology in the Middle East 11:15–20.

Altwegg, R., T. H. Ringsby, and B.-E. Sæther. 2000. Phenotypic correlates and consequences of dispersal in a metapopulation of house sparrows *Passer domesticus*. Journal of Animal Ecology 69:762–770.

American Ornithologists' Union. 1998. Check-list of North American Birds. 7th edition. American Ornithologists' Union, Washington, DC.

Ames, P. L. 1971. The morphology of the syrinx in passerine birds. Bulletin of the Peabody Museum of Natural History, Yale University 37:1–194.

Amr, Z. S., W. N. Al-Melhim, and M. A. Yosef. 1997. Mammal remains from pellets of the Eagle Owl, *Bubo bubo*, from Azraq Nature Reserve, Jordan. Zoology in the Middle East 14:5–9.

Anderson, T. R. 1977. Reproductive responses of sparrows to a super-abundant food supply. Condor 79:205–208.

Anderson, T. R. 1978. Population studies of North American sparrows, *Passer* spp. Occasional Papers of the Museum of Natural History, University of Kansas 70:1–58.

Anderson, T. R. 1979. Experimental synchronization of sparrow reproduction. Wilson Bulletin 91:317–319.

Anderson, T. R. 1980. Comparison of nestling diets of sparrows, *Passer* spp., within and between habitats. Acta XVII Congressus Internationalis Ornithologici 1162–1170.

Anderson, T. R. 1984. A comparative analysis of overlap in nestling diets of village populations of sparrows (*Passer* spp.) in Poland. Ekologia Polska 32:693–707.

Anderson, T. R. 1989. Determinate vs. indeterminate laying in the house sparrow. Auk 106:730–732.

Anderson, T. R. 1990. Excess females in a breeding population of house sparrow [*Passer domesticus* (L.)]. Pp. 87–93 in Granivorous Birds in the Agricultural Landscape (J. Pinowski and J. D. Summers-Smith, eds.). PWN–Polish Scientific Publishers, Warsaw.

Anderson, T. R. 1994. Breeding biology of house sparrows in northern lower Michigan. Wilson Bulletin 106:537–548.

Anderson, T. R. 1995. Removal indeterminacy and the proximate determination of clutch size in the house sparrow. Condor 97:197–207.

Anderson, T. R. 1997. Intermittent incubation during egg laying in house sparrows. Wilson Bulletin 109:324–328.

Anderson, T. R. 1998. Cessation of breeding in the multi-brooded house sparrow (*Passer domesticus*). International Studies on Sparrows 25:3–30.

Andersson, S., and M. Ahlund. 1991. Hunger affects dominance among strangers in house sparrows. Animal Behaviour 41:895–897.

Andrewartha, H. G., and L. C. Birch. 1954. The Distribution and Abundance of Animals. University of Chicago Press, Chicago, Illinois.

Angelici, F. M. 1993. House sparrow (*Passer domesticus domesticus*) as predator of Turkish gecko (*Hemidactylus turcicus*). Okologie der Vogel 15:119–120.

Angelici, F. M., L. Latella, L. Luiselli, and F. Riga. 1997. The summer diet of the little owl (*Athene noctua*) on the island of Astipalala (Dodecanese, Greece). Journal of Raptor Research 31:280–282.

Anwar, M. 1966. *Isospora lacazei* (Labbe, 1893) and *I. chloridis* sp. n. (Protozoa: Eimariidae) from the English sparrow (*Passer domesticus*), greenfinch (*Chloris chloris*) and chaffinch (*Fringilla coelebs*). Journal of Protozoology 13:84–90.

Aparicio, J. M. 1998. Patterns of fluctuating asymmetry in developing primary feathers: a test of the compensational growth hypothesis. Proceedings of the Royal Society of London B 265:2353–2357.

Ar, A., H. Ranh, and C. V. Paganelli. 1979. The avian egg: mass and strength. Condor 81:331–337.

Ar, A., and Y. Yom-Tov. 1978. The evolution of parental care in birds. Evolution 32:655–669.

Arens, J. R., and S. J. Cooper. 2005. Seasonal and diurnal variation in metabolism and ventilation house sparrows. Condor 107:433–444.

Arnold, T. W., F. C. Rohwer, and T. Armstrong. 1987. Egg viability, nest predation, and the adaptive significance of clutch size in prairie ducks. American Naturalist 130:643–653.

Arroyo, B. E. 1997. Diet of Montagu's harrier *Circus pygargus* in central Spain: analysis of temporal and geographic variation. Ibis 139:664–672.

Asch, A., and D. D. Roby. 1995. Some factors affecting precision of the total body electrical conductivity technique for measuring body composition in live birds. Wilson Bulletin 107:306–316.

Ash, J. S., and P. R. Colston. 1981. A house × Somali sparrow *Passer domesticus* × *P. castanopterus* hybrid. Bulletin of the British Ornithologists' Club 101:291–294.

Asnani, M. V. 1984. Comparative study on growth and metabolic function of the liver of house sparrow and house swift during post-hatching period. Pavo 22:45–52.

Assenmacher, I. 1973. The peripheral endocrine glands. Pp. 183–286 in Avian Biology. vol. III (D. S. Farner and J. R. King, eds.). Academic Press, New York.

Attiwill, A. R. 1970. On the spread of pines and bridal creeper by birds. South Australian Ornithologist 25:212.

Avise, J. C., and D. Walker. 1998. Pleistocene phylogeographic effects on avian opulations and the speciation process. Proceedings of the Royal Society of London B 265:457–463.

Babu, T. H. 1988. Effectiveness of certain chemicals and fungicides on the feeding behaviour of house sparrows. Pavo 26:17–23.

Bährmann, U. 1967. Bemerkungen zur Handschwingenmauser des Haussperlings (*Passer domesticus domesticus* L.). Beiträge zur Vogelkunde 12:363–366.

Baker, A. J. 1980. Morphometric differentiation in New Zealand populations of the house sparrow (*Passer domesticus*). Evolution 34:638–653.

Baker, J. R. 1938. The evolution of breeding seasons. Pp. 161–177 in Evolution: Essays on Aspects of Evolutionary Biology Presented to Prof. F. S. Goodrichon his 70th Birthday (C. R. de Beer, ed.). Clarendon Press, Oxford, UK.

Baker, M. 1995. Environmental component of clutch-size variation in house sparrows (*Passer domesticus*). Auk 112:249–252.

Balat, F. 1973. Die zwischenartlichen Brut beziehungen zwischen dem Haussperling, *Passer domesticus* (L.) und der Mehlschwalbe, Delichonurbica (L.). Zoologicke Listy 22:213–222.

Balat, F. 1974. Gelegegegrosse Brutverluste des Haussperlings, *Passer domesticus* (L.) im Mittelmahren. Zoologicke Listy 23:229–240.

Balat, F. 1977. Ortstreue und Ortswechsel des Haussperlings, *Passer domesticus*. Zoologicke Listy 26:237–244.

Baldaccini, N. E., S. Benvenuti, V. Fiaschi, and F. Papi. 1975. Pigeon navigation: effects of wind deflection at home cage on homing behaviour. Journal of Comparative Physiology 99:177–186.

Balmer, D., and J. Marchant. 1993. The sparrows fall. British Birds 86:631–633.

Baptista, L. F., and L. Petrinovich. 1984. Social interaction, sensitive phases and the song template hypothesis in the white-crowned sparrow. Animal Behaviour 32:172–181.

Barba, E.,and J. A. Gil-Delgado. 1990. Competition for nest-boxes among four vertebrate species: an experimental study in orange groves. Holarctic Ecology 13: 183–186.

Barba, E., J. A. Gil-Delgado, and J. S. Monrós. 1995. The costs of being late: consequences of delaying great tit *Parus major* first clutches. Journal of Animal Ecology 64:642–651.

Barfuss, D. W., and L. C. Ellis. 1971. Seasonal cycles in melatonin synthesis by the pineal gland as related to testicular function in the house sparrow (*Passer domesticus*). General and Comparative Endocrinology 17:183–193.

Barkowska, M., A. H. Kruszewicz, and A. Haman. 1995. The application of Richards' model as a measure of weight growth in house sparrow (*Passer domesticus*) and tree sparrow (*Passer montanus*) nestlings. Pp. 9–29 in Nestling Mortality of Granivorous Birds due to Microorganisms and Toxic Substances: Synthesis (J. Pinowski, B. P. Kavanagh, and B. Pinowska, eds.). PWN–Polish Scientific Publishers, Warsaw.

Barkowska, M., B. Pinowska, J. Pinowski, J. Romanowski, and K.-H. Hahm. 2000. Evaluation of the TOBEC method for calculating fat mass in tree sparrows *Passer montanus* and house sparrows *Passer domesticus*. Acta Ornithologica 35:135–145.

Barnard, C. J. 1980a. Equilibrium flock size and factors affecting arrival and departure in feeding house sparrows. Animal Behaviour 28:503–511.

Barnard, C. J. 1980b. Factors affecting flock size mean and variance in a winter population of house sparrows (*Passer domesticus*). Behaviour 74:114–127.

Barnard, C. J. 1980c. Flock feeding and time budgets in the house sparrow (*Passer domesticus* L.). Animal Behaviour 28:295–309.

Barnard, C. J. 1980d. Flock organization and feeding budgets in a field population of house sparrows (*Passer domesticus*). Acta XVII Congressus Internationalis Ornithologici 1117–1121.

Barnard, C. J., and R. M. Sibly. 1981. Producers and scroungers: a general model and its application to captive flocks of house sparrows. Animal Behaviour 29:543–550.

Barnett, L. B. 1970. Seasonal changes in temperature acclimatization of the house sparrow, *Passer domesticus*. Comparative Biochemistry and Physiology 33:559–578.

Barrows, W. B. 1889. The English sparrow (*Passer domesticus*) in North America. Bulletin 1. U. S. Department of Agriculture, Division of Economic Ornithology and Mammalogy.

Bartholomew, G. A., Jr. 1949. The effect of light intensity and day length on reproduction in the English sparrow. Bulletin of the Museum of Comparative Zoology 101:433–476.

Bates, J. M., and R. M. Zink. 1992. Seasonal variation in gene frequencies in the house sparrow (*Passer domesticus*). Auk 109:658–662.

Bauer, H. G., and G. Heine. 1992. Die Entwicklung der Brutvogebestande am Bodensee: Vergleich halbquantitativer Rasterkartierungen 1980/81 und 1990/91. Journal für Ornithologie 133:1–22.

Baumgart, W. 1984. Zur Charakterisierung von Haus- und Weidensperling, *Passer domesticus* und *Passer hispaniolensis*, als "zetdifferente Arten." Beiträge zur Vogelkunde 30:217–242.

Baumann, F. H., and R. Baumann. 1977. A comparative study of the respiratory properties of bird blood. Respiration Physiology 31:333–343.

Baverstock, P. R., M. Adams, R. W. Polkinghorne, and M. Gelder. 1982. A sex-linked enzyme in birds: Z-chromosome conservation but no dosage compensation. Nature 296:763–766.

Becker, D. M. 1985. Food habits of Richardson's merlins in southeastern Montana. Wilson Bulletin 97:226–230.

Beer, J. R. 1961. Winter feeding patterns in the house sparrow. Auk 78:63–71.

Behnke, J., P. McGregor, J. Cameron, I. Hartley, M. Shepherd, F. Gilbert, C. Barnard, J. Hurst, S. Gray, and R. Wiles. 1999. Semi-quantitative assessment of wing feather mite (Acarina) infestations on passerine birds from Portugal: evaluation of the criteria for accurate quantification of mite loads. Journal of Zoology, London 248:337–347.

Beimborn, D. A. 1967. Population ecology of the English sparrow *Passer domesticus domesticus* in North America. M. S. Thesis, University of Wisconsin, Milwaukee.

Beimborn, D. A. 1976. Sex ratios in the house sparrow: sources of bias. Bird-Banding 47:13–18.

Bell, B. D. 1994. House sparrows collecting feathers from live feral pigeons. Notornis 41:144–145.

Bello, R. E. 2000. *Anolis* sp. and *Gonatodes albogularis* (yellow-headed gecko) predation. Herpetological Review 31:239–240.

Bennett, W. A. 1990. Scale of investigation and the detection of competition: an example from the house sparrow and house finch introductions in North America. American Naturalist 135:725–747.

Bent, A. C. 1938. Life Histories of North American Birds of Prey. Part 2. Bulletin 170. United States National Museum Bulletin, Washington, DC.

Bent, A. C.. 1950. Life Histories of North American Wagtails, Shrikes, Vireos, and Their Allies. Bulletin 197. United States National Museum, Washington, DC.

Bentley, G. E., I. T. Moore, S. A. Sower, and J. C. Wingfield. 2004. Evidence for a novel gonadotropin-releasing hormone in hypothalamic and forebrain areas in songbirds. Brain, Behavior and Evolution 63:34–46.

Bentley, G. E., N. Perfito, K. Ukena, K. Tsutsui, and J. C. Wingfield. 2003. Gonadotropin-inhibitory peptide in song sparrows (*Melospiza*) in different reproductive conditions, and in house sparrows (*Passer domesticus*) relative to chicken-gonadotropin-releasing hormone. Journal of Neuroendocrinology 15:794–802.

Berger, A. J. 1957. Nesting behavior of the house sparrow. Jack-Pine Warbler 35:86–92.

Bergmann, H.-H., and H.-W. Helb. 1982. Stimmen der Vögel Europas. BLV Verlagsgesellschaft, München.

Bergtold, W. H. 1913. A study of the house finch. Auk 30:40–73.

Bergtold, W. H. 1921. The English sparrow (*Passer domesticus*) and the motor vehicle. Auk 38:244–250.

Beri, Y., P. M. Raizada, and R. K. Bhatnagar. 1972. Food in adults of common sparrow *Passer domesticus* Linnaeus. Entomologists' Newsletter 2:65–66.

Berman, S., M. Cibischino, P. Dellaripa, and L. Montren. 1990. Intraspecific variation in the hindlimb musculature of the house sparrow. Condor 92:199–204.

Bern, H. A., and R. S. Nishioka. 1965. Fine structure of the median eminence of some passerine birds. Proceedings of the Zoological Society, Calcutta 18:107–119.

Berndt, R. 1961. Haussperlinge (*Passer domesticus*) baden im Schnee. Ornithologische Mitteilungen 13:17.

Best, L. B. 1992. Characteristics of corn rootworm insecticide granules and the grit used by cornfield birds: evaluating potential avian risks. American Midland Naturalist 128:126–138.

Best, L. B., and J. P. Gionfriddo. 1991a. Characterization of grit use by cornfield birds. Wilson Bulletin 103:58–81.

Best, L. B., and J. P. Gionfriddo. 1991b. Integrity of five granular insecticide carriers in house sparrow gizzards. Environmental Toxicology and Chemistry 10:1487–1492.

Best, L. B., and J. P. Gionfriddo. 1994. Effects of surface texture and shape on grit selection by house sparrows and northern bobwhite. Wilson Bulletin 106:689–695.

Best, L. B., and J. P. Gionfriddo. 1995. Integrity of cellulose granular pesticide carrier in house sparrow gizzards. Environmental Toxicology and Chemistry 14:851–853.

Best, L. B., and J. P. Gionfriddo. 1996. Grit color selection by house sparrows and northern bobwhites. Journal of Wildlife Management 60:836–842.

Best, L. B., and T. R. Stafford. 2002. Influence of daily grit consumption rate and diet on gizzard grit counts. Journal of Wildlife Management 66:381–391.

Beven, G. 1947. Display of house-sparrow. British Birds 40:308–310.

Bezzel, E. 1985a. Birdlife in intensively used rural and urban environments. Ornis Fennica 62:90–95.

Bezzel, E. 1985b. Randzonen im Siedlungsgebiet des Haussperlings (*Passer domesticus*): Fallbeispiele aus Nordalpentälern. Garmischervogelkindliche Berichte 14:1–12.

Bhattacharyya, S. 1990. A survey of pseudoscorpions in the nests of *Passer domesticus* (Linnaeus) in West Bengal. Environment and Ecology 8:245–247.

Bhattacharyya, S. 1995. Survey of nest associated Acarina fauna in West Bengal. Environment and Ecology 13:547–564.

Bhattacharyya, T. K., and A. Ghosh. 1965. Seasonal histophysiologic study of the interrenal of the house sparrow. Acta Biologia Hungarica 16:69–77.

Biadun, W. 1994. The breeding avifauna of the parks and cemeteries of Lublin (SE Poland). Acta Ornithologica 29:1–13.

Binkley, S. 1974a. Computer methods of analysis for biorhythm data. Pp. 53–62 in Biological Rhythms in the Marine Environment (P. J. DeCoursey, ed.). University of South Carolina Press, Columbia, South Carolina.

Binkley, S. 1974b. Pineal and melatonin: circadian rhythms and body temperatures of sparrows. Pp. 582–585 in Chronobiology (F. Scheving, F. Halberg, and J. Pauly, eds.). Igaku Shom, Tokyo.

Binkley, S. 1976. Comparative biochemistry of the pineal gland of birds and mammals. American Zoologist 16:57–65.

Binkley, S. 1977. Constant light: effects on the circadian locomotor rhythm in the house sparrow. Physiological Zoology 50:170–181.

Binkley, S. 1983. Rhythms in ocular and pineal N-acetyltransferase: a portrait of an enzyme clock. Comparative Biochemistry and Physiology 75A:123–129.

Binkley, S. 1990. The Clockwork Sparrow: Time, Clocks, and Calendars in Biological Organisms. Prentice Hall, Englewood Cliffs, New Jersey.

Binkley, S., K. Adler, and D. H. Taylor. 1973. Two methods for using period length to study rhythmic phenomena. Journal of Comparative Physiology 83:63–71.

Binkley, S., E. Kluth, and M. Menaker. 1971. Pineal function in sparrows: circadian rhythms and body temperature. Science 174:311–314.

Binkley, S., E. Kluth, and M. Menaker. 1972. Pineal and locomotor activity: levels and arrhythmia in sparrows. Journal of Comparative Physiology 77:163–169.

Binkley, S., and K. Mosher. 1985a. Direct and circadian control of sparrow behavior by light and dark. Physiology and Behavior 35:785–797.

Binkley, S., and K. Mosher. 1985b. Oral melatonin produces arrhythmia in sparrows. Experientia 41:1615–1617.

Binkley, S., and K. Mosher. 1986. Photoperiod modifies circadian resetting responses in sparrows. American Journal of Physiology 251:R1156–R1162.

Binkley, S., and K. Mosher. 1987a. Circadian rhythm resetting in sparrows: early response to doublet light pulses. Journal of Biological Rhythms 2:1–11.

Binkley, S., and K. Mosher. 1987b. Sparrow circadian rhythm responses to rotated light-dark schedules. Physiology and Behavior 41:361–370.

Binkley, S., and K. Mosher. 1988. Two circadian rhythms in pairs of sparrows. Journal of Biological Rhythms 3:249–254.

Binkley, S., and K. Mosher. 1992. Activity rhythms in house sparrows exposed to natural lighting for one year. Journal of Interdisciplinary Cycle Research 23:17–33.

Binkley, S., K. Mosher, and K. B. Reilly. 1983. Circadian rhythms in house sparrows: lighting ad lib. Physiology and Behaviour 31:829–837.

Bircham, P. M. M., and W. J. Jordan. 1997. A consideration of some of the changes in distribution of "common birds" as revealed by The New Atlas of Breeding Birds in Britain and Ireland. Ibis 139:183–186.

Birkhead, T. R., J. P. Veiga, and F. Fletcher. 1995. Sperm competition and unhatched eggs in the house sparrow. Journal of Avian Biology 26:343–345.

Birkhead, T. R., J. P. Veiga, and A. P. Møller. 1994. Male sperm reserves and copulation behaviour in the house sparrow, *Passer domesticus*. Proceedings of the Royal Society of London B 256:247–251.

Bjordal, H. 1983. Effects of freezing, freeze-drying and skinning on body dimensions of house sparrows *Passer domesticus*. Cinclus 6:105–108.

Bjordal, H. 1984. Bill measurements of house sparrows *Passer domesticus* before and after skeletal preparation. Cinclus 7:21–23.

Blackmore, F. H. 1969. The effect of temperature, photoperiod and molt on the energy requirements of the house sparrow, *Passer domesticus*. Comparative Biochemistry and Physiology 30:433–444.

Blair, R. B. 1996. Land use and avian species diversity along an urban gradient. Ecological Applications 6:506–519.

Blem, C. 1968. Determination of caloric and nitrogen content of excreta voided by birds. Poultry Science 47:1205–1208.

Blem, C. R. 1973. Geographic variation in the bioenergetics of the house sparrow. Ornithological Monographs 14:96–121.

Blem, C. R. 1974. Geographic variation of thermal conductance in the house sparrow *Passer domesticus*. Comparative Biochemistry and Physiology 47A:101–108.

Blem, C. R. 1975a. Energetics of nestling house sparrows *Passer domesticus*. Comparative Biochemistry and Physiology 52A:305–312.

Blem, C. R. 1975b. Geographic variation in wing-loading of the house sparrow. Wilson Bulletin 87:543–549.

Blem, C. R. 1976. Efficiency of energy utilization of the house sparrow, *Passer domesticus*. Oecologia (Berlin) 25:257–264.

Blem, C. R. 1977. Reanalysis of geographic variation of house sparrow energetics. Auk 94:358–359.

Bo, M. S., S. M. Cicchino, and M. M. Martinez. 1996. Diet of long-winged harrier (*Circus buffoni*) in southeastern Buenos Aires Province, Argentina. Journal of Raptor Research 30:237–239.

Boag, P. T., and A. J. Van Noordwijk. 1987. Quantitative genetics. Pp. 45–78 in Avian Genetics: A Population and Ecological Approach (F. Cooke and P. A. Buckley, eds.). Academic Press, London.

Bocchi, G. D., D. Mainardi, and C. Orlando. 1960. Gruppi sanquigni eibridazione interspecifica in pseci e in uccelli. Istituto Lombardo (Rend. Sc.) B 94:63–74.

Bochkov, A. V., and T. D. Galloway. 2001. Parasitic cheyletoid mites (Acari: Cheyletoidea) associated with passeriform birds (Aves: Passerformes) in Canada. Canadian Journal of Zoology 79:2014–2028.

Bock, W. J. 1974. The avian skeleton: muscular system. Pp. 119–257 in Avian Biology, vol. IV (D. S. Farner and J. R. King, eds.). Academic Press, New York.

Bock, W. J. 1992. Methodology in avian macrosystematics. Bulletin of the British Ornithologists' Club 112A:53–72.

Bock, W. J., and J. J. Morony, JR. 1978a. Relationships of the passerine finches (Passeriformes: Passeridae). Bonner zoologische Beiträge 29:122–147.

Bock, W. J., and J. J. Morony. 1978b. The preglossale of Passer (Aves: Passeriformes): a skeletal neomorph. Journal of Morphology 155:99–110.

Bogliani, G., and A. Brangi. 1990. Abrasion of the status badge in the male Italian sparrow *Passer italiae*. Bird Study 37:195–198.

Bokotey, A. 1996. Preliminary results of work on the ornithological atlas of Lvov city (Ukraine). Acta Ornithologica 31:85–88.

Bonneaud, C., J. Mazuc, G. Gonzalez, C. Haussy, O. Chastel, B. Faivre, and G. Sorci. 2003. Assessing the cost of mounting an immune response. American Naturalist 161:367–379.

Bonneaud, C., G. Sorci, V. Morin, H. Westerdahl, R. Zoorob, and H. Wittzell. 2004. Diversity of Mhc class I and IIB genes in house sparrows (*Passer domesticus*). Immunogenetics 55:855–865.

Bordignon, L. 1985. Precoce nidificazione di Passera d'Italia *Passer domesticus italiae*. Gli Uccelli D'Italia 10:69–70.

Bosakowski, T. 1986. Winter population trends of the house finch and ecologically similar species in northeastern New Jersey. American Birds 40:1105–1110.

Bose, M., and F. Guidali. 2001. Seasonal and geographic differences in the diet of the barn owl in an agro-ecosystem in northern Italy. Journal of Raptor Research 35:240–246.

Bösenberg, D. R. 1957. Sperlinge und ihre Bekämpfung. Biologische Zentralanstalt der Deutschen Akademie der Landwirthschafts-wissenschaften zu Berlin 24:1–11.

Bosenberg, K. 1958. Geschlechterverhaltnis und Sterblichkeit der Nestlinge beim Haussperling (*Passer domesticus* L.). Ornithologische Mitteilungen 10:86–88.

Boughton, D. C. 1937a. Notes on avian coccidiosis. Auk 54:500–509.

Boughton, D. C. 1937b. Studies on oocyst production in avian coccidiosis. II. Chronic isosporan infections in the sparrow. American Journal of Hygiene 25:203–211.

Bouvier, D. G. 1968. La salmonellose chez les oiseaux sauvages, notammentchez les petits passereauxdes environs de Lausanne. Nos Oiseaux 29:293–295.

Bower, S. 1999. Fortpflanzungsaktivitat, Habitatnutzung und Populationsstruktur eines Schwarms von Haussperlingen (*Passer d. domesticus*) im Hamburger Stadtgebiet. Hamburger avifaunistische Beiträge 30:91–128.

Bowes, V. 1990. Parathyroid infection in English sparrows. Canadian Veterinary Journal 31:592.

Box, E. D. 1966. Blood and tissue protozoa of the English sparrow (*Passer domesticus domesticus*) in Galveston, Texas. Journal of Protozoology 13:204–208.

Box, E. D. 1967. Influence of Isospora infections on patency of avian Lankesterella (Atoxoplasma, Garnham, 1950). Journal of Parasitology 53:1140–1147.

Boyce, D. A., Jr. 1985. Prairie falcon prey in the Mojave Desert, California. Raptor Research 19:129–134.

Brackbill, H. 1960. Determinate laying by house sparrows. Condor 62:479.

Brandstätter, R. 2003. Encoding time of day and time of year by the avian circadian system. Journal of Neuroendocrinology 15:398–404.

Brandstätter, R., and U. Abraham. 2003. Hypothalamic circadianorganization in birds. I. Anatomy, functional morphology, and terminology of the suprachiasmatic region. Chronobiology International 20:637–655.

Brandstätter, R., U. Abraham, and U. Albrecht. 2001. Initial demonstration of rhythmic Per gene expression in the hypothalamus of a non-mammalian vertebrate, the house sparrow. NeuroReport 12:1167–1170.

Brandstätter, R., V. Kumar, U. Abraham, and E. Gwinner. 2000. Photoperiodic information acquired and stored in vivo is retained in vitro by a circadian oscillator, the avian pineal gland. Proceedings of the National Academy of Sciences, USA 97:12324–12328.

Brandstätter, R., V. Kumar, T. J. Van't Hof, and E. Gwinner. 2001. Seasonal variations of in vivo and in vitro melatonin production in a passeriform bird, the house sparrow (*Passer domesticus*). Journal of Pineal Research 31:120–126.

Brasil, M. De C., and S. B. Amato. 1992. [Faunistic analysis of the helmints of sparrows (*Passer domesticus* L., 1758) captured in Campo Grande, Rio de Janeiro, RJ.] Memorias. Instituto Oswaldo Cruz, Rio de Janeiro 87(Supplement 1):43–48. (In Portuguese, English summary).

Brasil, M. De C., S. B. Amato, and J. F. R. Amato. 1991. [Revision of the Brazilian species of the genus Leucochloridium carus, 1835 (Digenea, Leudochloridiidae).] Revista Brasileira de Biologia 51:537–543. (In Portuguese, English summary).

Breitwisch, R., and M. Breitwisch. 1991. House sparrows open an automatic door. Wilson Bulletin 103:725–726.

Breitwisch, R., and J. Hudak. 1989. Sex differences in risk-taking behavior in foraging flocks of house sparrows. Auk 106:150–153.

Breuner, C. W., D. H. Jennings, M. C. Moore, and M. Orchinik. 2000. Pharmacological adrenalectomy with mitotane. General and Comparative Endocrinology 120:27–34.

Breuner, C. W., and M. Orchinik. 2001. Seasonal regulation of membrane and intracellular corticosteroid receptors in the house sparrow brain. Journal of Neuroendocrinology 13:412–420.

Brichetti, P. 1992. Biometria delle uova e dimensione delle covate in alcune specie

di Charadriiformes e Passeriformes nidificanti in Italia. Rivista Italiana di Ornitologia, Milano 62:136–144.
Brichetti, P., M. Caffi, and S. Gandini. 1993. Biologia riprodutiva di ma popolazione di Passera d'Italia, *Passer italiae*, nidificante in una "colombaia" della pianura lombarda. Avocetta 17:65–71.
Brockway, B. P. 1965. Effects of capture and caging on thyroid activity of house sparrows (*Passer domesticus*). Ohio Journal of Science 65:130–137.
Brooke, R. K. 1973. House sparrows feeding at night in New York. Auk 90:206.
Brooker, M. G., and L. C. Brooker. 1989. Cuckoo host in Australia. Australian Zoological Reviews 2:1–67.
Broun, M. 1971. House sparrows feeding young at night. Auk 88:924–925.
Broun, M. 1972. Apparent migratory behavior in the house sparrow. Auk 89:187–189.
Brown, C. R. 1986. Cliff swallow colonies as information centers. Science 234:83–85.
Brown, N. S., and G. I. Wilson. 1975. A comparison of the ectoparasites of the house sparrow (*Passer domesticus*) from North America and Europe. American Midland Naturalist 154–165.
Brown, R. E., J. J. Baumel, and R. D. Klemm. 1995. Mechanics of the avian propatagium: flexion-extension mechanism of the avian wing. Journal of Morphology 255:91–105.
Brown, R. E., and A. C. Cogley. 1996. Contributions of the propatagium to avian flight. Journal of Experimental Zoology 276:112–124.
Brown, W. H. 1969. Winter bird-population study 52: city park. Audubon Field Notes 23:555–558.
Buchanan, K. L., M. R. Evans, and A. R. Goldsmith. 2003. Testosterone, dominance signaling and immunosuppression in the house sparrow, *Passer domesticus*. Behavioral Ecology and Sociobiology 55:50–59.
Buchanan, K. L., M. R. Evans, A. R. Goldsmith, D. M. Bryant, and L. V. Rowe. 2001. Testosterone influences basal metabolic rate in male house sparrows: a new cost of dominance signalling. Proceedings of the Royal Society of London B 268:1337–1344.
Buges, S. 1991. [Das ungewohnliche Nahrungsbehehmen des Haussperlings (*Passer domesticus* L.) und der Elster (*Pica pica* L.).] Zpravy Moravskeho ornitologickeho sdruzent 49:165. (In Czechoslovakian, German summary).
Bulatova, N. Sh., S. I. Radjabli, and E. N. Panov. 1972. Karylogical description of the species of the genus *Passer*. Experientia 28:1369–1371.
Bullard, R. W. 1988. Characteristics of bird-resistance in agricultural crops. Proceedings of the Vertebrate Pest Conference 13:305–309.
Bumpus, H. C. 1897. The variations and mutations of the introduced sparrow, *Passer domesticus*. Biological Lectures Delivered at the Marine Biological Laboratory of Wood's Holl, 1896, 1–15.
Bumpus, H. C. 1899. The elimination of the unfit as illustrated by the introduced sparrow, Passer domesticus. Biological Lectures Delivered at the Marine Biological Laboratory of Wood's Holl, 1897/1898, 209–228.

Bunning, E. 1960. Circadian rhythms and their time measurement in photoperiodism. Cold Spring Harbor Symposia on Quantitative Biology 25:249–256.

Burger, J. 1976. House sparrows usurp Hornero nests in Argentina. Wilson Bulletin 88:357–358.

Burke, T., and M. W. Bruford. 1987. DNA fingerprinting in birds. Nature 327:149–152.

Burrage, B. R. 1964. A nesting study of the house sparrow *Passer domesticus* in San Diego County, California. Transactions of the Kansas Academy of Science 67: 693–701.

Burton, A. N., J. R. Mclintock, J. Spalatin, and J. R. Rumpel. 1966. Western equine encephalitis in Saskatchewan birds and mammals 1962–1963. Canadian Journal of Microbiology 12:133–141.

Burtt, E. H., Jr., and J. M. Ichida. 1999. Occurrence of feather-degrading bacilli in the plumage of birds. Auk 116:364–372.

Busche, G. 1999. Bestandsentwicklung von Brutvögeln im Westen Schleswig-Holsteins1945–1995. Bilanzen im räumlich-zeitlichen Vergleich. Vogelwelt 120: 193–210.

Bush, F. M. 1967. Developmental and populational variation in electrophoretic properties of dehydrogenases, hydrolases and other blood proteins of the house sparrow, *Passer domesticus*. Comparative Biochemistry and Physiology 22:273–287.

Bush, F. M., and W. W. Fraser. 1969. Dissociation-reassociation of lactate dehydrogenase: reversed isozyme migration and kinetic properties. Proceedings of the Society for Experimental Biology and Medicine 131:13–15.

Buss, I. O. 1942. A managed cliff swallow colony in southern Wisconsin. Wilson Bulletin 54:153–161.

Buttemer, W. A. 1992. Differential overnight survival by Bumpus' house sparrows: an alternative interpretation. Condor 94:944–954.

Calhoun, J. B. 1947a. The role of temperature and natural selection in relation to the variations in the size of the English sparrow in the United States. American Naturalist 81:203–228.

Calhoun, J. B. 1947b. Variations in the plumages of the English Sparrow. Auk 64:305–306.

Caraco, T., and M. C. Bayham. 1982. Some geometric aspects of house sparrow flocks. Animal Behaviour 30:990–996.

Casotti, G. 2001a. Effect of season on kidney morphology in house sparrows. Journal of Experimental Biology 204:1201–1206.

Casotti, G. 2001b. Luminal morphology of the avian lower intestine: evidence supporting the importance of retrograde peristalsis for water conservation. Anatomical Record 263:289–296.

Casotti, G., and E. J. Braun. 2000. Renal anatomy in sparrows from different environments. Journal of Morphology 243:283–291.

Cassone, V. M. 1988. Circadian variation of [^{14}C]2-deoxyglucose uptake within the suprachismatic nucleus of house sparrow, *Passer domesticus*. Brain Research 459:178–182.

Cassone, V. M. 1990. Melatonin: time in a bottle. Oxford Reviews in Reproductive Biology 12:319–367.
Cassone, V. M., and D. S. Brooks. 1991. Sites of melatonin action in the brain of the house sparrow, *Passer domesticus*. Journal of Experimental Zoology 260:302–309.
Cassone, V. M., D. S. Brooks, D. B. Hodges, T. A. Kelm, L. Jun, and W. S. Warren. 1992. Integration of circadian and visual function in mammals and birds: brain imaging and the role of melatonin in biological clock regulation. Pp. 299–317 in Advances in Metabolic Techniques for Brain Imaging of Behavioral and Learning Functions (F. Gonzalez-Lima, Th. Finkenstadt, and H. Scheich, eds.). Kluwer Academic Publishers, Dordrecht.
Cassone, V. M., D. S. Brooks, and T. A. Kelm. 1995. Comparative distribution of 2[^{125}I]iodomelatonin binding in the brains of diurnal birds: outgroup analysis with turtles. Brain and Behavior Evolution 45:241–256.
Cassone, V. M., and M. Menaker. 1984. Is the avian circadian system a neuroendocrine loop? Journal of Experimental Zoology 232:539–549.
Cassone, V. M., and M. Menaker. 1985. Circadian rhythms of house sparrows are phase-shifted by pharmacological manipulation of brain serotonin. Journal of Comparative Physiology A 156:145–152.
Cassone, V. M., and R. Y. Moore. 1987. Retinohypothalamic projection and suprachiasmatic nucleus of the house sparrow, *Passer domesticus*. Journal of Comparative Neurology 266:171–182.
Casto, S. D. 1974. Molt schedule of house sparrows in northwestern Texas. Wilson Bulletin 86:176–177.
Castroviejo, J., L. C. Christian, and A. Gropp. 1969. Karotypes of four species of birds of the families Ploceidae and Paridae. Journal of Heredity 60:134–136.
Caughley, G., G. C. Grigg, J. Caughley, and G. J. E. Hill. 1980. Does dingo predation control the densities of kangaroos and emus? Australian Wildlife Research 7:1–12.
Caviedes-Vidal, E., D. Afik, C. Martinez Del Rio, and W. H. Karasov. 2000. Dietary modulation of intestinal enzymes of the house sparrow (*Passer domesticus*): testing an adaptive hypothesis. Comparative Biochemistry and Physiology A 125:11–24.
Caviedes-Vidal, E., and W. H. Karasov. 1995. Influences of diet composition on pancreatic enzyme activities in house sparrows. American Zoologist 35:78A.
Caviedes-Vidal, E., and W. H. Karasov. 1996. Glucose and amino acid absorption in house sparrow intestine and its dietary modulation. American Journal of Physiology 271:R561–R568.
Caviedes-Vidal, E., and W. H. Karasov. 2001. Developmental changes in digestive physiology of nestling house sparrows, *Passer domesticus*. Physiological and Biochemical Zoology 74:769–782.
Chabot, C., and M. Menaker. 1987. Environmental and endocrine control of feeding and perch hopping rhythmicity in the house sparrow. Abstracts of the Society of Neurosciences 13:1038.

Chabot, C. C., and M. Menaker. 1992. Circadian feeding and locomotor rhythms in piegeons and house sparrows. Journal of Biological Rhythms 7:287–299.

Chaffee, R. R. J., Y. Cassuto, and S. M. Horvath. 1965. Studies of the effects of cold acclimation on myoglobin levels in sparrows, mice, hamsters, and monkeys. Canadian Journal of Physiology and Pharmacology 43:1021–1025.

Chaffee, R. R. J., and W. W. Mayhew. 1964. Studies on chemical thermoregulation in the house sparrow (*Passer domesticus*). Canadian Journal of Physiology and Pharmacology 42:863–866.

Chaffee, R. R. J., W. W. Mayhew, M. Drebin, and Y. Cassuto. 1963. Studies on thermogenesis in cold-acclimated birds. Canadian Journal of Biochemistry and Physiology 41:2215–2220.

Chahal, B. S., G. S. Simwat, and H. S. Brar. 1973. Bird pests of crops and their control. Pesticides 7:18–20.

Chahota, R., and R. C. Katoch. 2000. Comparative efficacy of some current diagnostic techniques for diagnosis of chlamydiosis among domestic poultry and wild carriers. Indian Journal of Animal Sciences 70:11–13.

Chamberlain, D. E., R. J. Fuller, M. Shrubb, R. G. H. Bunce, J. C. Duckworth, D. G. Garthwaite, A. J. Impey, and A. D. M. Hart. 1999. The Effects of Agricultural Management on Farmland Birds. British Trust for Ornithology, Thetford, UK.

Chamberlain, R. W., R. E. Kissling, D. D. Stamm, D. B. Nelson, and R. K. Sikes. 1956. Venezuelan equine encephalomyelitis in wild birds. American Journal of Hygiene 63:261–273.

Chamberlain, R. W., R. E. Kissling, D. D. Stamm, and W. D. Sudia. 1957. Virus of St. Louis encephalitis in three species of wild birds. American Journal of Hygiene 65:110–118.

Chandra-Bose, D. A., and J. C. George. 1965. Studies on the structure and physiology of the flight muscles of birds. 13. Characterization of the avian pectoralis. Pavo 3:14–22.

Chang, M.-H., J. G. Chediack, E. Cavirdes-Vidal, and W. H. Karasov. 2004. L-Glucose absorption in house sparrows (*Passer domesticus*) is nonmediated. Journal of Comparative Physiology B 174:181–188.

Chappell, M. A., C. Bech, and W. A. Buttemer. 1999. The relationship of central and peripheral organ masses to aerobic performance variation in house sparrows. Journal of Experimental Biology 202:2269–2279.

Chastel, O., and M. Kersten. 2002. Brood size and body condition in the house sparrow *Passer domesticus*: the influence of brooding behaviour. Ibis 144:284–292.

Chastel, O., A. Lacroix, and M. Kersten. 2003. Pre-breeding energy requirements: thyroid hormone, metabolism and the timing of reproduction in house sparrows *Passer domesticus*. Journal of Avian Biology 34:296–306.

Chediack, J. G., E. Caviedes-Vidal, V. Fasulo, L. J. Yamin, and W. H. Karasov. 2003. Intestinal passive absorption of water-soluble compounds by sparrows: effect of molecular size and luminal nutrients. Journal of Comparative Physiology B 173:187–197.

Chediack, J. G., E. Caviedes-Vidal, W. H. Karasov, and M. Pestchanker. 2001. Passive absorption of hydrophilic carbohydrate probes by the house sparrow *Passer domesticus*. Journal of Experimental Biology 204:723–731.

Cheke, A. S. 1966. Sparrows in Corsica and Sardinia. Ibis 108:630–631.

Cheke, A. S. 1967. Notes on the ageing and sexing of juvenile house sparrows. Ringers' Bulletin 3(2):7–8.

Cheke, A. S. 1969. Mechanism and consequences of hybridization in sparrows *Passer*. Nature 222:179–180.

Cheke, A. S. 1972. Movements and dispersal among house sparrows, *Passer domesticus* (L.), at Oxford, England. Pp. 211–212 in Productivity, Population Dynamics and Systematics of Granivorous Birds (S. C. Kendeigh and J. Pinowski, eds.). PWN–Polish Scientific Publishers, Warsaw.

Cherian, K. M., and J. C. George. 1966. Histochemical demonstration of dihydrolipoic dehydrogenase activity in the red and white fibers of the avian pectoralis. Journal of Animal Morphology and Physiology 13:210–213.

Chilvers, B. L., P. E. Cowan, D. C. Waddington, P. J. Kelly, and T. J. Brown. 1998. The prevalence of infection of *Giardia* spp. and *Cryptosporidium* spp. in wild animals of farmland, southeastern North Island, New Zealand. International Journal of Environmental Health Research 8:59–64.

Christidis, L. 1986. Chromosomal evolution in finches and their allies (families: Ploceidae, Fringillidae, and Emberizidae). Canadian Journal of Genetics and Cytology 28:762–769.

Christidis, L. 1987. Biochemical systematics within palaeotropic finches (Aves: Estrildidae). Auk 104:380–392.

Churcher, P. B., and J. H. Lawton. 1987. Predation by domestic cats in an English village. Journal of Zoology, London 212:439–455.

Cink, C. L. 1976. The influence of early learning on nest site selection in the house sparrow. Condor 78:103–104.

Cink, C. L. 1977. Winter ecology and behavior of North American house sparrow populations. Ph. D. Dissertation, University of Kansas, Lawrence.

Cink, C. L. 1980. Ambush-like predation by a blue jay on fledgling house sparrows. Kansas Ornithological Society Bulletin 31:25–26.

Cink, C. L., and P. E. Lowther. 1987. Diferenciacion sexual anomala en gorriones sudamericanos. Revista de Ecologia Latinoamericos 1:20–24.

Clark, G. M., and C. E. Yunker. 1956. A new genus and species of dermanyssid (Acarina: Mesostigmata) from the English sparrow, with observations on its life cycle. Proceedings of the Helminthological Society of Washington 23:93–101.

Clark, J. H. 1903. A much mated house sparrow. Auk 20:306–307.

Clench, M. H. 1970. Variability in body pterylosis, with special reference to the genus *Passer*. Auk 87:650–691.

Clergeau, P. 1990. Mixed flocks feeding with starlings: an experimental field study in western Europe. Bird Behaviour 8:95–100.

Clergeau, P., J.-P. L. Savard, G. Mennechez, and G. Falardeau. 1998. Bird abundance and diversity along an urban-rural gradient: a comparative study between two cities on different continents. Condor 100:413–425.

Cobb, S. 1960. Observations on the comparative anatomy of the avian brain. Perspectives in Biology and Medicine 3:383–408.

Cockburn, T. A., C. A. Scoter, and A. D. Langmuir. 1957. Ecology of western equine and St. Louis encephalitis viruses: a summary of field investigations in Weld County, Colorado, 1949 to 1953. American Journal of Hygiene 65:130–146.

Cody, M. L. 1966. A general theory of clutch size. Evolution 20:174–184.

Cole, L. C. 1960. Competitive exclusion. Science 132:348–349.

Cole, L. J. 1917. Determinate and indeterminate laying cycles in birds. Anatomical Record 11:504–505.

Cole, S. R., and D. T. Parkin. 1981. Enzyme polymorphisms in the house sparrow, *Passer domesticus*. Biological Journal of the Linnean Society 15:13–22.

Cole, S. R., and D. T. Parkin. 1986. Adenosine deaminase polymorphism in the house sparrow, *Passer domesticus*. Animal Genetics 17:77–88.

Coleman, J. D. 1974. The use of artificial nest sites erected for starlings in Canterbury, New Zealand. New Zealand Journal of Zoology 1:349–354.

Collinge, W. E. 1914. Some observations on the food of nestling house sparrows. Journal of the Board of Agriculture 21(7):1–6.

Common, A. M. 1956. "Anting" by house sparrow. British Birds 49:155.

Conder, P. J. 1947. Sexual behaviour and call of the house-sparrow. British Birds 40:212–213.

Conradi, E. 1905. Song and call-notes of English sparrows when reared by canaries. American Journal of Psychology 16:190–198.

Constantini, C. 1996. Supposed hybrid house × tree sparrow in northern Italy. British Birds 89:457–458.

Cooke, C. H. 1947. Sexual behaviour of house-sparrow. British Birds 40:308.

Cooper, S. L. 1998. Reaction of female house sparrow to displaying male. British Birds 91:238–240.

Cordero, P. J. 1990a. Breeding success and behaviour of a pair of house and tree sparrow (*Passer domesticus*, *Passer montanus*) in the wild. Journal fur Ornithologie 131:165–167.

Cordero, P. J. 1990b. Phenotypes of juvenile offspring of a mixed pair consisting of a male house sparrow and a female tree parrow *Passer* spp. Ornis Fennica 67:52–55.

Cordero, P. J. 1991a. Phenotypes of adult hybrids between house sparrow *Passer domesticus* and tree sparrow *Passer montanus*. Bulletin of the British Ornithologists' Club 111:44–46.

Cordero, P. J. 1991b. Predation in house sparrow and tree sparrow (*Passer* spp.) nests. Pp. 111–120 in Nestling Mortality of Granivorous Birds due to Microorganisms and Toxic Substances (J. Pinowski, B. P. Kavanaugh, and W. Gorski, eds.). PWN–Polish Scientific Press, Warsaw.

Cordero, P. J. 1993. Factors influencing numbers of syntopic house sparrows and Eurasian tree sparrows in Spain. Auk 110:382–385.

Cordero, P. J. 2002. Hybrid fertility or intra-specific extra-pair fertilisations in mixed pairs of house and tree sparrows, *Passer domesticus* and *Passer montanus*? International Studies on Sparrows 29:5–10.

Cordero, P. J., S. C. Griffith, J. M. Aparicio, and D. T. Parkin. 2000. Sexual dimorphism in house sparrow eggs. Behavioral Ecology and Sociobiology 48:353–357.

Cordero, P. J., and J. D. Rodriguez-Teijeiro. 1988. Posicion y orientacion de nidos en arboles en el gorrion comun (*Passer domesticus*). Publicaciones del Departamento de Zoologia, Universidad de Barcelona 14:99–103.

Cordero, P. J., and J. D. Rodriguez-Teijeiro. 1990. Spatial segregation and interaction between house sparrows and tree sparrows (*Passer* spp.) in relation to nest site. Ekologia Polska 38:443–452.

Cordero, P. J., and J. C. Senar. 1990. Interspecific nest defence [sic] in European Sparrows: different strategies to deal with a different species of opponent? Ornis Scandinavica 21:71–73.

Cordero, P. J., and J. C. Senar. 1994. Persistent tree sparrows *Passer montanus* can counteract house sparrow *P. domesticus* competitive pressure. Bird Behaviour 10:7–13.

Cordero, P. J., and J. D. Summers-Smith. 1993. Hybridization between house and tree sparrow (*Passer domesticus*, *P. montanus*). Journal für Ornithologie 134:69–77.

Cordero, P. J., J. H. Wetton, and D. T. Parkin. 1999a. Extra-pair paternity and male badge size in the house sparrow. Journal of Avian Biology 30:97–102.

Cordero, P. J., J. H. Wetton, and D. T. Parkin. 1999b. Within-clutch pattern of egg viability and paternity in the house sparrow. Journal of Avian Biology 30:103–107.

Cornelius, L. W. 1969. Field notes on Salmonella infection in greenfinches and house sparrows. Bulletin of the Wildlife Disease Association 5:142–143.

Cottam, C. 1929. The fecundity of the English sparrow in Utah. Wilson Bulletin 41:193–194.

Cowie, R. J., and J. R. Simons. 1991. Factors affecting the use of feeders by garden birds. I. The positioning of feeders with respect to cover and housing. Bird Study 38:145–150.

Craggs, J. D. 1967. Population studies of an isolated colony of house sparrows (*Passer domesticus*). Bird Study 14:53–60.

Craggs, J. D. 1976. An isolated colony of house sparrows. Bird Study 23:281–284.

Cramp, S., and C. M. Perrins (eds.). 1994. The Birds of the Western Palearctic, vol. VIII. Oxford University Press, UK.

Crawford, R. L. 1974. Bird casualties at a Leon County, Florida TV tower: October 1966–September 1973. Bulletin of Tall Timbers Research Station 18:1–27.

Crespi, B. J. 1990. Measuring the effect of natural selection on phenotypic interaction systems. American Naturalist 145:32–47.

Crespi, B. J., and F. L. Bookstein. 1989. A path-analytic model for the assessment of selection on morphology. Evolution 43:18–28.

Crespo, F. O. 1977. La presencia del gorrion europeo, *Passer domesticus* L., en el Ecuador. Revista de la Universidad Catolica 5:193–197.

Crewe, M. 1997. Hybrid house × tree sparrow at Timworth. Suffolk Birds 45:159–160.

Crossner, K. A. 1977. Natural selection and clutch size in the European starling. Ecology 58:885–892.

Cyprich, D., J. Pinowski, and M. Krumpal. 2002. Seasonal changes in numbers of fleas (Siphonaptera) in nests of the house sparrow (*Passer domesticus*) and tree sparrow (*P. montanus*) in Warsaw surroundings (Poland). Acta Parasitologica 47:58–65.

Daanje, A. 1941. Uber der Verhalten des Haussperlings (*Passer d. domesticus* [L.]). Ardea 50:1–42.

Danilov, N. N., E. S. Nekrasov, L. N. Dobrinskij, and K. I. Kopbin. 1969. Studies on the variability of *Passer domesticus* L. and *P. montanus* populations. International Studies on Sparrows 3:24–27.

Darwin, C. 1871. The Descent of Man, and Selection in Relation to Sex, vol. I. John Murray, London.

Da Silva, J. M. C., and D. C. Oren. 1990. Introduced and invading birds in Belem, Brazil. Wilson Bulletin 102:309–313.

Dathe, H. 1988. Über die ErhÑhrung einer Waldohreule, Asio otus, inmitten der Großstadt Berlin. BeitrÑge zur Vogelkunde 34:41–46.

Davidson, A. H. 1994. Common grackle predation on adult passerines. Wilson Bulletin 106:174–175.

Davis, E. A., Jr. 1955. Seasonal changes in the energy balance of the English sparrow. Auk 72:385–411.

Davis, J. 1953. Precocious sexual development in the juvenal English sparrow. Condor 55:117–120.

Davis, J. 1954. Seasonal changes in bill length of certain passerine birds. Condor 56:142–149.

Davis, J., and B. S. Davis. 1954. The annual gonad and thyroid cycles of the English sparrow in sourthern California. Condor 56:328–345.

Davis, M. 1944. Purple grackle kills English sparrow. Auk 61:139–140.

Davis, M. 1945. English sparrow anting. Auk 62:641.

Dawson, A. 1991. Photoperiodic control of testicular regression and moult in male house sparrows *Passer domesticus*. Ibis 133:312–316.

Dawson, A. 1998a. Photoperiodic control of the termination of breeding and the induction of moult in house sparrows *Passer domesticus*. Ibis 140:35–40.

Dawson, A. 1998b. Thyroidectomy of house sparrows (*Passer domesticus*) prevents photo-induced testicular growth but not increased hypothalamic gonadotrophin-releasing hormone. General and Comparative Endocrinology 110:196–200.

Dawson, D. G. 1964. The eggs of the house sparrow. Notornis 11:187–189.

Dawson, D. G. 1967. Roosting sparrows (*Passer domesticus*) killed by rainstorm, Hawke's Bay, New Zealand. Notornis 14:208–210.

Dawson, D. G. 1968. An intraspecific attack in house sparrows. Notornis 15:267.

Dawson, D. G. 1970. Estimation of grain loss due to sparrows (*Passer domesticus*) in New Zealand. New Zealand Journal of Agricultural Research 13:681–688.

Dawson, D. G. 1972a. House sparrow, *Passer domesticus* (L.), breeding in New Zealand. Pp. 129–130 in Productivity, Population Dynamics and Systematics of

Granivorous Birds (S. C. Kendeigh and J. Pinowski, eds.). PWN–Polish Scientific Publishers, Warsaw.

Dawson, D. G. 1972b. The breeding biology of house sparrows. D. Phil. Thesis, Oxford University, UK, vi + 82.

Dawson, D. G., and P. C. Bull. 1970. A questionnaire survey of bird damage to fruit. New Zealand Journal of Agricultural Research 13:362–371.

Dearborn, N. 1912. The English sparrow as a pest. Farmers' Bulletin 493. United States Department of Agriculture, Washington, DC.

Deckert, G. 1969. Zur Ethologie und Okologie des Haussperlings (*Passer d. domesticus* L.). Beiträge zue Vogelkunde 15:1–84.

Degraaf, R. M., and T. J. Maier. 2001. Obtaining and storing house sparrow eggs in quantity for nest-predation studies. Journal of Field Ornithology 72:124–130.

Degraaf, R. M., T. J. Maier, and T. K. Fuller. 1999. Predation of small eggs in artificial nests: effects of nest position, edge, and potential predator abundance in extensive forest. Wilson Bulletin 111:236–242.

Desrochers, A. 1992. Age-related differences in reproduction by European blackbirds: restraint or constraint? Ecology 73:1128–1131.

De Wavrin, H. 1991. Observation d'un Moineau domestique (*Passer domesticus*) en migration. Aves 28:229.

Dexter, C. 1999. The Remorseful Day. Fawcett Books, New York.

Dexter, R. W. 1959. Two 13–year-old age records for the house sparrow. Bird-Banding 30:182.

Dhindsa, M. S., H. S. Toor, and P. S. Sandhu. 1984. Community structure of birds damaging pearl millet and sorghum and estimation of grain loss. Indian Journal of Ecology 11:154–159.

Diaz, M. 1990. Interspecific patterns of seed selection among granivorous passerines: effects of seed size, seed nutritive value and bird morphology. Ibis 132:467–476.

Diaz, M. 1994. Variability in seed size selection by granivorous passerines: effects of bird size, bird size variability, and ecological plasticity. Oecologia (Berlin) 99:1–6.

Dobson, A. P. 1987. A comparison of seasonal and annual mortality for both sexes of fifteen species of common British birds. Ornis Scandinavica 18:122–128.

Dobson, A. 1990. Survival rates and their relationship to life-history traits of some common British birds. Current Ornithology 7:115–146.

Dolnik, V. R. 1972. The water storation by the migratory fat deposition in *Passer domesticus bactrianus* Zar. et kud.—the arid zone migrant. Pp. 103–109 in Productivity, Population Dynamics and Systematics of Granivorous Birds (S. C. Kendeigh and J. Pinowski, eds.). PWN–Polish Scientific Publishers, Warsaw.

Dolnik, V. R., and T. I. Blyumental. 1967. Autumnal premigratory and migratory periods in the chaffinch (*Fringilla coelebs coelebs*) and some other temperate-zone passerine birds. Condor 69:435–468.

Dolnik, V. R., and V. M. Gavrilov. 1975. A comparison of the seasonal and daily variations of bioenergetics, locomotor activities and major body composition in the sedentary house sparrow (*Passer d. domesticus* [L.]) and the migratory

"Hindian" sparrow (*Passer d. bactrianus* Dar. et Kudash). Ekologia Polska 23: 211–226.

Donald, P. F., R. E. Green, and M. F. Heath. 2001. Agricultural intensification and the collapse of Europe's farmland bird populations. Proceedings of the Royal Society of London B 268:23–29.

Donham, R. S., J. C. Wingfield, P. W. Mattocks, Jr., and D. S. Farner. 1982. Changes in testicular and plasma androgens with photoperiodically induced increase in plasma LH in the house sparrow. General and Comparative Endocrinology 48:342–347.

Doughty, R. 1978. The English sparrow in the American landscape: a paradox in nineteenth century early wildlife conservation. Research Paper 19. School of Geography, University of Oxford, UK.

Dozsa, I. 1962/63. Die Haussperling als Salmonella typhi murium reservoir. Aquila 69/70:225–229.

Draber-Molko, A. 1997. Protocalliphora azurea (Fall.) (Diptera: Calliphordiae) and other insects found in nests of sparrows, *Passer domesticus* (L.) and *Passer montanus* (L.) in the vicinity of Warsaw. International Studies on Sparrows 22/23:3–10.

Drent, R. H., and S. Daan. 1980. The prudent parent: energetic adjustments in avian breeding. Ardea 68:225–252.

Dubach, M. 1981. Quantitative analysis of the respiratory system of the house sparrow, budgerigar and violet-eared hummingbird. Respiration Physiology 46:43–60.

Duncan, R. A. 1996. House sparrow (*Passer domesticus*) trends in coastal northwest Florida-Alabama based on Christmas Bird Count data. Alabama Birdlife 42(2):1–2.

Dutenhoffer, M. S., and D. L. Swanson. 1996. Relationship of basal to summit metabolic rate in passerine birds and the aerobic capacity model for the evolution of endothermy. Physiological Zoology 69:1232–1254.

Dyer, M. I., J. Pinowski, and B. Pinowska. 1977. Population dynamics. Pp. 53–105 in Granivorous Birds in Ecosystems (J. Pinowski and S. C. Kendeigh, eds.). Cambridge University Press, Cambridge, UK.

Earle, R. A. 1988. Reproductive isolation between urban and rural populations of Cape sparrows and house sparrows. Acta XIX Congressus Internationalis Ornithologici 1778–1786.

Easterbrook, T. G. 1999. Population trends of wintering birds around Banbury, Oxfordshire, 1975–96. Bird Study 46:16–24.

Eastzer, D., P. R. Chu, and A. P. King. 1980. The young cowbird: average or optimal nestling? Condor 82:417–425.

Eaton, W. F. 1924. Decrease of the English sparrow in eastern Massachusetts. Auk 41:604–606.

Edgar, D. R., and G. P. Kershaw. 1994. The density and diversity of the bird populations in three residential communities in Edmonton, Alberta. Canadian Field-Naturalist 108:156–161.

Edwards, H. H., G. D. Schnell, R. L. Debois, and V. H. Hutchison. 1992. Natural and induced remanent magnetism in birds. Auk 109:43–56.

Elcavage, P., and T. Caraco. 1983. Vigilance behaviour in house sparrow flocks. Animal Behaviour 31:303–304.

Elgar, M. A. 1986a. House sparrows establish foraging flocks by giving chirrup calls if the resources are divisible. Animal Behaviour 34:169–174.

Elgar, M. A. 1986b. The establishment of foraging flocks in house sparrows: risk of predation and daily temperature. Behavioral Ecology and Sociobiology 19:433–438.

Elgar, M. A. 1987. Food intake rate and resource availability: flocking decisions in house sparrows. Animal Behaviour 35:1168–1176.

Elgar, M. A., P. J. Burren, and M. Posen. 1984. Vigilance and perception of flock size in foraging house sparrows (*Passer domesticus*). Behaviour 90:215–223.

Elgar, M. A., and C. P. Catterall. 1981. Flocking and predator surveillance in house sparrows: test of an hypothesis. Animal Behaviour 29:868–872.

Elgar, M. A., and C. P. Catterall. 1982. Flock size and feeding efficiency in house sparrows. Emu 82:109–111.

Elgar, M. A., H. McKay, and P. Woon. 1986. Scanning, pecking and alarm flights in house sparrows. Animal Behaviour 34:1892–1894.

Elliott, J. J., and R. S. Arbib, Jr. 1953. Origin and status of the house finch in the eastern United States. Auk 70:31–37.

Ellis, D. H., C. H. Ellis, B. A. Sabo, A. M. Rea, J. Dawson, J. K. Fackler, C. T. Larue, T. G. Grubb, J. Schmitt, D. G. Smith, and M. Kery. 2004. Summer diet of the peregrine falcon in faunistically rich and poor zones of Arizona analyzed with capture-recapture modeling. Condor 106:873–886.

Ely, C. A., and T. J. Bowman. 1969. Storm mortality at a house sparrow roost. Kansas Ornithological Society Bulletin 20:6–7.

Encke, F.-W. 1965a. Nahrungsuntersuchungen an Nestlingen des Haussperlings (*Passer d. domesticus*) in verschiedenen Biotopen, Jahreszeiten und Altersstufen. Beiträge zur Vogelkunde 11:153–184.

Encke, F.-W. 1965b. Über Gelege-, Schlupf- und Ausflugsstärken das Haussperlings (*Passer d. domesticus*) in Abhängigkeit von Biotop und Brutperiode. Beiträge zur Vogelkunde 10:268–287.

Engler, B., and H.-G. Bauer. 2002. Dokumentation eines starken Bestandsrückgangs beim Haussperling (*Passer domesticus*) in Deutschland auf Basis von Literaturangaben von 1850–2000. Die Vogelwarte 41:196–210.

Erdoğan, A., and I. Kiziroğlu. 1995. Brutbiologische Untersuchungen am Feld—*Passer montanus* und Haussperling *P. domesticus* in Beytepe/Ankara. Ornithologische Verhandlungen 25:211–218.

Ericson, P. G. P., T. Tyrberg, A. S. Kjellberg, L. Jonnson, and I. Ullen. 1997. The earliest record of house sparrows (*Passer domesticus*) in northern Europe. Journal of Archaeological Science 24:183–190.

Errington, P. L. 1933. Food habits of southern Wisconsin raptors. Part II. Hawks. Condor 35:19–29.

Errington, P. L. 1946. Predation and vertebrate populations. Quarterly Review of Biology 21:144–177, 221–245.

Escobar, J. V., and J. A. Gil-Delgado. 1984. Estrategias de nidificaci¢n en *Passer domesticus*. Donana, Acta Vertebrata 11:65–78.

Eskin, A. 1971. Some properties of the system controlling the circadian rhythm of sparrows. Pp. 55–78 in Biochronometry (M. Menaker, ed.). National Academy of Sciences, Washington, DC.

Etzold, F. 1891. Die Entwicklung der Testikel von Fringilla domestica von Winterruhe bis zum Eintritt der Brunft. Zeitschrift für wissenschaftliche Zoologie 52:46–84.

Evans, M. R., A. R. Goldsmith, and S. R. A. Norris. 2000. The effects of testosterone on antibody production and plumage coloration in male house sparrows (*Passer domesticus*). Behavioral Ecology and Sociobiology 47:156–163.

Evans, T. R., and L. C. Drickamer. 1994. Flight speeds of birds determined using Doppler radar. Wilson Bulletin 106:154–156.

Evenden, F. G. 1957. Observations on the nesting behavior of the house finch. Condor 59:112–117.

Everett, M. J. 1992. Black kite feeding on house sparrows. British Birds 85:495–496.

Ewins, P. J., M. J. R. Miller, M. E. Barker, and S. Postupalsky. 1994. Birds breeding in or beneath osprey nests in the Great Lakes basin. Wilson Bulletin 106:743–749.

Falconer, D. S. 1981. Introduction to Quantitative Genetics. 2nd edition. Longman, London.

Farner, D. S., R. S. Donham, R. A. Lewis, P. W. Mattocks, Jr., T. R. Darden, and J. P. Smith. 1977. The circadian component in the photoperiodic mechanism of the house sparrow, *Passer domesticus*. Physiological Zoology 50:247–268.

Farner, D. S., A. Oksche, F. I. Kamemoto, J. R. King, and H. E. Cheyney. 1961. A comparison of the effect of long daily photoperiods on the pattern of energy storage in migratory and non-migratory finches. Comparative Biochemistry and Physiology 2:125–142.

Farner, D. S., and J. C. Wingfield. 1980. Reproductive endocrinology of birds. Annual Review of Physiology 42:457–472.

Favaloro, N. 1942. The usurpation of nests, nest sites and materials. Emu 41:268–276.

Fernandez-Juricic, E. 2001. Avian spatial segregation at edges and interiors of urban parks in Madrid, Spain. Biodiversity and Conservation 10:1303–1316.

Fevold, H. R., and K. B. Eik-Nes. 1962. Progesterone metabolism by testicular tissue of the English sparrow (*Passer domesticus*) during the annual reproductive cycle. General and Comparative Endocrinology 2:506–515.

Fevold, H. R., and K. B. Eik-Nes. 1963. Progesterone metabolism by testicular tissue of the English sparrow (*Passer domesticus*). General and Comparative Endocrinology 3:335–345.

Fiuczynski, D., and D. Nethersole-Thompson. 1980. Hobby studies in England and Germany. British Birds 73:275–295.

Fleischer, R. C. 1982. Clutch size in Costa Rican house sparrows. Journal of Field Ornithology 53:280–281.

Fleischer, R. C. 1983. A comparison of theoretical and electrophoretic assessments of genetic structure in populations of the house sparrow (*Passer domesticus*). Evolution 37:1001–1009.

Fleischer, R. C., and R. F. Johnston. 1982. Natural selection on body size and proportions in house sparrows. Nature 298:747–749.

Fleischer, R. C., and R. F. Johnston. 1984. The relationships between winter climate and selection on body size of house sparrows. Canadian Journal of Zoology 62:405–410.

Fleischer, R. C., R. F. Johnston, and W. J. Klitz. 1983. Allozymic heterozygosity and morphological variation in house sparrows. Nature 304:628–630.

Fleischer, R. C., P. E. Lowther, and R. F. Johnston. 1984. Natal dispersal in house sparrows: possible causes and consequences. Journal of Field Ornithology 55:444–456.

Fleischer, R. C., and M. T. Murphy. 1992. Relationships among allozyme heterozygosity, morphology and lipid levels in house sparrows during winter. Journal of Zoology, London 226:409–419.

Flemban, H. M., and T. D. Price. 1997. Morphological differences among populations of house sparrows from different altitudes in Saudi Arabia. Wilson Bulletin 109:539–544.

Flux, J. E. C., and C. F. Thompson. 1986. House sparrows taking insects from car radiators. Notornis 33:190–191.

Foley, J. O. 1928. A note on the spermatogenetic wave in the testes of the adult English sparrow (*Passer domesticus*). Anatomical Record 41:367–371.

Folk, C., and I. Novotny. 1970. Variation in body weight and wing length in the house sparrow, *Passer domesticus* L., in the course of a year. Zoologicke Listy 19:333–342.

Folstad, I., and A. J. Karter. 1992. Parasites, bright males, and the immunocompetence handicap. American Naturalist 139:603–622.

Forschner, D. C. 1990. Effects of brood-size manipulations on parental investment of house sparrows (*Passer domesticus*) in northern lower Michigan. Denison Journal of Biological Science 26:20–29.

Forsyth, M. I. "The English Sparrow."

Foster, N. H. 1917. Measurements and weights of birds' eggs. Irish Naturalist 26:41–47.

Fraga, R. M. 1980. The breeding of Rufous horneros (*Furnarius rufus*). Condor 82:58–68.

Francy, D. B., T. G. T. Jaenson, J. O. Lundstrom, E.-B. Schildt, A. Espmark, B. Hendriksson, and B. Niklasson. 1989. Ecologic studies of mosquitoes and birds as hosts of Ockelbo virus in Sweden and isolation of Inkoo and Batai viruses from mosquitoes. American Journal of Tropical Medicine and Hygiene 41:355–363.

Frank, R., G. J. Sirons, and D. Wilson. 1981. Residues of 4–aminopyridine in poisoned birds. Bulletin of Environmental Contamination and Toxicology 26:389–392.

Fretwell, S. D., and H. L. Luas, Jr. 1970. On territorial behavior and other factors influencing habitat distribution in birds. Acta Biotheoretica 19:16–36.

Friedmann, H. 1929. The cowbirds: a study in the biology of social parasitism. Charles C. Thomas Publishers, Springfield, Illinois.

Friedmann, H. 1963. Host relations of the parasitic cowbirds. United States National Museum Bulletin. Smithsonian Institution, Washington, DC.

Frimer, O. 1989. Food and predation in suburban sparrowhawks *Accipiter nisus* during the breeding season. Dansk Ornitologisk Forenings Tidsskrift 83:35–44.

Fryday, S. L., and P. W. Greig-Smith. 1994. The effects of social learning on the food choice of the house sparrow (*Passer domesticus*). Behaviour 128:281–300.

Fryday, S. L., A. D. M. Hart, and N. J. Dennis. 1994. Effects of exposure to an organophosphate on the seed-handling efficiency of the house sparrow. Bulletin of Environmental Contamination and Toxicology 53:869–876.

Fryer, J. C. F. 1939. The destruction of buds of trees and shrubs by birds. British Birds 33:90–94.

Fulgione, D., G. Aprea, M. Milone, and G. A. Odierna. 2000. Chromosomes and heterochromatin in the Italian sparrow, *Passer italiae*, a taxon of presumed hybrid origin. Folia Zoologica 49:199–204.

Fulgione, D., A. Esposito, C. E. Rusch, and M. Milone. 2000. Song clinal variability in *Passer italiae*, a species of probable hybrid origins. Avocetta 24:107–112.

Fulgione, D., and M. Milone. 1998. On the enigmatic populations of the Italian sparrow. Biologia e Conservacione della Fauna 102:183–191.

Fulgione, D., C. E. Rusch, A. Esposito, and M. Milone. 1998. Dynamics of weight, fat and moult in the Italian sparrow *Passer domesticus*. Acta Ornithologica 33:93–98.

Gahr, M., H.-R. Guttinger, and D. E. Kroodsma. 1993. Estrogen receptors in the avian brain: survey reveals general distribution and forebrain areas unique to songbirds. Journal of Comparative Neurology 327:112–122.

Gallelli, G. 1948. Anormale zoofagia in un passero (*P. italiae*) in Milano. Rivista Italiana di Ornitologia 18:194–196.

Garcia, J. M. F. 1994. House sparrow learning to exploit food of caged goldfinch. British Birds 87:276.

Gaston, S. 1971. The influence of the pineal organ on the circadian activity rhythm of birds. Pp. 541–546 in Biochronometry (M. Menaker, ed.). National Academy of Sciences, Washington, DC.

Gaston, S., and M. Menaker 1968. Pineal function: the biological clock in the sparrow? Science 160:1125–1127.

Gates, J. M. 1972. Red-tailed hawk populations and ecology in east-central Wisconsin. Wilson Bulletin 84:421–433.

Gavett, A. P., and J. S. Wakeley. 1986a. Blood constituents and their relation to diet in urban and rural house sparrows. Condor 88:279–284.

Gavett, A. P., and J. S. Wakeley. 1986b. Diets of house sparrows in urban and rural habitats. Wilson Bulletin 98:137–144.

Gavrilov, E. I. 1965. On hybridisation of Indian and house sparrows. Bulletin of the British Ornithologists' Club 85:112–114.

Gavrilov, E., S. Erkoohov, A. Grjaznov, S. Brokhovich, and A. Goloshchapov. 1995. Number evaluation of migratory sparrows inhabiting south-eastern Kazakhstan and northern Kirgizstan. Pp. 365–380 in Nestling Mortality of Granivorous Birds due to Microorganisms and Toxic Substances: Synthesis (J. Pinowski, B. P. Kavanagh, and B. Pinowska, eds.). PWN–Polish Scientific Publishers, Warsaw.

Gavrilov, E. I., and A. B. Goloshchapov. 1992. Age characters of Spanish and Indian sparrows males. International Studies on Sparrows 19:23–26.

Gavrilov, E. I., and M. N. Korelov. 1968. [The Indian sparrow as a distinct good species.] Byulleten' Moskovskogo Obshchestva Ispytateley Prirody Otdel biologicheskiy 73:115–122. (In Russian, English summary).

Gavrilov, E. I., and B. Stephan. 1980. Zwitter von *Passer italiae* und *Passer hispaniolensis* aus Kasachstan. Mitteilungen aus dem zoologische Museum in Berlin 56. Supplementheft. Annalem für Ornithologie 4:29–31.

Gavrilov, V. M. 1979. [Molt characteristics of sedentary and migrative subspecies of the chaffinch (*Fringilla coelebs*) and house-sparrow (*Passer domesticus*).] Ornitologiya 14:158–163. (In Russian).

Gaymer, R., R. A. A. Blackman, P. G. Dawson, M. Penny, and M. Penny. 1969. The endemic birds of Seychelles. Ibis 111:157–176.

Gehlbach, F. R. 1994. The eastern screech owl. Texas A&M University Press, College Station, Texas.

Geiler, H. 1959. Geschlechterverhaltnis, Korpergewicht und Flugellange der Individuen einer mittdeutschen Sperlingspopulation. Beiträge zur Vogelkunde 6:359–366.

George, J. C., and R. M. Naik. 1960. Intramuscular fat store in the pectoralis of birds. Auk 77:216–217.

George, J. C., A. K. Susheela. and N. V. Vallyathan. 1964. Histochemical for biochemical differentiation and regional specialization in the pectoralis muscle of the house sparrow. Pavo 2:115–119.

George, T. L. 1987. Greater land bird densities on island vs. mainland: relation to nest predation level. Ecology 68:1393–1400.

Gering, J. C., and R. B. Blair. 1999. Predation on artificial bird nests along an urban gradient: predatory risk or relaxation in urban environments? Ecography 22:532–541.

Geyikoğlu, F., and A. Özkaral. 2000. A histochemical study of the origin regions of the triceps muscle of the sparrow (*Passer domesticus*). Turkish Journal of Zoology 24:107–111.

Geyikoğlu, F., A. Temelli, and A. Özkaral. 2002. Muscle fiber types of the tunica muscularis externa of the upper part of the sparrow (*Passer domesticus*) esophagus. Turkish Journal of Zoology 26:217–221.

Gier, L. J., and O. Grounds. 1944. Histological study of the digestive system of the English sparrow. Auk 61:241–243.

Giesbrech, D. S., and C. D. Ankney. 1998. Predation risk and foraging behaviour: an experimental study of birds at feeders. Canadian Field-Naturalist 112:668–675.

Gil-Delgado, J. A., R. Pardo, J. Bellot, and I. Lucas. 1979. Avifauna del naranjal valenciano, II. el gorrion comun (*Passer domesticus* L.). Mediterranea 3:69–99.

Ginn, H. B., and D. S. Melville. 1983. Moutling in Birds. British Trust for Ornithology, Tring, England.

Gionfriddo, J. P., and L. B. Best. 1995. Grit use by house sparrows: effects of diet and grit size. Condor 97:57–67.

Gionfriddo, J. P., L. B. Best, and B. J. Giesler. 1995. A saline-flushing technique for determining the diet of seed-eating birds. Auk 112:780–782.

Gistsov, A. P., and E. I. Gavrilov. 1984. Constancy in the dates and routes of spring migrations in the Spanish and Indian sparrows in the foothills of western Tien Shan. International Studies on Sparrows 11:22–33.

Glue, D. 1994. Siskins arrive early on orange peanut bags as house sparrow numbers decline. BTO News 194:14–15.

Godsey, M. S., Jr., N. S. Blackmore, N. A. Panella, K. Burkhalter, K. Gottfried, L. A. Halsey, R. Rutledge, S. A. Langevin, R. Gates, K. M. Lamonte, A. Lambert, R. S. Lanciotti, C. G. M. Blackmore, T. Loyless, L. Stark, R. Oliveri, L. Conti, and N. Komar. 2005. West Nile virus epizootiology in the southeastern United States, 2001. Vector-borne and Zoonotic Diseases 5:82–89.

Goecke, C. S., and D. L. Goldstein. 1997. Renal glomerular and tubular effects of antidiuretic hormone and two antidiuretic hormone analogues in house sparrows (*Passer domesticus*). Physiological Zoology 70:283–291.

Goés, R. M., and H. Dolder. 2002. Cytological steps during spermiogenesis in the house sparrow (*Passer domesticus*, Linnaeus). Tissue and Cell 34:273–282.

Goglia, I., and M. Milone. 1995. Le popolazioni di *Passer italiae* e *Passer montanus* in Campania nel periodo 1985–1990. Avocetta 19:144.

Goldstein, D. L. 1993. Influence of dietary sodium and other factors on plasma aldosterone concentrations and in vitro properties of the lower intestine in house sparrows (*Passer domesticus*). Journal of Experimental Biology 176:159–174.

Goldstein, D. L., and E. J. Braun. 1986a. Lower intestinal modification of ureteral urine in hydrated house sparrows. American Journal of Physiology 250:R89–R95.

Goldstein, D. L., and E. J. Braun. 1986b. Proportions of mammalian-type and reptilian-type nephrons in the kidneys of two passerine birds. Journal of Morphology 187:173–179.

Goldstein, D. L., and E. J. Braun. 1988. Contributions of the kidneys and intestines to water conservation, and plasma levels of antidiuretic hormone, during dehydration in house sparrows (*Passer domesticus*). Journal of Comparative Physiology B 158:353–361.

Goldstein, D. L., L. Guntle, and C. Flaugher. 2001. Renal response to dietary protein in the house sparrow *Passer domesticus*. Physiological and Biochemical Zoology 74:461–467.

Goldstein, D. L., V. Reddy, and K. Plaga. 1999. Second messenger production in avian medullary nephron segments in response to peptide hormones. American Journal of Physiology 276:R847–R854.

Goldstein, D. L., and A. Zahedi. 1990. Variation in osmoregulatory parameters of captive and wild house sparrows (*Passer domesticus*). Auk 107:533–538.

Gonzalez, G., G. Sorci, and F. De Lope. 1999. Seasonal variation in the relationship between cellular immune response and badge size in male house sparrows (*Passer domesticus*). Behavioral Ecology and Sociobiology 46:117–122.

Gonzalez, G., G. Sorci, A. P. Møller, P. Ninni, C. Haussy, and F. De Lope. 1999. Immunocompetence and condition-dependent sexual advertisement in male house sparrows (*Passer domesticus*). Journal of Animal Ecology 68:1225–1234.

Gonzalez, G., G. Sorci, L. C. Smith, and F. De Lope. 2001. Testosterone and sexual signalling in male house sparrows (*Passer domesticus*). Behavioral Ecology and Sociobiology 50:557–562.

Gonzalez, G., G. Sorci, L. C. Smith, and F. De Lope. 2002. Social control and physiological cost of cheating in status signaling male house sparrows (*Passer domesticus*). Ethology 108:289–302.

Goodridge, A. G. 1964. The effect of insulin, glucagon and prolactin on lipid synthesis and related metabolic activity in migratory and non-migratory finches. Comparative Biochemistry and Physiology 13:1–26.

Gorska, E. 1975. [The investigations on the common roostings of the sparrow, *Passer domesticus* (L.) and starling, *Sturnus vulgaris* L. in Poznan in winters 1970/71 and 1971/72.] Przeglad Zoologiczny 19:230–238. (In Polish, English summary).

Gorska, E. 1990. Seasonal patterns in diurnal activity for the house sparrow [*Passer domesticus* (L.)]. Pp. 43–57 in Granivorous Birds in the Agricultural Landscape (J. Pinowski and J. D. Summers-Smith, eds.). PWN–Polish Scientific Publishers, Warsaw.

Gorska, E. 1991. [Annual rhythm of starting and finishing of the daily activity in the urban populations of collared dove (*Streptopelia decaocto*), house sparrow (*Passer domesticus*), blackbird (*Turdus merula*), starling (*Sturnus vulgaris*) and jackdaw (*Corvus monedula*) in Slupsk.] Notatki Ornitologiczne 32:37–54. (In Polish, English summary).

Gorzelski, W., J. Pinowski, P. Kaminski, and A. Kruszewicz. 1995. Lipid and protein contents and heavy metals in relation to survival of house sparrow (*Passer domesticus*) and tree sparrow (*Passer montanus*) nestlings. Pp. 203–221 in Nestling Mortality of Granivorous Birds due to Microorganisms and Toxic Substances: Synthesis (J. Pinowski, B. P. Kavanagh, and B. Pinowska, eds.). PWN–Polish Scientific Publishers, Warsaw.

Gotmark, F., and P. Post. 1996. Prey selection by sparrowhawks, *Accipiter nisus*: relative predation risk for breeding passerine birds in relation to their size, ecology and behaviour. Philosophical Transactions of the Royal Society of London B 351:1559–1577.

Gould, S. J., and R. C. Lewontin. 1979. The spandrels of San Marco and the Panglossian paradigm: a critique of the adaptationist programme. Proceedings of the Royal Society of London B 205:581–598.

Gowanlock, J. N. 1914. The grackle as a nest-robber. Bird-Lore 16:187–188.

Gowaty, P. A. 1984. House sparrows kill eastern bluebirds. Journal of Field Ornithology 55:378–380.

Graber, R. R., and J. W. Graber. 1965. Variation in avian brain weights with special reference to age. Condor 67:300–318.

Grant, P. R. 1972. Centripetal selection and the house sparrow. Systematic Zoology 21:23–30.

Gray, R. D. 1994. Sparrows, matching and the ideal free distribution: can biological and psychological approaches be synthesized? Animal Behaviour 48:411–423.

Greenman, C. G., L. B. Martin II, and M. Hau. 2005. Reproductive state, but not testosterone, reduces immune function in male house sparrows (*Passer domesticus*). Physiological and Biochemical Zoology 78:60–68.

Gregory, R. D. 1999. Broad-scale habitat use of sparrows, finches and buntings in Britain. Vogelwelt 120(Supplement):163–173.

Greig-Smith, P. W. 1987. Aversions of starlings and sparrows to unfamiliar, unexpected or unusual flavours and colours in food. Ethology 74:155–163.

Greig-Smith, P. W., and C. M. Rowney. 1987. Effects of colour on the aversions of starlings and house sparrows to five chemical repellents. Crop Protection 6:402–409.

Gress, B. 1985. House sparrow found feeding western kingbird nestlings. Kansas Ornithological Society Bulletin 36:25–26.

Griffith, S. C. 2000. A trade-off between reproduction and a condition-dependent sexually selected ornament in the house sparrow, *Passer domesticus*. Proceedings of the Royal Society of London B 267:1115–1119.

Griffith, S. C., I. P. F. Owens, and T. Burke. 1999a. Environmental determination of a sexually selected trait. Nature 400:358–360.

Griffith, S. C., I. P. F. Owens, and T. Burke. 1999b. Female choice and annual reproductive success favour less-ornamented male house sparrows. Proceedings of the Royal Society of London B 266:765–770.

Griffith, S. C., I. R. K. Stewart, D. A. Daeson, I. P. F. Owens, and T. Burke. 1999. Contrasting levels of extra-pair paternity in mainland and island populations of the house sparrow (*Passer domesticus*): is there an "island effect"? Biological Journal of the Linnean Society 68:303–316.

Griffiths, R., M. C. Double, K. Orr, and R. J. G. Dawson. 1998. A DNA test to sex most birds. Molecular Ecology 7:1071–1075.

Grimm, H. 1954. Biometrische Bemerkungen uber mitteldeutsche und westdeutsche Sperlingspopulationen. Journal für Ornithologie 95:306–318.

Gross, A. O. 1944. Food of the snowy owl. Auk 61:1–18.

Gross, A. O. 1965. The incidence of albinism in North American birds. Bird-Banding 36:67–71.

Grozczynski, J., P. Jablonski, G. Lesinski, and J. Romanowski. 1993. Variation in diet of tawny owl *Strix aluco* L. along an urbanization gradient. Acta Ornithologica 27:113–123.

Grubb, T. C., Jr. 1989. Ptilochronology: feather growth bars as indicators of nutritional status. Auk 106:314–320.

Grubb, T. C., Jr., and L. Greenwald. 1982. Sparrows and a brushpile: responses to different combinations of predation risk and energy cost. Animal Behaviour 30:637–640.

Grubb, T. C., Jr., and V. V. Pravosudov. 1994. Ptilochronology: follicle history fails to influence growth of an induced feather. Condor 96:214–217.

Grun, G. 1975. Die Erhanrung der Sperlinge *Passer domesticus* (L.) und *Passer montanus* (L.) unter verschiedenen Umweltbedingungen. International Studies on Sparrows 8:24–103.

Gruwell, J. A., C. L. Fogarty, S. G. Bennett, G. L. Challet, K. S. Vanderpool, M. Jozan, and J. P. Webb, Jr. 2000. Role of peridomestic birds in the transmission of St. Louis encephalitis virus in southern California. Journal of Wildlife Diseases 36:13–34.

Guillory, H. D., and J. H. Deshotels. 1981. House sparrows flushing prey from trees and shrubs. Wilson Bulletin 93:554.

Gupta, Y. K., and P. N. Saxena. 1999. Effect of mexacarbate on liver and serum phosphatases of *Passer domesticus* (Linnaeus). Pavo 37:77–80.

Gwinner, E. 1986. Circannual Rhythms: Endogenous Annual Clocks in the Organization of Seasonal Processes. Springer-Verlag, Berlin.

Gwinner, E. 1989. Melatonin in the circadian system of birds: model of internal resonance. Pp. 127–145 in Circadian Clocks and Ecology (T. Hiroshage and K. Honma, eds.). Hokkaido University Press, Sapporo, Japan.

Gwinner, E., and R. Brandstätter. 2001. Complex bird clocks. Philosophical Transactions of the Royal Society of London B 356:1801–1810.

Haarakangas, H., H. Hyvarinen, and M. Ojanen. 1974. Seasonal variations and the effects of nesting and moulting on liver mineral content in the house sparrow (*Passer domesticus* L.). Comparative Biochemistry and Physiology 47A:153–163.

Haase, E. 1975. The effects of testosterone propionate on secondary sexual characters and testes of house sparrows, *Passer domesticus*. General and Comparative Endocrinology 26:248–252.

Haffer, J. 1989. Parapatrische Vogelarten der palarktischen Region. Journal für Ornithologie 130:475–512.

Haftorn, S. 1994. Diurnal rhythm of passerines during the polar night in Pasvik, North Norway, with comparative notes from South Norway. Fauna Norvegicus, Series C, Cinclus 17:1–8.

Hahn, T. P., and G. F. Ball. 1995. Changes in brain GnRH associated with photorefractoriness in house sparrows (*Passer domesticus*). General and Comparative Endocrinology 99:349–363.

Hamerstrom, F. N., Jr., and F. Hamerstrom. 1951. Food of young raptors on the Edwin S. George Reserve. Wilson Bulletin 63:16–25.

Hamilton, S., and R. F. Johnston. 1978. Evolution in the House Sparrow. VI. Variability and niche breadth. Auk 95:313–323.

Hamilton, W. D., and M. Zuk. 1982. Heritable true fitness and bright birds: a role for parasites? Science 218:384–387.

Hammer, M. 1948. Investigations on the feeding-habits of the house-sparrow (*Passer*

domesticus) and the tree-sparrow (*Passer montanus*). Danish Review of Game Biology 1(2):1–59.
Hammon, W. McD., W. C. Reeves, and G. E. Sather. 1951a. Japanese B encephalitis virus in the blood of experimentally inoculated birds: epidemiological implications. American Journal of Hygiene 53:249–261.
Hammon, W. McD., W. C. Reeves, and G. E. Sather. 1951b. Western equine and St. Louis encephalitis viruses in the blood of experimentally infected wild birds and epidemiological implications of findings. Journal of Immunology 67:357–367.
Hankinson, M. D. 1999. Male house sparrows behave as if a fertilization window exists. Auk 116:1141–1144.
Hanotte, O., E. Cairns, T. Robson, M. C. Double, and T. Burke. 1992. Cross-species hybridization of a single-locus minisatellite probe in passerine birds. Molecular Ecology 1:127–130.
Hardy, E. 1932. Courtship of house-sparrow. British Birds 25:301.
Harper, D. G. C. 1984. Moult interruption in passerines resident to Britain. Ringing and Migration 5:101–104.
Harris, J. A. 1911. A neglected paper on natural selection in the English sparrow. American Naturalist 45:314–319.
Harrison, C. J. O. 1963. "Industrial" discoloration of house sparrow and other birds. British Birds 56:296–297.
Harrison, J. G. 1960. A comparative study of the method of skull pneumatisation in certain birds: part one. Bulletin of the British Ornithologists' Club 80:167–172.
Harrison, J. G. 1961. A comparative study of the method of skull pneumatisation in certain birds: part two. Bulletin of the British Ornithologists' Club 81:12–17.
Harrison, J. M. 1961a. The significance of some plumage phases of the house-sparrow, *Passer domesticus* (Linnaeus) and the Spanish sparrow, *Passer hispaniolensis* Temminck: part one. Bulletin of the British Ornithologists' Club 81:96–103.
Harrison, J. M. 1961b. The significance of some plumage phases of the house-sparrow, *Passer domesticus* (Linnaeus) and the Spanish sparrow, *Passer hispaniolensis* Temminck: part two. Bulletin of the British Ornithologists' Club 81:119–124.
Hart, J. W. 1949. Observations on blood parasites of birds in South Carolina. Journal of Parasitology 35:79–82.
Hartman, F. A. 1946. Adrenal and thyroid weights in birds. Auk 63:42–64.
Hartup, B. K., A. A. Dhondt, K. V. Syndenstricker, W. M. Hochachka, and G. V. Kollias. 2001. Host range and dynamics of mycoplasmal conjunctivitis among birds in North America. Journal of Wildlife Diseases 37:72–81.
Hartwig, H. G. 1974. Electron microscopic evidence for a retinohypothalamic projection to the suprachiasmatic nucleus of *Passer domesticus*. Cell and Tissue Research 153:89–99.
Hau, M., and E. Gwinner. 1992. Circadian entrainment by feeding cycles in house sparrows, *Passer domesticus*. Journal of Comparative Physiology A 170:403–409.

Hau, M., and E. Gwinner. 1994. Melatonin facilitates synchronization of sparrow circadian rhythms to light. Journal of Comparative Physiology A 173:343–347.

Hau, M., and E. Gwinner. 1995. Continuous melatonin administration accelerates resynchronization following phase shifts of a light-dark cycle. Physiology and Behavior 58:89–95.

Hau, M., and E. Gwinner. 1996. Food as a circadian zeitgeber for house sparrows: the effect of different food access durations. Journal of Biological Rhythms 11:196–207.

Hau, M., and E. Gwinner. 1997. Adjustment of house sparrow circadian rhythms to a simultaneously applied light and food zeitgeber. Physiology and Behavior 62:973–981.

Haukioja, E., and J. Reponen. 1969. [The movements of the house sparrow (*Passer domesticus*) during a yearcycle.] Eripainos Porin Lintutieteellinen Yhdistys ry:n vuosikirjasta 1968:23–26. (In Finnish, English summary).

Hauser, D. C. 1973. Comparison of anting records from two locations in North Carolina. Chat 37:91–102.

Havlín, J. 1974. Von Haussperling (*Passer domesticus*) und Feldsperling (*P. montanus*) an reifenden Getreidepflanzen verursachte Schäden. Zoologicke Listy 23:241–259.

Hayes, R. O., L. C. Lamotte, and P. Holden. 1967. Ecology of arboviruses in Hale County, Texas, during 1965. American Journal of Tropical Medicine and Hygiene 16:675–687.

Haywood, S. 1993. Sensory and hormonal control of clutch size in birds. Quarterly Review of Biology 68:33–59.

Hegner, R. E., and J. C. Wingfield. 1984. Social facilitation of gonadal recrudescence. Journal of Steroid Biochemistry 20:1549.

Hegner, R. E., and J. C. Wingfield. 1986a. Behavioral and endocrine correlates of multiple brooding in the semicolonial house sparrow *Passer domesticus*. I. Males. Hormones and Behavior 20:294–312.

Hegner, R. E., and J. C. Wingfield. 1986b. Gonadal development during autumn and winter in house sparrows. Condor 88:269–278.

Hegner, R. E., and J. C. Wingfield. 1986c. Behavioral and endocrine correlates of multiple brooding in the semicolonial house sparrow *Passer domesticus*. II. Females. Hormones and Behavior 20:313–326.

Hegner, R. E., and J. C. Wingfield. 1987a. Effects of brood-size manipulations on parental investment, breeding success, and reproductive endocrinology of house sparrows. Auk 104:470–480.

Hegner, R. E., and J. C. Wingfield. 1987b. Effects of experimental manipulation of testosterone levels on parental investment and breeding success in male house sparrows. Auk 104:462–469.

Hegner, R. E., and J. C. Wingfield. 1987c. Social status and circulating levels of hormones in flocks of house sparrows. Ethology 76:1–14.

Hegner, R. E., and J. C. Wingfield. 1990. Annual cycle of gonad size, reproductive hormones, and breeding activity of free-living house sparrows [*Passer domesticus*

(L.)] in rural New York. Pp. 123–135 in Granivorous Birds in the Agricultural Landscape (J. Pinowski and J. D. Summers-Smith, eds.). PWN–Polish Scientific Publishers, Warsaw.

Heigl, S., and E. Gwinner. 1994. Periodic melatonin in the drinking water synchronizes circadian rhythm in sparrows. Naturwissenschaften 81:83–85.

Heigl, S., and E. Gwinner. 1995. Synchronization of circadian rhythms of house sparrows by oral melatonin: effects of changing period. Journal of Biological Rhythms 10:225–233.

Heigl, S., and E. Gwinner. 1999. Periodic food availability synchronizes locomotor and feeding activity in pinealectomized house sparrows. Zoology 102:1–9.

Heij, C. J. 1986. Nest of house sparrows, *Passer domesticus* (L.). composition and occupants. International Studies on Sparrows 13:28–34.

Heij, C. J., and C. W. Moeliker. 1990. Population dynamics of Dutch house sparrows in urban, suburban and rural habitats. Pp. 59–85 in Granivorous Birds in the Agricultural Landscape (J. Pinowski and J. D. Summers-Smith, eds.). PWN–Polish Scientific Publishers, Warsaw.

Hein, W. K., D. S. Westneat, and J. P. Poston. 2003. Sex of opponent influences response to a potential status signal in house sparrows. Animal Behaviour 65: 1211–1221.

Heisler, C. 1978. Die Bedeutung der Beine für die Temperaturregulation bei Haussperling (*Passer domesticus*) und Zebrafink (*Taeniopygia guttata castanotis*). Die Vogelwarte 29:261–268.

Hellebrekers, W. PH. J. 1950. Measurements and weights of eggs of birds on the Dutch list. E. J. Brill, Leiden, Netherlands.

Hendel, R. C., and F. W. Turek. 1978. Suppression of locomotor activity in sparrows by treatment with melatonin. Physiology and Behavior 21:275–278.

Hendricks, P. 1991. Repeatability of size and shape of American pipit eggs. Canadian Journal of Zoology 69:2624–2628.

Hibbert-Ware, A. 1937. Report of the little owl food inquiry: 1936–37. British Birds 31:162–187, 205–229.

Hill, E. F. 1971. Toxicity of selected mosquito larvicides to some common avian species. Journal of Wildlife Management 35:757–762.

Hill, E. F. 1972. Avoidance of lethal dietary concentrations of insecticide by house sparrows. Journal of Wildlife Management 36:635–639.

Hill, G. E. 1993. House finch (*Carpodacus mexicanus*). No. 46 in The Birds of North America (A. Poole and F. Gill, eds.). Academy of Natural Sciences, Philadelphia.

Hinds, D. S., and W. A. Calder. 1971. Tracheal dead space in the respiration of birds. Evolution 25:429–440.

Hodson, K. 1978. Prey utilized by merlins nesting in shortgrass prairies of southern Alberta. Canadian Field-Naturalist 92:76–77.

Hohn, E. O. 1956. Seasonal recrudescence of thymus in adult birds. Canadian Journal of Biochemistry and Physiology 34:90–101.

Hoi, H., R. Václav, and D. Slobodová. 2003. Postmating sexual selection in house sparrows: can females estimate "good fathers" according to their early paternal effort? Folia Zoologica 52:299–308.

Holden, P. 1955. Recovery of western equine encephalomyelitis virus from naturally infected English sparrows of New Jersey, 1953. Proceedings of the Society for Experimental Biology and Medicine 88:490–492.

Holden, P., D. B. Francy, C. J. Mitchell, R. O. Hayes, J. S. Lazuick, and T. B. Hughes. 1973. House sparrows, *Passer domesticus* (L.), as hosts of arboviruses in Hale County, Texas. II. Laboratory studies with western equine encephalitis virus. American Journal of Tropical Medicine and Hygiene 22:254–262.

Holden, P., R. O. Hayes, C. J. Mitchell, D. B. Francy, J. S. Lazuick, and T. B. Hughes. 1973. House sparrows, *Passer domesticus* (L.), as hosts of arboviruses in Hale County, Texas. I. Field studies, 1965–1969. American Journal of Tropical Medicine and Hygiene 22:244–253.

Hole, D. G., M. J. Whittingham, R. B. Bradbury, G. Q. A. Anderson, P. L. M. Lee, J. D. Wilson, and J. R. Krebs. 2002. Widespread local house-sparrow extinctions. Nature 418:931–932.

Honza, M. 1992. Seasonal changes in a bird community in the vicinity of agricultural farms. Folia Zoologica 41:139–149.

Hopkins, A. D. 1938. Bioclimatics: a science of life and climate relations. Miscellaneous Publications of the United Sates Department of Agriculture 280:1–188.

Hopla, C. E., D. B. Francy, C. H. Calisher, and J. S. Lazuick. 1993. Relationship of cliff swallows, ectoparasites, and an alphavirus in west-central Oklahoma. Journal of Medical Entomology 30:267–262.

Houde, P. 1987. Critical evaluation of DNA hybridization studies in avian systematics. Auk 104:17–32.

Hoyt, D. F. 1979. Practical methods of estimating volume and fresh weight of bird eggs. Auk 96:73–77.

Howard, W. J. 1937. Bird behavior as a result of emergence of seventeen year locusts. Wilson Bulletin 49:43–44.

Huber, G. C., and E. C. Crosby. 1929. The nuclei and fiber paths of the avian diencephalon, with consideration of telencephalic and certain mesencephalic centers and connections. Journal of Neurology 48:1–225.

Hubregste, V. 1992. House sparrows operating electronic doors. The Australia Bird Watcher 14:241.

Hudson, J. W., and S. L. Kimzey. 1964. Body temperature and metabolism cycles in the house sparrow, *Passer domesticus*, compared with the white-throated sparrow, *Zonotrichia albicollis*. American Zoologist 4:294–295.

Hudson, J. W., and S. L. Kimzey. 1966. Temperature regulation and metabolic rhythms in populations of the house sparrow, *Passer domesticus*. Comparative Biochemistry and Physiology 17:203–217.

Huggins, R. A. 1941. Egg temperatures of wild birds under natural conditions. Ecology 22:148–157.

Hull, S. L. 1998. Alarm calls and predator discrimination in populations of the house sparrow *Passer domesticus* in Leeds. Naturalist 123:19–24.

Hume, R. A. 1983. Hybrid tree X house sparrow paired with house sparrow. British Birds 76:234–235.

Hunninen, A. V., and M. D. Young. 1950. Blood protozoa of birds at Columbia, South Carolina. Journal of Parasitology 36:258–260.

Hunt, K. A., D. M. Bird, P. Mineau, and L. Shutt. 1992. Selective predation of organophosphate-exposed prey by American kestrels. Animal Behaviour 43:971–976.

Hussain, I., S. Ahmad, S. Munir, and A. A. Khan. 1993. Laboratory evaluation of strychnine treated cereal grains against house sparrow, *Passer domesticus*. Pakistan Journal of Zoology 25:121–125.

Il'Enko, A. I. 1958. [Factors determining the start of multiplication in the population of house sparrows (*Passer domesticus* L.) in Moscow.] Zoologicheskii Zhurnal 37:1867–1873. (In Russian, English translation at Alexander Library).

Illescas-Gomez, P., and R. Lopez-Roman. 1980. *Choanotaenia passerina* (Fuhrmann, 1907) Fuhrmann, 1932; primera cita Espana, parasito del *Passer domesticus* L. Revista Iberia de Parasitologia 40:399–405.

Il'Yenko, A. I. 1965. [Competition for nesting places in house sparrow population.] Zoologicheskii Zhurnal 44:1874–1878. (In Russian, English translation at Alexander Library).

Immelmann, K. 1971. Ecological aspects of periodic reproduction. Pp. 341–389 in Avian Biology, vol. I (D. S. Farner and J. R. King, eds.). Academic Press, New York.

Indykiewicz, P. 1990. Nest-sites and nests of house parrow [*Passer domesticus*] in an urban environment. Pp. 95–121 in Granivorous Birds in the Agricultural Landscape (J. Pinowski and J. D. Summers-Smith, eds.). PWN–Polish Scientific Publishers, Warsaw.

Indykiewicz, P. 1991. Nests and nest-sites of the house sparrow *Passer domesticus* (Linnaeus, 1758) in urban, suburban and rural environments. Acta Zoologica Cracoviensia 34:475–495.

Ion, I. 1971. Studiu asupra compozitiei hranei consumata de puii vrabiei de casa—*Passer domesticus* L. si vrabiei de cimp—*Passer montanus* L. Muzeul de Stiintele Naturii Bacau, Studii di Comunicari-1971: 263–276.

Ion, I. 1992. New investigations on food of house sparrow—*Passer domesticus* L. and tree sparrow—*Passer montanus* L. in Roumania. International Studies on Sparrows 19:37–42.

Ion, I., and R. Ion. 1978. Observation during the breeding season on the house sparrow, *Passer domesticus* L. and the tree sparrow, *Passer montanus* L. (Aves, Passeriformes). Trav. Mus. Hist. nat. "Grigore Antipa" 19:329–333.

Ivanitzky, V. V. 1996. [Behavior of males and females of Indian sparrow (*Passer indicus*) during nesting.] Zoologicheskii Zhurnal 75:249–255. (In Russian, English summary).

Ivanov, B. E. 1987. Productivity due to reproduction of house sparrow, *Passer domesticus* (L.), populations inhabiting animal farms. Ekologia Polska 35:699–721.

Ivanova, S. 1935. Über den Mechanismus der Wirkung von Licht auf die Hoden der Vögel (*Passer domesticus*). Archiv for experimentale Pathologie und Pharmakologie 179:349–359.

Jackson, J. A., and B. J. Schardien Jackson. 1985. Interactions between house sparrows and common ground-doves on Walker's Cay, Bahamas. Wilson Bulletin 97:379–381.

Jacob, J., and A. Zeman. 1970. Die Burzeldrüsenlipide des Haussperlings *Passer domesticus*. Zeitschrift Naturforschung 25b:984–988.

Jacob, J., and V. Ziswiler. 1982. The uropygial gland. Pp. 199–324 in Avian Biology, vol. VI (D. S. Farner and J. R. King, eds.). Academic Press, New York.

Jacobsen, E. M. 1995. [Point count censuses of birds breeding in urban areas, 1976–1994.] Dansk Ornitologisk Forenings Tidsskrift 89:111–118. (In Danish: English summary).

Jadhav, B. V., A. V. Bosale, A. B. Gavhane, and A. P. Jadhav. 1990. A new cestode from a house sparrow, *Passer domesticus* Hyderabad. Rivista di Parassitologia 7(51):77–80.

Jain, N. C., N. K. Chandiramani, and I. P. Singh. 1971. Studies on avian pleuropneumonia-like organisms. 2. Occurrence of Mycoplasma in wild birds. Indian Journal of Animal Sciences 41:301–305.

Jaksic, F. M., E. F. Pavez, J. E. Jimenez, and J. C. Torres-Mura. 2001. The conservation status of raptors in the metropolitan region, Chile. Journal of Raptor Research 35:151–158.

James, F. C., and C. Nesmith. 1988. Nongenetic effects in geographic differences among nestling populations of red-winged blackbirds. Acta XIX Congressus Internationalis Ornithologici 1424–1433.

James, P. C., and A. R. Smith. 1987. Food habits of urban-nesting merlins, *Falco columbarius*, in Edmonton and Fort Saskatchewan, Alberta. Canadian Field-Naturalist 101:592–594.

Jani, M. B., G. K. Menon, and R. V. Shah. 1984. Incubation patch formation in the house sparrow: distribution pattern of phosphomonoesterases. Pavo 22:73–79.

Jani, M. B., G. K. Menon, and R. V. Shah. 1985a. Incubation patch formation in house sparrow: histochemical localization of glucose-6–phosphate, alpha glycerophosphate, and B [beta]-hydroxybuterate dehydrogenases and lipids in the skin. Pavo 23:93–100.

Jani, M. B., G. K. Menon, and R. V. Shah. 1985b. Lactate, succinate and malate dehydrogenases in incubation patch of house sparrow: a histochemical study. Pavo 23:85–92.

Janik, D., J. Dittami, and E. Gwinner. 1992. The effect of pinealectomy on circadian plasma levels in house sparrows and European starlings. Journal of Biological Rhythms 7:277–286.

Janssen, R. B. 1983. House sparrows build roost nests. Loon 55:64–65.

Jasinski, A. 1973. Fine structure of capillaries in the pecten ocelli of the sparrow, *Passer domesticus*. Zeitscrift für Zellforschung und mikroskopische Anatomie 146:281–292.

Jawor, J. M., and R. Breitwisch. 2003. Melanin ornaments, honesty, and sexual selection. Auk 120:249–265.

Jenkins, A. R., and G. M. Avery. 1999. Diets of breeding peregrine and Lanner falcons in South Africa. Journal of Raptor Research 33:190–206.

Jenni, L., and U. Schaffner. 1984. Herbstbewegungen von Haus- und Feldsperling *Passer domesticus* und *P. montanus* in der Schweiz. Der Ornitholgische Beobachter 81:61–67.

Jensen, H., B.-E. Sæther, T. H. Ringsby, J. Tufto, S. C. Griffith, and H. Ellegran. 2003. Sexual variation in heritability and genetic correlations of morphological traits in house sparrow (*Passer domesticus*). Journal of Evolutionary Biology 16:1296–1307.

Jentzsch, M. 1988. Vogelbeute der Schleiereule (*Tyto alba*) im Helme- Unstrut-Gebiet. Beiträge zur Vogelkunde 34:221–229.

Johnson, C. A., L.-A. Giraldeau, and J. W. A. Grant. 2001. The effect of handling time on interference among house sparrows foraging at different seed densities. Behaviour 138:597–614.

Johnson, O. W. 1968. Some morphological features of avian kidneys. Auk 85:216–228.

Johnston, R. F. 1965. Nestsite sitting by breeding house sparrows. Kansas Ornithological Society Bulletin 16:17–18.

Johnston, R. F. 1966. Colorimetric studies of soil color-matching by feathers of house sparrows from the central United States. Kansas Ornithological Society Bulletin 17:19–23.

Johnston, R. F. 1967a. Sexual dimorphism in juvenile house sparrows. Auk 84:275–277.

Johnston, R. F. 1967b. Some observations on natural mass mortality of house sparrows. Kansas Ornithological Society Bulletin 18:9–10.

Johnston, R. F. 1969a. Aggressive foraging behavior in house sparrows. Auk 86:558–559.

Johnston, R. F. 1969b. Character variation and adaptation in European sparrows. Systematic Zoology 18:206–231.

Johnston, R. F. 1969c. Taxonomy of house sparrows and their allies in the Mediterranean basin. Condor 71:129–139.

Johnston, R. F. 1972. Color variation and natural selection in Italian sparrows. Bollettino de Zoologia 39:351–362.

Johnston, R. F. 1973. Evolution in the house sparrow. IV. Replicate studies in phenetic covariation. Systematic Zoology 22:219–226.

Johnston, R. F. 1975. Studies in phenetic and genetic covariation. Pp. 333–353 in Proceedings of the 8th International Conference of Numerical Taxonomy (G. Estabrook, ed.). Freeman, San Francisco.

Johnston, R. F. 1976a. Estimating variation in bony characters and a comment on the Kluge-Kerfoot effect. Occasional Papers of the Museum of Natural History, University of Kansas 53:1–8.

Johnston, R. F. 1976b. Evolution in the house sparrow. V. Covariation of skull and hindlimb sizes. Occasional Papers of the Museum of Natural History, University of Kansas 56:1–8.

Johnston, R. F., and R. C. Fleischer. 1981. Overwinter mortality and sexual size dimorphism in the house sparrow. Auk 98:503–511.

Johnston, R. F., and W. J. Klitz. 1977. Variation and evolution in a granivorous bird: the house sparrow. Pp. 15–51 in Granivorous Birds in Ecosystems (J. Pinowski and S. C. Kendeigh, eds.). Cambridge University Press, Cambridge, UK.

Johnston, R. F., D. M. Niles, and S. A. Rohwer. 1972. Hermon Bumpus and natural selection in the house sparrow *Passer domesticus*. Evolution 26:20–31.

Johnston, R. F., and R. K. Selander. 1963. Further remarks on discolouration in house sparrows. British Birds 56:469.

Johnston, R. F., and R. K. Selander. 1964. House sparrows: rapid evolution of races in North America. Science 144:548–550.

Johnston, R. F., and R. K. Selander. 1971. Evolution in the house sparrow. II. Adaptive differentiation in North American populations. Evolution 25:1–28.

Johnston, R. F., and R. K. Selander. 1972. Variation, adaptation, and evolution in the North American house sparrows. Pp. 301–326 in Productivity, Population Dynamics and Systematics of Granivorous Birds (S. C. Kendeigh and J. Pinowski, eds.). PWN–Polish Scientific Publishers, Warsaw.

Johnston, R. F., and R. K. Selander. 1973. Evolution in the house sparrow. III. Variation in size and sexual dimorphism in Europe and North and South America. American Naturalist 107:373–390.

Jokimäki, J., and M.-L. Kaisanlahti-Jokimäki. 2003. Spatial similarity of urban bird communities: a multiscale approach. Journal of Biogeography 30:1183–1193.

Jokimäki, J., J. Suhonen, K. Inki, and S. Jokinen. 1996. Biogeographical comparison of winter bird assemblages in urban environments in Finland. Journal of Biogeography 23:379–386.

Jones, D. N., and J. Wieneke. 2000. The suburban bird community of Townsville revisited: changes over 16 years. Corella 24:53–60.

Jones, M. M. 1980a. Diurnal variation in the distribution of lipid in the pectoralis muscle of the house sparrow (*Passer domesticus*). Journal of Zoology, London 191:475–486.

Jones, M. M. 1980b. Nocturnal loss of muscle protein from house sparrows (*Passer domesticus*). Journal of Zoology, London 192:33–39.

Jones, M. M. 1982. Growth of the pectoralis muscle of the house sparrow (*Passer domesticus*). Journal of Anatomy 135:719–731.

Jones, M. M. 1990. Muscle protein loss in laying house sparrows *Passer domesticus*. Ibis 133:193–198.

Jopek, Z., E. Kucharczak, and J. Pinowski. 1995. The concentration of iron, zinc, copper and lead in sparrow (*Passer* spp.) eggs. Pp. 181–201 in Nestling Mortality of Granivorous Birds due to Microorganisms and Toxic Substances: Synthesis (J. Pinowski, B. P. Kavanagh, and B. Pinowska, eds.). PWN–Polish Scientific Publishers, Warsaw.

Judd, S. D. 1898. The food of shrikes. United States Department of Agriculture Bulletin 9:15–26.

Juricova, Z., I. Literak, and J. Pinowski. 1999. Antibodies to arboviruses in house sparrows (*Passer domesticus*) in Czech Republic. Preliminary report. International Studies on Sparrows 26:67–68.

Juricova, Z., I. Literak, and J. Pinowski. 2000. Antibodies to arboviruses in house sparrows (*Passer domesticus*) in the Czech Republic. Acta Veterinaria Brno 69:213–215.

Juricova, Z., J. Pinowski, I. Literak, K. H. Haham, and J. Romanowski. 1998. Antibodies to alphavirus, flavivirus, and bunyavirus arboviruses in house sparrows (*Passer domesticus*) and tree sparrows (*P. montanus*) in Poland. Avian Diseases 42:182–185.

Jurik, M. 1974. Bionomics of fleas in birds' nests in the territory of Czechoslovakia. Acta Scientiarum Naturalium Brno 8(10):1–54.

Kaczmarek, S. 1991. [Fleas from the nests of *Passer domesticus* and *Passer montanus*.] Wiadomosci Parazytologiczne 37:67–70. (In Polish, English summary).

Kadhim, A.-H. H., K. Y. Al-Dabbagh, M. N. Al-Nakash, and I. N. Waheed. 1987. The annual cycle of male house sparrow *Passer domesticus* in central Iraq. Journal of Biological Sciences Research 18:1–9.

Kaleta, E. F. 2002. Foot-and-mouth disease: susceptibility of domestic poultry and free-living birds to infection and to disease—a review of the historical and current literature concerning the role of birds in spread of foot-and-mouth disease viruses. Deutsche Tierärztliche Wochenschrift 109:391–399.

Kalinoski, R. 1975. Intra- and interspecific aggression in house finches and house sparrows. Condor 77:375–384.

Kalmbach, E. R. 1940. Economic status of the English sparrow in the United States. Technical Bulletin No. 711. United States Department of Agriculture, Washington, DC.

Kalmus, H. 1984. Wall clinging: energy saving by the house sparrow *Passer domesticus*. Ibis 126:72–74.

Karasov, W. H., and J. M. Diamond. 1983. Adaptive regulation of sugar and amino acid transport by vertebrate intestine. American Journal of Physiology 245: G443–G462.

Karolewski, M. A., A. B. Łukowski, J. Pinowski, and J. Trojanowski. 1991. Chlorinated hydrocarbons in eggs and nestlings of *Passer montanus* and *P. domesticus* from urban and suburban areas of Warsaw. Preliminary report. Pp. 189–195 in Nestling Mortality of Granivorous Birds due to Microorganisms and Toxic Substances (J. Pinowski, B. P. Kavanaugh, and W. Gorski, eds.). PWN–Polish Scientific Press, Warsaw.

Kaul, H. N., A. C. Mishar, V. Dhanda, S. M. Kulkarni, and S. N. Guttikar. 1978. Ectoparasitic arthropods of birds and mammals from Rajasthan State, India. Indian Journal of Parasitology 2:19–25.

Keck, W. N. 1932a. Control of the sex characters in the English sparrow, *Passer domesticus* (Linnaeus). Anatomical Record 54(Supplement):77.

Keck, W. N. 1932b. Control of the sex characters in the English sparrow, *Passer domesticus* (Linnaeus). Proceedings of the Society for Experimental Biology and Medicine 30:158–159.

Keck, W. N. 1933. Control of the bill color of the male English sparrow by injection of male hormone. Proceedings of the Society for Experimental Biology and Medicine 30:1140–1141.

Keck, W. N. 1934. The control of the secondary sex characters in the English sparrow, *Passer domesticus* (Linnaeus). Journal of Experimental Zoology 67:315–347.

Kedar, H., M. A. Rodriguez-Gironés, S. Yedvab, D. W. Winkler, and A. Lotem. 2000. Experimental evidence for offspring learning in parent-offspring communication. Proceedings of the Royal Society of London B 267:1723–1727.

Keil, W. 1972. Investigations on food of house- and tree sparrows in a cereal-growing area during winter. Pp. 253–262 in Productivity, Population Dynamics and Systematics of Granivorous Birds (S. C. Kendeigh and J. Pinowski, eds.). PWN–Polish Scientific Publishers, Warsaw.

Kendeigh, S. C. 1941. Length of day and energy requirements for gonad development and egg-laying in birds. Ecology 22:237–248.

Kendeigh, S. C. 1944. Effect of air temperature on the rate of energy metabolism in the English sparrow. Journal of Experimental Zoology 96:1–16.

Kendeigh, S. C. 1949. Effect of temperature and season on energy resources of the English sparrow. Auk 66:113–127.

Kendeigh, S. C. 1961. Energy of birds conserved by roosting in cavities. Wilson Bulletin 73:140–147.

Kendeigh, S. C. 1969. Energy responses of birds to their thermal environments. Wilson Bulletin 81:441–449.

Kendeigh, S. C. 1976. Latitudinal trends in the metabolic adjustments of the house sparrow. Ecology 57:509–519.

Kendeigh, S. C., and C. R. Blem. 1974. Metabolic adaptation to local climate in birds. Comparative Biochemistry and Physiology 48A:175–187.

Kendeigh, S. C., V. R. Dol'Nik, and V. M. Gavrilov. 1977. Avian energetics. Pp. 127–204 in Granivorous Birds in Ecosystems (J. Pinowski and S. C. Kendeigh, eds.). Cambridge University Press, Cambridge, UK.

Kendeigh, S. C., J. E. Kontogiannis, A. Mazac, and R. R. Roth. 1969. Environmental regulation of food intake by birds. Comparative Biochemistry and Physiology 31:941–957.

Kendeigh, S. C., and J. Pinowski (eds.). 1972. Productivity, Population Dynamics and Systematics of Granivorous Birds. PWN–Polish Scientific Publishers, Warsaw.

Kendeigh, S. C., and H. E. Wallin. 1966. Seasonal and taxonomic differences in the size and activity of the thyroid glands in birds. Ohio Journal of Science 66:369–379.

Kendra, P. E., R. R. Roth, and D. W. Tallamy. 1988. Conspecific brood parasitism in the house sparrow. Wilson Bulletin 100:80–90.

Kennedy, E. D. 1991. Determinate and indeterminate egg-laying patterns: a review. Condor 93:106–124.

Kimball, R. T. 1996. Female choice for male morphological traits in house sparrows, *Passer domesticus*. Ethology 102:639–648.

Kimball, R. T. 1997. Male morphology and nest-site quality in house sparrows. Wilson Bulletin 109:711–719.

Kirschbaum, A. 1933. The experimental modification of the seasonal sexual cycle of the English sparrow, *Passer domesticus*. Anatomical Record 57(Supplement):62.

Kirschbaum, A., and C. A. Pfeiffer. 1941. Deposition of melanin in sparrow bill following local action of testosterone propionate in alcoholic solution. Proceedings of the Society for Experimental Biology and Medicine 46:649–651.

Kirschbaum, A., C. A. Pfeiffer, J. Van Heuverswyn, and W. U. Gardner. 1939. Studies on gonad-hypophyseal relationship and cyclic osseous changes in the English sparrow, *Passer domesticus* L. Anatomical Record 75:249–263.

Kirschbaum, A., and A. R. Ringoen. 1936. Seasonal sexual activity and its experimental modification in the male sparrow, *Passer domesticus* Linnaeus. Anatomical Record 64:453–473.

Kissling, R. E., R. W. Chamberlain, R. K. Sikes, and M. E. Eidson. 1954. Studies of North American arthropod-borne encephalitides. III. Eastern equine encephalitis in wild birds. American Journal of Hygiene 60:251–265.

Kiziroğlu, I., M. N. Sisli, and Ü. Alp. 1987. Zur interspezifischen Konkurrenz in der Besiedlung von Nistkästen bei Ankara/Türkei mit Angaben zur Brutbiologie verschiedener Höhlenbrüter-Arten. Vogelwelt 198:169–175.

Klein, S. E. B., S. Binkley, and K. Moser. 1985. Circadian phase of sparrows: control by light and dark. Photochemistry and Photobiology 41:453–457.

Klem, D. Jr., S. A. Finn, and J. H. Nave, Jr. 1983. Gross morphology and general histology of the ventriculus, intestinum, caeca and cloaca of the house sparrow (*Passer domesticus*. Proceedings of the Pennsylvania Academy of Science 57:27–32.

Klem, D. Jr., C. R. Brancato, J. F. Catalano, and F. L. Kuzmin. 1982. Gross morphology and general histology of the esophagus, ingluvies and proventriculus of the house sparrow (*Passer domesticus*). Proceedings of the Pennsylvania Academy of Science 56:141–146.

Kleven, S. H., and W. O. Fletcher. 1983. Laboratory infection of house sparrows (*Passer domesticus*) with Mycoplasma gallisepticum and Mycoplasma synoviae. Avian Diseases 27:308–311.

Klimkiewicz, M. K., and A. G. Futcher. 1987. Longevity records of North American birds: Coerebinae to Estrildidae. Journal of Field Ornithology 58:318–333.

Klitz, W. 1972. Genetic consequences of colonization with subsequent expansion: the house sparrow in North America. Ph. D. Dissertation, University of Kansas, Lawrence.

Klomp, H. 1970. The determination of clutch-size in birds: a review. Ardea 58:1–124.

Kobayashi, H., and M. Wada. 1973. Neuroendocrinology in birds. Pp. 287–347 in Avian Biology, vol. III (D. S. Farner and J. R. King, eds.). Academic Press, New York.

Kocian, L., D. Nemethová, D. Melicherová, and A. Matusková. 2003. Breeding bird communities in three cemeteries in the city of Bratislava (Slovakia). Folia Zoologica 52:177–188.

Kohler, L. S. 1930. Polygamy in house sparrows. Oologist 47:104.

Kokernot, R. H., J. Hayes, R. L. Will, C. H. Tempelis, D. H. M. Chan, and B. Radivojevic. 1969a. Arbovirus studies in the Ohio-Mississippi basin, 1964–1967. II. St. Louis encephalitis virus. American Journal of Tropical Medicine and Hygiene 18:750–761.

Kokernot, R. H., J. Hayes, R. L. Will, B. Radivojevic, K. R. Boyd, and D. H. M. Chan. 1969b. Arbovirus studies in the Ohio-Mississippi basin, 1964–1967. III. Flanders virus. American Journal of Tropical Medicine and Hygiene 18:762–767.

Komar, N., S. Langavin, S. Hinten, N. Nemeth, E. Edwards, D. Hettler, B. Davis, R. Bowen, and M. Bunning. 2003. Experimental infection of North American birds with the New York 1999 strain of West Nile virus. Emerging Infectious Diseases 9:311–322.

Korf, H. T-W., N. H. Zimmerman, and A. Oksche. 1982. Intrinsic neurons and neural connections of the pineal organ of the house sparrow, *Passer domesticus*, as revealed by anterograde and retrograde transport of horseradish peroxidase. Cell and Tissue Research 222:243–260.

Korpimaki, E. 1988. Diet of breeding Tengmalm's owls Aegolius funereus: long-term changes and year-to-year variation under cyclic food conditions. Ornis Fennica 65:21–30.

Koteja, P. 1986. Maximum cold-induced oxygen consumption in the house sparrow *Passer domesticus* L. Physiological Zoology 59:43–48.

Kozłowski, P. 1992. [Nest-boxes as a site of bird broods in Warsaw urban parks.] Acta Ornithologica 27:21–33. (In Polish, English summary).

Kozłowski, S., E. Małyszko, J. Pinowski, and A. Kruszewicz. 1991. The effect of microorganisms on the mortality of house sparrow (*Passer domesticus*) and tree sparrow (*Passer montanus*) embryos. Pp. 121–128 in Nestling Mortality of Granivorous Birds due to Microorganisms and Toxic Substances (J. Pinowski, B. P. Kavanaugh, and W. Gorski, eds.). PWN–Polish Scientific Press, Warsaw.

Krebs, J. R., N. S. Clayton, R. R. Hampton, and S. J. Shettleworth. 1995. Effects of photoperiod on food-storing and the hippocampus of birds. NeuroReport 6:1701–1704.

Krebs, J. R., J. D. Wilson, R. B. Bradbury, and G. M. Siriwardena. 1999. The second Silent Spring? Nature 400:611–612.

Krementz, D. G., and C. D. Ankney. 1986. Bioenergetics of egg production by female house sparrows. Auk 103:299–305.

Krementz, D. G., and C. D. Ankney. 1988. Changes in lipid and protein reserves and in diet of breeding house sparrows. Canadian Journal of Zoology 66:950–956.

Krementz, D. G., and C. D. Ankney. 1995. Changes in total body calcium and diet of breeding house sparrows. Journal of Avian Biology 26:162–167.

Kricher, J. C. 1983. Correlation between house finch increase and house sparrow decline. American Birds 37:358–360.

Krogstad, S., B.-E. Sæther, and E. J. Solberg. 1996. Environmental and genetic determinants of reproduction in the house sparrow: a transplant experiment. Journal of Evolutionary Biology 9:979–991.

Kruger, C. 1944. En Undersógelse af Graaspurvens (*Passer d. domesticus*) og Skovspurvens (*Passer m. montanus*) Traek. Dansk Ornithologisk Forenings Tidsskrift 38:105–114.

Kruszewicz, A. G. 1989. Blood parasites from passerine birds collected in northern India. International Studies on Sparrows 16:29–31.

Kruszewicz, A. 1991. The effect of Isospora lacazei on the development of nestling house sparrow (*Passer domesticus*). Preliminary report. Pp. 171–171 in Nestling Mortality of Granivorous Birds due to Microorganisms and Toxic Substances (J. Pinowski, B. P. Kavanaugh, and W. Gorski, eds.), PWN–Polish Scientific Press, Warsaw.

Kruszewicz, A. G. 1994. Factors affecting moult in the house sparrow *Passer domesticus*. Journal für Ornithologie 135(Sonderheft):54.

Kruszewicz, A. G. 1995a. The epizootic role of the house sparrow (*Passer domesticus*) and tree sparrow (*Passer montanus*). Literature review. Pp. 339–351 in Nestling Mortality of Granivorous Birds due to Microorganisms and Toxic Substances: Synthesis (J. Pinowski, B. P. Kavanaugh, and B. Pinowska, eds.). PWN–Polish Scientific Press, Warsaw.

Kruszewicz, A. G. 1995b. The occurrence of Isospora lacazei (Coccidia: Eimeridae) and its influence on nestling growth in house sparrows (*Passer domesticus*) and tree sparrows (*Passer montanus*). Pp. 291–305 in Nestling Mortality of Granivorous Birds due to Microorganisms and Toxic Substances: Synthesis (J. Pinowski, B. P. Kavanaugh, and B. Pinowska, eds.). PWN–Polish Scientific Press, Warsaw.

Kruszewicz, A. G., A. H. Kruszewicz, R. Pawiak, and M. Mazurkiewicz. 1995. Bacteria in house sparrow (*Passer domesticus*) and tree sparrow (*Passer montanus*) nestlings: occurrence and influence on growth and mortality. Pp. 267–282 in Nestling Mortality of Granivorous Birds due to Microorganisms and Toxic Substances: Synthesis (J. Pinowski, B. P. Kavanaugh, and B. Pinowska, eds.). PWN–Polish Scientific Press, Warsaw.

Kruszewicz, A. G., J. Pinowski, A. H. Kruszewicz, M. Mazurkiewicz, R. Pawiak, and E. Małyszko. 1995. Occurrence of fungi in house sparrow (*Passer domesticus*) and tree sparrow (*Passer montanus*) nestlings. Pp. 283–290 in Nestling Mortality of Granivorous Birds due to Microorganisms and Toxic Substances: Synthesis (J. Pinowski, B. P. Kavanaugh, and B. Pinowska, eds.). PWN–Polish Scientific Press, Warsaw.

Kuenzel, W. J., and A. Van Tienhoven. 1982. Nomenclature and location of avian hypothalamic nuclei and associated circumventricular organs. Journal of Comparative Neurology 206:293–313.

Kuhn, T. S. 1970. The structure of scientific revolutions. 2nd edition. University of Chicago Press, Chicago, Illinois.

Kulczycki, A., and M. Mazur-Gierasinska. 1968. Nesting of house sparrow *Passer domesticus* (Linnaeus, 1758). Acta Zoologica Cracoviensia 13:231–250.

Kumar, M. 1995. Spermatogenesis in the testis of house sparrow, *Passer domesticus*: histological observations. Pavo 33:1–4.
Kumudanathan, K., N. Shivanarayan, and A. Banu. 1983. Breeding biology of house sparrow *Passer domesticus* at Rajendranagar, Hyderabad (A. P.). Pavo 21:1–11.
Kutzscher, C.. 1993. Ein interessanter Flohfund am Haussperling (*Passer domesticus* L.). Entomologische Nachrichten und Berichte 37:138–139.
Lack, D. 1940. Variation in the introduced English sparrow. Condor 42:239–241.
Lack, D. 1947. The significance of clutch-size. Ibis 89:302–352.
Lack, D. 1954a. The Natural Regulation of Animal Numbers. Clarendon Press, Oxford, UK.
Lack, D. 1954b. Visible migration in S. E. England, 1952. British Birds 47:1–15.
Lack, D., and E. Lack. 1951. The breeding biology of the swift *Apus apus*. Ibis 93:501–546.
Łącki, A. 1962. [Observations of the biology of clutches of house-sparrow, *Passer domesticus* (L.).] Acta Ornithologica 6:195–207. (In Polish, English summary).
Lainson, R. 1959. Atoxoplasma garnham, 1950, as a synonym for Lankesterella labbé, 1899. Its life cycle in the English sparrow (*Passer domesticus domesticus*, Linn.). Journal of Protozoology 6:360–371.
Laitman, R. S., and F. W. Turek. 1979. The effect of pinealectomy on entrainment of the locomotor activity rhythm in sparrows maintained on various short days. Journal of Comparative Physiology B 134:339–343.
Lal, P., and J. P. Thapliyal. 1982. Role of thyroid in the response of bill pigmentation to male hormone of the house sparrow, *Passer domesticus*. General and Comparative Endocrinology 48:135–142.
Lande, R., and S. J. Arnold. 1983. The measurement of selection on correlated characters. Evolution 37:1210–1226.
Lazarus, J., and M. Symonds. 1992. Contrasting effects of protective and obstructive cover on avian vigilance. Animal Behaviour 43:519–521.
Leck, C. F. 1973. A house sparrow roost in Lima, Peru. Auk 90:888.
Lehto, H. 1993. House sparrow X tree sparrow hybrids in Finland. Dutch Birding 15:264–265.
Lendvai, Á., Z. Barta, A. Liker, and V. Bókony. 2004. The effect of energy reserves on social foraging: hungry sparrows scrounge more. Proceedings of the Royal Society of London B 271:2467–2472.
Lenz, M. 1990. The breeding bird communities of three Canberra suburbs. Emu 90:145–153.
Lepage, D., and C. M. Francis. 2002. Do feeder counts reliably indicate bird population changes? 21 years of winter bird counts in Ontario, Canada. Condor 104: 255–270.
Lepczyk, C. A, and W. H. Karasov. 2000. Effect of ephemeral food restriction on growth of house sparrows. Auk 117:164–174.
Lesher, S. W., and S. C. Kendeigh. 1941. Effect of photoperiod on molting of feathers. Wilson Bulletin 53:169–180.

Lessells, C. M., and P. T. Boag. 1987. Unrepeatable repeatabilities: a common mistake. Auk 104:116–121.

Lever, C. 1987. Naturalized Birds of the World. John Wiley & Sons, New York.

Lifjeld, J. T. 1994. Do female house sparrows copulate with extra-pair males to enhance their fertility? Journal of Avian Biology 25:75–76.

Liker, A., and Z. Barta. 2001. Male badge size predicts dominance against females in house sparrows. Condor 103:151–157.

Liker, A., and Z. Barta. 2002. The effects of dominance on social foraging tactic use in house sparrows. Behaviour 139:1061–1076.

Lima, S. L. 1987. Distance to cover, visual obstructions, and vigilance in house sparrows. Behaviour 102:231–238.

Lindell, C. 1996. Patterns of nest usurpation: when should species converge on nest niches? Condor 98:464–473.

Literak, I., J. Pinowski, M. Anger, Z. Juricova, H. Kyu-Hwang, and J. Romanowski. 1997. Toxoplasma gondii antibodies in house sparrows (*Passer domesticus*) and tree sparrows (*P. montanus*). Avian Pathology 26:823–827.

Literak, I., K. Sedlak, Z. Juricova, and I. Pavlasek. 1999. Experimental toxoplasmosis in house sparrows (*Passer domesticus*). Avian Pathology 28:363–368.

Lockie, J. D. 1952. A comparison of some aspects of the retinae of the Manx shearwater, fulmar petrel, and house sparrow. Quarterly Journal of Microscopical Science 93:347–356.

Lockley, A. K. 1992. The position of the hybrid zone between the house sparrow *Passer domesticus domesticus* and the Italian sparrow *Passer domesticus italiae* in the Alpes Maritimes. Journal für Ornithologie 133:77–82.

Lockley, A. K. 1996. Changes in the position of the hybrid zone between the house sparrow *Passer domesticus domesticus* and the Italian sparrow *Passer domesticus italiae* in the Alpes Maritimes. Journal für Ornithologie 137:243–248.

Lofts, B., and A. J. Marshall. 1956. The effects of prolactin administration on the internal rhythm of reproduction in male birds. Journal of Endocrinology 13:101–106.

Lofts, B., R. K. Murton, and R. J. P. Thearle. 1973. The effects of testosterone propionate and gonadotropins on the bill pigmentation and testes of the house sparrow (*Passer domesticus*). General and Comparative Endocrinology 21:202–209.

Löhrl, H. 1963. Zur Hohenverbreitung einiger Vogel in den Alpen. Journal für Ornithologie 104:62–68.

Löhrl, H., and R. Bohringer. 1957. Untersuchungen an einer sudwestdeutschen Populationdes Haussperlings (*Passer d. domesticus*). Journal für Ornithologie 98:229–240.

Loisel, G. 1900. Etudes sur la spermatogenese chez le moineau domestique. Journal de l'Anatomie et de la Physiologie Normales et Pathologoqies de l'Homme et Animaux 36:160–185.

Long, J. L. 1981. Introduced Birds of the World. Universe Books, New York.

Lord, R. D. Jr. 1956. A comparative study of the eyes of some falconiform and passeriform birds. American Midland Naturalist 56:325–344.

Lord, R. D., C. H. Calisher, W. A. Chappell, W. R. Metzger, and G. W. Fischer. 1974. Urban St. Louis encephalitis surveillance through wild birds. American Journal of Epidemiology 99:360–363.
Lord, R. D., C. H. Calisher, and W. P. Doughty. 1974. Assessment of bird involvement in three urban St. Louis encephalitis epidemics. American Journal of Epidemiology 99:364–367.
Lord, R. D., T. H. Work, P. H. Coleman, and J. G. Johnston, Jr. 1973. Virological studies of avian hosts in the Houston epidemic of St. Louis encephalitis, 1964. American Journal of Tropical Medicine and Hygiene 22:662–671.
Lo Valvo, F., and G. Lo Verde. 1987. Studio Variabilita' fenotipica delle Popolazioni Italiane di Passere e loro Posizione Tassonomica. Rivista Italiana di Ornitologia, Milano 57:97–110.
Lowther, P. E. 1977a. Bilateral size dimorphism in house sparrow gynandromorrphs. Auk 94:377–380.
Lowther, P. E. 1977b. Selection intensity in North American house sparrows. Evolution 31:649–656.
Lowther, P. E. 1979a. Growth and dispersal of nestling house sparrows: sexual differences. Inland Bird Banding 51:23–29.
Lowther, P. E. 1979b. Overlap of house sparrow broods in the same nest. Bird-Banding 50:160–162.
Lowther, P. E. 1979c. The nesting biology of house sparrows in Kansas. Kansas Ornithological Society Bulletin 30:23–28.
Lowther, P. E. 1983. Breeding biology of house sparrows: intercolony variation. Occasional Papers of the Museum of Natural History, University of Kansas 107:1–17.
Lowther, P. E. 1985. Nest mortality in house sparrows. Kansas Ornithological Society Bulletin 36:27–32.
Lowther, P. E. 1988. Spotting pattern of the last laid egg of the house sparrow. Journal of Field Ornithology 59:51–54.
Lowther, P. E. 1990. Breeding biology of house sparrows: patterns of intra-clutch variation in egg size. Pp. 138–149 in Granivorous Birds in the Agricultural Landscape (J. Pinowski and J. D. Summers-Smith, eds.). PWN–Polish Scientific Publishers, Warsaw.
Lowther, P. E. 1996. Breeding biology of house sparrows: observations of a suburban colony. Meadowlark 5:2–7.
Lowther, P. E., and C. L. Cink. 1992. House sparrow. No. 12 in The Birds of North America (A. Poole, P. Stettenheim, and F. Gill, eds.). Philadelphia Academy of Sciences, Philadelphia.
Lu, J., and V. M. Cassone. 1993a. Daily melatonin administration synchronizes circadian patterns of basal metabolism and behavior in pinealectomized house sparrows, *Passer domesticus*. Journal of Comparative Physiology A 173:775–782.
Lu, J., and V. M. Cassone. 1993b. Pineal regulation of circadian rhythms of 2-deoxyl[^{14}C]glucose uptake and 2[^{125}I]iodomelatonin binding in the visual system of the house sparrow, *Passer domesticus*. Journal of Comparative Physiology A 173:765–774.

Ludwig, G. V., R. S. Cook, R. G. McLean, and D. B. Francy. 1986. Viremic enhancement due to transovarially acquired antibodies to St. Louis encephalitis virus in birds. Journal of Wildlife Diseases 22:326–334.

Lumsden, H. G. 1989. Test of nest box preferences of eastern bluebirds, *Sialia sialis*, and tree swallows, *Tachycineta bicolor*. Canadian Field-Naturalist 103:595–597.

Lund, Hj. M.-K. 1956. [The house-sparrow in north Norway, its history, methods of dispersal and notes on its life-history.] Dansk Ornitologisk Forenings Tidsskrift 50:67–76. (In Norwegian, English summary).

Lundstrom, J. O., and B. Niklasson. 1996. Ockelbo virus (Togaviridae: Alphavirus) neutralizing antibodies in experimentally infected Swedish birds. Journal of Wildlife Diseases 32:87–93.

Lundstrom, J. O., M. J. Turell, and B. Niklasson. 1992. Antibodies to Ockelbo virus in three orders of birds (Anseriformes, Galliformes and Passeriformes) in Sweden. Journal of Wildlife Diseases 28:144–147.

Luniak, M. 1996. Inventory of the avifauna of Warsaw: species composition, abundance and habitat distribution. Acta Ornithologica 31:67–80.

Luniak, M., A. Haman, P. Kozłowski, and T. Mizera. 1992. Wyniki leg¢w ptak¢w gniezdzacych sie w skrzynkach w parkach miejskich Warszawy I Poznania. Acta Ornithologica 27:49–63.

Lutz, P. L., I. S. Longmuir, and K. Schmidt-Nielsen. 1974. Oxygen affinity of bird blood. Respiration Physiology 20:325–330.

MacDonald, J. W. 1977. Cutaneous salmonellosis in a house sparrow. Bird Study 25:59.

MacDonald, J. W., and L. W. Cornelius. 1969. Salmonellosis in wild birds. British Birds 62:28–30.

Machacek, P. 1977. Ektoparaziti vrabce domaciho [*Passer domesticus* (L.)] a vrabce polniho [*Passer montanus* (L.)]. Scripta Fac. Sci. Nat. Ujep Brunensis, Biologia 2 7:71–86.

Macke, T. 1965. Gelungene Kreuzung von *Passer domesticus* und *hispaniolensis*. Journal für Ornithologie 106:461–462.

Mackowicz, R., J. Pinowski, and M. Wieloch. 1970. Biomass production by house sparrow (*Passer d. domesticus* L.) and tree sparrow (*Passer m. montanus* L.) populations in Poland. Ekologia Polska 18:465–501.

Macmillan, B. W. H. 1981. Food of house sparrows and greenfinches in a mixed farming district, Hawke's Bay, New Zealand. New Zealand Journal of Zoology 8:93–104.

Macmillan, B. W. H., and B. J. Pollock. 1985. Food of nestling house sparrows (*Passer domesticus*) in mixed farmland of Hawke's Bay, New Zealand. New Zealand Journal of Zoology 12:307–317.

Madej, C. W., and K. Clay. 1991. Avian seed preference and weight loss experiments: the effect of fungal endophyte-infected tall fescue seeds. Oecologia (Berlin) 88:296–302.

Mahabal, A., and D. B. Bastawade. 1985. Population ecology and communal roosting behaviour of pariah kite *Milvus migrans govinda* in Pune (Maharashtra). Journal of the Bombay Natural History Society 82:337–346.

Maina, J. N. 1984. Morphometrics of the avian lung. 3. The structural design of the passerine lung. Respiration Physiology 55:291–307.

Maldonado, A., A. Arenas, M. C. Tarradas, J. Carranza, I. Luque, A. Miranda, and A. Perea. 1994. Prevalence of antibodies to avian paramyxoviruses 1, 2 and 3 in wild and domestic birds in southern Spain. Avian Pathology 23:145–152.

Małyszko, E., J. Pinowski, S. Kozłowski, B. Bernacka, W. Pepiński, and A. Kruszewicz. 1991. Auto- and allochtonous flora and fauna of the intestinal tract of *Passer domesticus* and *Passer montanus* nestlings. Pp. 129–137 in Nestling Mortality of Granivorous Birds due to Microorganisms and Toxic Substances (J. Pinowski, B. P. Kavanaugh, and W. Gorski, eds.). PWN–Polish Scientific Press, Warsaw.

Manganaro, A. 1997. Dati sull'alimentazione del Guto comune, *Asio otus*, nella Laguna di Orbetello (Grosseto, Italia centrale). Rivista italiana di Ornitologia, Milano 67:151–157.

Manganaro, A., L. Ranazzi, R. R. Anazzi, and A. Sorace. 1990. La dieta dell'Alloco, *Strix aluco*, nel parco di Villa Dona Pamphili (Roma). Rivista Italiana di Ornitologia, Milano 60:37–52.

Manly, B. F. J. 1976. Some examples of double exponential fitness functions. Heredity 36:229–234.

Mann, G. S. 1986. House sparrows (*Passer domesticus* [Linn.]) causing severe damage to peach buds in residential areas at Ludhiana, Punjab. Tropical Pest Management 32:43.

Manwell, C., and C. M. A. Baker. 1975. Molecular genetics of avian proteins. XIII. Protein polymorphism in three species of Australian passerines. Australian Journal of Biological Science 28:545–557.

Manwell, R. D. 1957. Blood parasitism in the English sparrow, with certain biological implications. Journal of Parasitology 43:428–433.

Marchant, J. H., R. Hudson, S. P. Carter, and P. Whittington. 1990. Population trends of British breeding birds. British Trust for Ornithology, Tring, Hertfordshire.

Marcos, J., and J. S. Monros. 1994. Variacion intrappuesta del peso del huevo del Gorrion Comun en el naranjal valenciano. Actas de las XII Jornadas Ornitologicas Espanolas 267–272.

Markus, M. B. 1964. Premaxillae of the fossil *Passer predomesticus* Tchernov and the extant South African Passerinae. Ostrich 35:245–246.

Marquez, J., and E. J. Braun. 2002. Muscle fiber typing of the distal thoracic limb of *Passer domesticus*. Federation of American Societies for Experimental Biology Journal 16:A397.

Marshall, A. J. 1952. The interstitial cycle in relation to autumn and winter sexual behaviour in birds. Proceedings of the Zoological Society of London 121:727–740.

Martin, L. B. II, M. Pless, J. Svoboda, and M. Wikelski. 2004. Immune activity in temperate and tropical house sparrows: a common-garden experiment. Ecology 85:2323–2331.

Martin, L. B. II, A. Scheuerlein, and M. Wikelski. 2003. Immune activity elevates energy expenditure of house sparrows: a link between direct and indirect costs. Proceedings of the Royal Society of London B 270:153–158.

Mason, P. 1986. Brood parasitism in a host generalist, the shiny cowbird. I. The quality of different species as hosts. Auk 103:52–60.

Massa, B. 1989. Comments on *Passer italiae* (Vieillot 1817). Bulletin of the British Ornithologists' Club 109:196–198.

Master, T. L. 1979. An incident of blue jay predation on a house sparrow. Wilson Bulletin 91:470.

Mathew, K. L. 1985. Sequence of primary, secondary and rectrix moult in the house sparrow, *Passer domesticus*. Pavo 23:53–62.

Mathew, K. L., and C. Lukose. 1995. Five-striped squirrel *Funambulus pennanti* (Wroughton) feeding on fledgling house sparrow *Passer domesticus*. Journal of the Bombay Natural History Society 92:256.

Mathew, K. L., and R. M. Naik. 1986. Interrelation between moulting and breeding in a tropical population of the house sparrow *Passer domesticus*. Ibis 128:260–265.

Mathew, K. L., and R. M. Naik. 1993. Seasonal variation in diet of house sparrow nestlings under tropical regime. Pavo 31:27–34.

Mathew, K. L., and R. M. Naik. 1994. Seasonal variation in the energy reserve of nestlings of house sparrow *Passer domesticus*. Pavo 32:73–80.

Mathew, K. L., and R. M. Naik. 1998. A comparative study of the breeding biology of the house sparrow in rural and urban habitats in tropics. Pavo 36:19–26.

Mathew, K. L., and V. V. Padmavat. 1985. Effect of captivity on moult of the house sparrow, *Passer domesticus*. Pavo 23:63–76.

Matuhin, A. 1994. Polygyny, infanticide and fledging cleptoparasitism in the Indian sparrow in southern Kazakhstan. Journal für Ornithologie 135 (Sonderheft): 124.

Mayes, W. E. 1927. House-sparrows' winter nest-building. British Birds 20:273–274.

Mayr, E. 1940. Speciation phenomena in birds. American Naturalist 74:249–278.

Mayr, E. 1949. Enigmatic sparrows. Ibis 91:304–306.

Mayr, E. 1956. Geographical character gradients and climatic adaptation. Evolution 10:105–108.

Mayr, E. 1963. Animal Species and Evolution. Belknap Press, Cambridge, Massachusetts.

Mayr, E. 1982. The Growth of Biological Thought: Diversity, Evolution and Inheritance. Belknap Press, Cambridge, Massachusetts.

Mazuc, J., C. Bonneaud, O. Chastel, and G. Sorci. 2003. Social environment affects female and egg testosterone levels in the house sparrow (*Passer domesticus*). Ecology Letters 6:1084–1090.

Mazuc, J., O. Chastel, and G. Sorci. 2003. No evidence for differential maternal allocation to offspring in the house sparrow (*Passer domesticus*). Behavioral Ecology 14:340–346.

McCanch, N. V. 1992. A gynandromorphic house sparrow. British Birds 85:673–674.
McClure, H. E.. 1962. Ten years and 10,000 birds. Bird-Banding 33:1–21, 69–84.
McGillivray, W. B. 1978. House sparrows nesting near a Swainson's hawk nest. Canadian Field-Naturalist 92:201–202.
McGillivray, W. B. 1980a. Communal nesting in the house sparrow. Journal of Field Ornithology 51:571–572.
McGillivray, W. B. 1980b. Nest grouping and productivity in the house sparrow. Auk 97:396–399.
McGillivray, W. B. 1981. Climatic influences on productivity in the house sparrow. Wilson Bulletin 93:196–206.
McGillivray, W. B. 1983. Intraseasonal reproductive costs for the house sparrow (*Passer domesticus*). Auk 100:25–32.
McGillivray, W. B., and R. F. Johnston. 1987. Differences in sexual size dimorphism and body proportions between adult and subadult house sparrows in North America. Auk 104:681–687.
McGillivray, W. B., and E. C. Murphy. 1984. Sexual differences in longevity of house sparrows at Calgary, Alberta. Wilson Bulletin 96:456–458.
McGraw, K. J., J. Dale, and E. A. Mackillop. 2003. Social environment during molt and the expression of melanin-based plumage pigmentation in male house sparrows (*Passer domesticus*). Behavioral Ecology and Sociobiology 53:116–122.
McGroarty, D. L., and R. C. Dobson. 1974. Ectoparasite populations on house sparrows in northwestern Indiana. American Midland Naturlaist 91:479–486.
McInnes, P. F., D. E. Andersen, D. J. Hoff, M. J. Hooper, and L. L. Kinkel. 1996. Monitoring exposure of nestling songbirds to agricultural application of an organophosphorus insecticide using cholinesterase activity. Environmental Toxicology and Chemistry 15:544–552.
McLean, I. C. 1988. Breeding behaviour of the long-tailed cuckoo on Little Barrier Island. Notornis 35:89–98.
McLean, R. G., G. Frier, G. L. Parham, D. B. Francy, T. P. Monath, E. G. Campos, A. Therrien, J. Kerschner, and C. H. Calisher. 1985. Investigations of the vertebrate hosts of eastern equine encephalitis during an epizootic in Michigan, 1980. American Journal of Tropical Medicine and Hygiene 34:1190–1202.
McLean, R. G., L. J. Kirk, R. B. Shriner, and M. Townsend. 1993. Avian hosts of St. Louis encephalitis virus in Pine Bluff, Arkansas, 1991. American Journal of Tropical Medicine and Hygiene 49:46–52.
McLean, R. G., J. Mullenix, J. Kerschner, and J. Hamm. 1983. The house sparrow (*Passer domesticus*) as sentinel for St. Louis encephalitis virus. American Journal of Tropical Medicine and Hygiene 32:1120–1129.
McLean, R. G., J. P. Webb, E. G. Campos, J. Gruwell, D. B. Francy, D. Womeldorf, C. M. Myers, T. H. Work, and M. Jozan. 1988. Antibody prevalence of St. Louis encephalitis virus in avian hosts in Los Angeles, California, 1986. Journal of the American Mosquito Control Association 4:524–528.
McLeod, B. J., and E. G. Thompson. 2002. Predation on house sparrows (*Passer*

domesticus) and hedge sparrows (*Prunella modularis*) by brushtail possums (*Trichosurus vulpecula*) in captivity. Notornis 49:95–99.

McMillan, J. P., J. A. Elliott, and M. Menaker. 1975a. On the role of eyes and brain photoreceptors in the sparrow: arrhythmicity in constant light. Journal of Comparative Physiology 102:263–268.

McMillan, J. P., J. A. Elliott, and M. Menaker. 1975b. On the role of eyes and brain photoreceptors in the sparrow: Aschoff's rule. Journal of Comparative Physiology 102:257–262.

McMillan, J. P., H. C. Keatts, and M. Menaker. 1975. On the role of eyes and brain photoreceptors in the sparrow: entrainment to light cycles. Journal of Comparative Physiology 102:251–256.

McMillan, J. P., H. A. Underwood, J. A. Elliott, M. H. Stetson, and M. Menaker. 1975. Extraretinal light perception in the sparrow. IV. Further evidence that the eyes do not participate in photoperiodic photoreception. Journal of Comparative Physiology 97:205–213.

McVean, A., and P. Haddlesey. 1980. Vigilance schedules among house sparrows *Passer domesticus*. Ibis 122:533–536.

Medeiros, F. 1995. Morphometric variation of the house sparrow in the Azores. Boletim do Museu Municipal do Funchal (Historia Natural) Sup. No. 4:421–431.

Medeiros, F. M. 1997a. Genetic variation in the Azorean populations of the house sparrow (*Passer domesticus*) relative to the original populations from the mainland. International Studies on Sparrows 24:3–17.

Medeiros, F. M. 1997b. Population structure, abandance and philopatry of the house sparrow (*Passer domesticus*) in the Azores. International Studies on Sparrows 22–23:23–30.

Medeiros, F. M. 1998. Population dynamics of the house sparrow: an approach. Boletim de Museu Municipal do Funchal (Historia Natural) Supplement 5:219–224.

Meeke, C. A., and P. A. Melrose. 1994. Seasonal changes in the morphology of gonadotropin-releasing hormone (GnRH) neurons in the house sparrow. Society of Neuroscience Abstracts 20:90.

Mehrotra, S. N. 1951. Diurnal mitosis in the testes of Indian birds. Proceedings of the National Academy of Sciences, India 21B:33–38.

Meier, A. H., and J. W. Dusseau. 1973. Daily entrainment of the photoinducible phases for photostimulation of the reproductive system in the sparrows, *Zonotrichia albicollis* and *Passer domesticus*. Biology of Reproduction 8:400–410.

Meier, A. H., D. M. Martin, and R. MacGregor, III. 1971. Temporal synergism of corticosterone and prolactin controlling gonadal growth in sparrows. Science 173:1240–1242.

Meier, A. H., and A. C. Russo. 1985. Circadian organization of the avian annual cycle. Pp. 303–343 in Current Ornithology, vol. 2 (R. F. Johnston, ed.). Plenum Press, New York.

Meinertzhagen, R. 1955. The speed and altitude of bird flight (with notes on other animals). Ibis 97:81–117.

Meise, W. 1936. Zur Systematik und Verbreitungsgeschichte der Haus- und Weidensperlinge, *Passer domesticus* (L.) und *hispaniolensis* (T.). Jounal für Ornithologie 84:631–672.

Meise, W. 1937. Sperlingsmischgebiete und Artentstehung durch Kreuzung. Forschungen und Fortschrifte 13:286–287.

Menaker, M. 1965. Circadian rhythms and photoperiodism in *Passer domesticus*. Pp. 385–395 in Circadian Clocks (J. Aschoff, ed.). North-Holland Publishing, Amsterdam.

Menaker, M. 1968a. Extraretinal light perception in the sparrow. I. Entrainment of the biological clock. Proceedings of the National Academy of Sciences, USA 59:414–421.

Menaker, M. 1968b. Light reception by extra-retinal receptors in the brain of the sparrow. Pp. 299–300. Proceedings 76th Annual Convention, American Psychological Association.

Menaker, M. 1971. Rhythms, reproduction, and photoreception. Biology of Reproduction 4:295–308.

Menaker, M. 1982. The search for principles of physiological organization in vertebrate circadian systems. Pp. 1–12 in Vertebrate Circadian Systems (J. Aschoff, S. Daan, and G. Groos, eds.). Springer-Verlag, Berlin.

Menaker, M., and A. Eskin. 1966. Entrainment of circadian rhythm by sound in *Passer domesticus*. Science 155:1579–1581.

Menaker, M., and H. Keatts. 1968. Extraretinal light perception in the sparrow. II. Photoperiodic stimulation of testis growth. Proceedings of the National Academy of Sciences, USA 60:146–151.

Menaker, M., R. Roberts, J. Elliott, and H. Underwood. 1970. Extraretinal light perception in the sparrow. III. The eyes do not participate in photoperiodic photoreception. Proceedings of the National Academy of Sciences, USA 67:320–325.

Menaker, M., J. S. Takahashi, and A. Eskin. 1978. The physiology of circadian pacemakers. Annual Review of Physiology 40:501–526.

Menon, C. K., M. B. Jani, and R. V. Shah. 1978. Feather papillae in the incubation patches of house sparrows. Condor 80:101.

Menon, G. K., and B. Pilo. 1983. House sparrows nesting in an unusual two-story nest. Pavo 21:101–103.

Meredev, A. N., and V. I. Golovkova. 1978. [Helminth fauna of the birds of the Murgab River Valley]. Izv. Akademia Nauk Turkmenskoi SSR 1978:38–43. (In Russian, English translation at Alexander Library).

Metzdorf, R., M. Gahr, and L. Fusadi. 1999. Distribution of aromatase, estrogen receptor, and androgen receptor mRNA in the forebrain of songbirds and nonsongbirds. Journal of Comparative Neurology 407:115–129.

Metzmacher, M. 1986. Moineaux domestiques, *Passer domestics*, et espagnols, *Passer hispaniolensis*, dans une region l'ouest Algerien: analyse comparative de leur morphologie externe. Gerfaut 76:317–334.

Micks, D. W. 1949. Malaria in the English sparrow. Journal of Parasitology 35:543–544.

Middleton, J. 1965. Testicular responses of house sparrows and white-crowned sparrows to short daily photoperiods with low intensities of light. Physiological Zoology 38:255–266.
Miller, D. S. 1935. Effects of thyroxin on plumage of the English sparrow, *Passer domesticus* (Linnaeus). Journal of Experimental Zoology 71:293–308.
Miller, D. S. 1939. A study of the physiology of the sparrow thyroid. Journal of Experimental Zoology 80:259–285.
Miller, W. 1970. Factors affecting the status of eastern and mountain bluebirds in southwestern Manitoba. Blue Jay 28:38–46.
Mills, S. H., and J. E. Heath. 1970. Thermoresponsiveness of the preoptic region of the brain in house sparrows. Science 168:1008–1009.
Minock, M. E. 1969. Salinity tolerance and discrimination in house sparrows (*Passer domesticus*). Condor 71:79–80.
Mirza, Z. B. 1972. Study of the fecundity, mortality, numbers, biomass and food of a population of house sparrows in Lahore, Pakistan. Pp. 141–150 in Productivity, Population Dynamics and Systematics of Granivorous Birds (S. C. Kendeigh and J. Pinowski, eds.). PWN–Polish Scientific Publishers, Warsaw.
Mitchell, C. J., and R. O. Hayes. 1973. Breeding house sparrows, *Passer domesticus*, in captivity. Ornithological Monographs 14:39–48.
Mitchell, C. J., R. O. Hayes, P. Holden, and T. B. Hughes, Jr. 1973. Nesting activity of the house sparrow in Hale County, Texas, during 1968. Ornithological Monographs 14:49–59.
Mitchell, C. J., R. O. Hayes, and T. B. Hughes. 1984. Relative abundance of birds along transects in an endemic zone of western equine encephalitis virus activity in west Texas. Bulletin of the Society of Vector Ecology 9:30–36.
Mitchell, C. J., and T. B. Hughes, Jr. 1972. Homing in house sparrows, *Passer domesticus*. Bird-Banding 43:213–214.
Mitschke, A., and S. Baumung. 2001. Brutvogel-Atlas Hamburg. Hamburger Avifaunistische Beiträge 31:299–301.
Mitschke, A., H.-H. Geisler, S. Baumung, and L. Andersen. 1999. Ornithologischer Jahresbericht 1996 und 1997 für das Hamburger Berichtsgebiet. Hamburger avifaunistische Beiträge 30:129–204.
Moens, L., and R. Coessens. 1970. Seasonal variation in the adrenal cortex cells of the house sparrow, *Passer domesticus* (L.), with special reference to a possible zonation. General and Comparative Endocrinology 15:95–100.
Møller, A. P. 1987a. House sparrow, *Passer domesticus*, communal displays. Animal Behaviour 35:203–210.
Møller, A. P. 1987b. Variation in badge size in male house sparrows *Passer domesticus*: evidence for status signaling. Animal Behaviour 35:1637–1644.
Møller, A. P. 1988. Badge size in the house sparrow *Passer domesticus*: Effects of intra- and intersexual selection. Behavioral Ecology and Sociobiology 22:373–378.
Møller, A. P. 1989. Natural and sexual selection on a plumage signal of status and

on morphology in house sparrows, *Passer domesticus*. Journal of Evolutionary Biology 2:125–140.

Møller, A. P. 1990. Sexual behavior is related to badge size in the house sparrow *Passer domesticus*. Behavioral Ecology and Sociobiology 27:23–29.

Møller, A. P. 1991. Clutch size, predation, and distribution of avian unequal competitors in a patchy environment. Ecology 72:1336–1349.

Møller, A. P. 1992. Frequency of female copulations with multiple males and sexual selection. American Naturalist 139:1089–1101.

Møller, A. P. 1994. Directional selection on directional asymmetry: testes size and secondary sexual characters in birds. Proceedings of the Royal Society of London B 258:147–151.

Møller, A. P., and J. Erritzoe. 1988. Badge, body and testes size in house sparrows *Passer domesticus*. Ornis Scandinavia 19:72–73.

Møller, A. P., and J. Erritzoe. 1992. Acquisition of breeding coloration depends on badge size in male house sparrows *Passer domesticus*. Behavioral Ecology and Sociobiology 31:217–277.

Møller, A. P., and J. Erritzoe. 2000. Predation against birds with low immunocompetence. Oecologia 122:500–504.

Møller, A. P., R. T. Kimball, and J. Errtzoe. 1996. Sexual ornamentation, condition, and immune defence in the house sparrow *Passer domesticus*. Behavioral Ecology and Sociobiology 39:317–322.

Moltoni, E. 1954. Ulteriore notizia di passero (*Passer italiae*) divoratore di lucertole. Rivista Italiana di Ornitologia 24:217.

Molzahn, A. 1997. Spätbrut vom Haussperling (*Passer domesticus*) 1997 in Hamburg. Hamburger avifaunistische Beiträge 29:182.

Moore, R. G., and L. M. Hanks. 2000. Avian predation on the evergreen bagworm (Lepidoptera: Psychidae). Prodeedings of the Entomological Society of Washington 102:350–352.

Moreau, R. E. 1931. An Egyptian sparrow-roost. Ibis 13:204–208.

Moreno-Rueda, G., and M. Soler. 2001. Reconocimento de huevos en el gorrión común *Passer domesticus*, una especie con parasitismo de cría intraespecífico. Ardeola 48:225–231.

Moreno-Rueda, G., and M. Soler. 2002. Cría en cautividad del gorrión común *Passer domesticus*. Ardeola 49:11–17.

Morris, J. G. 1969. The control of feral pigeons and sparrows associated with intensive animal production. Australian Journal of Science 32:9–14.

Morsy, T. A., S. A. M. Mazyad, and M. S. Younis. 1999. Feather and nest mites of two common resident birds in two ecologically different Egyptian governorates. Journal of the Egyptian Society of Parasitology 29:417–430.

Mostini, L. 1987. Nidificazione tardiva di passera d'Italia, *Passer domesticus italiae*. Rivista Italiana di Ornithologia, Milano 57:149–150.

Moulton, M. P., and D. K. Ferris. 1991. Summer diets of some introduced Hawaiian finches. Wilson Bulletin 103:286–292.

Mountford, M. D. 1968. The significance of litter-size. Journal of Animal Ecology 37:363–367.

Mueller, N. S. 1977. Control of sex differences in the plumage of the house sparrow, *Passer domesticus*. Journal of Experimental Zoology 202:45–48.

Mueller, N. S. 1986. Abrupt change in food preference in fledgling house sparrows. Journal of the Elishu Mitchell Scientific Society 102:7–9.

Mullerova-Franekova, M., and L. Kocian. 1995. Structure and dynamics of breeding bird communities in three parks in Bratislava. Folia Zoologica 44:111–121.

Murakami, N., T. Kawano, K. Nakahara, T. Nasu, and K. Shiota. 2001. Effect of melatonin on circadian rhythm, locomotor activity and body temperature in the intact house sparrow, Japanese quail and owl. Brain Research 889:220–224.

Murakami, N., H. Nakamura, R. Nishi, N. Marumoto, and T. Nasu. 1994. Comparison of circadian oscillation of melatonin release in pineal cells of house sparrow, pigeon and Japanese quail, using cell perfusion systems. Brain Research 651:209–214.

Murphy, E. C. 1978a. Breeding ecology of house sparrows: spatial variation. Condor 80:180–193.

Murphy, E. C. 1978b. Seasonal variation in reproductive output of house sparrows: the determination of clutch size. Ecology 59:1189–1199.

Murphy, E. C. 1980. Body size of house sparrows: reproductive and survival correlates. Acta XVII Congressus Internationalis Ornithologici 1155–1161.

Murphy, E. C. 1985. Bergmann's rule, seasonality, and geographic variation in body size of house sparrows. Evolution 39:1327–1334.

Murphy, E. C., and E. Haukioja. 1986. Clutch-size in nidicolous birds. Current Ornithology 4:141–180.

Murphy, M. E., and J. R. King. 1982. Semi-synthetic diets as a tool for nutritional ecology. Auk 99:165–167.

Murphy, M. E., and J. R. King. 1991. Ptilochronology: a critical evaluation of assumptions and utility. Auk 108:695–704.

Murton, R. K., B. Lofts, and A. H. Orr. 1970. The significance of circadian based photosensitivity in the house sparrow *Passer domesticus*. Ibis 112:448–456.

Murton, R. K., B. Lofts, and N. J. Westwood. 1970. Manipulation of photo- refractoriness in the house sparrow *Passer domesticus* by circadian light regimes. General and Comparative Endocrinology 14:107–113.

Murton, R. K., and N. J. Westwood. 1974. An investigation of photo-refractoriness in the house sparrow by artificial photoperiods. Ibis 116:298–313.

Myrcha, A., J. Pinowski, and T. Tomek. 1972. Energy balance of nestlings of tree sparrows, *Passer m. montanus* (L.), and house sparrows, *Passer d. domesticus* (L.). Pp. 59–83 in Productivity, Population Dynamics and Systematics of Granivorous Birds (S. C. Kendeigh and J. Pinowski, eds.). PWN– Polish Scientific Publishers, Warsaw.

Naik, R. M. 1974. Recent studies on the granivorous birds in India. International Studies on Sparrows 7:21–25.

Naik, R. M., and L. Mistry. 1972. Breeding season and reproductive rate of *Passer domesticus* (L.) in Baroda, India. Pp. 133–140 in Productivity, Population Dynamics and Systematics of Granivorous Birds (S. C. Kendeigh and J. Pinowski, eds.). PWN–Polish Scientific Publishers, Warsaw.

Naik, R. M., and L. Mistry. 1980. Breeding season in a tropical population of the house sparrow. Journal of the Bombay Natural History Society 75:1118–1142.

Nandy, P. K., and C. K. Manna. 1996. Changes in some testicular biochemical components during peak breeding and nonbreeding phases of two common Indian pest birds, the rose-ringed parakeet (*Psittacula krameri*) and the house sparrow (*Passer domesticus*). Pavo 34:65–70.

Nankinov, D. N. 1984. Nesting habits of the tree sparrow, *Passer montanus* (L.), in Bulgaria. International Studies on Sparrows 11:47–70.

Navarro, C., A. Marzal, F. De Lope, and A. P. Møller. 2003. Dynamics of an immune response in house sparrows *Passer domesticus* in relation to time of day, body condition and blood parasite infection. Oikos 101:291–298.

Neff, J. A. 1959. Controlling English sparrows by means of poisoned baits. Technical Release No. 9–59. Bureau of Sport Fisheries and Wildlife, Denver, Colorado.

Neff, M. 1973. Untersuchungen über das embryonale und postembryonale Organwachstum bei Vogelarten mit verschiedenem Ontogenesemodus. Revue Suisse de Zoologie 79:1471–1597.

Nemec, F., D. Cyprich, and M. Krumpal. 1995. The occurrence of fleas (Siphonaptera) in the nests of the house sparrow (*Passer domesticus* L., 1758) and the tree sparrow (*Passer montanus* L., 1758) in Plzen (western Czech Republic). International Studies on Sparrows 20–21:21–25.

Nero, R. W. 1951. Pattern and rate of cranial "ossification" in the house sparrow. Wilson Bulletin 63:84–88.

Nero, R. W. 1987. House sparrow killed by red squirrel. Blue Jay 45:180–181.

Neumann, K., and J. H. Wetton. 1996. Highly polymorphic microsatellites in the house sparrow *Passer domesticus*. Molecular Ecology 5:307–309.

Newman, J. A., and T. Caraco. 1989. Co-operative and non-co-operative bases of food-calling. Journal of Theoretical Biology 141:197–209.

Newton, I. 1986. The Sparrowhawk (396 p.). T. & A. D. Poyser, Calton, Staffordshire, UK.

Newton, I. 1992. Experiments on the limitation of bird numbers by territorial behaviour. Biological Reviews 67:129–173.

Newton, I. 1998. Population Limitation in Birds. Academic Press, San Diego.

Nhlane, M. E. D. 2000. The breeding biology of the house sparrow *Passer domesticus* at Blantyre, Malawi. Ostrich 71:80–82.

Nice, L. B., M. M. Nice, and R. M. Kraft. 1935. Erythrocites and hemoglobin in the blood of some American birds. Wilson Bulletin 47:120–124.

Nicholls, T. J., A. R. Goldsmith, and A. Dawson. 1988. Photorefractoriness in birds and comparison with mammals. Physiological Reviews 68:133–176.

Nichols, J. T. 1934. Sex ratio in the house sparrow. Bird-Banding 5:188–189.

Nichols, J. T. 1935. Seasonal and individual variations in house sparrows. Bird-Banding 6:11–15.

Nicolai, B. 1971. Beitrag zur Nistweise des Haussperlings (*Passer domesticus*). Beiträge zur Vogelkunde 17:78–79.

Niethammer, G. 1953. Gewicht und Flugellange beim Haussperling (*Passer d. domesticus*). Journal für Ornithologie 94:282–289.

Niethammer, G. 1958. Das Mischgebiet zwischen *Passer d. domesticus* und *Passer d. italiae* in Süd-Tirol. Journal für Ornithologie 99:431–437.

Niethammer, G. 1969. Vergleich der Renthendorfer Haussperlinge von heute mit einer von C. L. Brehm vor 110 Jahren gesammelten Seite. Journal für Ornithologie 110:265–268.

Niethammer, G. 1971. Some problems connected with the house sparrow's colonisation of the world. Ostrich Supplement 8:445–458.

Niethammer, G., and K. Bauer. 1960. Das Mischgebiet zwischen *Passer d. domesticus* und *Passer d. italiae* im Tessin. Ornithologische Beobachter 5:241–242.

Niewiadowska, A., M. Barkowska, Z. Juricova, I. Literak, J. Pinowski, B. Pinowska, and J. Romanowski. 1998. Chlorinated aromatic hydrocarbons in the brains of sparrows (*Passer domesticus*, *P. montanus*) from suburban areas of Warsaw. Proceedings of the Latvian Academy of Sciences, Section B 52(Supplement):48–52.

Niles, D. M. 1973. Geographic and seasonal variation in the occurrence of incompletely pneumatized skulls in the house sparrow. Condor 75:354–356.

Nir, Y., R. Goldwasser, Y. Lasowski, and A. Avivi. 1967. Isolation of arboviruses from wild birds in Israel. American Journal of Epidemiology 86:372–378.

Nir, Y., Y. Lasowski, A. Avivi, and R. Goldwasser. 1969. Survey of antibodies to arboviruses in the serum of various animals in Israel during 1965–1966. American Journal of Tropical Medicine and Hygiene 18:416–422.

Nivison, J. J. 1978. The vocal behavior and displays of the house sparrow, *Passer domesticus* L., in the United States. Ph. D. Dissertation, Wayne State University, Detroit, Michigan.

Nordmeyer, A., H. Oekle, and E. Plagemann. 1970. Biometrische Untersuchungen an nordwestdeutschen Haussperling-Populationen. International Studies on Sparrows 4:50–54.

Nordmeyer, A., H. Oekle, and E. Plagemann. 1972. Biometrical studies of house sparrow, *Passer domesticus* (L.), populations in northwestern Germany. Pp. 337–350 in Productivity, Population Dynamics and Systematics of Granivorous Birds (S. C. Kendeigh and J. Pinowski, eds.). PWN–Polish Scientific Publishers, Warsaw.

Norris, R. A. 1963. A preliminary study of avian blood groups with special reference to the Passeriformes. Bulletin of Tall Timbers Research Station 4:1–71.

North, C. A. 1968. A study of house sparrow populations and their movements in the vicinity of Stillwater, Oklahoma. Ph. D. Dissertation, Oklahoma State University, Stillwater.

North, C. A. 1969. Preliminary report on house sparrow reproductivity and population fluctuations in Coldspring, Wisconsin, 1969. International Studies on Sparrows 3:43–66.

North, C A. 1972. Population dynamics of the house sparrow, *Passer domesticus* (L.), in Wisconsin, USA. Pp. 195–212 in Productivity, Population Dynamics and Sys-

tematics of Granivorous Birds (S. C. Kendeigh and J. Pinowski, eds.). PWN–Polish Scientific Publishers, Warsaw.
North, C. A. 1973. Movement patterns of the house sparrow in Oklahoma. Ornithological Monographs 14:79–91.
North, C. A. 1980. Attentiveness and nesting behavior of the male and female house sparrow (*Passer domesticus*) in Wisconsin. Acta XVII Congressus Internationalis Ornithologici 1122–1128.
Novikov, B. G. 1946. Intracellular determination of the dimorphism of color of the plumage in *Passer domesticus* (L.). Comptes Rendus (Doklady) de l'Academis des Sciencesde l'URSS 52:453–455.
Novotny, I. 1970. Breeding bionomy, growth and development of young house parrow (*Passer domesticus*, Linné 1758). Acta Scientiarum Naturalium Brno 4(7):1–57.
Nowikow, B. G. 1935. Die Analyse der sekindären Geschlechtsunterschiede in der Gefiederfärbung bei den Sperlingsvögeln (Passeres). I. Biologisches Zentralblatt 55:285–293.
Nyland, K. B., M. P. Lombardo, and P. A. Thorpe. 2003. Left-sided directional bias of cloacal contacts during house sparrow copulations. Wilson Bulletin 115:470–473.
O'Connor, R. J. 1975a. Growth and metabolism in nestling passerines. Symposium of the Zoological Society of London 35:277–306.
O'Connor, R. J. 1975b. Initial size and subsequent growth in passerine nestlings. Bird-Banding 46:329–340.
O'Connor, R. J. 1975c. Nestling thermolysis and developmental change in body temperature. Comparative Biochemistry and Physiology 52A:419–422.
O'Connor, R. J. 1975d. The influence of brood size upon metabolic rate and body temperature in nestling blue tits *Parus caeruleus* and house sparrows *Passer domesticus*. Journal of Zoology, London 175:391–403.
O'Connor, R. J. 1977. Differential growth and body composition in altricial passerines. Ibis 119:147–166.
O'Donald, P. 1973. A further analysis of Bumpus' data: the intensity of natural selection. Evolution 27:398–404.
Odum, E. P. 1935. The heart rate of small birds. Science 101:153–154.
Oehmke, H.-J., J. Priedkalns, M. Vaupel-Von Harnack, and A. Aksche. 1969. Fluoreszenz- und elektronemikroskopische Untersuchungen am Zwischenhern Hypophysensystem von *Passer domesticus*. Zeitschrift für Zellforschung und mikroskopische Anatomie 95:109–133.
O'Hara, R. J. 1991. Phylogeny and classification of birds: a study in molecular evolution (book review). Auk 108:990–994.
Ojanen, M., and M. Tynjala. 1983. [Occurrence of house sparrow (*Passer domesticus*) at Tauvo Bird Observatory.] Aureola 8:113–118. (In Finnish, English summary).
Oksche, A., and M. Vaupeel-Von Harnack. 1966. Elektronenmikroskopische Untersuchungen zur Frage der Sinneszellen im Pinealorgan der Vögel. Zeitschrift für Zellforschung und mikroskopische. Anatomie 69:41–60.

Oliphant, L. W., and S. McTaggart. 1977. Prey utilized by urban merlins. Canadian Field-Naturalist 91:190–192.

Opdam, P. 1979. Feeding ecology of a sparrowhawk population (*Accipiter nisus*). Ardea 66:137–155.

Owen, O. S. 1957. Observations on territorial behavior in the English sparrow. Bulletin of the Ecological Society of America 38:101–102.

Owens, I. P. F., and I. R. Hartley. 1991. "Trojan sparrows": evolutionary consequences of dishonest invasion for the badge-of-status model. American Naturalist 138:1187–1205.

Packard, G. C. 1967a. House sparrows: evolution of populations from the Great Plains and Colorado Rockies. Systematic Zoology 16:73–89.

Packard, G. C. 1967b. Seasonal variation in bill length of house sparrows. Wilson Bulletin 79:345–346.

Paganelli, C. V., A. Olszowka, and A. Ar. 1974. The avian egg: surface area, volume, and density. Condor 76:319–325.

Palokangas, R., I. Nuuja, and J. Koivusaari. 1975. Seasonal changes in some thermoregulatory variables of the house sparrow (*Passer domesticus* L.). Comparative Biochemistry and Physiology 52A:299–304.

Palomeque, J., L. Palacios, and J. Planas. 1980. Comparative respiratory functions of blood in some passeriform birds. Comparative Biochemistry and Physiology 66A:619–624.

Panicker, R. G., and H. R. Acharya. 1994. Histomorphology of proventriculus, ventriculus and intestine of developing and adult house sparrow, *Passer domesticus*. Pavo 32:145–152.

Panov, E. N., and S. I. Radjabli. 1972. [Relationships between the house sparrow (*Passer domesticus* L.) and the spanish sparrow (*Passer hispaniolensis* Temm.) in Tajikistan and the possible isolating mechanisms.] Problemy Evolyutsii 2:263–275. (In Russian, English summary).

Paperna, I., and H. Gill. 2003. Schizogenic stages of Haemoproteus from Wenyon's Baghdad sparrows are also found in *Passer domesticus biblicus* in Israel. Parasitology Research 91:486–490.

Paradis, E., S. R. Baillie, W. J. Sutherland, and R. D. Gregory. 1998. Patterns of natal and breeding dispersal in birds. Journal of Animal Ecology 67:518–536.

Paralkar, V. K. 1995. A common garden snake (*Calotes versicolor*) killing an adult house sparrow (*Passer domesticus*). Journal of the Bombay Natural History Society 92:426.

Parfitt, D. E. 1984. Relationship of morphological plant characteristics of sunflower to bird feeding. Canadian Journal of Plant Science 64:37–42.

Parker, J. W. 1982. Additional records of house sparrows nesting on raptor nests. Southwestern Naturalist 27:240–241.

Parkin, D. T. 1987. Evolutionary genetics of house sparrows. Pp. 381–406 in Avian Genetics: A Population and Ecological Approach (F. Cooke and P. A. Buckley, eds.). Academic Press, London.

Parkin, D. T. 1988. Genetic variation in the house sparrow (*Passer domesticus*). Acta XIX Congressus Internationalis Ornithologici 1652–1657.

Parkin, D. T., and S. R. Cole. 1984. Genetic differentiation and rates of evolution in some introduced populations of the house sparrow, *Passer domesticus* in Australia and New Zealand. Heredity 54:15–23.

Parkin, D. T., and S. R. Cole. 1985. Genetic variation in the house sparrow, *Passer domesticus*, in the East Midlands of England. Biological Journal of the Linnean Society 23:287–301.

Parr, S. J. 1985. The breeding ecology and diet of the hobby *Falco subbuteo* in southern England. Ibis 127:60–73.

Parrish, J. W., and M. M. Mote. 1984. Serum chemical levels in captive female house sparrows. Wilson Bulletin 96:138–141.

Parrish, J. W., J. A. Ptacek, and K. L. Will. 1984. The detection of near-ultraviolet light by nonmigratory and migratory birds. Auk 101:53–58.

Parshad, R. K., and S. Bhutani. 1987. Ovarian steroidogenesis during the reproductive cycle of house sparrow (*Passer domesticus*). Pavo 25:79–84.

Paterson, A. M., R. L. Palma, and R. D. Gray. 1999. How frequently do avian lice miss the boat? Implications for coevolutionary studies. Systematic Biology 48:214–223.

Pawiak, R., M. Mazurkiewicz, J. Molenda, J. Pinowski, and A. Wieliczko. 1991. The occurrence of Escherichia coli strains pathogenic to humans and animals in the eggs and nestlings of *Passer* spp. Pp. 139–151 in Nestling Mortality of Granivorous Birds due to Microorganisms and Toxic Substances (J. Pinowski, B. P. Kavanaugh, and W. Gorski, eds.). PWN–Polish Scientific Press, Warsaw.

Pawlina, I. M., and G. Proulx. 1996. Study of house sparrow (*Passer domesticus*) feeding preference to natural color and guard coat blue coated seeds. Crop Protection 15:143–146.

Payne, R. B. 1961. Growth rate of the lens of the eye of house sparrows. Condor 63:358–360.

Payne, R. B. 1972. Mechanisms and control of molt. Pp. 103–155 in Avian Biology, vol. II (D. S. Farner and J. R. King, eds.). Academic Press, New York.

Paz Nava, M., J. P. Veiga, and M. Puerta. 2001. White blood cell counts in house sparrows (*Passer domesticus*) before and after moult and after testosterone treatment. Canadian Journal of Zoology 79:145–148.

Pearse, E. A., and G. Smith. 1990. The Times Books World Weather Guide. Random House, New York.

Pearse, T. 1940. Polygamy in the English sparrow. Condor 42:124–125.

Péczely, P. 1976. Etude cirannuelle de la fonction corticosurrenalienne chez les expeces de passereaux migrants et non migrants. General and Comparative Endocrinology 30:1–11.

Peirce, M. A. 1980. Haematozoa of British birds: post-mortem and clinical findings. Bulletin of the British Ornithologists' Club 100:158–160.

Peirce, M. A. 1981. Distribution and host-parasite check-list of the haematozoa of birds in Western Europe. Journal of Natural History 15:419–458.

Penhallurick, R. D. 1993. House sparrows nesting in cliffs in Scilly. British Birds 86:435–436.
Pennycott, T. W., R. E. Cough, A. M. Wood, and H. W. Reid. 2002. Encephalitis of unknown aetiology in young starlings (*Sturnus vulgaris*) and house sparrows (*Passer domesticus*). Veterinary Record 151:213–214.
Perkins, L. E. L., and D. E. Swayne. 2003. Varied pathogenicity of a Hong Kong-origin H5N1 avian influenza virus in four passerine species and budgerigars. Veterinary Pathology 40:14–24.
Perrins, C. M., and D. Moss. 1975. Reproductive rates in the great tit. Journal of Animal Ecology 44:695–706.
Peterjohn, B. G., J. R. Sauer, and W. A. Link. 1994. The 1992 and 1993 summary of the North American Breeding Bird Survey. Bird Populations 2:46–61.
Petretti, F. 1991. Italian sparrows (*Passer italiae*) breeding in black kite (*Milvus migrans*) nests. Avocetta 15:15–17.
Pfeiffer, C. A. 1947. Gonadotrophic effects of exogenous sex hormones on the testes of sparrows. Endocrinology 41:92–104.
Pfeiffer, C. A., and A. Kirschbaum. 1941. Secretion of androgen by the sparrow ovary following stimulation with pregnant mare serum. Yale Journal of Biology and Medicine 13:315–322.
Pfeiffer, C. A., and A. Kirschbaum. 1943. Relation of interstitial cell hyperplasia to secretion of male hormone in the sparrow. Anatomical Record 85:211–227.
Pfeiffer, C. A., A. Kirschbaum, and W. U. Gardner. 1940. Relation of estrogen to ossification and the levels of serum calcium and lipoid in the English sparrow, *Passer domesticus*. Yale Journal of Biology and Medicine 43:279–284.
Philipson, W. R. 1938. House-sparrows excavating nest-hole. British Birds 32:17.
Phillips, J. C. 1915. Notes on American and Old World English sparrows. Auk 32:51–59.
Phillis, W. 1972. Seasonal abundance of *Dermanyssus hirundinis* and *D. americanus* (Mesostigmata: Dermanyssidae) in the nests of house sparrow. Journal of Medical Entomology 9:111–112.
Piechocki, R. 1954. Statisctische Festellungen an 20 000 Sperlingen (*Passer d. domesticus*). Journal für Ornithologie 95:297–305.
Pilgrim, R. L. C., and R. L. Palma. 1982. A list of the chewing lice (Insecta: Mallophaga) from birds in New Zealand. Notornis 29(Supplement):1–32.
Pinowska, B. 1975. Food of female house sparrows (*Passer domesticus* L.) in relation to stages of the nesting cycle. Polish Ecological Studies 1:211–225.
Pinowska, B. 1979. The effect of energy and building resources of females on the production of house sparrow (*Passer domesticus* [L.]) populations. Ekologia Polska 27:383–396.
Pinowska, B., G. Chylinski, and B. Gonder. 1976. Studies on the transmitting of Salmonellae by house sparrows (*Passer domesticus* L.) in the region of Zulawy. Polish Ecological Studies 2:113–121.
Pinowska, B., and K. Krasnicki. 1985a. Changes in the content of magnesium, cop-

per, calcium, nitrogen and phosphorus in female house sparrows during the breeding cycle. Ardea 73:175–182.

Pinowska, B., and K. Krasnicki. 1985b. Quantity of gastroliths and magnesium and calcium contents in the body of female house sparrows during egg-laying period. Zeszyty Naukowe Filli UW, 48, Biologia 10:125–130.

Pinowska, B., and J. Pinowski. 1977. Fecunity, mortality, numbers and biomass dynamics of a population of the house sparrow, *Passer domesticus* (L.). International Studies on Sparrows 10:26–41.

Pinowski, J. 1967. Introduction. International Studies on Sparrows 1:5–8.

Pinowski, J., M. Barkowska, A. H. Kruszewicz, and A. G. Kruszewicz. 1994. The causes of the mortality of eggs and nestlings of *Passer* spp. Journal of Biosciences 19:441–451.

Pinowski, J., M. Barkowska, and B. Pinowska. 1995. Interaction of microorganisms, heavy metals and pesticides from liver and their effect on the development and mortality of *Passer* spp. nestlings. Pp. 307–338 in in Nestling Mortality of Granivorous Birds due to Microorganisms and Toxic Substances: Synthesis (J. Pinowski, B. P. Kavanaugh, and B. Pinowska, eds.). PWN–Polish Scientific Press, Warsaw.

Pinowski, J., B. P. Kavanagh, and W. Gorski (eds.). 1991. Nestling Mortality of Granivorous Birds due to Microorganisms and Toxic Substances. PWN–Polish Scientific Publishers, Warsaw.

Pinowski, J., B. P. Kavanagh, and B. Pinowska (eds.). 1995. Nestling Mortality of Granivorous Birds due to Microorganisms and Toxic Substances: Synthesis. PWN–Polish Scientific Publishers, Warsaw.

Pinowski, J., and S. C. Kendeigh (eds.). 1977. Granivorous Birds in Ecosystems. Cambridge University Press, Cambridge, UK.

Pinowski, J., A. Łukowski, R. Szczepanowski, A. Haman, and P. Kaminski. 1995. Accumulation of organochlorine insecticides and polychlorinated biphenyls in eggs and nestlings (*Passer* spp.) and their possible health effects. Pp. 223–249 in Nestling Mortality of Granivorous Birds due to Microorganisms and Toxic Substances: Synthesis (J. Pinowski, B. P. Kavanaugh, and B. Pinowska, eds.). PWN–Polish Scientific Press, Warsaw.

Pinowski, J., M. Mazurkiewicz, E. Małyszko, R. Pawiak, S. Kozłowski, A. Kruszewicz and P. Indykiewicz. 1988. The effect of micro-organisms on embryo and nestling mortality in house sparrow (*Passer domesticus*) and tree sparrow (*Passer montanus*). Prod. Int. 100 Do-G Meeting. Current Topics Avian Biology 273–282.

Pinowski, J., M. Mazurkiewicz, R. Pawiak, and A. Haman. 1995. Lethal effects of microorganisms on embryos of house sparrows (*Passer domesticus*) and tree sparrows (*Passer montanus*). Pp. 251–265 in Nestling Mortality of Granivorous Birds due to Microorganisms and Toxic Substances: Synthesis (J. Pinowski, B. P. Kavanaugh, and B. Pinowska, eds.). PWN–Polish Scientific Press, Warsaw.

Pinowski, J., A. Niewiadowska, Z. Juricova, I. Literak, and J. Romanowski. 1999.

Chlorinated aromatic hydrocarbons in the brains and lipids of sparrows (*Passer domesticus* and *Passer montanus*) from rural and suburban areas near Warsaw. Environmental Contamination and Toxicology 63:736–743.

Pinowski, J., B. Pinowska, and J. Truszkowski. 1972. Escape from the nest and brood desertion by the tree sparrow, *Passer m. montanus* (L.), the house sparrow, *Passer d. domesticus* (L.), and the great tit, *Parus m. major* L. Pp. 397–405 in Productivity, Population Dynamics and Systematics of Granivorous Birds (S. C. Kendeigh and J. Pinowski, eds.). PWN–Polish Scientific Publishers, Warsaw.

Pinowski, J., J. Romanowski, M. Barkowska, K. Sawicka-Kapusta, P. Kaminski, and A. G. Kruszewicz. 1993. Lead and cadmium in relation to body weight and mortality of the house sparrow *Passer domesticus* and tree sparrow *Passer montanus* nestlings. Acta Ornithologica 28:63–68.

Pinowski, J., K. Sawicka-Kapusta, M. Barkowska, J. Romanowski, B. Pinowska, and P. Kaminski. 1995. Heavy metals in nestlings of *Passer* spp. in urban and suburban environments. Archiwum Ochrony Srodowiska 2:73–82.

Pinowski, J., and J. D. Summers-Smith (eds.). 1990. Granivorous Birds in the Agricultural Landscape. PWN–Polish Scientific Publishers, Warsaw.

Pinowski, J., and M. Wieloch. 1972. Energy flow through nestlings and biomass production of house sparrow, *Passer d. domesticus* (L.), and tree sparrow, *Passer m. montanus* (L.), populations in Poland. Pp. 151–163 in Productivity, Population Dynamics and Systematics of Granivorous Birds (S. C. Kendeigh and J. Pinowski, eds.). PWN–Polish Scientific Publishers, Warsaw.

Pitman, C. R. 1961. Unusual nesting behaviour of the house sparrow, *Passer domesticus* (L.). Bulletin of the British Ornithologists' Club 81:148–149.

Pitts, T. D. 1979. Nesting habits of rural and suburban house sparrows in northwest Tennessee. Journal of the Tennessee Academy of Science 54:145–148.

Pochop, P. A., and R. J. Johnson. 1993. Pentagon milk-box nest box. Journal of Field Ornithology 64:239–243.

Pocock. T. N. 1966. Contributions to the osteology of African birds. Ostrich Supplement 6:83–94.

Pogue, D. W., and W. A. Carter. 1995. Breeding biology of secondary cavity-nesting birds in Oklahoma. Southwestern Naturalist 40:167–173.

Polani, A., A. R. Goldsmith, and M. R. Evans. 2000. Ectoparasites of house sparrows (*Passer domesticus*): an experimental test of the immuno-competence handicap hypothesis and a new model. Behavioral Ecology and Sociobiology 47:230–242.

Poltz, J., and J. Jacob. 1974. Burzeldrusensekrete Ammern (Emberizidae), Finken (Fringillidae) und Webern (Ploceidae). Journal für Ornithologie 115:119–127.

Popp, J. W. 1988. Selection of horse dung pats by foraging house sparrows. Journal of Field Ornithology 59:385–388.

Potter, E. F. 1970. Anting in wild birds, its frequency and probable purpose. Auk 87:692–713.

Potter, E. F., and D. C. Hauser. 1974. Relationship of anting and sunbathing to molting in wild birds. Auk 91:537–563.

Prasad, R., and S. C. Patnaik. 1977. Karyotypes of five passerine birds belonging to family Ploceidae. Caryologia 30:361–368.

Preiser, F. 1957. Untersuchungen uber die Ortsstetigkeit und Wanderung der Sperlinge (*Passer domesticus domesticus* L.) als Grundlage fur die Bekampfung. Ph. D. Thesis, Landwirtschaftlichen Hochschule Hohenheim, Stuttgart-Hohenheim, Germany.

Priedkalns, J., and A. Oksche. 1969. Ultrastructure of synaptic terminals in nucleus infundibularis and nucleus supraopticus of *Passer domesticus*. Zeitschrift für Zellforschung und Mikroskopische Anatomie 98:135–147.

Prys-Jones, R. P., L. Sshifferli, and D. W. McDonald. 1974. The use of an emetic in obtaining food samples from passerines. Ibis 116:90–94.

Przygodda, W. 1960. Berinungen von Haussperlingen in Bonn. Ornithologische Mitteilungen 12:21–25.

Puerta, M., M. P. Nava, C. Venero, and J. P. Veiga. 1995. Hematology and plasma chemistry of house sparrows (*Passer domesticus*) along the summer months and after testosterone treatment. Comparative Biochemistry and Physiology 110A: 303–307.

Pugesek, B. H., and A. Tomer. 1996. The Bumpus house sparrow data: a reanalysis using structural equation models. Evolutionary Ecology 10:387–404.

Pulliam, H. R. 1973. On the advantages of flocking. Journal of Theoretical Biology 38:419–422.

Pulliam, H. R. 1985. Foraging efficiency, resource partitioning, and the coexistence of sparrow species. Ecology 66:1829–1836.

Pulliam, H. R. 1988. Sources, sinks, and population regulation. American Naturalist 132:652–661.

Pulliam, H. R., G. H. Pyke, and T. Caraco. 1982. The scanning behavior of juncos: a game-theoretical approach. Journal of Theoretical Biology 95:89–103.

Purcell, K. L., and J. Verner. 1999. Nest predators of open and cavity nesting birds in oak woodlands. Wilson Bulletin 111:251–256.

Pyke, G. H., H. R. Pulliam, and E. L. Charnov. 1977. Optimal foraging: a selective review of theory and tests. Quarterly Review of Biology 52:137–154.

Quay, W. B. 1967. Comparative survey of the anal glands of birds. Auk 84:379–389.

Quay, W. B. 1987. Spontaneous continuous release of spermatozoa and its predawn surge in male passerines. Gamete Research 16:83–92.

Quay, W. B., and A. Renzoni. 1963. Comparative and experimental studies of pineal structure and cytology in passeriform birds. Rivista di Biologia 56:363–407.

Quay, W. B., and A. Renzoni. 1966. Studies on the "commisuro-pineal neurosecretory cells" of birds. Rivista di Biologia 59:231–266.

Quiring, D. P., and P. H. Bade. 1943a. Metabolism of the English sparrow. Growth 7:309–315.

Quiring, D. P., and P. H. Bade. 1943b. Organ and gland weights of the English sparrow. Growth 7:299–307.

Rademacher, B. 1951. Beringungsversuche uber die Ortstreue der Sperlinge (*Passer d. domesticus* L. und *Passer m. montanus* L.). Zeitschrift für Pflanzenkrankheiten und Pflanzenschutz 58:416–426.

Radke, W. J., and M. J. Frydendall. 1974. A survey of emetics for use in stomach contents recovery in the house sparrow. American Midland Naturalist 92:164–172.

Radunzel, L. A., D. M. Muschitz, V. M. Bauldry, and P. Arcese. 1997. A long-term study of the breeding success of eastern bluebirds by year and cavity type. Journal of Field Ornithology 68:7–18.

Radwan, J. 1993. Are dull birds still dull in UV? Acta Ornithologica 27:125–130.

Ralph, C. L., and D. C. Dawson. 1968. Failure of the pineal body of two species of birds (*Coturnix coturnix japonica* and *Passer domesticus*) to show electrical responses to illumination. Experientia 24:147–148.

Rana, B. D. 1989a. Population ecology of *Passer domesticus* in the Indian arid zone. International Studies on Sparrows 16:1–7.

Rana, B. D. 1989b. Some observations on neophobic behaviour among house sparrow, Passer domesticus. Pavo 27:35–38.

Rana, B. D. 1991. Comparative repellency of five fungicide toxicants to the house sparrow (*Passer domesticus indicus*). Pavo 29:55–59.

Rana, B. D. 1992. Evaluation of certain stupefying chemicals against house sparrow (*Passer domesticus*). Pavo 30:61–66.

Rana, B. D., and M. Idris. 1986. Population structure of the house sparrow, *Passer domesticus indicus* in western Rajastham Desert. Pavo 24:91–96.

Rana, D. B., and M. Idris. 1987. Food habits of the house sparrow *Passer domesticus indicus* in an arid environment. Japanese Journal of Ornithology 35:125–128.

Rana, B. D., and M. Idris. 1989. Breeding biology of *Passer domesticus indicus* in the Indian desert. International Studies on Sparrows 16:18–28.

Rana, B. D., and M. Idris. 1991a. Evaluation of bird depredations to standing crops in an arid environment. Pavo 29:61–66.

Rana, B. D., and M. Idris. 1991b. The effect of predation on egg and nestling mortality among *Streptopelia decaocto* and *Passer domesticus indicus* in an arid environment. Pp. 55–60 in Nestling Mortality of Granivorous Birds due to Microorganisms and Toxic Substances (J. Pinowski, B. P. Kavanaugh, and W. Gorski, eds.). PWN–Polish Scientific Press, Warsaw.

Rand, A. L. 1956. Changes in English sparrow population densities. Wilson Bulletin 68:69–70.

Rappole, J. H., and Z. Hubalek. 2003. Migrant birds and West Nile virus. Journal of Applied Microbiology 94:47–58.

Ravikumar, G., B. Ssenthilkimaran, P. D. Tewary, and A. K. Goel. 1995. Circadian aspect of photoperiodic time measurement in a female house sparrow, *Passer domesticus*. Journal of Biological Rhythms 10:319–323.

Ravikumar, G., and P. D. Tewary. 1990. Photorefractoriness and its termination in the subtropical house sparrow, *Passer domesticus* involvement of circadian rhythm. Chronobiology International 7:187–191.

Ravikumar, G., and P. D. Tewary. 1991. Endogenous circadian rhythm in the photoperiodic ovarian response of the subtropical sparrow, *Passer domesticus*. Physiology and Behavior 50:L637–L639.

Ray-Chaudhuri, R. 1976. Cytotaxonomy and chromosome evolution in Passeriformes (Aves): a comparative karotype study of seventeen species. Zeitschrift für Zoologische Systematik und Evolutions-Forschung 14:299–320.

Reebs, S. G. 1989. Acoustical entrainment of circadian activity rhythms in house sparrows: constant light is not necessary. Ethology 82:172–181.

Reeves, W. C., R. E. Bellamy, A. F. Gelb, and R. P. Scrivani. 1964. Analysis of the circumstances to abortion of a western equine encephalitis epidemic. American Journal of Hygiene 80:205–220.

Reeves, W. C., R. E. Bellamy, and R. P. Scrivani. 1958. Relationship of mosquito vectors to winter survival of encephalitis viruses. I. Under natural conditions. American Journal of Hygiene 67:78–89.

Reeves, W. C., W. McD. Hammon, D. P. Furman, H. E. McClure, and B. Brookman. 1947. Recovery of western equine enceophalomyelitis virus from wild bird mites (*Liponyssus silvarium*) in Kern County, California. Science 105:411–412.

Reeves, W. C., W. McD. Hammon, S. Lazarus, B. Brookman, H. E. McClure, and W. H. Doetschman. 1952. The changing picture of encephalitis in the Yakima Valley, Washington. Journal of Infectious Diseases 90:291–301.

Rehkamper, G., H. D. Frahm, and K. Zilles. 1991. Quantitative development of brain and brain structures in birds (Galliformes and Passeriformes) compared to that in mammals (Insectivores and Primates). Brain Behavior and Evolution 37:125–143.

Rekasi, J. 1968. Data on the food biology of *Passer d. domesticus* (L.). International Studies on Sparrows 2:26–39.

Reyer, H.-U., W. Fischer, P. Steck, T. Nabulon, and P. Kessler. 1998. Sex-specific nest defense in house sparrows (*Passer domesticus*) varies with badge size of males. Behavioral Ecology and Sociobiology 42:93–99.

Reynolds, J., and F. G. Stiles. 1982. Distribucion y densidad de poblaciones del gorrion comun (*Passer domesticus*; Aves: Ploceidae) en Costa Rica. Revista de Biologia Tropical 30:65–71.

Rich, E. L., and L. M. Romero. 2001. Daily and photoperiod variations of basal and stress-induced corticosterone concentrations in house sparrows (*Passer domesticus*). Journal of Comparative Physiology B 171:543–547.

Ricklefs, R. E. 1969. Preliminary models for growth rates in altricial birds. Ecology 50:1031–1039.

Ridgway, R. 1889. The Ornithology of Illinois, vol. I. Natural History Survey of Illinois, Springfield, Illinois.

Riley, G. M. 1936a. Factors affecting the diurnal spermatogenic cycle of the male sparrow (*Passer domesticus*). Anatomical Record 64(Supplement):40.

Riley, G. M. 1936b. Light regulation of sexual activity in the male sparrow (*Passer domesticus*). Proceedings of the Society for Experimental Biology and Medicine 34:331–332.

Riley, G. M. 1937. Experimental studies on spermatogenesis in the house sparrow, *Passer domesticus* (Linnaeus). Anatomical Record 67:327–346.

Riley, G. M. 1938. Cytological studies on spermatogenesis in the house sparrow, *Passer domesticus*. Cytologia 9:165–176.

Riley, G. M. 1940. Light versus activity as a regulator of the sexual cycle in the house sparrow. Wilson Bulletin 52:73–86.

Riley, G. M., and E. Witschi. 1937. Comparative effects of light stimulation and administration of gonadotropic hormones on female sparrows. Anatomical Record 70(Supplement):50.

Riley, G. M., and E. Witschi. 1938. Comparative effects of light stimulation and administration of gonadotropic hormones on female sparrows. Endocrinology 23:618–624.

Ringoen, A. R. 1940. The effects of theelin administration upon the reproductive system of the female English sparrow, *Passer domesticus* (Linnaeus). Journal of Experimental Zoology 83:379–389.

Ringoen, A. R. 1942. Effects of continuous green and red light illumination on gonadal response in the English sparrow, *Passer domesticus* (Linnaeus). American Journal of Anatomy 71:99–116.

Ringoen, A. R., and A. Kirschbaum. 1937a. Correlation between ocular stimulation and spermatogenesis in the English sparrow (*Passer domesticus*). Proceedings of the Society for Experimental Biology and Medicine 36:111–113.

Ringoen, A. R., and A. Kirschbaum. 1937b. Daily light ration and gonadal activity in the English sparrow, *Passer domesticus* Linnaeus. Anatomical Record 67(Supplement):41–42.

Ringoen, A. R., and A. Kirschbaum. 1939. Factors responsible for the sexual cycle in the English sparrow, *Passer domesticus* (Linnaeus): ocular stimulation and spermatogenesis; effect of increased light ration on ovarian development. Journal of Experimental Zoology 80:173–191.

Ringsby, T. H., B.-E. Sæther, R. Altwegg, and E. J. Solberg. 1999. Temporal and spatial variation in survival rates of a house sparrow, *Passer domesticus*, metapopulation. Oikos 85:419–425.

Ringsby, T. H., B.-E. Sæther, and E. J. Solberg. 1998. Factors affecting juvenile survival in house sparrow *Passer domesticus*. Journal of Avian Biology 29:241–247.

Ringsby, T. H., B.-E. Sæther, J. Tufto, H. Jensen, and E. J. Solberg. 2002. Asynchronous spatiotemporal demography of a house sparrow metapopulation in a correlated environment. Ecology 83:561–569.

Rising, J. D. 1972. Age and seasonal variation in dimensions of house sparrows, *Passer domesticus* (L.), from a single population in Kansas. Pp. 327–336 in Productivity, Population Dynamics and Systematics of Granivorous Birds (S. C. Kendeigh and J. Pinowski, eds.). PWN–Polish Scientific Publishers, Warsaw.

Ritchison, G. 1985. Plumage variability and social status in captive male house sparrows. Kentucky Warbler 61:39–42.

Riters, L. V., D. P. Teague, and M. B. Schroeder. 2004. Social status interacts with badge size and neuroendocrine physiology to influence sexual behavior in male house sparrows (*Passer domesticus*). Brain, Behavior and Evolution 63:141–150.

Robbins, C. S. 1973. Introduction, spread, and present abundance of the house sparrow in North America. Ornithological Monographs 14:3–9.

Robinson, R. A., J. D. Wilson, and H. Q. P. Crick. 2001. The importance of arable habitat on farmland birds in grassland landscapes. Journal of Applied Ecology 38:1059–1069.

Rocio, L., I. Acosta, I. Navarrete, and P. N. Gutierrez. 1981. Primera cita en Espana de Capillaria tridens (Nematoda, Trichuridae), parasitio del gorrion comun *Passer domesticus*. Revista Iberica de Parasitologia 41:316–318.

Rodriguez-Estrealla, R., and L. B. R. Rodriguez. 1997. Crested Caracara food habits in the Cape region off Baja California, Mexico. Journal of Raptor Research 31:228–233.

Rohwer, S. 1975. The social significance of avian plumage variability. Evolution 29:593–610.

Romanowski, J., J. Pinowski, K. Sawicka-Kapusta, and T. Wlostowski. 1991. The effect of heavy metals upon development and mortality of *Passer domesticus* and *Passer montanus* nestlings. Preliminary report. Pp. 197–204 in Nestling Mortality of Granivorous Birds due to Microorganisms and Toxic Substances (J. Pinowski, B. P. Kavanaugh, and W. Gorski, eds.). PWN–Polish Scientific Press, Warsaw.

Romero, L. M., and R. C. Romero. 2002. Corticosterone responses in wild birds: the importance of rapid initial sampling. Condor 104:129–135.

Root, T. 1988. Atlas of Wintering North American Birds: an Analysis of Christmas Bird Count Data. University of Chicago Press, Chicago, Illinois.

Rose, L. N. 1983. House sparrows sunning in glass jars. British Birds 76:316.

Rosen, S., A. Hadami, and K. Davidov. 1984. The occurrence of the northern and tropical fowl mites (*Ornithonyssus sylvarum* and *O. bursa*) in house sparrows (*Passer domesticus*) in a poultry farm. Refuah Veterinarith 41:37–39.

Ross, C. C. 1963. Albinism among North American birds. Cassinia 47:2–21.

Rosser, B. W. C., and J. C. George. 1986. The avian pectoralis: histochemical characterization and distribution of muscle fiber types. Canadian Journal of Zoology 64:1174–1185.

Rossetti, K. 1983. House sparrows taking insects from spiders' webs. British Birds 76:412.

Roudneva, L. M. 1970. Contribution a l'analyse du mecanisme de la phase refractaire des gonades chez les oiseaux. Annales d'Endocrinologie, Paris 31:1065–1069.

Rowan, W. 1925. Relation of light to bird migration and developmental changes. Nature 115:494–495.

Rowan, W. 1938. Light and seasonal reproduction in animals. Biological Reviews 13:374–402.

Royall, W. C., Jr. 1969. Trapping house sparrows to protect experimental grain crops. Wildlife Leaflet 484. Fish and Wildlife Service, United States Department of the Interior, Washington, DC.

Ruprecht, A. L. 1967. A hybrid house sparrow × tree sparrow. Bulletin of the British Ornithologists' Club 87:78–81.

Ruprecht, A. L. 1968. [The morphological variability of the *Passer domesticus* (L.) skull in postnatal development.] Acta Ornithologica 18:27–44. (In Polish, English summary).

Rytkonen, S., P. Kuokkanen, M. Hukkanen, and K. Huntala. 1998. Prey selection by sparrowhawks *Accipiter nisus* and characteristics of vulnerable prey. Ornis Fennica 75:77–87.

Saarela, S., J. S. Keith, E. Hohhola, and P. Trayhurn. 1991. Is the "mammalian" brown fat-specific mitochondrial uncoupling protein present in adipose tissues in birds? Comparative Biochemistry and Physiology 100B:45–49.

Sæther, B.-E., T. H. Ringsby, O. Bakke, and E. J. Solberg. 1999. Spatial and temporal variation in demography of a house sparrow metapopulation. Journal of Animal Ecology 68:628–637.

Sage, B. L. 1957. Remarks on the taxonomy, history and distribution of the house sparrow introduced to Australia. Emu 57:349–352.

Sage, B. L. 1963. The incidence of albinism and melanism in British birds. British Birds 56:409–416.

Sahin, R. 1996. Besetung von Nestern in einer neu entstandenen Mehlschwalben-Kolonie (*Delichon urbica*) durch Haussperlinge (*Passer domesticus*) in Samsun/Türkei. Ökologie der Vögel 18:45–54.

Saini, H. K., and M. S. Dhindsa. 1991. Diet of the house sparrow in an intensively cultivated area. Japanese Journal of Ornithology 39:93–100.

Saini, H. K., M. S. Dhindsa, and H. S. Toor. 1989. Morphometric variation in an Indian population of the house sparrow, *Passer domesticus indicus*. Gerfaut 79:69–79.

Saldanha, C. J., N. S. Clayton, and B. A. Schlinger. 1999. Androgen metabolism in the juvenile Oscine forebrain: a cross-species analysis of neural sites implicated in memory function. Journal of Neurobiology 40:397–406.

Saldanha, C. J., P. Popper, P. E. Micevych, and B. A. Schlinger. 1998. The passerine hippocampus is a site of high aromatase: inter- and intraspecies comparisons. Hormones and Behavior 34:85–97.

Sanchez-Aguado, F. J. 1986. Sobre la alimentacion de los Gorriones Molinero y Comun (*Passer montanus* L. y *P. domesticus* L.) en invierno y primavera. Ardeola 33:17–33.

Sanders, C. W., and W. H. Elder. 1976. Oral chemosterilisation of the house sparrow. International Pest Control 18:4–6.

Sandhu, P. S., M. S. Dhindsa, and H. S. Toor. 1985. Evaluation of brodifacoum, an anticoagulant, against house sparrow. Indian Journal of Ecology 12:173–174.

Sano, M. 1990. [First record of the house sparrow in Japan]. Japanese Journal of Ornithology 39:31–33. (In Japanese, English summary).

Sappington, J. N. 1975. Cooperative breeding in the house sparrow (*Passer domesticus*). Ph. D. Dissertation, Mississippi State University.

Sappington, J. N. 1977. Breeding biology of house sparrows in north Mississippi. Wilson Bulletin 89:300–309.

Sarwar, H. A., and K. N. Murty. 1982. Destruction of pearl millet nursery by sparrows *Passer domesticus* (Linnaeus) and its avoidance. Journal of the Bombay Natural History Society 79:200–201.

Sawicka-Kapusta, K., J. Pinowski, M. Barkowska, J. Romanowski, and P. Kaminski. 1995. The concentration of heavy metals (Cd, Fe, Pb and Zn) in the livers of house sparrow (*Passer domesticus*) and tree sparrow (*Passer montanus*) nestlings from parks and suburban areas of Warsaw. Pp. 117–138 in Nestling Mortality of Granivorous Birds due to Microorganisms and Toxic Substances: Synthesis (J. Pinowski, B. P. Kavanagh, and B. Pinowska, eds.). PWN–Polish Scientific Publishers, Warsaw.

Saxena, M., and R. S. Mathur. 1976. Annual male reproductive cycle of house sparrow *Passer domesticus* (Linnaeua). Current Science 45:103–105.

Schafer, E. W., Jr., R. B. Bunton, N. F. Lockyer, and J. W. De Grazio. 1973. Comparative toxicity of seventeen pesticides to the quelea, house sparrow, and red-winged blackbird. Toxicology and Applied Pharmacology 26:154–157.

Scherner, E. R. 1974. Untersuchungen zur popularen Variabilitat des Haussperlings (*Passer domesticus*). Vogelwelt 95:41–60.

Schifferli, L. 1976. Factors affecting weight and condition in the house sparrow particularly when breeding. D. Phil. Thesis, Oxford University, UK.

Schifferli, L. 1977. Bruchstucke von Schneckenhauschen als Calciumquelle für die Bildung des Eischale beim Haussperling *Passer domesticus*. Ornithologische Beobachter 74:71–74.

Schifferli, L. 1978a. Die Rolle des Männchens während der Bebrütung der Eier beim Haussperling *Passer domesticus*. Ornithologische Beobachter 75:44–47.

Schifferli, L. 1978b. Experimental modification of brood size among house sparrows *Passer domesticus*. Ibis 120:365–369.

Schifferli, L. 1979. Warum legen Singvögel (Passeres) ihre Eier am frühen Morgan? Ornithologische Beobachter 76:33–36.

Schifferli, L. 1980a. Changes in the fat reserves in female house sparrows *Passer domesticus* during egg laying. Acta XVII Congressus Internationalis Ornithologici 1129–1135.

Schifferli, L. 1980b. Growth and mortality of male and female nestling house sparrow *Passer domesticus* in England. Avocetta 4:49–62.

Schifferli, L. 1980c. Juvenile house sparrows parasitizing nestling house sparrows. British Birds 73:189–190.

Schifferli, L. 1981. Federgewicht des Haussperlings *Passer domesticus* im Jahresverlauf. Ornithologische Beobachter 78:113–115.

Schifferli, L., and A. Schifferli. 1980. Die Verbreitung des Haussperlings *Passer domesticus domesticus* und des Italiensperlings *Passer domestticus italiae* im Tesson und im Misox. Ornithologischer Beobachter 77:21–26.

Schmidt, E. 1968. Der Haussperling (*Passer domesticus* [L.]) und der Feldsperling (*Passer montanus* [L.]) aus Nahrung der Schleiereule (*Tyto alba* [Scop.]) in Ungarn. International Studies on Sparrows 2:96–101.

Schmidt, G. 1966. Haussperling (*Passer domesticus*) als Felsholenbruter am Nordmeer. Vogelwelt 87:91–92.

Schneider, R. 1995. Der Wanderfalke Falco peregrinus als Brutvogel in der Grossstadt —Neue Chancen für eine vom Aussterben bedrohte Tierart? Ornithologischer Beobachter 92:313–319.

Schnell, G. D. 1965. Recording the flight-speed of birds by Doppler radar. Living Bird 4:79–87.

Schnell, G. D., and J. J. Hellack. 1978. Flight speeds of brown pelicans, chimney swifts, and other birds. Bird-Banding 49:108–112.

Schnetter, M., H. K. Englert, K. Haas, and J. Schneider. 1968. Über eine Salmonellen-Epidemie unter den Singvögeln an den Futterplätzen des Schwarzwaldes im Winter 1966/67. Angewandte Ornitologie 3:89–90.

Schöll, R. W. 1959. Über das Vorkommen von Sperlingen am Brenner-Paß (Tirol). Journal für Ornithologie 100:439–440.

Schöll, R. W. 1960. Die Sperlingsbesiedlung des Pustertales/Südtirol und seiner nördlichen Seitentäler. Anzeiger der Ornitholoischen Gesellschat in Bayen 5:591–596.

Schwabel, H. 1997. The contents of maternal testosterone in house sparrow *Passer domesticus* eggs vary with breeding conditions. Naturwissenschaften 84:406–408.

Schwagmeyer, P. L., and D. W. Mock. 2003. How constantly are good parents good parents? Repeatability of parental care in the house sparrow, *Passer domesticus*. Ethology 109:303–313.

Schwagmeyer, P. L., D. W. Mock, and G. A. Parker. 2002. Biparental care in house sparrows: negotiation or sealed bid? Behavioral Ecology 13:713–721.

Scott, D. M. 1988. House sparrow and chipping sparrow feed the same fledgling brown-headed cowbird. Wilson Bulletin 100:323–324.

Secker, H. L. 1975. Communal display of house sparrow. Emu 75:229–230.

Seel, D. C. 1960. The behaviour of a pair of house sparrows while feeding young. British Birds 53:303–310.

Seel, D. C. 1966. Further observations on the behaviour of a pair of house sparrows rearing young. Bird Study 13:207–209.

Seel, D. C. 1968a. Breeding seasons of the house sparrow and tree sparrow *Passer* spp. at Oxford. Ibis 110:129–144.

Seel, D. C. 1968b. Clutch-size, incubation and hatching success in the house sparrow and tree sparrow *Passer* spp. at Oxford. Ibis 110:270–282.

Seel, D. C. 1969. Food, feeding rates and body temperature in the nestling house sparrow *Passer domesticus* at Oxford. Ibis 111:36–47.

Seel, D. C. 1970. Nestling survival and nestling weights in the house sparrow and tree sparrow *Passer* spp. at Oxford. Ibis 112:1–14.

Seibert, H. C. 1949. Differences between migrant and non-migrant birds in food and water intake at various temperatures and photoperiods. Auk 66:128–153.

Selander, R. K., and R. F. Johnston. 1967. Evolution in the house sparrow. I. Intrapopulation variation in North America. Condor 69:217–238.

Selander, R. K., R. F. Johnston, and T. H. Hamilton. 1964. Colorimetric methods in ornithology. Condor 66:491–495.

Selander, R. K., and S. Y. Yang. 1966. The incubation patch of the house sparrow, *Passer domesticus* Linnaeus. General and Comparative Endocrinology 6:325–333.

Selas, V. 1993. Selection of avian prey by breeding sparrowhawks *Accipiter nisus* in southern Norway: the importance of size and foraging behaviour of prey. Ornis Fennica 70:144–154.

Senar, J. C., and J. L. Copete. 1995. Mediterranean house sparrows (*Passer domesticus*) are not used to freezing temperatures: an analysis of survival rates. Journal of Applied Statistics 22:1069–1074.

Sengupta, S. 1981. Adaptive significance of the use of margosa leaves in nests of house sparrows *Passer domesticus*. Emu 81:114–115.

Sengupta, S., and Shrilata. 1997. House sparrow *Passer domesticus* uses Krrishnachura leaves as an antidote to malarial fever. Emu 97:248–249.

Sergio, F., and G. Bogliani. 1999. European hobby density, nest area occupancy, diet, and productivity in relation to intensive agriculture. Condor 101:806–817.

Sergio, F., and A. Boto. 1999. Nest dispersion, diet, and breeding success of black kites (*Milvus migrans*) in the Italian pre-Alps. Journal of Raptor Research 33:207–217.

Seymour, C., R. G. Lewis, M. Kim, D. F. Gagnon, J. G. Fox, F. E. Dewhirst, and B. J. Paster. 1994. Isolation of Helicobacter strains from wild bird and swine feces. Applied and Environmental Microbiology 60:1025–1028.

Shafi, M. M., L. Hussain, A. A. Khan, S. Ahmed, and M. S. Ahmed. 1984. Laboratory evaluation of 4-aminopyridine against house sparrows (*Passer domesticus*). Tropical Pest Management 30:302–305.

Shah, R. V., G. K. Menon, and M. B. Jani. 1979. Incubation patch formation in the house sparrow: possible role of integumentary glycosaminoglycans. Monitore Zoologico Italiano 13:1–9.

Sharma, S. K. 1995. An unusual nesting site of house sparrow *Passer domesticus* (Linn.). Journal of the Bombay Natural History Society 92:422.

Sharpe, C., D. Ascanio, and R. Restall. 1997. Three species of exotic passerine in Venezuela. Cotinga 7:43–44.

Sharrock, J. T. R. 1976. The Atlas of Breeding Birds of Britain and Ireland. British Trust for Ornithology, Tring, UK.

Shields, G. F. 1982. Comparative avian cytogenetics: a review. Condor 84:45–58.

Shields, W. M. 1977. The social significance of avian winter plumage variability: a comment. Evolution 31:905–907.

Shrubb, M. 1980. Farming influences on the food and hunting of kestrels. Bird Study 27:109–115.

Shugart, G. W. 1988. Uterovaginal sperm-storage glands in sixteen species with comments on morphological differences. Auk 105:379–384.

Sibley, C. G. 1970. A comparative study of the egg-white proteins of passerine birds. Peabdoy Museum of Natural History Bulletin 32:1–131.

Sibley, C. G., and J. E. Ahlquist. 1985. The phylogieny and classification of passerine birdsbased on comparisons of the genetic material, DNA. Acta XVIII Congressus Internationalis Ornithologici 83–121.

Sibley, C. G., and J. E. Ahlquist. 1990. Phylogeny and Classification of Birds: a Study in Molecular Evolution. Pp. 675–683 in Passeridae. Yale University Press, New Haven, Connecticut.

Sibley, C. G., and B. L. Monroe, Jr. 1990. Distribution and Taxonomy of Birds of the World. Yale University Press, New Haven, Connecticut.

Siki, M. 1992. [Researches on the breeding biology of house sparrow (*Passer domesticus* L.)]. Doga, Turkish Journal of Zoology 16:243–247. (In Turkish, English summary).

Simeonov, S. D. 1964. [Über die Nahrung des Haussperlings in der Umgebung von Sofia.] Annuaire de l'Universite de Sofia 61:239–275. (In Bulgarian, German summary).

Simmons, K. E. L. 1951. Display of house-sparrow. British Birds 44:18–19.

Simmons, K. E. L. 1952. Some notes on the behaviour of house-sparrows. British Birds 45:323–325.

Simmons, K. E. L. 1957. The taxonomic significnce of the head-scratching method of birds. Ibis 99:178–181.

Simms, E. A. 1948. Unusual behaviour of house-sparrow. British Birds 41:344.

Simwat, G. S. 1977. Studies on the feeding habits of house sparrow *Passer domesticus* (L.) and its nestling in Punjab. Journal of the Bombay Natural History Society 74:175–179.

Singer, R.,and Y. Yom-Tov. 1988. The breeding biology of the house sparrow *Passer domesticus* in Israel. Ornis Scandinavica 19:139–144.

Singh, J., C. P. Nair, and A. David. 1951. Five years' observation on the incidence of blood Protozoa in house sparrows (*Passer comesticus* Linnaeus) and in pigeons (Columba livia Gmelin) in India. Indian Journal of Malariology 5:229–233.

Siriwardena, G. M., S. R. Baillie, and J. D. Wilson. 1999. Temporal variation in the annual survival rates of six granivorous birds with contrasting population trends. Ibis 141:621–636.

Sitdolphi, R. H. D. 1974a. House sparrow plucking Barbary dove. Notornis 21:283–284.

Sitdolphi, R. H. D. 1974b. The adaptable house sparrow. Notornis 21:88.

Slonaker, J. R. 1918. A physiological study of the anatomy of the eye and its accessory parts of the English sparrow (*Passer domesticus*). Journal of Morphology 31:351–459.

Slonaker, J. R. 1921. The development of the eye and its accessory parts in the English sparrow (*Passer domesticus*). Journal of Morphology 33:263–357.

Smal, C. M. 1987. The diet of the barn owl *Tyto alba* in southern Ireland, with reference to a recently introduced prey species—the Bank Vole *Cleithrionomys glareolus*. Bird Study 34:113–125.

Smith, G. T., E. A. Brenowitz, and G. S. Prins. 1996. Use of PG-21 immuno- cytochemistry to detect androgen receptors in the songbird brain. Journal of Histochemistry and Cytochemistry 44:1075–1080.

Smith, J. P. 1982. Changes in blood levels of thyroid hormones in two species of passerine birds. Condor 84:160–167.

Smith, N. J. H. 1973. House sparrows (*Passer domesticus*) in the Amazon. Condor 75:242–243.

Smith, N. J. H. 1980. Further advances of house sparrows into the Brazilian Amazon. Condor 82:109–111.

Snow, D. W. 1955. The abnormal breeding of birds in the winter 1953/54. British Birds 48:120–126.

Sodhi, N. S. 1992a. Comparison between urban and rural bird communities in prairie Saskatchewan: urbanization and short-term population trends. Canadain Field-Naturalist 106:210–215.

Sodhi, N. S. 1992b. Sex and age differences in risk-taking behavior in house sparrows. Condor 94:293–294.

Sodhi, N. S. 1993. Correlates of hunting range size in breeding merlins. Condor 95:316–321.

Sodhi, N. S., A. Didiuk, and L. W. Oliphant. 1990. Differences in bird abundance in relation to proximity to merlin nests. Canadian Journal of Zoology 68:852–854.

Sodhi, N. S., P. C. James, I. G. Warkentin, and L. W. Oliphant. 1992. Breeding ecology of urban merlins (*Falco columbarius*). Canadian Journal of Zoology 70:1477–1483.

Sodhi, N. S., and L. W. Oliphant. 1993. Prey selection by urban-feeding merlins. Auk 110:727–735.

Solberg, E. J., and T. H. Ringsby. 1996. Hybridisation between house sparrow *Passer domesticus* and tree sparrow *Passer montanus*. Journal für Ornithologie 137:525–528.

Solberg, E. J., and T. H. Ringsby. 1997. Does male badge size signal status in small island populations of house sparrows, *Passer deomesticus*? Ethology 103:177–186.

Solberg, E. J., T. H. Ringsby, A. Altwegg, and B.-E. Sæther. 2000. Fertile house sparrow × tree sparrow (*Passer domesticus* × *Passer montanus*) hybrids. Journal für Ornithologie 141:102–104.

Solonen, T. 1997. Effect of sparrowhawk *Accipiter nisus* predation on forest birds in southern Finland. Ornis Fennica 74:1–14.

Sooter, C. A., M. Schaeffer, R. Gorrie, and T. A. Cockburn. 1954. Transovarian passage of antibodies following naturally acquired encephalitis infection in birds. Journal of Infectious Diseases 95:165–167.

Sorace, A. 1992. Some data on egg size in Italian sparrow *Passer italiae*. International Studies on Sparrows 19:31–36.

Sorace, A. 1993. Breeding time and clutch size of Italian sparrow, *Passer italiae*, in some localities of central Italy. Rivista Italiana di Ornitologia, Milano 63:64–68.

Southern, H. N. 1945. The economic importance of the house sparrow, *Passer domesticus* L.: a review. Annals of Applied Biology 32:57–67.

Stahl, R. S., T. W. Custer, P. A. Pochop, and J. L. Johnston. 2002. Improved method for quantifying the avicide 3-chlorop-toluidine hydrochloride in bird tissues using a deuterated surrogate/GC/MS method. Journal of Agricultural and Food Chemistry 20:732–738.

Stainton, J. M. 1982. Timing of bathing, dusting and sunning. British Birds 75:65–86.
Stamm, D. D. 1958. Studies on the ecology of equine encephalomyelitis. American Journal of Public Health 48:328–335.
Stamm, D. D. 1963. Susceptibility of bird populations to eastern, western, and St. Louis encephalitis viruses. Proceedings of the XIII International Ornithological Congress 591–603.
Stangel, P. W. 1986. Lack of effects from sampling blood from small birds. Condor 88:244–245.
Steinbacher, J. 1952. Jahreszeitliche Veränderungen am Schnabel des Haussperlings (*Passer domesticus* L.). Bonner Zoologische Beiträge 3:23–30.
Steinbacher, J. 1954. Über die Sperlings-Formen von Sardinien und Sizilien. Senckenbergiana 34:307–310.
Stenhouse, J. H. 1928. Remarkable decrease of the house-sparrow in Fair Isle and Shetland. Scottish Naturalist 169:162–163.
Stephan, B. 1982. Zur Mauser dreier Passer-Arten in Kazachstan. Mitteilungen aus dem zoologische Museum in Berlin 58. Supplementheft. Annalem für Ornitholgie 6:91–100.
Stephan, B. 1986. Die Evolutionstheorie und der taxoniomsche Status des Italiensperlings. Mitteilungen aus dem zoologische Museum in Berlin 62. Supplementheft. Annalem für Ornitholgie 10:25–68.
Stephan, B., and E. I. Gavrilov. 1980. Passer-Hybriden aus Kasachstan. Mitteilungen aus dem zoologische Museum in Berlin 56. Supplementheft. Annalem für Ornitholgie 4:25–28.
Stepniewski, J. 1992. Legi pary miezanej: wrobel (*Passer domesticus*) × mazurek (*Passer montanus*). Notatki Ornitologiczne 33:334.
Stettenheim, P. 1972. The integument of birds. Pp. 1–63 in Avian Biology, vol. II (D. S. Farner and J. R. King, eds.). Academic Press, New York.
Stewart, P. A. 1973. Replacement of cavity-hunting starlings and house sparrows after removal. Wilson Bulletin 85:291–294.
Stewart, R., and T. B. Rambo. 2000. Cloacal microbes in house sparrows. Condor 102:679–684.
Stoddard, H. L., Jr., and R. A. Norris. 1967. Bird casualties at a Leon County, Florida TV tower: an eleven-year study. Bulletin of Tall Timbers Research Station 8:1–104.
Stoner, D. 1939. Parasitism of the English sparrow on the northern cliff swallow. Wilson Bulletin 51:221–222.
Storer, R. W. 1966. Sexual dimorphism and food habits in three North American accipiters. Auk 83:423–436.
Strasser, R., and H. Schwabel. 2004. Yolk testosterone organizes behavior and male plumage coloration in house sparrows (*Passer domesticus*). Behavioral Ecology and Sociobiology 56:491–497.
Studd, M., R. D. Montgomerie, and R. J. Robertson. 1983. Group size and predator surveillance in foraging house sparrows (*Passer domesticus*). Canadian Journal of Zoology 61:226–231.

Suffern, C. 1951. Robbery of nest material by house-sparrow. British Birds 44:18.

Summers-Smith, D. 1954a. Colonial behaviour in the house sparrow. British Birds 47:249–265.

Summers-Smith, D. 1954b. The communal display of the house-sparrow *Passer domesticus*. Ibis 96:116–128.

Summers-Smith, D. 1955. Display of the house-sparrow *Passer domesticus*. Ibis 97:296–305.

Summers-Smith, D. 1956. Movements of house sparrows. British Birds 49:465–488.

Summers-Smith, D. 1957. Mortality of the house sparrow. Bird Study 4:265–270.

Summers-Smith, D. 1958. Nest-site selection, pair formation and territory in the house-sparrow *Passer domesticus*. Ibis 100:190–203.

Summers-Smith, D. 1963. The House Sparrow. Collins, London.

Summers-Smith, J. D. 1988. The Sparrows. T. & A. D. Poyser, Calton, Staffordshire, UK.

Summers-Smith, J. D. 1993. New distributional records for house sparrow *Passer domesticus*. Bulletin of the British Ornithologists' Club 113:62.

Summers-Smith, J. D. 1995. The Tree Sparrow. Bath Press, Bath, UK.

Summers-Smith, J. D. 1999. Current status of the house sparrow in Britain. British Wildlife 10:381–386.

Summers-Smith, J. D. 2000. Decline of house sparrows in large towns. British Birds 93:256–257.

Summers-Smith, D., and J. D. R. Vernon. 1972. The distribution of *Passer* in northwest Africa. Ibis 114:259–262.

Sushkin, P. P. 1927. On the anatomy and classication of the weaverbirds. Bulletin of the American Museum of Natural History 57:1–32.

Sustek, Z., and J. Kristofik. 2003. Beetles (Coleoptera) in nests of house and tree sparrows (*Passer domesticus* and *P. montanus*). Biologia, Bratislawa 58:953–965.

Sweazea, K. L., and E. J. Braun. 2003. Glucose transport in skeletal muscle of *Passer domesticus*, the common house sparrow. Federation of American Societies for Experimental Biology Journal 17:A863.

Szabo, K., A. Szalmás, A. Liker, and Z. Barta. 2002. Effects of haematophagous mites on nestling house sparrows (*Passer domesticus*). Acta Parasitologia 47:318–322.

Takahashi, J. S., and M. Menaker. 1982a. Entrainment of the circadian system of the house sparrow: a population of oscillators in pinealectomized birds. Journal of Comparative Physiology 146:245–253.

Takahashi, J. S., and M. Menaker. 1982b. Role of the suprachiasmatic nuclei in the circadian system of the house sparrow, *Passer domesticus*. Journal of Neuroscience 2:815–828.

Takahashi, J. S., C. Norris, and M. Menaker. 1978. Circadian photoperiodic regulation of testis growth in the house sparrow: is the pineal gland involved? Pp. 153–156 in Comparative Endocrinology (P. J. Gaillard and H. H. Boer, eds.). Elsevier/North Holland Biomedical Press, Amsterdam.

Tatner, P. 1983. The diet of urban magpies *Pica pica*. Ibis 125:90–107.

Tatschl, J L. 1968. Unusual nesting site for house sparrows. Auk 85:514.

Taylor, D. 1994. House sparrow × tree sparrow hybrids in Kent, England. Dutch Birding 16:122–123.

Taylor, R. M., T. H. Work, H. S. Hurlbt, and Farag Rizk. 1956. A study of the ecology of West Nile virus in Egypt. American Journal of Tropical Medicine and Hygiene 5:579–620.

Tchernov, E. 1962. Paleolithic avifauna in Palestine. Bulletin of the Research Council of Israel 11B:95–151.

Teal, J. M. 1969. Direct measurement of CO_2 production during flight in small birds. Zoologica 54:17–23.

Tewary, P. D., and G. Ravikumar. 1989a. Effect of exogenous 17 B-estradiol on gonadal photostimulation in house sparrow, *Passer domesticus* (Linn.). Current Science 58:721–723.

Tewary, P. D., and G. Ravikumar. 1989b. Photoperiodically induced ovarian growth in the subtropical house sparrow (*Passer domesticus*). Environmental Control in Biology 27:105–111.

Thearle, R. J. P. 1968. Urban bird problems. Pp. 181–197 in The Problems of Birds as Pests (R. K. Murton and E. N. Wright, eds.). Academic Press, London.

Thomas, D. H. 1997. The ecophysiological role of the avian lower gastrointestinal tract. Comparative Biochemistry and Physiology 118A:247–255.

Thomas, M. 1982. Magpie chasing and probably catching house sparrow. British Birds 75:36–37.

Thompson, H. M., K. A. Tarrant, and A. D. M. Hart. 1992. Exposure of starlings, house sparrows and skylarks to pesticides. Pp. 194–199 in Pesticides, Cereal Farming and the Environment (P. Greig-Smith, G. Frampton, and T. Hardy, eds.). HMSO, London.

Threadgold, L. T. 1958. Photoperiodic response of the house sparrow, *Passer domesticus*. Nature 182:407–408.

Threadgold, L. T. 1960a. A study of the annual cycle of the house sparrow at various latitudes. Condor 62:190–201.

Threadgold, L. T. 1960b. Testicular response of the house sparrow, *Passer domesticus*, to short photoperiods and low intensities. Physiological Zoology 33:190–205.

Thurber, W. A. 1972. House sparrows in Guatemala. Auk 89:200.

Thurber, W. A. 1986. Range expansion of the house sparrow through Guatemala and El Salvador. American Birds 40:341–350.

Tiemeier, O. W. 1950. The os opticus of birds. Journal of Morphology 86:25–46.

Tinbergen, L. 1946. [The sparrow-hawk (*Accipiter nisus* L.) as a predator of Passerine birds.] Ardea 34:1–213. (In Dutch, English summary).

Tipton, K. W., E. H. Floyd, J. G. Marshall, and J. B. McDevitt. 1970. Resistance of certain grain sorghum hybrids to bird damage in Louisiana. Agronomy Journal 62:211–213.

Tiwari, J. 1990. Five-striped squirrel *Funambulus pennanti* (Wroughton) killing birds. Journal of the Bombay Natural History Society 87:137.

Tixier-Vidal, A., and B. K. Follett. 1973. The adenohypophysis. Pp. 109–182 in Avian Biology, vol. III (D. S. Farner and J. R. King, eds.). Academic Press, New York.

Townsend, C. W., and J. H. Hardy, Jr. 1909. A note on the English sparrow (*Passer domesticus*). Auk 36:78–79.

Trivers, R. L. 1972. Parental investment and sexual selection. Pp. 136–179 in Sexual Selection and the Descent of Man, 1871–1971 (B. Campbell, ed.). Aldine, Chicago.

Tucker, D. 1965. Electrohysiological evidence for olfactory function in birds. Nature 207:34–36.

Tucker, V. A. 1968. Respiratory physiology of house sparrows in relation to high-altitude flight. Journal of Experimental Biology 48:55–66.

Turček, F. J. 1972. A micropopulation of the house sparrow, *Passer comesticus* (L.), in urban habitat (observations). International Studies on Sparrows 6:24–30.

Turek, F. W., C. Desjardins, and M. Menaker. 1976. Antigonadal and progonadal effects of testosterone in male house sparrows. General and Comparative Endocrinology 28:395–402.

Turek, F. W., J. P. McMillan, and M. Menaker. 1976. Melatonin: effects on the circadian locomotor rhythm of sparrows. Science 194:1441–1443.

Turner, E. R. A. 1965. Social feeding in birds. Behaviour 24:1–46.

Ueck, M. 1970. Weitere Untersuchungen zur Feinstruktur und Innervation des Pinealorgans von *Passer domesticus* L. Zeitschrift für Zellforschung und mikroskopische Anatomie 105:276–302.

UK Raptor Working Group. 2000. Report of the UK Raptor Working Group. Department of the Environment, Transport and the Regions, Bristol, UK.

Underwood, H.,and M. Menaker. 1970. Photoperiodically significant photoreception in sparrows: is the retina involved? Science 167:298–301.

Universal Reference Book. 2001. Weather America: a thirty-year summary of statistical weather data and rankings. Grey House Publishing, Millerton, New York.

Vaclav, R., and H. Hoi. 2002a. Different reproductive tactics in house sparrows signalled by badge size: is there a benefit to being average? Ethology 108:569–582.

Vaclav, R.,and H. Hoi. 2002b. Importance of colony size and breeding synchrony on behaviour, reproductive success and paternity in house sparrows *Passer domesticus*. Folia Zoologica 51:35–48.

Vaclav, R., H. Hoi, and D. Blomqvist. 2003. Food supplementation affects extrapair paternity in house sparrows (*Passer domesticus*). Behavioral Ecology 14:730–735.

Vaisanen, R. A., and O. Hilden. 1993. [Long-term trends of winter populations in the Greenfinch, Bullfinch, Yellowhammer and House Sparrow in Finland.] Linnut 28(3):21–24. (In Finnish, English summary).

Van Brink, J. M. 1959. L'expression morphologique de la digametie chez les Saurosides et les Monotremes. Chromosomia (Berlin) 10:1–72.

Van Den Bosch, R. H., and J. A. J. Metz. 1992. Analyzing the velocity of animal range expansion. Journal of Biogeography 19:135–150.

Van Riper, C. III, S. G. Van Riper, M. L. Goff, and M. Laird. 1986. The epizootiology and ecological significance of malaria in Hawaiian land birds. Ecological Monographs 56:327–344.

Van Riper, C. III, S. G. Van Riper, and W. R. Hansen. 2002. Epizootiology and effect of avian pox on Hawaiian forest birds. Auk 119:929–942.

Vaurie, C. 1949. Notes on some Ploceidae from western Asia. American Museum Novitates 1406:1–41.

Vaurie, C. 1956. Systematic notes on Palearctic birds, No. 24. Ploceidae: the genera *Passer*, *Petronia*, and *Montifringilla*. American Museum Novitates 1814:1–27.

Vaux, G. 1962. Beobactungen über Zugbewegungen und Wiederansiedlung des Haussperlings (*Passer d. domesticus* L.) auf Helgoland. Schriften des Naturwissenschaftlichen Vereins für Schleswig-Holstein 33:33–36.

Veiga, J. P. 1990a. A comparative study of reproductive adaptations in house and tree sparrows. Auk 107:45–59.

Veiga, J. P. 1990b. Infanticide by male and female house sparrows. Animal Behaviour 39:496–502.

Veiga, J. P. 1990c. Sexual conflict in the house sparrow: interference between polygynously mated females versus asymmetric male investment. Behavioral Ecology and Sociobiology 27:345–350.

Veiga, J. P. 1992a. Hatching asynchrony in the house sparrow: a test of the egg-viability hypothesis. American Naturalist 139:669–675.

Veiga, J. P. 1992b. Why are house sparrows predominantly monogamous? A test of hypotheses. Animal Behaviour 43:361–370.

Veiga, J. P. 1993a. Badge size, phenotypic quality, and reproductive success in the house sparrow: a study on honest advertisement. Evolution 47:1161–1170.

Veiga, J. P. 1993b. Does brood heat loss influence seasonal patterns of brood size and hatching asynchrony in the house sparrow? Ardeola 40:163–168.

Veiga, J. P. 1993c. Prospective infanticide and ovulation retardation in free-living house sparrows. Animal Behaviour 45:43–46.

Veiga, J. P. 1995. Honest signaling and the survival cost of badges in the house sparrow. Evolution 49:570–572.

Veiga, J. P. 1996a. Mate replacement is costly to males in the multi-brooded house sparrow: an experimental study. Auk 113:664–671.

Veiga, J. P. 1996b. Permanent exposure versus facultative concealment of sexual traits: an experimental study in the house sparrow. Behavioral Ecology and Sociobiology 39:342–352.

Veiga, J. P. 2003. Infanticide by male house sparrows: gaining time or manipulating females? Proceedings of the Royal Society of London B 270(Supplement): 587–589.

Veiga, J. P., and L. Boto. 2000. Low frequency of extra-pair fertilisations in house sparrows breeding at high density. Journal of Avian Biology 31:237–244.

Veiga, J. P., and M. Puerta. 1995. Nutritional constraints determine the expression of a sexual trait in the house sparrow, *Passer domesticus*. Proceedings of the Royal Society of London B 263:229–234.

Vierke, J. 1970. Die Besiedlung Südafrikas durch den Haussperling (*Passer domesticus*). Journal für Ornithologie 111:94–103.

Voltura, K. M., P. L. Schwagmeyer, and D. W. Mock. 2002. Parental feeding rates

in the house sparrow, *Passer domesticus*: are larger-badged males better fathers? Ethology 108:1011–1022.
Waddington, D. C., and J. F. Cockrem. 1987. Homing ability of the house sparrow. Notornis 34:57–58.
Waghray, P., and H. Taher. 1993. Unusual nesting site of house sparrow *Passer domesticus* (Linn.) in Hyderabad. Journal of the Bombay Natural History Society 90:98.
Wagner, H. O. 1959. Die Einwanderung des Haussperlings in Mexiko. Zeitschrift für Tierpsychologie 16:584–592.
Walsberg, G. E., and B. O. Wolf. 1995. Variation in the respiratory quotient of birds and implications for indirect calorimetry using measurements of carbon dioxide production. Journal of Experimental Biology 196:213–219.
Ward, P. 1965. Feeding ecology of the black-faced dioch *Quelea quelea* in Nigeria. Ibis 107:173–214.
Ward, P., and A. Zahavi. 1973. The importance of certain assemblages of birds as "information-centers" for food-finding. Ibis 115:517–534.
Warkentin, I. G., and L. W. Oliphant. 1990. Habitat use and foraging behaviour of urban merlins (*Falco columbarius*) in winter. Journal of Zoology (London) 221:539–563.
Washington, D. 1973. Breeding the house sparrow. Avicultural Magazine 79:109–115.
Washington, D. 1990. Birds bathing in deep water. British Birds 83:166–197.
Watson, D. 1977. The Hen Harrier. T. & A. D. Poyser, Berkhamsted, UK.
Watson, J. R. 1970. Dominance-subordination in caged groups of house sparrows. Wilson Bulletin 82:268–278.
Weatherbee, D. K. 1951. Air-sacs in the English sparrow. Auk 68:242–244.
Weatherbee, D. K., and N. S. Weatherbee. 1961. Artificial incubation of eggs of various bird species and some attributes of neonates. Bird-Banding 32:141–159.
Weaver, R. L. 1939a. The northern distribution and status of the English sparrow in Canada. Canadian Field-Naturalist 53:95–99.
Weaver, R. L. 1939b. Winter observations and a study of the nesting of English sparrows. Bird-Banding 10:73–79.
Weaver, R. L. 1942. Growth and development of English sparrows. Wilson Bulletin 54:183–191.
Weaver, R. L. 1943. Reproduction in English sparrows. Auk 60:62–74.
Webb, D. M. and J. Zhang. 2005. FoxP2 in song-learning birds and vocal-learning mammals. Journal of Heredity 96:212–216.
Weddle, C. B. 2000. Effects of ectoparasites on nestling body mass in the house sparrow. Condor 102:684–687.
Weiner, J. 1972. Energy requirements of house sparrows, *Passer d. domesticus* (L.), in southern Poland. Pp. 45–57 in Productivity, Population Dynamics and Systematics of Granivorous Birds (S. C. Kendeigh and J. Pinowski, eds.). PWN–Polish Scientific Publishers, Warsaw.
Weitzel, N. H. 1988. Nest-site competition between the European starling and native breeding birds in northwestern Nevada. Condor 90:515–517.

Wells, J. V., K. V. Rosenberg, E. H. Dunn, D. L. Tessaglia-Hymes, and A. A. Dhondt. 1998. Feed counts as indicators of spatial and temporal variation in winter abundance of resident birds. Journal of Field Ornithology 69:577–585.

Werler, E., and E. C. Franks. 1975. Some unusual nest sites of the house sparrow. Wilson Bulletin 87:113.

Wessels, T. 1976. Connecticut house sparrow nesting in December. Auk 93:837.

Westneat, D. F., I. R. K. Stewart, E. H. Woeste, J. Gipson, L. Abdulkadir, and J. P. Poston. 2002. Patterns of sex ratio variation in house sparrows. Condor 104:598–609.

Wetmore, A. 1936. The number of contour feathers in passeriform and related birds. Auk 53:159–169.

Wetmore, P. W. 1941. Blood parasites of birds of the District of Columbia and Patuxent Research Refuge vicinity. Journal of Parasitology 26:379–393.

Wetton, J. H., T. Burke, D. T. Parkin, and E. Cairns. 1995. Single-locus DNA fingerprinting reveals that male reproductive success increases with age through extra-pair paternity in the house sparrow (*Passer domesticus*). Proceedings of the Royal Society of London B 260:91–98.

Wetton, J. H., R. E. Carter, D. T. Parkin, and D. Walters. 1987. Demographic study of a wild house sparrow population by DNA fingerprinting. Nature 327:147–149.

Wetton, J. H., and D. T. Parkin. 1991. Sperm competition and fertility in the house Sparrow. Acta XX Congressus Internationalis Ornithologici 2435–2441.

Wetton, J. H., D. T. Parkin, and R. E. Carter. 1992. The use of genetic markers for parentage analysis in *Passer domesticus* (house sparrows). Heredity 69:243–254.

Wettstein, O. 1959. Ergänzende Nachricten über des süd-alpine Mischgebiet der Haussperlinge. Journal für Ornithologie 100:103–104.

Wever, R. 1966. A mathematical model for circadian rhythms. Pp. 47–63 in Circadian Clocks (J. Aschoff, ed.). North-Holland Publishing, Amsterdam.

Whitfield-Rucker, M. G., and V. M. Cassone. 1996. Melatonin binding in the house sparrow song control system: sexual dimorphism and the effect of photoperiod. Hormones and Behavior 30:528–537.

Whitfield-Rucker, M., and V. M. Cassone. 2000. Photoperiodic regulation of the male house sparrow song control center: gonadal dependent and independent mechanisms. General and Comparative Endocrinology 118:173–183.

Wickler, W. 1982. Immanuel Kant and the song of the house sparrow. Auk 99:590–591.

Wideman, R. F., Jr. 1987. Renal regulation of avian calcium and phosphorus metabolism. Journal of Nutrition 117:808–815.

Wieloch, M. 1975. Food of nestling house sparrows, *Passer domesticus* L. and tree sparrows, *Passer montanus* L. in agrocenoses. Polish Ecological Studies 1(3):227–242.

Wieloch, M., and A. Fryska. 1975. Biomass production and energy requirements in populations of house sparrows (*Passer d. domesticus* L.) and tree sparrow (*Passer m. montanus* L.) during the breeding season. Polish Ecological Studies 1(3):243–252.

Wieloch, M., and S. Strawinski. 1976. [Production of the populations of the sparrow *Passer domesticus* (L.) and *Passer montanus* (L.) and a trial at assessing the role of the house sparrows as consumers during their breeding season in agrocenoses.] Pp. 7–15 in Ekologie Ptakow Wybiezeza (S. Strawinski, ed.). Gdanskie Towarzystwo Naukowe, Gdansk, Poland. (In Polish, English summary).

Wiens, J. A., and M. I. Dyeer. 1977. Assessing the potential impact of granivorous birds in ecosystems. Pp. 205–266 in Granivorous Birds in Ecosystems (J. Pinowski and S. C. Kendeigh, eds.). Cambridge University Press, Cambridge, UK.

Will, R. L. 1969. Fecundity, density, and movements of a house sparrow population in southern Illinois. Ph. D. Dissertation, University of Illinois, Urbana.

Will, R. L. 1973. Breeding success, numbers, and movements of house sparrows at McLeansboro, Illinois. Ornithological Monographs 14:60–78.

Williams, G. C. 1966. Natural selection, the costs of reproduction, and a refinement of Lack's principle. American Naturalist 100:687–690.

Williams, J. B. 1985. Validation of the doubly labeled water technique for measuring energy metabolism in starlings and sparrows. Comparative Biochemistry and Physiology 80A:349–353.

Williams, J. E., O. P. Young, D. M. Watts, and T. J. Reed. 1971. Wild birds as eastern (EEE) and western (WEE) equine encephalitis sentinels. Journal of Wildlife Diseases 7:188–194.

Willner, G. R., J. E. Gates, and W. J. Devlin. 1983. Nest box use by cavity-nesting birds. American Midland Naturalist 109:194–201.

Wilson, J. E., and J. W. MacDonald. 1967. Salmonella infection in wild birds. British Veterinary Journal 123:212–219.

Wilson, P. W., and E. M. Grigsby. 1979. Combined nesting of red-tailed hawks and house sparrows. Inland Bird Banding 51:75.

Wing, L. 1943. Spread of the starling and the English sparrow. Auk 60:74–87.

Wingfield, J. C., K. S. Matt, and D. S. Farner. 1984. Physiologic properties of steroid hormone-binding proteins in avian blood. General and Comparative Endocrinology 53:281–292.

Wingfield, J. C., and D. S. Farner. 1993. Endocrinology of reproduction in wild birds. Pp. 163–327 in Avian Biology, vol. IX (D. S. Farner, J. R. King, and K. C. Parkes, eds.). Academic Press, New York.

Winterbottom , J. M. 1959. Expansion of the range of the house sparrow. Dept. Nat. Cons. Rept. 16:92–94.

Witherby, H. F., F. C. R. Jourdain, N. F. Ticehurst, and B. W. Tucker. 1943. The Handbook of British Birds, vol. I. Witherby, London.

Witschi, E. 1935. Seasonal sex characters in birds and their hormonal control. Wilson Bulletin 47:177–188.

Witschi, E. 1936. The bill of the sparrow as an indicator for the male sex hormone. I. Sensitivity. Proceedings of the Society for Experimental Biology and Medicine 33:484–486.

Witschi, E., and R. P. Woods. 1936. The bill of the sparrow as an indicator for the male sex hormone. II. Structural basis. Journal of Experimental Zoology 73:445–459.

Witt, H.-H., J. Crespo, E. De Juana, and J. Varela. 1981. Comparative feeding ecology of Audouin's gull *Larus audouinii* and the herring gull *L. argentatus* in the Mediterranean. Ibis 123:519–526.

Witt, K. 2000. Situation der Vögel in städtischen Bereich: Beispiel Berlin. Vogelwelt 121:107–128.

Wojtusiak, R. J., H. Wojtusiak, and B. Ferens. 1946. Homing experiments on birds. VI. Investigations on the tree and house sparrows (*Passer arboreus* Bewick and *P. domesticus* L.). Bulletin of the International Academy of Poland 2:99–106.

Wolfson, A., and H. A. Stahlecker, Jr. 1950. Gonadal response to injections of testosterone propionate in the male English sparrow and white-throated sparrow. Anatomical Record 108:593–594.

Wood, K. A. 1998. Seasonal changes in diet of pied currawongs *Strepera graculina* at Wollongong, New South Wales. Emu 98:157–180.

Wood, R. H. 1986. Water rail feeding on passerines in garden. British Birds 79:397–400.

Woolfenden, G. E., and W. B. Robertson, Jr. 1975. First nesting of the house sparrow at Dry Tortugas. Florida Field Naturalist 3:23–24.

Wootton, J. T. 1987. Interspecific competition between introduced house finch populations and two associated passerine species. Oecologia (Berlin) 71:325–331.

Wyllie, I. 1981. The Cuckoo. Universe Books, New York.

Yakobi, V. E. 1979. [On the species independence of the Indian sparrow.] Zoologeskii Zhurnal 58:136–137. (In Russian, English summary).

Yalden, D. W. 1980. Notes on the diets of urban kestrels. Bird Study 27:235–238.

Yalden, D. W., and P. E. Yalden. 1985. An experimental investigation of examining kestrel diet by pellet analysis. Bird Study 32:50–55.

Yokoyama, K. 1980. The possible role of the pineal in photoperiodic time measurement in two species of passerine birds. Acta XVII Congressus Internationalis Ornithologici 439–443.

Yom-Tov, Y. 2001. Global warming and body mass decline in Israeli passerine birds. Proceedings of the Royal Society of London B 268:947–952.

Yom-Tov, Y., R. McCleary, and D. Purchase. 1992. The survival rate of Australian passerines. Ibis 134:374–379.

Yosef, R. 1991. Foraging habits, hunting and breeding success of Lanner falcons (*Falco biarmicus*) in Israel. Journal of Raptor Research 25:77–81.

Yosef, R. 1996. Raptors feeding in migration at Eilat, Israel: opportunistic behavior or migratory strategy? Journal of Raptor Research 30:242–245.

Zahavi, A. 1975. Mate selection—selection for a handicap. Journal of Theoretical Biology 53:205–214.

Zalewski, A. 1994. Diet of urban and suburban tawny owls (*Strix aluco*) in the breeding season. Journal of Raptor Research 28:246–252.

Zar, J. H. 1977. Environmental temperature and the fatty acid compositions of house sparrow (*Passer domesticus*) muscle and brain. Comparative Biochemistry and Physiology 57A:127–131.

Zeidler, K. 1966. Untersuchungen über Flügelbefiederung und Mauser des Haussperlings (*Passer domesticus* L.). Journal für Ornithologie 107:113–153.
Zimmerman, N. H., and M. Menaker. 1975. Neural connections of sparrow pineal: role in circadian control of activity. Science 190:477–479.
Zimmerman, N. H., and M. Menaker. 1979. The pineal gland: a pacemaker within the circadian system of the house sparrow. Proceedings of the National Academy of Sciences, USA 76:999–1003.
Zink, R. M., and M. C. McKitrick. 1995. The debate over species concepts and its implications for ornithology. Auk 112:701–719.
Ziswiler, V. 1965. Zur Kenntnis des Samenöffnens und der Struktur des hörnernen Gaumens bei körnerfressenden Oscines. Journal für Ornithologie 106:1–48.
Zukal, J. 1992. [Mammals in the food of the long-eared owl, *Asio otus* L.] Lynx (Praha) 26:21–26. (In Czech, English summary).

INDEX